TELECOMMUNICATION BY SPEECH

TELECOMMUNICATION BY SPEECH

The Transmission Performance of Telephone Networks

D. L. RICHARDS B.Sc. (Eng), C.Eng., F.I.E.E.

Post Office Research Station, London

LONDON BUTTERWORTHS

THE BUTTERWORTH GROUP

ENGLAND

Butterworth & Co (Publishers) Ltd
London: 88 Kingsway, WC2B 6AB

AUSTRALIA

Butterworths Pty Ltd
Sydney: 586 Pacific Highway, NSW 2067
Melbourne: 343 Little Collins Street, 3000
Brisbane: 240 Queen Street, 4000

CANADA

Butterworth & Co (Canada) Ltd
Toronto: 14 Curity Avenue, 374

NEW ZEALAND

Butterworths of New Zealand Ltd
Wellington: 26–28 Waring Taylor Street, 1

SOUTH AFRICA

Butterworth & Co (South Africa) (Pty) Ltd
Durban: 152–154 Gale Street

First published 1973

ISBN 0 408 70344 X

Printed in Hungary

CONTENTS

GLOSSARY OF SYMBOLS

The attempt has been made, as far as possible, to use symbols that are likely to be familiar to the reader, that is to use those that have become established in the literature. This has resulted in the need to use the same symbol in more than one meaning; where no confusion seems likely to result (e.g. use of Z for sensation level and for impedance) this practice has not been avoided. In some cases, spectrum density is required as well as power and this has been indicated by, for example, W' and W; it is convenient to use logarithmic scales in many cases ($10 \log_{10}(\quad)$ or $20 \log_{10}(\quad)$, as the case may be) and the units of these have been called decibels (dB). Logarithmic scales of spectrum density have also been distinguished from the corresponding logarithmic scale of 'power' by use of the ($'$). Except in a few instances (mainly for parameters), Greek letters have been avoided.

Various suffixes have been used, for example to distinguish listening from conversation testing conditions, e.g. x_L and x_C or to define a value of a stimulus corresponding to a given value of a response (assessment score), e.g. $x_{2.5}$.

(a) Mathematical symbols

f	= frequency, Hz
Δf	= width of a given band of frequency
$(\Delta f)_c$	= width of a 'critical band' of frequency as defined by Fletcher
K	= $10 \log_{10} (\Delta f)_c$
I	= acoustic intensity, W/m^2
p	= sound pressure, N/m^2
I_0	= a datum of intensity, 1 pW/m^2
p_0	= a datum of sound pressure, 20 μN/m^2

p_{in} = sound pressure, at a given frequency, of an acoustical test signal applied to an acoustical path, N/m^2

p_{out} = sound pressure, at a given frequency, measured at the output from an acoustical path, N/m^2

β = intensity level, $10 \log_{10} I/I_0$, or sound pressure level, $20 \log_{10} p/p_0$, dB

β'_S = level of spectrum density (long term mean intensity) of speech at a mouth reference point, dB relative to 1 pW/m^2 or to 20 $\mu N/m^2$

β_0 = hearing threshold for pure tones referred to an ear reference point, dB relative to 1 pW/m^2 or to 20 $\mu N/m^2$

β'_N = level of spectrum density of a continuous spectrum sound other than speech (e.g. acoustical noise), dB relative to 1 pW/m^2 or to 20 $\mu N/m^2$

N' = an abbreviation for β'_N

λ = loudness of sound

Λ = loudness level of sound

l = length or distance (Section 2.2.3.1 and Section 3.2.4.1)

α, δ, γ, ϕ, θ = parameters that define the relative locations of mouth and ear on a human head and the manner in which a handset is held

Z = general symbol for impedance (a complex quantity), Ω

Z_M = impedance looking into the electrical terminals of a microphone

Z_E = impedance looking into the electrical terminals of an earphone

Z_C = impedance looking into the line terminals of a telephone set

Z_{S0} = impedance that, when it is connected across the line terminals of a telephone set, causes sidetone to vanish

Z_L = impedance of the line seen by a telephone set

Z_{LTC} = impedance of a local telephone circuit seen from the junction

Z_J = impedance looking from an LTC into a junction or trunk

Z_B = balance impedance in a telephone set (Z_B governs, but is not necessarily equal to, Z_{S0})

k = a function of Z_C, Z_{S0} and Z_L that governs suppression of sidetone in a telephone set

x = transmission loss, electrical, acoustical or nominal overall reference equivalent, dB (used with various subscripts)

w = transmission loss of a trunk circuit or a chain of trunk circuits (usually having a 2 wire interface at least at one end), dB

y = transmission loss of a junction, dB

z = transmission loss of a 4 wire circuit, dB

L_{ins} = insertion loss of a transmission item, or group of items, dB (the order of the items and the terminating impedances must be indicated, e.g. L_{ins}(SL + FB; Z_C/Z_J) for a subscriber's line and feeding bridge)

L_{mis} = mismatch (or reflection) loss at an interface, dB (the impedances looking each way must be indicated, e.g. $L_{mis}(Z_C/Z_L)$)

L_{ret} = return loss due to impedance mismatch at an interface, dB (the impedances looking each way must be indicated, e.g. $L_{ret}(Z_{S0}/Z_L)$)

b = balance return loss at a 4 wire/2 wire termination, dB

g = transmission loss of an electrical portion of a crosstalk path, dB

L_S = transmission loss (sending) of the electrical circuit of a subscriber's set, dB

L_R = transmission loss (receiving) of the electrical circuit of a subscriber's set

n, r = ratios of the numbers of turns on a telephone induction coil

η_S = theoretical efficiency (sending) of electrical circuit of a subscriber's set (Section 5.1.5.1)

η_R = theoretical efficiency (receiving) of electrical circuit of a subscriber's set (Section 5.1.5.1)

L_{ST} = transmission loss of the electrical part of the sidetone path of a subscriber's set

x_S = sending reference equivalent of a local telephone circuit, dB

x_R = receiving reference equivalent of a local telephone circuit, dB

s = loudness or other rating of a sending part of a telephone connection, dB

r = loudness or other rating of a receiving part of a telephone connection, dB

S, with various subscripts,
= general symbol for acoustical/electrical or electrical/acoustical sensitivities, for example:

S_M = sensitivity of a microphone, dB relative to 1 V per N/m²

S_E = sensitivity of an earphone, dB relative to 1 N/m^2 per V
S_{MG} = sending sensitivity of part of a speech path from a mouth reference point to a defined point G in the circuit, units as S_M
S_{GE} = receiving sensitivity of part of a speech path from a point G in the circuit to an ear reference point, units as S_E
L_{ME} = air to air transmission loss from a mouth reference point to an ear reference point via a speech path, dB
L_{RME} = the value of L_{ME} when a given reference speech path is used at its 'zero' setting, dB
L_{MEST} = the value of L_{ME} for a sidetone path for speech
L_{RNST} = air to air transmission loss encountered by room noise in reaching a telephone user's ear through the sidetone path
Z = sensation level at a given frequency, dB
Z_{R0} = the value of Z, at a given frequency, when $L_{ME} = 0$ dB
Z_R = the value of Z, at a given frequency, when a reference system is used at its zero setting, dB
M = threshold shift due to the presence of a masking sound (M is a function of frequency), dB
or the masking audiogram of a sound, dB
$P(Z)$ = a function of Z related to articulation or listening effort
$Q(Z)$ = a function of Z related to loudness or audibility
C_k, m = parameters used to define $Q(Z)$
k = a parameter used to define certain properties of $P(Z)$ or $Q(Z)$ (Section 4.2.1.5)
B' = a frequency weighting factor used for averaging $P(Z)$ or $Q(Z)$
$\overline{P(Z)}$ = a weighted average of $P(Z)$
$\overline{Q(Z)}$ = a weighted average of $Q(Z)$
$\overline{Z}$ = a weighted average of Z, derived from $\overline{Q(Z)}$, dB
C = a constant introduced to enable a convenient scale to be used for λ when it is derived from $\overline{Q(Z)}$
G = a frequency weighting factor used for averaging transformed values of L_{ME}
$\overline{L_{ME}}$ = a weighted average of L_{ME}, dB
$\overline{L_{RME}}$ = a weighted average of L_{RME}, dB
$\overline{L}$ = an abbreviation used to denote $\overline{L_{ME}}$ or $\overline{L_{RME}}$, dB
$\phi(\ )$ = a function relating $\overline{P(Z)}$ to a measurable score such as articulation or listening effort
Y = an assessment score
Y_{LP} = mean opinion score, loudness preference scale

Y_{LE} = mean opinion score, listening effort scale
Y_C = mean opinion score, conversation
A, B = parameters defining Y_{LP} as a function of transmission loss, x
α, β, γ = parameters defining Y_{LE} or Y_C as a function of x
p, P = probability (used with various suffixes)
σ = standard deviation of a probability distribution
m, k = parameters defining a probability function (Section 3.7.2)
v = instantaneous value of a quantity varying in time
v_{rms} = root mean square value of a quantity which is varying in time
x = a standardised variable equal to v/v_{rms} (Section 2.1.3.2)
l, k = parameters specifying a gamma distribution that defines the probability density function of x (Section 2.1.3.2)
V = voltage generally, with suffixes to denote location, V also used with various suffixes to represent level of speech voltage, dBV
E = electromotive force (open circuit voltage), with suffixes to denote location, V
W = mean power of a signal (usually speech or a component of speech), mW or μW
W_s = mean power of a syllabic segment of speech
W_0 = mean power of noise or the noise equivalent to an audibility threshold
W_d = mean power of distortion products within the duration of a syllabic segment
W' = spectrum density of a continuous spectrum signal (usually speech averaged over the duration of activity), mW/Hz or μW/Hz
W'_0 = spectrum density of a noise or the noise equivalent to an audibility threshold
S = level of speech power at a given point in a telephone connection, dBm
s = level of the mean power (W_s) of a syllabic segment of speech, dBm
R = decibel equivalent of the ratio of the power of a Gaussian test signal to that of the products of nonlinear distortion, dB (Section 6.2)
L = effective transmission loss of a nonlinear transmission device when a Gaussian test signal is used, dB (Section 6.2)

N = level of psophometrically weighted circuit noise at a given point in a connection, dBmp (also used for unweighted level of circuit noise)

U = unaffected level of a compandor, dBm, referred to a given point in a connection (Section 4.2.2.2)

G_n = decibel equivalent of ratio of speech power to that of injected circuit noise, dB (Section 4.2.2.1 and 6.2)

Q = setting of modulated noise reference unit, dB or decibel equivalent of a weighted average of the ratio of the mean power (W_s) of each syllabic segment to the mean power (W_d) of the associated distortion products in a non-linear system

B = number of quantised amplitudes each side of zero provided by a quantised transmission system (Section 6.2)

G = companding advantage provided by non-uniform quantising in a quantised transmission system, dB (Section 6.2)

T_{max} = level of a sinusoidal test tone used to define the extreme of the working range of a quantised transmission system, dBm

T_{min} = level of a sinusoidal test tone used to define the minimum level of the working range of a quantised transmission system, dBm

M, r, D = quantities used in reference to stability of a circuit (Section 4.2.3)

(b) Abbreviations for special units

dBm = the absolute (power) level in decibels

dBmO = the absolute (power) level in decibels referred to a point of zero relative level on a circuit

dBr = the relative (power) level in decibels

dBmOp = the absolute psophometric power level in decibels referred to a point of zero relative level on a circuit

pWp = the absolute psophometric power in pW

pWOp = the absolute psophometric power in pW referred to a point of zero relative level on a circuit

dBV = 20 $\log_{10}$ (voltage)

dBA = the absolute level, measured with 'A' frequency weighting, of a sound measured with a sound level meter complying with IEC Recommendation Publication 179. The datum is 20 $\mu N/m^2$ at 1000 Hz

SRE = sending reference equivalent

RRE = receiving reference equivalent
ORE = overall reference equivalent
STRE = sidetone reference equivalent

(c) Other abbreviations used frequently

BMSP = British modal speaking position
GSC = group switching centre
LTC = local telephone circuit
RVL = reference vocal level
SV5B = speech voltmeter type 5B
13-2P-27 and 16-4T-31 are used to identify the principal transmission parts of the United Kingdom Post Office telephone sets, respectively of the 300 and 700 types. The first group (13 or 16) represents the type No of the microphone; the second group (2P or 4T) represents the type No of the earphone; the third group (27 or 31) represents the type No of the electrical circuit. Corresponding descriptions are used for other types of telephone set. The description is not exhaustive but requires supplementing when, for example, a regulator is included.

FOREWORD

At the present time there are some 14 million telephones in use in the United Kingdom, and more than 250 million telephones in the world. By the year 2000 these figures may well have increased to 65 million and 1 400 million respectively.

The world's telephones are interconnected by a vast and ever increasing network of transmission links of many different types, including wire lines, inland and submarine coaxial cables, microwave radio relay systems and earth orbiting communication satellites, and new systems such as waveguides and optical fibres are being developed. A variety of modulation and multiplexing methods, including amplitude, frequency/phase and pulse code modulation, frequency division and time division multiplexing, are employed. Several different types of exchange switching system, for example Strowger, Crossbar and electronic, are used to establish connections between telephone customers. Each of these transmission and switching systems introduces its distinctive impairments to telephonic conversation in the form of noise, distortion, crosstalk and in some cases delay. The network itself is in a state of almost continuous growth and change, with new technologies appearing and old ones disappearing.

And yet the basic requirements for satisfactory voice communication are enduring and unchanging.

Against this complex and evolving network situation the telephone transmission engineer has had to evolve methods for translating the basic requirements for voice communication into performance standards and rules for system design that not only take account of the variety of transmission and switching systems, but also the very important economic factors involved. A reduction by a few decibels of the noise allowed in a long transoceanic telephone transmission system could increase the cost by a million pounds or more.

Since nation needs to speak to nation the standards of telephone transmission and the rules by which they are applied to system design must also be agreed internationally—The International Telecommunication Union and its Consultative Committees play an important role in achieving such agreement.

This book, which presents a comprehensive and coordinated account of the principles and practice of telephone communication system design from the viewpoint of transmission performance, is the outcome of 25 years work by an engineer in the British Post Office. He would, I am sure, be the first to acknowledge that it also represents in considerable measure a bringing together and consolidation of the work and achievements of telephone transmission engineers in many organisations and several countries.

To such engineers, and their contributions to the improvement of human communication and a better ordered world, this book could well be dedicated.

W. J. BRAY
DIRECTOR OF RESEARCH, POST OFFICE

PREFACE

This book was originally intended to contain descriptions of experimental and theoretical assessment methods used in studying telephone transmission performance problems and to associate with these a comprehensive body of fundamental information suitable for guiding designers of telephone equipment and planners of telephone networks. This intention has, however, had to be modified. First, such a treatment would merely transfer material already available in accessible sources without reducing the confusion that confronts the newcomer to this type of work; secondly, even a superficial treatment on these lines would occupy considerable bulk which could not be justified; thirdly, it seemed to the author that what was needed in this field was a vade-mecum rather than a new thesis. This book therefore represents the author's attempt to satisfy the requirements by: (a) providing an orderly classification of knowledge in the field concerned, (b) clothing the framework so created with sufficient information to satisfy those readers who do not want to delve too deeply and (c) offering those more studiously inclined guidelines to the very extensive literature by means of a bibliography classified according to the same system as that used in the text.

The arrangement used is based on the system developed over more than 20 years by the author for personal use in indexing references to publications concerned with the transmission performance of telephone communication systems. Chapters 3 and 4, together with their associated bibliographies, represent fulfilment of the author's original intentions and these form the core of the book; reference should accordingly be made there for information on methods of assessment and on the effects of various factors potentially harmful to conversations over telephone connections. Chapters 1 and 2 provide introductory matter arranged in a manner that has proved useful to the author and will, it is hoped, be found convenient to other telephone engineers. The later chapters give information on the characteristics of equipment and on the manner in which the design processes may proceed.

The bibliography is such an essential adjunct to the text that it has been presented in a special form to display the development of the subject and to guide the enquirer to the information he requires. To these ends, the entries are placed in chronological order within each subdivision which is numbered in the same way as the associated text.

Planning the transmission performance of telephone networks is an extremely complex subject and so definitions are particularly important if confused discussion is to be avoided. Terms that can be defined by simple statements are given in the text and can be located with the aid of the index. When rather esoteric ideas have to be dealt with, it has been the author's dictum that: 'When you have formulated proper definition, you have stated almost everything worth saying on the matter'. By the converse, the reader is invited to treat the whole book as largely composed of definitions.

The complexity of the subject arises mainly from two sources. First, it is concerned with interactions between humans using telephone connections as part of the communication medium; secondly, the items of plant and equipment concerned are extremely numerous (even in a given connection), perform very repetitive functions, have relatively long economic lives and are installed piecemeal rather than by changing the entire system.

Consequences of these facts are first that telephone networks have to be treated as evolving organisms and secondly that evolution of their characteristics is slow, although the technologies used in individual items may be very advanced and rapidly changing. Like biological organisms, telephone networks respond well only when modifications are properly conceived but exhibit signs analogous to disease if changes are allowed to take place in inappropriate forms or manners. Because network characteristics respond in this way, historical perspective is needed to foresee the likely consequences of actions taken to improve a telephone network or to make it more economic. Chapters 1, 7 and 8 contain historical material that seems of value here and further reading is possible by following up references in the associated parts of the bibliography.

In view of these complexities, philosophy of network design

principles is very important if real progress is to be made and Chapter 7 is devoted to this topic. Examples of certain applications of these principles, using the information assembled in Chapter 4, are given in Chapter 8. Descriptions of telephone hardware are contained in Chapters 5 and 6 and their associated bibliographical references.

Because of his affiliation, the author has taken his examples mainly from past and current practice in the Post Office telephone network of the United Kingdom. The telephonic needs of persons are, however, much the same in all countries and so developments of all national networks are subject to constraints in accordance with the same fundamental principles. Telephony has long been a subject for international cooperation and so equipment practices, as well as network characteristics, have tended to become alike, especially under the aegis of the ITU.

The reader will certainly not find all his questions answered within this book but the author will be content if his efforts help to formulate proper questions.

'It is a great and necessary proof of wisdom and sagacity to know what questions may be reasonably asked. For if a question is absurd in itself and calls for an answer where there is no answer, it does not only throw disgrace on the questioner, but often tempts an incautious listener into absurd answers, thus presenting, as the ancients said, the spectacle of one person milking a he-goat, and of another holding the sieve.'

Critique of Pure Reason, Immanuel Kant (1724–1804)

ACKNOWLEDGEMENTS

The author gratefully acknowledges help given, knowingly and unknowingly, by numerous persons. It has been possible to assemble the material for this book only because of the fortunate circumstances in which the author has found himself and the personal contacts he has been able to make in his day-to-day work for the Post Office. Far too many persons have been involved to list them individually and so the author trusts that those who have contributed and who read this will recognise themselves as included in one or more of the following categories. The author's obligations may be classed as follows:

1. Exposure to the ideas of past workers, especially those of the Post Office whose names appear in Sections 1.1, 7.1 and 8.1 of the Bibliography.

2. Working associations over many years with colleagues in the Post Office and co-workers in various United Kingdom Government Establishments, commercial telephone manufacturing companies and their laboratories and British Universities.

3. Opportunities for meeting co-workers of other countries in and out of international conferences, symposia, seminars and other gatherings. Numerous countries are included but especially the following: Australia, Belgium, Canada, Czechoslovakia, Denmark, Federal Germany, France, Italy, Japan, Netherlands, Sweden, Switzerland and the United States of America. The author has visited most of these countries and is very grateful for the fruitful discussions resulting from these visits. Naturally, included in this category are the many contacts with past and present members of the CCITT Secretariat and Laboratory.

4. Practical encouragement by the present Director of Research of the Post Office, Mr. W. J. Bray, and permission by the Senior Director of Development of the Post Office, Professor J. H. H.

Merriman, to make use of information contained in the book. Use has been made of access to records, notebooks and other unpublished material reporting work carried out in the Post Office Research Department.

5. Figures 2.4, 2.5, 2.6, 2.8, 2.9, 2.11, 2.20, 2.21, 2.22, 2.23, 3.1, 3.2, 3.11, 3.12, 3.13, 3.15 and 3.32 have appeared in the following articles published by the Institution of Electrical Engineers, London: *Proceedings IEE*, **103B**, 679–691 (1956)
International Conference on Satellite Communication (IEE, London, 1962), 247–251
Proceedings IEE, **111**, 941–949 (1964)
Proceedings IEE, **111**, 931–940 (1964)
Proceedings IEE, **116**, 955–964 (1969)
Proceedings IEE, **118**, 423–436 (1971)

Acknowledgement is made for permission by the Institution to reproduce these figures.

Figures 3.33, 3.37, 4.21, 4.22 and 4.28 have been prepared from Post Office drawings which appeared in two papers presented by the author at the 4th International Symposium on Human Factors in Telephony (1968). The papers were later reproduced in a volume published in 1970 by VDE-Verlag GmbH, Berlin, reporting the Symposium.

The views expressed, comments made and conclusions drawn are on the authority only of the author and do not necessarily represent those of the Post Office or any other organisation.

1. GENERAL BACKGROUND

1.1 TELEPHONE NETWORKS

1.1.1 Historical

The telephone was invented by the Scottish-born Alexander Graham Bell in the United States of America in 1876, and demonstrated at a meeting of the British Association in Glasgow the same year. It was first offered to the British public in 1878 by The Telephone Company, Limited, which was established to exploit the Bell patents in the United Kingdom. At first only point to point connections were possible and then only between stations not more than a few miles apart. The first telephone sets used Bell receivers as microphones as well as using them as earphones and so the transmission performance was rather poor, probably corresponding to an overall reference equivalent of at least 50 dB associated with an extremely narrow and peaky frequency response (see Section 8.1 of Chapter 8). Telephone exchanges were then opened to interconnect customers of the company and three exchanges had been opened by the Telephone Company before the end of 1879 and these then served about 200 customers by overhead wires.

The carbon microphone was invented by Thomas Alva Edison in 1877 in the United States of America and, in 1879, the Edison Telephone Company of London, Limited was formed to exploit the invention. This company used telephone sets containing Edison carbon microphones and induction coils with local batteries and opened three exchanges serving some 170 subscribers by 1880. The Edison carbon microphone was much more sensitive than the Bell receiver used as a microphone and, in fact, with its aid (and of Bell receivers as earphones) conversation was demonstrated between London and Norwich (115 miles) using one of the Great Eastern Railway Company's telegraph lines.

In 1880, these two telephone companies merged to form the United Telephone Company which remained in business until 1889 when it was absorbed by the National Telephone Company, Limited which had been formed in 1881 and had built up a large telephone business mainly in the north of Britain.

In 1880, it was established by legal process that telephony was embraced by telegraphy of which the Post Office held a government monopoly since 1869. The Post Office was concerned by the competition of telephony with telegraphy and so, from 1881 it allowed the telephone companies to continue under licence to operate only on local bases. Additional companies were also licensed and later the operating areas were increased but all the licences were drafted to expire at the end of 1911. Longer distance facilities were also licensed from 1884 but the Post Office already had a very extensive countrywide telegraph network* and the Government wished to reserve the provision of the longer distance facilities so that they could be established on a more rational basis than would be possible purely commercially.

In 1881, the United Telephone Company had 914 exchange subscribers and 186 private wires. Other companies, and the Post Office, operated local networks but the opportunities for interconnecting subscribers on different exchanges were then few. Exchanges owned by the same company could be treated as other subscribers and interconnection made but subscribers had to rent lines from more than one company if they wanted wider interconnection. Such interconnection as was available was at first confined to limited areas and interurban interconnection was very limited until the Post Office established a backbone trunk system, which was formally opened in 1895.

Post Office interest in local telephony grew gradually from its telegraph customers who had Wheatstone ABC instruments and could be interconnected over a great deal of the country via Post Office telegraph lines. The Post Office at first supplied some of its telegraph customers with telephone sets to use as an alternative to the telegraph instruments and gradually the greater convenience caused customers to abandon their telegraph facilities. Without actively stimulating telephony business on its own account, the Post Office had, by 1892, opened 35 telephone exchanges serving 4691 subscribers' lines and 691 Government telephone lines.

The first interurban trunk line dedicated to telephony in the United Kingdom was opened in 1882 between Leeds and Bradford. London was not connected telephonically to any other town until

* In 1872 the Post Office had 5000 telegraph offices and 15 million telegrams were handled per year; many business houses rented lines and instruments. By 1900 there were over 12 000 telegraphs offices.

1884 when a circuit to Brighton was opened. In 1890 four London–Birmingham circuits were opened using 150 lb/mile hard drawn copper conductors on poles; these circuits were then the longest (130 miles) in the United Kingdom. Two London–Paris circuits (275 miles) were opened in 1891.* In 1892, in preparation for taking over all long distance telephone circuits, the Post Office started planning a backbone trunk system of 800 lb/mile conductors with 12 in spacing on main routes, such as London–Glasgow, 600 lb/mile on secondary routes such as Cardiff–Birmingham and 400 lb/mile on shorter routes such as London–Brighton; most of these circuits were in service by 1896, including circuits between Glasgow and Belfast via a submarine cable opened in 1893. Trunk and junction circuits under 100 miles in length were usually provided at that time on conductors at least 100 lb/mile, cable being used only for rather short lengths and in towns. In 1896, the Post Office also took over the trunk lines operated by companies and these amounted to 29 000 wire miles.

In 1891, no overall network transmission plan was feasible but the need for some transmission aim for telephony was recognised. Mr. A. R. Bennett addressed the British Association in Cardiff in a lecture entitled: 'On the telephoning of great cities'; he formulated four 'conditions indispensable to the establishment of an effective and popular telephone exchange service in a large town.' The first related to transmission performance and is worth quoting: 'Speech must be loud and distinct and privacy of communication complete; and a subscriber's line and apparatus must be adapted equally well to speaking to another across the street or to one 500 miles away. There must furthermore be absence of all disturbing sounds.' The other three conditions related respectively to switching, tariffs and ability to extend the network.

At that time local networks extended only a few miles and were mostly of overhead construction, much galvanised iron wire being used. Loop resistances were however not excessive, partly because earth returns were common. The telephone sets in use in 1891 were rather varied although most incorporated carbon granule microphones of types appreciably better in performance than the original Edison type, which used compressed lampblack as the active element.

* Feasibility was proved in advance by construction of artificial lines and laboratory speech tests by the United Kingdom Post Office; naturally the 21 miles span of the English Channel by submarine cable was the most difficult technical feature of the route.

1891 also saw the introduction of dry (ie not filled with paraffin wax) paper core underground cables with impervious lead sheaths and this revolutionised the local distribution networks which were then becoming extremely congested and the operating companies were encountering considerable wayleave difficulties. The conductors of the underground cables used for subscribers' lines were, until 1905, almost exclusively 20 lb/mile.

By 1905, the development of telephone sets had reached a stable phase. Magneto exchanges (which were the most common type until 1903) used a handset type of local battery telephone set which remained in service until the early 1930s. With the accelerating introduction of Common battery exchanges, the 'solid back' type of carbon granule microphone, typified by the Post Office Transmitter No 1 (fitted in the Post Office Telephones No 1, a wall type, and No 2, the candlestick type) and the Western Electric 2617A, became standard as also was the Bell receiver used as the earphone. The microphones used in the local battery ((LB) sets were unsuitable for Common battery (CB) working and those used for CB exchanges were not entirely suitable for LB areas. After 1903 most new exchanges were of the CB type, including the Automatic exchanges first opened in 1912.

1.1.2 Transmission planning, standards of performance and methods of assessment

It is worth pausing at this stage in describing the development of telephony in the United Kingdom to consider the problems of network transmission planning, standards of performance and method of assessment of the communication efficiency of telephone connections. Until the first decade of the 20th century, no properly coordinated telephone network could be said to exist enabling any two telephone subscribers in the United Kingdom to communicate with each other.

Potential users of telephone connections are dispersed unevenly and certain concentrations of mutual interest affect the pattern of traffic. Figure 1.1 shows in diagrammatical form the topographical arrangement of a possible telephone network. Each subscriber's station is connected by an individual line (termed the subscriber's line) to a local exchange which allows any pair of subscribers' lines on that exchange to be connected together. Subscribers served by different local exchanges which are nearby (say in the same town area) can be interconnected over junction circuits between the local exchanges. Because switching takes place at both ends of each junction, these circuits are available as common paths for many individual subscribers (unlike the subscribers' lines which connect only particular subscribers to their local exchanges). The junction circuits are therefore used more economically than the subscribers' lines

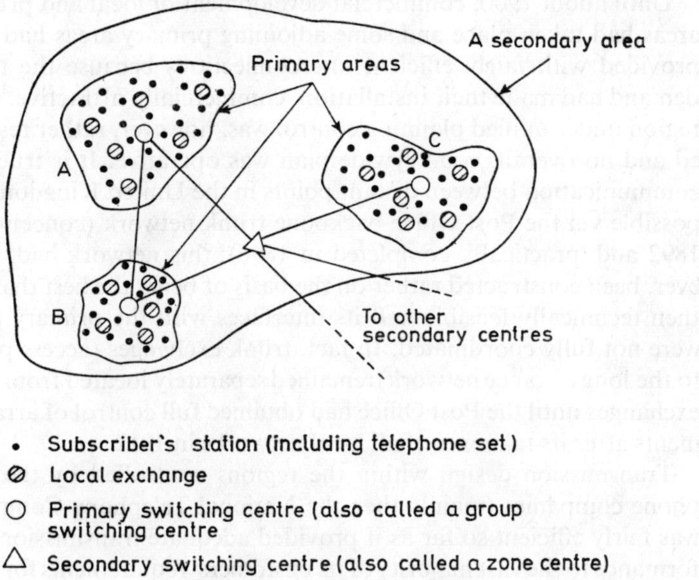

Figure 1.1. A Telephone Network

***Notes:** 1. Tertiary and quaternary switching centres may exist; 2. In dense primary areas some pairs of local exchanges will be interconnected through a third (called a Tandem exchange, see Figure 1.4.)*

and the number of junctions required to interconnect two local exchanges depends on the telephone traffic between the respective local exchange areas.

Individual local exchanges are normally connected directly by junctions only within a certain area, termed in Figure 1.1 a primary area. Calls between subscribers in different primary areas make use of one or more primary switching centres which form a higher stage in a hierarchical system of switching centres by which longer distance circuits (called trunk circuits) are used to interconnect primary switching centres in the primary areas concerned. Higher stages in the hierarchy usually also exist involving secondary switching centres as indicated in Figure 1.1. Naturally, most of the total traffic is carried by the lowest parts of the hierarchy and it becomes unattractive economically to provide full interconnection between all pairs of switching centres at the highest stage of the hierarchy. For this reason, rather infrequently required connections often involve tandem connection of trunk circuits resulting in inferior transmission performance.

Until about 1900, commercial development of local and primary areas had taken place and some adjoining primary areas had been provided with fairly efficient interconnections because the traffic demand had made their installation commercially attractive. Each region under unified planning control was, however, rather restricted and no overall, countrywide plan was operative. It is true that communication between distant points in the United Kingdom was possible via the Post Office backbone trunk network (conceived in 1892 and practically completed in 1896); this network had, however, been constructed rather on the basis of being the best that was then technically feasible and its interfaces with the primary areas were not fully coordinated. In fact, trunk exchanges (access points to the long distance network) remained separately located from local exchanges until the Post Office had obtained full control of arrangements after its takeover of all local networks in 1912.

Transmission design within the regions controlled by the telephone companies (mainly then the National Telephone Company) was fairly efficient so far as it provided adequate transmission performance to the vast majority of its customers' requirements for minimum expenditure on lines. The National Telephone Company expended considerable effort in devising practical methods for measuring and calculating the relative transmission performances of all the types of lines then in common use and this was complete in 1904. These results were expressed in terms of equivalent lengths of 'standard cable', the values being referred to as 'standard cable equivalent' (s.c.e.) per mile. Not only were the purely electrical losses of lines attended to but also the subjective effects of changing the type of telephone set, the effects of varying the feeding current and the differing losses of various types of feeding bridge were also evaluated. The subjective determinations required in this work were performed by means of loudness comparisons with a transmission test circuit, shown in Figure 1.2, which had been standardised and continued to form the basis for such assessments until about 1930 when the telephone sets it contained became obsolete and therefore no longer obtainable for maintenance replacements. The arrange-

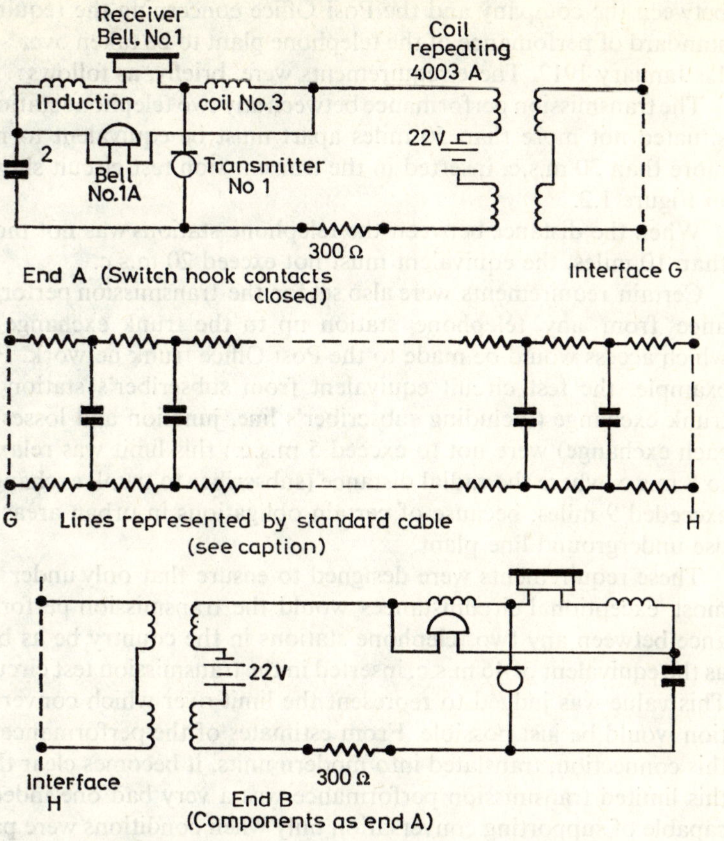

Figure 1.2. Transmission test circuit. Standardised in 1905 and used until about 1933.

Note: The Post Office and National Telephone Company used lengths of real cable wound on drums. The specification of such lengths of standard cable therefore included inductance and leakance as well as the resistance and capacitance shown

ment consisted of standardised telephone sets, selected from the type then in current use, 300 Ω resistors to represent the subscriber's line at each end, representative feeding bridges as used in the 22 V CB exchanges and an adjustable length of 'standard cable' (low capacitance, 20 lb/mile) to represent the equivalents of the junction and trunk lines, including any intervening switching points, between the two terminal exchange feeding bridges.

The results of this work formed the basis for agreement in 1905 between the company and the Post Office concerning the required standard of perfomance of the telephone plant to be taken over* on 1st January 1912. These requirements were, briefly, as follows:

The transmission performance between any two telephone stations situated not more than 50 miles apart must be equivalent to not more than 30 m.s.c. inserted in the transmission test circuit shown in Figure 1.2.

When the distance between the telephone stations was not more than 10 miles, the equivalent must not exceed 20 m.s.c.

Certain requirements were also set for the transmission performance from any telephone station up to the trunk exchange at which access would be made to the Post Office trunk network. For example, the test circuit equivalent from subscriber's station to trunk exchange (including subscriber's line, junction and losses in each exchange) were not to exceed 5 m.s.c.; this limit was relaxed to 8 m.s.c. where the radial distance (subscriber to trunk exchange) exceeded 9 miles, because of certain obligations in urban areas to use underground line plant.

These requirements were designed to ensure that only under the most exceptional circumstances would the transmission performance between any two telephone stations in the country be as bad as the equivalent of 46 m.s.c. inserted in the transmission test circuit. This value was judged to represent the limit over which conversation would be just possible. From estimates of the performance of this connection, translated into modern units, it becomes clear that this limited transmission performance was a very bad one indeed, capable of supporting conversation only when conditions were particularly favourable and, even then, powerful and close talking would be required and very attentive listening in the presence of negligible disturbance from room and circuit noises.

Although the foregoing standards of performance were not very effective in ensuring adequate transmission performance under relatively unfavourable conditions, the methods fulfilled the very important need of indicating where economies could be made with negligible degradation in performance. For example, until 1905, subscribers' underground cables used 20 lb/mile conductors exclusively but, thanks to the information assembled on the relative perform-

* The plant compulsorily transferred was that operated by commercial companies under the 1881 licences which were issued on a 30 year basis. A few municipalities operated their own local systems under licences of 1889; most of these handed their systems over voluntarily to the Post Office but two still remain independent, namely the City of Kingston upon Hull and the States of Jersey, Guernsey, Alderney and Sark.

ances of different types of line, it became possible to show that 10 lb/mile conductors in underground cables would be quite adequate for subscribers' lines up to about $1\frac{1}{2}$ miles in route length. Such lengths of subscribers' lines constituted a very large proportion of the local cable network and so, on introduction of 10 lb/mile cables in 1905, almost half the new copper could be saved, the heavier gauge conductors being reserved for continuing the lines beyond about one mile from the exchange. In 1909, the use of $6\frac{1}{2}$ lb/mile conductors in underground cables for subscribers' lines was introduced with still more saving in copper.

The first few years of the present century found the Post Office busily engaged in studies of how to economise in expenditure on copper in the trunk network without sacrifice in performance. Oliver Heaviside had shown in 1877 that advantage could be theoretically obtained by adding inductance to a line that had appreciable mutual capacitance between wires. The loading coil, invented by Prof. M. I. Pupin in the United States of America in 1899 provided the means of exploiting Heaviside's discovery and the Post Office began experimenting with loading from about 1902. The first schemes for loading junction cables were completed in 1908 resulting in the equivalent lengths of the cables being reduced, from the speech transmission point of view, to about 0.4 of their real lengths. Later, reduction in transmission equivalent to as little as 0.25 of the real length was obtainable: By 1915, the art of cable balancing to secure uniformity of impedance had enabled loading of long distance cables to be so effective that these could have transmission equivalents comparable with those of the heavy overhead open wire lines hitherto needed for long trunk circuits although the conductors of the cables were often somewhat lighter than those of the overhead lines they replaced. The most important advantage of the cables was their greater reliability and freedom from the effects of storms, etc. Unfortunately most of the long distance loaded cable schemes then contemplated were not carried out because of the war (1914–1918) and were rendered obsolete after the war by the prospects of amplification with valves which had been demonstrated as feasible in 1916*.

* The amplifying valve had been invented in 1905, by Lee de Forest, but practical amplifying equipment for telephony was still being developed in the laboratory in 1916. However, in that year important circuits to Ireland were disrupted by storm damage to the overhead sections and were restored over cables by use of experimental amplifying equipment supplied by the Research Section of the Post Office Engineering Department.

By 1915, the Post Office was fully in control of practically the whole telephone system of the United Kingdom and so it was possible to formulate comprehensive network transmission plans. By this time most trunk exchanges had been united with local

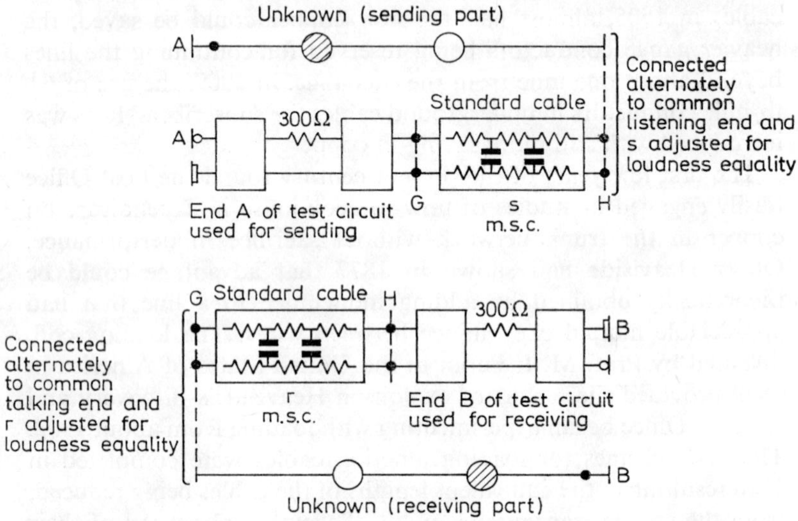

Figure 1.3. Method of expressing test circuit equivalents for part of a connection up to the zone centre.

Note: Transmission from A to H in the 'unknown' is said to be equivalent (for sending) to s m.s.c. inserted in the transmission test circuit. This may be described as a test circuit equivalent (sending) of s m.s.c.

Transmission from G to B in the 'unknown' is said to be equivalent (for receiving) to r m.s.c. inserted in the transmission test circuit. This may be described as a test circuit equivalent (receiving) of r m.s.c.

In general, sending and receiving equivalents will be different from each other for any given combination of telephone set, subscriber's line and feeding bridge (local telephone circuit) and whichever is worse (greater value in m.s.c.) must be considered.

By these methods local line allowances can be determined for any local telephone circuit. It was customary to plot the loop resistance of the subscriber's line as a function of s for sending and the standard cable equivalent of the subscriber's line as a function of r for receiving

exchanges and the 43 Zones (secondary areas, *see* Figure 1.1) that existed in 1908 (having been located largely for commercial reasons) were reduced in number to 9. So far as was practicable, the 9 Zone switching centres, which were located at the chief cities, were fully interconnected by relatively low loss trunk circuits intended not to exceed 15 m.s.c. in equivalents. Belfast was a difficult case and was obtainable from England, Scotland and Wales only indirectly by circuits in tandem; from London, for example, the equivalent to Belfast was about 25 m.s.c. From each Zone centre to a subscriber on any exchange in the home Zone, the transmission was intended to be equivalent to no worse than 10 m.s.c. for both sending and receiving (*see* Figure 1.3 for explanation); this maximum applied whether the terminal local exchange had direct access to the Zone centre or through a 'controlling centre' (which would now be called a Group centre). Taking these intended values for a possible overall telephone connection as illustrated in Figure 1.4a, the total equivalent, in terms of the length of standard cable inserted in the transmission test circuit of Figure 1.2, becomes a maximum of 35 m.s.c. The improvement from 46 m.s.c. contemplated in 1905 was certainly necessary and, in 1915, was deemed very desirable; even 35 m.s.c. provides only a connection that would be judged 'bad' by at least 30% of users accustomed to modern telephony. Its suitability may well have been supported at the time by the fact that about 80–90% of words would be understood over such a connection in an articulation test; it is now known that articulation tests are not reliable for judging suitability of telephone connections for public telephony.

The equivalent of 15 m.s.c. between Zone centres was to apply only for interconnecting the more remote pairs and 10 m.s.c. was to be the limit for Zone–Zone circuits used in any connection between stations not more than 300 but more than 200 miles apart; 5 m.s.c. was to apply when the distance between stations did not exceed 200 miles.

Statements of requirements as given above would now be embodied in diagrams such as those shown in Figure 1.4. Figure 1.4a shows a switching transmission and routing plan for longer distance traffic in accordance with the foregoing requirements. When the terminal exchanges are within the same Zone the arrangement would be as in Figure 1.4b or c; when the traffic justifies the provision, direct circuits will be used to interconnect Group Switching Centres without passing through a switching point at the Zone Centre. Shorter connections (approximately intraurban) are shown in Figure 1.4d and this illustrates the arrangement within a primary area (*see* Figure 1.1). A primary area planned in accordance with Figure 1.4d might appear topographically as shown in simplified form in Figure 1.5.

During the period covered by the foregoing discussion, planning was based on the assumption that overall transmission equivalent

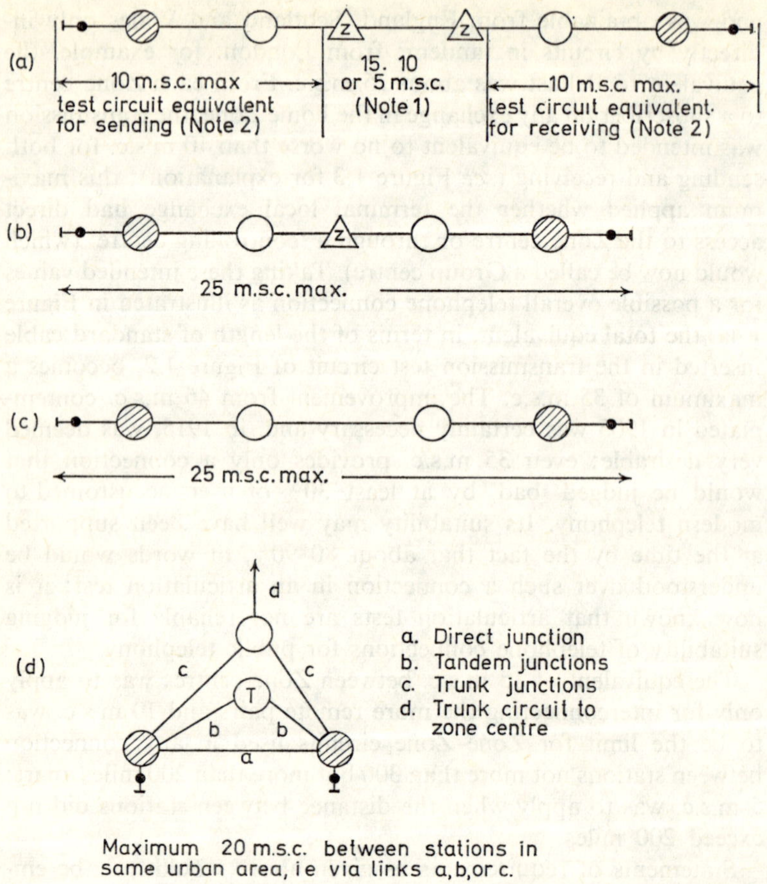

Figure 1.4. An early transmission plan (1915). (a) Between stations in different zones (b) Between stations in same zone via the zone centre (c) Between stations in same zone (d) Shorter distance traffic.

Notes: 1. 15 m.s.c. was the general maximum for Z–Z circuits and corresponded to an overall standard of 35 m.s.c.

10 m.s.c. applied for Z–Z circuits used to connect subscribers' stations in adjacent zones when the stations were more than 200 miles apart; or in non-adjacent zones and not more than 300 miles apart. Also applied for stations on central urban exchanges in any two large commercial centres 300–400 miles apart. The overall standard was then 30 m.s.c.

5 m.s.c. applied for stations in adjacent zones and not more than 200 miles apart. Applied also for stations in any two large commercial centres when traffic justified direct lines between local exchanges. The overall standard was then 25 m.s.c.

2. 10 m.s.c. test circuit equivalent was applied as the maximum for any exchange in the zone; for certain exchanges, called 'Controlling Centres', the limit was restricted to a maximum of 4 m.s.c.

(the equivalent length of a standardised type of cable) represented the degradation introduced by every transmission feature requiring attention in planning telephone networks. This assumption was certainly justified until amplification became practicable (in the 1920s); at this stage overall transmission loss (expressed as equivalent distortionless attenuation) replaced overall transmission equivalent by substituting decibels for miles of standard cable in

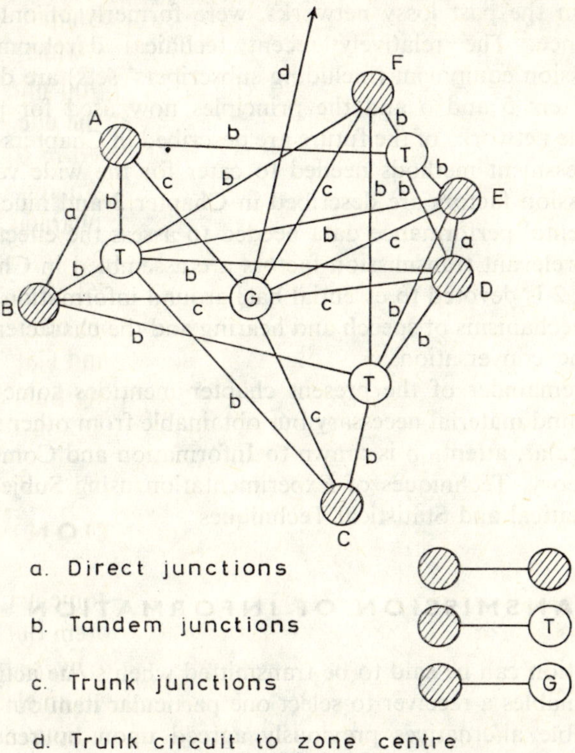

Figure 1.5. A primary area using tandem exchanges

the relationship 1 m.s.c. = 0.92209 dB (in the United States of America, the standard cable was slightly different, because inductance was omitted, and 1 USA m.s.c. = 0.94670 dB).

Use of overall loss has remained a workable basis for setting transmission standards until comparatively modern plant became widespread. The increasing variety of plant and equipment in telephone networks has now necessitated a gradual change in the criterion for planning from one based purely on ensuring sufficiently small losses of loudness of received speech to the modern assessment criteria which must take into account the wide variety of transmission factors now requiring to be controlled in planning (*see* Chapters 3 and 4).

The present stage in this evolutionary process will not be final; already the likelihood of extensive digital transmission networks being realised has shown the need for much more attention to be given to certain factors (e.g. attenuation/frequency distortion) which, in the past lossy networks, were formerly of only minor importance. The relatively recent technical developments in transmission equipment (including subscribers' sets) are described in Chapters 5 and 6 and the principles now used for planning telephone networks of the future are described in Chapters 7 and 8. The assessment methods needed to cater for the wide variety of transmission factors are described in Chapter 3 and much of the fundamental performance data needed to assess the effects of the various relevant transmission factors are assembled in Chapter 4. Chapter 2 is devoted to essential background information relating to the mechanisms of speech and hearing and the characteristics of telephone conversation.

The remainder of the present chapter mentions some general background material necessary but obtainable from other sources; in particular, attention is drawn to Information and Communication Theory, Techniques of Experimentation using Subjects, and Mathematical and Statistical Techniques.

1.2 TRANSMISSION OF INFORMATION

Information can be said to be transmitted when some action by a sender enables a receiver to select one particular item out of a set of possible alternatives previously agreed upon by sender and receiver as the alphabet to be used. Figure 1.6 shows an elementary communication system in which the information source can be regarded as operating a process in which items are selected from a defined ensemble of symbols, each representing a message element. They are coded in the sending apparatus as signal elements suitable

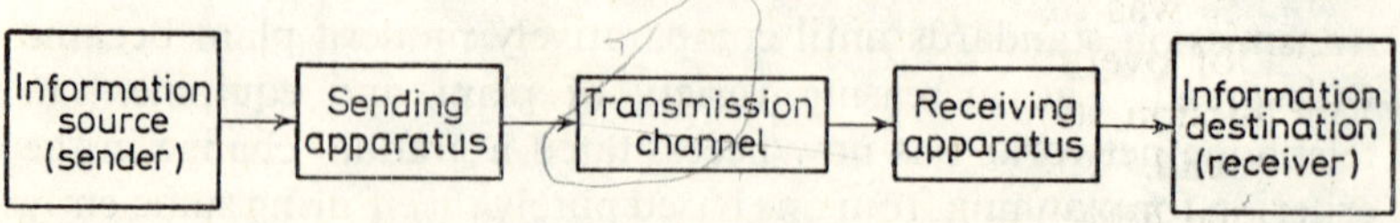

Figure 1.6. Elementary communication system

for transmission over the channel. The receiving apparatus decodes the signal elements, thereby restoring them to the same form as in the original ensemble of symbols representing message elements. An important feature of the situation assumed here is that the defined set of symbols (potential message elements) is known in advance at the destination as well as at the source. Uncertainty at the receiver is resolved whenever a message element is correctly decoded. Thus, the amount of information emitted by the source is governed by the statistics of the ensemble of potential message elements and the rate at which the selections take place.

This principle has been elaborated into a theory of communication which defines certain necessary properties of the transmission channel if it is to convey signals sufficient to permit selections to be made by the receiver at a given rate.

This theoretical model is of some value in studying the perform ance of speech transmission systems and certain methods of assessment are consistent with the theory. For example the articulation method (*see* Section 3.3 of Chapter 3) attempts to measure the capability of the system to reproduce sounds from a defined population, emitted by a talker (sender) and recognised as specific members of that population by a listener (receiver).

Two distinct quantities have to be considered: the rate at which the sender emits symbols of which each has a certain selective potential; and the capacity of the transmission channel to carry signals representing the messages embodied in the sequence of symbols. A very naive interpretation of the speech transmission process might equate the symbols emitted by the sender to the speech sounds in a language which serve to distinguish the meaning of one word from another; in English there are about 40 such sounds. That is, any spoken sentence can be transcribed by writing a sequence of signs drawn from a set of 40 different signs, and the written material can be read aloud by another person to produce the same meaning to a listener as he would have received by hearing the original talker. The rate at which the signs are written down, together with the statistical description of their probabilities, governs the information rate of the talker as the source. If the signs were equiprobable and independent, the information per sound would amount to $\log_2 40 = 5.3$ bits; assuming a rate of ten signs per second, the information rate would be 53 bits/s. In fact speech sounds are neither equiprobable nor independent (*see* Section 2.1 of Chapter 2) and taking account of the zero order probabilities and the digram transitions sets the maximum information rate of speech at about 43 bits/s. If account is taken of redundancy between sounds within words, the rate is reduced to, say, 35 bits/s. In sentences, redundancy between words must be considered and this reduces the rate of emitting information to an estimated value of about 12 bits/s. When the precise topic of conversation has been defined and is known to both participants, it has been estimated* that the information rate is no greater than about 0.2 bit/s.

The second quantity, channel capacity, can be considered first as it relates to the hearing mechanism of the listener. It is true that listeners can make very fine distinctions between individual sounds of a set and so the total number of distinguishably different sounds is rather large, perhaps equivalent to 2^{5000}. It requires application of a sound for about 50 to 200 ms for a discrimination to be made; taking an average of 100 ms, it might be stated that the ear can receive at 50 000 bits/s. This statement is however only true in the sense that a transmission capacity of at least this amount is necessary to ensure that the listener will never be able to distinguish the difference between sounds entering a transmission system and those arriving at the other end. In fact, there are very severe limitations in true information capacity of a listener; no human activity, whatever the senses or combinations of senses, can be continued at information rates exceeding about 43 bits/s and it has been shown that the limitation is in internal processing in the brain and not in indicating responses by muscle activity.

Turning now to communication systems interposed between talker and listener, it might be supposed that an information capacity of less than 50 bits/s would suffice not only to transmit all the information in the speech but also to satisfy the rate at which the listener can operate. Unfortunately this is far from being so as a rough calculation of the information capacity of a definite telephone speech path will show. A rather poor telephone connection might have a bandwidth W of 3000 Hz and the ratio of speech signal power S to noise power N might be 30 dB; by the Shannon formula, $C = W \log_2 (1 + S/N)$, the information transmission capacity C would be about 30 000 bits/s. The ratio of nearly 1000/1 implies that the efficiency with which the channel is used by a talker and listener is only 0.1% and constitutes a challenge to improve upon this state of affairs; the challenge has certainly been accepted

* This estimate was based on the average time taken to identify one pictorial sign out of a set when they can be distinguished by reference to a finite number of distinguishing features.

(*see* Section 6.5 of Chapter 6) but, even when very sophisticated methods are used, the efficiency, in the above terms, is improved only to about 1%.

In theory, the rate at which information is transmitted by a channel can equal the information rate of the source provided the channel capacity is not reduced below a value equal to the information rate of the source; when the channel capacity is less than this value, the rate of transmission cannot exceed that of the channel. The ratio of the rate of information transmission to the

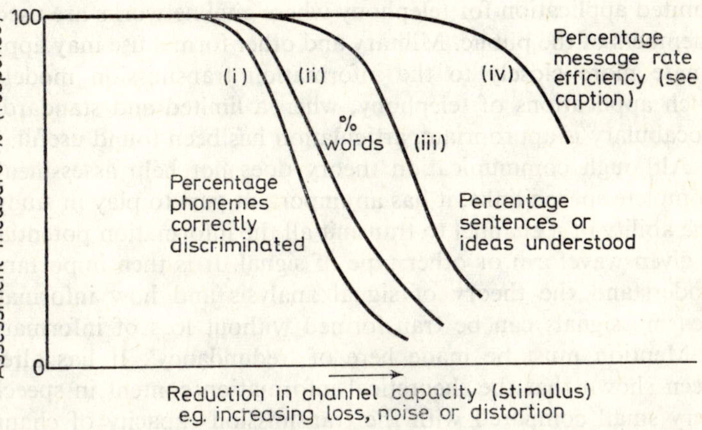

Figure 1.7. Performance of speech communication system according to various criteria.

Note: Message rate efficiency is the ratio of time required to complete a defined conversational task under perfect transmission conditions to that necessary using the speech link under consideration

information rate of the source is an obvious measure for assessing the transmission performance of the system and curve i in Figure 1.7 shows the form of such a measure plotted as a function of reduction in channel capacity. The figure includes also some practical assessment measures based on similar principles, but not, of course, to the same scale of channel capacity as curve i.

Unfortunately, the communication model discussed above does not provide a very useful practical description of the speech communication process on which reliable assessment methods can be based. The basic assumption of a definite and agreed ensemble is certainly not true for many ordinary telephone conversations, where a considerable part of the call is devoted to establishing some common basis for discussion; where the original ensemble cannot be defined, there can be no basis for employing the communication theory model. A more practical objection is that the participants are normally much more concerned with the effort required in conversing and in any difficulty encountered, than in any counting of errors committed in interpretation; if any such errors are at all likely the channel will be completely useless for ordinary telephone conversations.

For these reasons assessment methods based on the principle of measuring the rate of transmission of information have rather limited application for telephony where civil networks are used by members of the public. Military and other formal use may approximate more closely to the information transmission model; in such applications of telephony, when a limited and standardised vocabulary is appropriate, articulation has been found useful.

Although communication theory does not help assessment of complete speech links, it has an important part to play in studying the ability of a channel to transmit all the information potential of a given waveform or other type of signal. It is then important to understand the theory of signal analysis and how information bearing signals can be transformed without loss of information.

Mention must be made here of 'redundancy'. It has already been shown that the theoretical information content in speech is very small compared with the transmission capacity of channels usually considered necessary in telephony; this disparity must not, however, be considered as wholly wasteful. Communication theory teaches that messages can be transmitted without error at rates up to the information capacity of the channel but, if anything like the full channel capacity is exploited, very elaborate coding is necessary to match the signals to the characteristics of the medium, involving long sequences of message symbols; this requires time for organisation and so delay results. Therefore, if practically instantaneous transmission is required, a margin must remain between channel capacity and theoretical information content of the messages. Another use to which surplus channel capacity is put is the avoidance of error when, perhaps only temporarily, the channel capacity is reduced. Speech sounds having many different waveforms will be accepted, even by a discriminating listener, as the same phoneme; each phoneme thus has many aliases any one of which is acceptable. In recognising a speech sound only a few of the clues which distinguish it from others in the language are actually used; when the channel capacity is reduced, some clues will become obscured but others will be relatively unaffected by the particular form of channel degradation that occurs and it is only when practically all the clues have been obliterated that failure occurs. This property of message symbols to contain theoretically unnecessary elements is called redundancy; its presence results in the assessment problem becoming one of measuring the effort needed to take advantage of the redundancy to avoid error rather than of counting the occurrences of errors (which are very few indeed even when a very bad telephone channel is being used).

The redundant structure of speech and the active role adopted by listeners in interpreting what reaches their ears results in the need for multicriterion methods for assessment when a very wide range of transmission quality is to be covered; clues treated as redundant according to one criterion may be essential elements according to another (*see* Section 4.1.5 of Chapter 4).

Information theory and communication theory have, for the reasons mentioned above, proved of only limited value for defining and measuring telephony performance and no successful assessment method has yet been based on principles leading to scores in terms of, say, bits/s.

1.3 SUBJECTIVE ASSESSMENT

Taken as a whole and excluding fault conditions, the telephone connections provided for the customers to use are sufficient to enable those who are accustomed to the system to communicate with each other without excessive effort or undue risk of error. The effort required to converse on the telephone is, however, by no means negligible as is made evident by the appreciable difficulty commonly experienced by those who have never, or only rarely, used it before. It is also much more frequently necessary than in direct face-to-face conversation to spell out a word where context does not help; this is true even of experienced users of the telephone.

There is therefore an appreciable quality interval between a speech link set up to simulate direct air-to-air conditions, by using the best quality microphones and earphones, and even the best ordinary telephone connection*; this is revealed quantitatively by

* That is one free from faults and in accordance with well established telephone engineering standards; it should also be assumed that ambient conditions are reasonably favourable, say not more than 50 dBA level of room noise.

laboratory tests and studies of customers' opinions. This does not imply that telephone connections are unsatisfactory or that an appreciable volume of complaints is received; rather that human adaptability is such that the need for modest amounts of effort will be willingly accepted as a quite reasonable price to pay for the advantages. Laboratory tests reveal that considerable additional degradation will be tolerated before a substantial proportion of users will, when questioned, express unfavourable opinions. Even greater degradation would be accepted before any appreciable volume of direct complaints is generated by any transmission defects.

The total range of communication efficiency to be studied therefore extends, at its upper extremity, to such high quality that a voice would immediately be recognisable, although not previously heard over the telephone, and nuances, such as voice quality, inflection, etc. would be so well reproduced that no more spelling would be needed than in natural air-to-air conversation. The range extends to a lower extremity where the performance is so poor that conversation is only possible when a great deal of effort is used. The communication efficiency of ordinary telephone connections occupies only a portion of this range, although their range of performance is still fairly large.

Telephone connections are used by customers as aids in the performance of certain tasks involving conversation. To compensate for imperfections of ordinary telephone connections, the users are willing to expend some effort in recognising received speech sounds by exploitation of the multiple clues present as redundancy in the speech; use of effort in employing these faculties to decode the speech signals to recognise the meaning intended is likely to be acceptable as reasonable when the conversation task is light, the appreciable margin of intellectual effort available then being willingly drawn upon, probably without the telephone user being conscious of it. If, however, the business to be transacted during the conversation is such that the participants need to devote almost all their intellectual faculties to these affairs, the effort required for overcoming the communication deficiences of the connection will be irksome and may give rise to unfavourable opinions of the connection.

As the amount of any physical degradation becomes greater, the effect on such statistics as the proportions of subjects (or users) who express unfavourable opinions will be only gradual. There is no uniquely identifiable point on any scale of physical degradation that can be said to distinguish sharply between connections that are satisfactory and those that are not. The limiting value of any degrading feature must be decided in the light of all circumstances, economic, state of technology, etc. as well as the resulting subjective effects quoted above. Account must also be taken of the possibility that the effect of any degradation under consideration may, in other circumstances, be aggravated by additional factors, such as incorporation of the equipment in question into a more complicated telephone connection. It follows that, when considering any particular cause of inefficiency, the information required to make an engineering decision must show the relationship over a wide range of values between the subjective effects and a suitable physical measure of that cause of inefficiency. It is also extremely desirable to use a common scale for subjective effects so that the equivalent intensities of different causes can be clearly revealed. At the present time such comparisons are greatly complicated by use of an excessive number of scales used for expressing subjective effects caused by different physical agents. A major objective of future studies must be to express all causes in terms of one, or a very few, common scales of subjective response.

An important part of telephone engineering is concerned with human factors where the object is to design the item to be best adapted for the human purpose concerned. The question then arises, how to measure the fitness for purpose? Laboratory tests may be artificial and results may be in terms of somewhat irrelevant quantities; field tests, however, are imprecise and many important factors are largely unknown as well as uncontrollable. The method adopted is an iterative one: the physical and other conditions in the field are first surveyed; laboratory tests are conducted under conditions shown to be relevant by the field survey and using what seem to be relevant assessment criteria; finally assessment is made in the field using, as far as is possible the same assessment criteria as were used in the laboratory. This network design process is explained in Chapter 7 but the importance of integrating field and laboratory studies is an essential thread running through all the material presented here which is concerned in detail mainly with laboratory assessment studies from which fairly precise statements of human performance and opinions can be made.

Much of the laboratory testing covered in Chapter 3 follows the pattern in principle of measuring a relationship between a response and a physically definable stimulus. Determinations of response/stimulus relationships are well documented in the fields of psychology, biology and agriculture where the response is commonly rather difficult to define; consequently much of the methodology used in telephony assessment has been adapted from these disciplines. In telephony, the stimuli to be studied can be defined in terms of transmission losses, sound pressure (or intensity), levels of signals and noise, certain measures of waveform distortion, transmission times, etc. The responses that are relevant are of various kinds. In some cases direct physical measurement may be possible, such as of speech level, but more commonly only comparative judgements are possible such as louder/quieter, better/worse, undetectable/detectable, not objectionable/objectionable, etc. Such judgements can also be made on multicategory scales provided care is taken to avoid mixtures of different dimensions.

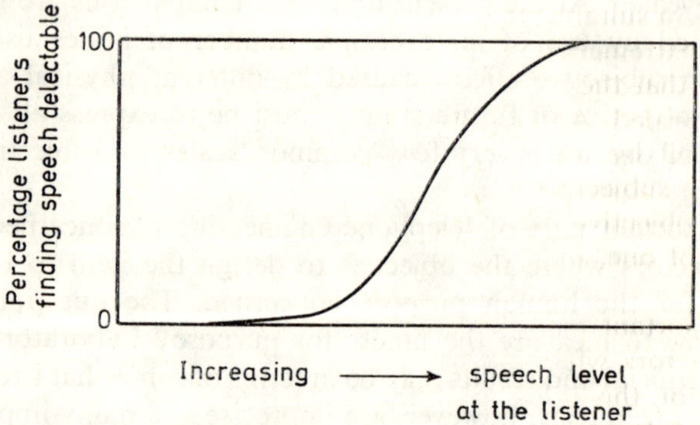

Figure 1.8. Results of a speech detectability experiment

In some cases observation of subjects' behaviour when performing some defined task is possible, such as counting the number of errors committed by a listener in an articulation test. The response/stimulus relationship then takes the form of percentage errors (or correct responses) as a function of the stimulus, say, loss. Some examples were included in Figure 1.7; this figure also included (curve iv) a function derived from measurements of time to complete a task (*see also* Section 3.6.2 of Chapter 3).

Possibly the commonest practical form of response is that to a question made after performance of some task requiring judgement. A simple example is determination of the threshold of detectability of speech heard over a simulated crosstalk path; Figure 1.8 shows the percentage of responses 'yes' to the question:

Can you detect the presence of any speech? Naturally, in such an experiment the subject must be told when to listen, for example by means of a light signal, and, to guard against excessive guessing, he may be given a switch that positively eliminates any speech so that he is assisted in resolving doubtful judgements and discouraged from making false positive responses.

Figure 1.9 shows some results of a more complicated experiment

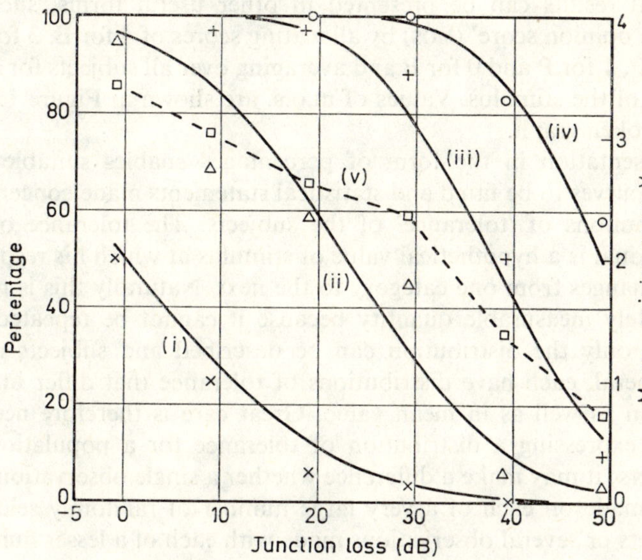

Figure 1.9. Free conversation transmission assessments. Percentage 'satisfied' subjects as functions of junction loss

(i) Percentage of subjects expressing the opinion 'Excellent'

(ii) Percentage of subjects expressing the opinion 'Excellent' or 'Good'

(iii) Percentage of subjects expressing the opinion 'Excellent', 'Good' or 'Fair'

(iv) Percentage of subjects expressing the opinion 'Excellent', 'Good', 'Fair' or 'Poor'

(v) Mean Opinion Score, Y_C

Notes: **1.** *The graphs shown in the figure apply when no circuit noise is present;* **2.** *Curves (i), (ii), (iii), and (iv) are fitted Gaussian probability integral functions; (ii) and (iii) are defined by the following parameters; (ii) mean 23.0 dB; standard deviation 12.8 dB (iii) mean 38.0 dB; standard deviation 10.0 dB:* **3.** *When circuit noise at a level of −60 dBmp was present (at the inputs to receive ends of +4 dB receive reference equivalent), the parameters of the curves corresponding to (ii) and (iii) in the figure became: (ii) mean 20.4 dB; standard deviation 11.6 dB, (iii) mean 34.2 dB; standard deviation 9.3 dB:* **4.** *Room noise at a level of 50 dBA was present at each end:* **5.** *The nominal overall reference equivalent was equal to the junction loss +14 dB*

in which the subjects completed a defined task requiring conversational cooperation. Conversations are made over each of a number of settings of some adjustable feature in a telephone connection; these settings constitute the values of the 'stimulus'. After each conversation the subjects voted according to a five category scale of 'excellent' (E), 'good' (G), 'fair' (F), 'poor' (P) and 'bad' (B). The figure shows the percentages of E, E+G, E+G+F and E+G+F+P which constitute the 'responses'. The same experimental results can be presented in other useful forms, such as 'mean opinion score' (MOS) by allocating scores of 4 for E, 3 for G, 2 for F, 1 for P and 0 for B and averaging over all subjects for each value of the stimulus. Values of m.o.s. are shown in Figure 1.9 by the broken graph.

Presentation in the form of percentages enables suitable sigmoid curves to be fitted and statistical statements made concerning distributions of 'tolerance' of the subjects. The tolerance of an individual is a hypothetical value of stimulus at which his response just changes from one category to the next. Naturally this is not a definitely measurable quantity because it cannot be repeated exactly; only the distribution can be described and subjects may, in general, each have distributions of tolerance that differ in dispersion as well as in mean value. Great care is therefore needed when expressing a distribution of tolerance for a population of subjects; it may make a difference whether a single observation has been made on each of a very large number of randomly selected subjects or several observations made with each of a lesser number of such subjects. Furthermore, populations of subjects may not be homogeneous and so likely sources of heterogeneity must be taken into account. The treatment of such situations and methods of analysis are described elsewhere.

Many subjective tests use only a binary decision procedure and some of the disadvantages associated with such subjective responses can be avoided by use of a 'rating' procedure in which the numerical result is obtained in terms of the setting of a reference device to correspond to the intercategory boundary. Measurement of loudness of sounds in phons and determinations of loudness ratings of telephone speech paths are examples of this method; in the former, the reference device varies the level of a tone at 1000 Hz until it is judged equally loud to the 'unknown' sound; in the latter a defined reference speech path is used containing an adjustable and calibrated loss element. Precise procedures for arriving at the balance setting are numerous and their relative merits can be argued; several examples are described in Sections 3.1, 3.2 and 3.4 of Chapter 3.

When assembling transmission performance information on telephone speech links, there is always a considerable excess of desired test conditions over the number that can possibly be treated with the resources available. One consequence is a temptation to take short cuts by simplifying the test procedures—this is particularly dangerous because procedures chosen for ease in experimentation and the saving of time tend to include undesirable artificialities unrepresentative of real telephone conditions. It is usually much safer to perform as many carefully chosen and properly designed tests as facilities will allow always ensuring that relevant situations are represented. The gap between the results obtainable in this manner and the coverage required for planning can then be filled by a careful selection of simplified tests and by use of plausible theoretical models which have been based on results from properly representative test conditions and have been verified as reliable for the purposes intended. This advice amounts to application of the true iterative scientific method, namely: 1. setting up a hypothesis; 2. deducing therefrom certain consequences; 3. performing experiments to test the hypothesis by checking the deductions; 4. modifying the hypothesis in the light of the new knowledge; 5. deducing further consequences; etc.

The wide variety of characteristics of telephone connections concerning which fundamental information is required for planning telephone networks arises because there is such a large number of different kinds of degradation and these can be combined in almost all possible ways. The different kinds of factor include loss, circuit and room noise, various distortions, sidetone, echo, propagation time and the effects of various voice switching devices and the effects of all practical combinations of these must be considered. In each laboratory experiment, only a few different degradations can be included and so there is need to present results in a form that facilitates estimation of the effects of combinations of degradations. Studies in assessment of still pictures on television screens has reached an advanced stage in respect of predicting the results of combinations of different impairments from those obtained with each impairment present separately. The corresponding problem in speech transmission assessment has not yet been resolved and more study is needed.

Much subjective experimentation in other fields is confined to describing the behaviour of an 'average' or 'normal' subject; in telephony, however, the problem is not to provide satisfactory service to 'average' customers but to satisfy all but a tiny minority. Consequently, it is just as important in the associated subjective studies to estimate dispersions as to determine mean values for the various quantities. This precludes use of special teams of trained testers except for certain specialised purposes such as determining loudness ratings or ultimate thresholds of detectability. Most of the work must be done with subjects as representative as possible of ordinary telephone customers and they must be employed in tasks relevant to normal use of the telephone (*see* Section 3.6 of Chapter 3).

In laboratory experiments, subjects can be influenced by factors additional to those controlled by the experimenter for particular observations. First, they enter the experiment with some background of exposure to contemporary telephone connections; this may blunt their critical faculties concerning degradations similar to those already experienced but may accentuate them so far as novel experiences are concerned; in either case some bias may be present. Secondly, in a sequence of presentations of different test conditions, the result from one condition may be influenced by conditions to which they have previously been exposed; this can sometimes be mitigated by judicious interposition of 'conditioning' treatments. Thirdly, novel experiences require some time for adaptation and too brief an exposure to a test condition may not yield an answer suitable for predicting the result of continual exposure to the same physical condition.

From the foregoing notes it should be clear that considerable care, patience and persistence are required when assembling telephony transmission information suitable for making network planning decisions.

1.4 MATHEMATICS AND STATISTICS

Little need be said about the mathematical equipment needed for telephone assessment except to note that the arts of curve fitting (including probit analysis, use of the logistic function and other biological assay techniques (*see* Section 3.7 of Chapter 3)) and estimation of parameters of mathematical models are necessary together with some knowledge of signal analysis and transmission theory (already referred to in Section 1.2). It is, perhaps, more important to point out the need for proper discipline in all measurement work, especially in the design of experiments according to logical principles that will permit proper statistical analysis to be made including tests of significance and estimates of confidence intervals.

Although there is not space for including details of the experimental designs used to obtain the information presented, it must be noted that, with such variable material as humans, little of value would be obtained from measurements conducted without proper attention to the planning of the observations.

Large experiments may be more efficient in certain respects than a series of separate small ones, but the work discussed here does not usually lend itself to experimentation in large units. Much of the subjective experimentation is exploratory rather than for the purpose of gathering data on a subject already well understood and so a sequential strategy is usually desirable. Small experiments incur less costly mistakes if wrong test conditions are chosen or administrative errors made. Each subject must not be used excessively, partly because he has other work to do and partly to avoid excessive practice effects and too prolonged exposure to testing procedures. Typically, experiments using 12 subjects (or 12 pairs when conversation is involved) and 12 test conditions are suitable. Arrangements in the form of Latin squares permit several factors to be included as well as the test conditions themselves, thereby reducing the experimental error so far as intra-experiment comparisons between test conditions are concerned. A very simple Latin square design is illustrated in Figure 1.10. Four test conditions (treatments) are to be compared and these are allocated at random to the four symbols A, B, C and D. There are four subjects and these are allocated to the four numbers 1, 2, 3 and 4 which are represented by the four rows of the design. Each subject is to be given each of the four treatments but the sequence is so chosen that each treatment is administered first, second, third and fourth as shown in the figure where the columns represent the order of administration of the treatments. Figure 1.10a shows an unrandomised Latin square and it will be seen that in three rows out of four treatment B follows immediately after treatment A and that other regularities are present. This feature is undesirable in experiments on subjects because residual effects may be present from previously experienced treatments and can be avoided by randomising rows and columns for every experiment. One example of such randomisation is shown in the figure; Figure 1.10b shows the rows arranged in the (random) order 1432 and Figure 1.10c shows the columns of this square arranged in the (random) order 3124. Statistical analysis will enable the effects of the three factors, treatments, subjects and sequence to be separated and analysis of variance will enable tests of significance to be made to indicate whether the mean values for the four treatments differ by more

(a)

A	B	C	D
B	C	D	A
C	D	A	B
D	A	B	C

(b)

A	B	C	D
D	A	B	C
C	D	A	B
B	C	D	A

(c)

C	A	B	D
B	D	A	C
A	C	D	B
D	B	C	A

Figure 1.10. Randomisation of a Latin Square design of experiment. (a) Original, standard, Latin square. (b) Rows interchanged at random (1432). (c) Columns interchanged at random (3124)

than could be accounted for by the operation of purely random incidence of errors. The analysis will also assign confidence intervals to the treatment mean values, namely statements of the range within which the true value lies. 95% confidence means that the statements will, in the long run of repetitions of such statements, be true in 95% of cases and in 5% of cases the true value will lie outside the assigned range. Naturally 99% confidence intervals will be wider than 95%.

Detailed descriptions are available in the literature for designing and analysing experiments suitable for particular problems; those used for loudness rating and articulation tests are included in CCITT Recommendations P.72 and P.45 (Appendices 1 and 2 in Red Book).

Experimental designs can become very sophisticated when special properties are required, such as to cope with a relatively large number of treatments without exposing each subject to every treatment. Whenever possible, however, complicated designs should be avoided; special properties are obtained only when certain assumptions can be made, such as absence of interaction between certain factors, and these cannot always be relied upon. Furthermore, missing values sometimes have to be fitted due to accidental circumstances, and this is a very difficult process with other than simple designs. Complicated designs also require more difficult analysis than simple ones.

Examples will be found of the use of various response metameters (transformation functions applied to observed quantities) used either to render variances independent of the magnitude of the observed quantity or to secure a linear relationship between the observed quantity and some physical magnitude associated with the stimulus or applied treatment.

Although the distribution functions of many quantities used here can be treated as though Gaussian, Poisson or binomial according to their nature, it is sometimes necessary to deal with continuous distribution functions that are appreciably skew; in such cases lognormal or gamma distributions are commonly found useful. In other cases, where Poisson or binomial distributions might be expected to apply*, it may be found that the observed variances are greater than would be expected indicating the presence of some heterogeneity; this may be dealt with by use of some compound distribution function† in which the probability of an event must be taken to possess some dispersion which may be definable in terms of a suitable continuous distribution function. The observed variance may also be found to be less than expected; this can occur when samples are taken systematically from a set of populations having different mean values and the samples pooled together, such as when scoring logatoms in an articulation test by

* Such as 'events' during a conversation (Poisson) and correct responses in an articulation test (binomial).

† If a particular cause can be assigned, for example to differences between individual members of a testing team, the experiment can, of course, be arranged so that the heterogeneity attributable to this factor is separated in the analysis of variance.

allocating marks for initial consonant, vowel and final consonant separately, and then adding all the marks together.

References to literature explaining the various mathematical and statistical techniques will be found in the relevant sections of the bibliography.

2. MECHANISM OF HUMAN COMMUNICATION

2.1 SPEECH

2.1.1 The production mechanism (including artificial mouths)

The organs by which speech is produced are illustrated in Figure 2.1. The main vocal tract extends from the larynx to the lips, this distance being about 170 mm for men and about 135 mm for women. For certain sounds the velum (soft palate) is lowered thereby coupling the nasal passages to the main vocal tract at about its centre. When speaking, the passage from pharynx to esophagus is closed and air passes from the lungs through the trachaea to the larynx, which provides a narrow slit, about 18 mm long and extending about 3 mm in depth. The width of this slit (termed the glottis) is variable; during the production of voiced sounds, the folds of the

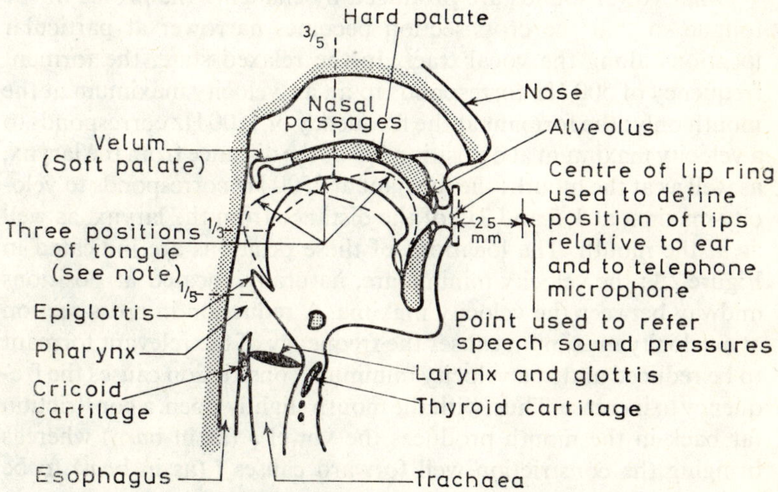

Figure 2.1. Median section of a mouth and vocal tract.
Note: The solid line shows the relaxed condition. The broken lines show tongue positions that give constrictions at the back (—-—) and front (– – – –) of the mouth

tissue of the larynx are stretched between thyroid and cricoid cartilages and the air forced through the glottis causes them to vibrate. (These folds are often called the vocal cords.) Air is thereby admitted to the vocal tract as a rapid sequence of pulses, the rate of pulsation (denoted by F_0) averaging about 130 per second for men and about 250 for women and ranging, for any given person, over perhaps ± half an octave.

For non-nasal sounds, with the muscles controlling the articulatory organs relaxed and the mouth slightly open, the vocal tract consists of a tube, practically constant in cross-section, open at one end (the mouth) and having a fundamental resonant frequency of 500 Hz and other resonances at frequencies of 1500, 2500, etc. Hz. These resonances are termed 'formants' and their frequencies are denoted by F_1, F_2, F_3, etc. The damping is such that the bandwidths of the formants are about 50 Hz for F_1 and F_2 and about 100 Hz for F_3. When excited by vibration of the folds of the larynx, such conditions result in emission of the neutral vowel ɜ (as in b*ir*d). Greater opening of the lips causes the vowel *a* (as in b*a*rn) to be emitted whereas nearly closing the lips produces *u* (as in boot); opening the mouth effectively shortens the length of the resonant tract and raises the resonant frequencies whereas partial closure increases the air velocity at the lips and thereby lowers the resonant frequencies.

Other vowel sounds are produced by changing the profile of the tongue so that the cross-section becomes narrower at particular locations along the vocal tract. In the relaxed state, the formant frequency of 500 Hz corresponds to an air velocity maximum at the mouth only; the formant at the frequency of 1500 Hz corresponds to a velocity maximum at a position 1/3 of the distance from the larynx, as well as at the mouth; the formant at 2500 Hz corresponds to velocity maxima at 1/5 and 3/5 of the distance from the larynx, as well as at the mouth. The locations of these positions are indicated in Figure 2.1; the velocity minima are, naturally, located at positions midway between the velocity maxima. A reduction in cross-section at a velocity maximum causes the frequency of the relevant formant to be reduced and at a velocity minimum constriction causes the frequency to increase. Thus, with the mouth slightly open, a constriction far back in the mouth produces the vowel ɔ (as in b*or*n) whereas bringing the constriction well forward causes *i* (as in b*ea*t) to be emitted.

Emission of speech from the vocal organs and the complementary process of recognition by the hearing mechanism cannot be treated as a simple process of transmitting information coded as a fixed set of sound patterns whose physical characteristics are precisely associated with elements in a language system. It is important to distinguish *phonetic* analysis, which is concerned with descriptions of the sounds themselves, from *phonemic* analysis, which is concerned with definitions of the distinctive elements of a particular language system. On the one hand, closely related dialects of the same language system can have quite different phonetic structures, although the phonemic system is the same; on the other hand, two languages, differing vastly in phonemic structure, can employ rather similar phonetic systems. An example of this distinction in English is provided by the sound *l* which has two differing phonetic forms, namely as in Wi*ll*iam, when the sound is formed near the front of the mouth and is sometimes called the clear form, and as in o*l*d, when it is formed further back in the mouth and sometimes called the dark form. These distinctly different pronunciations have no phonemic significance in English, their occurrence being dictated purely by rules of the language. In Russian and Spanish, however, it is possible to change the meaning of a word by using the wrong form. Another example is the phonemic distinction between aspirated and unaspirated plosives which is important in Eastern Armenian whereas, in English, both are used but only according to rules of pronunciation such as that the aspirated form in initial positions, like *p*ill but the unaspirated form follows a sibilant like s*p*ill.

The material given in this section is based mainly on phonemic distinctions, in the most part relating to the English language. Naturally, telephone transmission engineers deal directly with physical waveforms of signals and so, perhaps, their first interest is in phonetic descriptions of the signals a telephone system must handle most often in order to transmit speech faithfully. Information on such matters is supplied here by giving (necessarily rather crudely) physical descriptions of some of the waveforms typically associated with the various phonemes when emitted by a limited, and probably rather standardised, selection of talkers.

The phonemic descriptions are useful in the study of redundancy of messages, interpretation of the results of articulation tests and in understanding linguistic factors in the probabilities of certain confusions.

Table 2.1 gives a list of vowels and vowel combinations and includes indications of their relative frequencies of occurrence in English conversational speech over telephone connections. Three degrees of stress have been distinguished in the sample and the frequencies with which the vowels occurred in syllables bearing each degree are shown. Stress is the linguistic emphasis used, for example, to distinguish such different words as 'content' (that which is included) and 'content' (satisfied)—the first word has stress on its first syllable and, in the other word, the second syllable is stressed. Complete words may have stress laid upon them to denote their relatively greater importance. A syllable spoken with stress may have its vowel increased in intensity or its duration may be increased.

The various vowels are produced by particular combinations of the location of the constriction along the vocal tract and the opening or closing of the mouth (by lowering or raising the jaw). Figure 2.2 shows many of the vowels arranged according to their mode of formation and according to the frequencies of their first and second formants; a striking resemblance can be seen between the two parts of the figure.

The excitation for vowels is provided solely by modulation of the air stream from the lungs by the glottis; the production of consonants, however, involves other types of excitation either without 'voicing' or with additional modulation at the larynx by vibration of the vocal cords. The consonant sounds are listed in Table 2.2. Plosive, or stop, consonants are produced by the explosive emission of sound on opening a complete obstruction at the lips, alveolus, velum or pharynx; these (and many other consonants) exist as unvoiced/voiced pairs. The fricative consonants involve excitation by creating turbulence at particular locations along the vocal tract, near the teeth, alveolus or other parts; these consonants also exist as unvoiced/voiced pairs and, except for *h*, are represented by different symbols. *h* is unvoiced except when preceded by and followed by a vowel as in the word a*h*ead. The remaining consonants are always voiced in English although unvoiced counterparts of some exist in certain languages; for example the unvoiced counterpart of *j* exists in German in the word *ich* and the *ll* sound in Welsh represents the unvoiced counterpart of *l*.

The nasals are produced by lowering the velum, so opening the passage past the uvula and coupling the main vocal tract to the path through the nose. The semivowels are formed by various constrictions made by the tongue or lips. Absence of any sound before or after a vowel can affect the meaning and so blanks must be included as significant occurrences. The relative frequencies of occurrence of consonant and consonant combinations in English conversational speech are shown in Table 2.3.

The order in which the consonants are shown in Table 2.2 within each class corresponds to the places of articulation increasingly

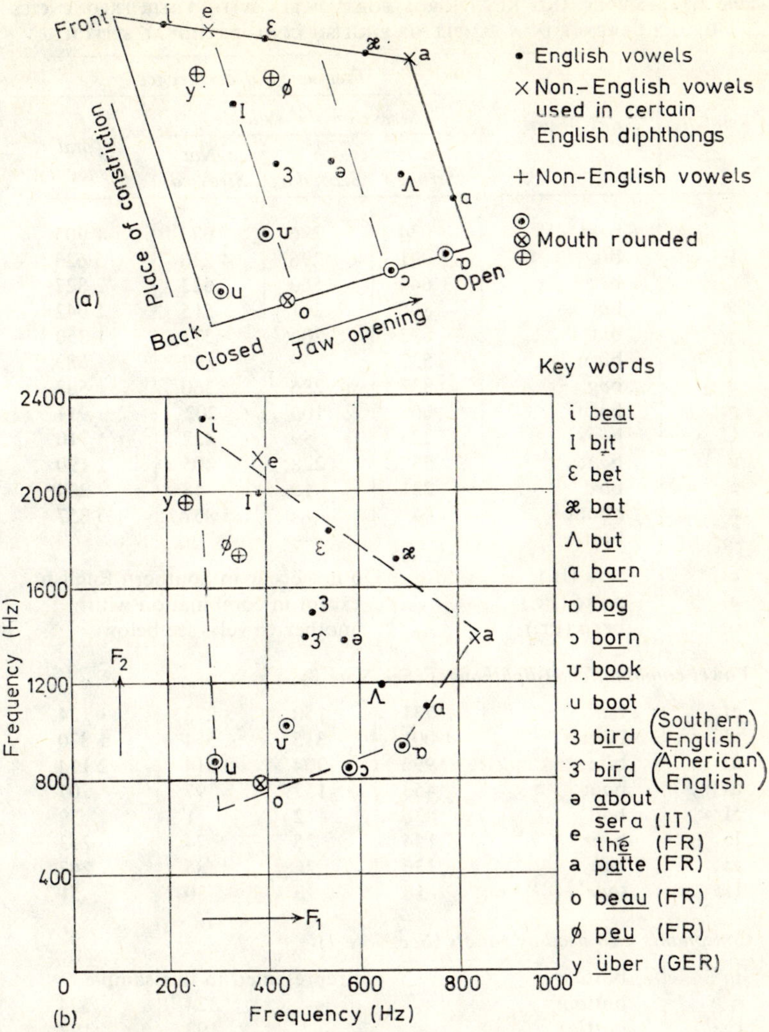

Figure 2.2. The vowel diagram and frequencies F_1 and F_2. (a) Vowel diagram (b) Frequencies of first and second formants

removed from the mouth towards the larynx and their formant frequencies falling.

Some languages make use of other types of speech sound; examples are nasalised vowels (French and Portuguese), clicks (Hottentot and other languages of Southern Africa) and implosives

Table 2.1. SYMBOLS AND KEY WORDS FOR VOWELS WITH THEIR FREQUENCIES OF OCCURRENCE IN A SAMPLE OF ENGLISH CONVERSATIONAL SPEECH

Symbol	Key Word	*Frequency of occurrence* — *Stress on the syllable* — *a* Strong	*b* Secondary	*c* Not Stressed	*f* Total Frequency
i	b*ea*t	934	296	675	1 905
I	b*i*t	1 091	296	4 236	5 623
ɛ	b*e*t	1 644	560	623	2 827
æ	b*a*t	866	261	515	1 642
ʌ	b*u*t	873	206	271	1 350
a	b*ar*n	537	49	97	683
ɒ	b*o*g	937	286	360	1 583
ɔ	b*or*n	609	100	202	911
U	b*oo*k	229	59	422	710
u	b*oo*t	557	228	405	1 190
ɜ	b*ir*d	231	15	48	294
ə	*a*bout butt*er*	—	—	5 857	5 857
e	s*e*ra (It.)	Do not occur in Southern English except in combination with another vowel—see below			
a	p*a*tte (Fr.)				
o	b*eau* (Fr.)				
Vowel combinations (diphthongs) (See Note 2)					
eI	b*ai*t	731	150	273	1 154
əU	b*oa*t	1 001	315	554	1 870
aI	b*i*te	996	304	814	2 114
aU	b*ou*t	453	157	97	707
ɔI	b*oi*l	36	2	1	39
Iə	h*ear*	154	25	74	253
eə	h*air*	136	26	85	247
Uə	t*our*	18	3	30	51
Consonants also used as vowels (See Note 1)					
m	bott*om*	not represented in this sample			
n	butt*on*	—	—	324	324
l	bott*le*	—	—	192	192
	TOTAL	12 033	3 338	16 155	31 526

Note 1 The semivowels w and j (key words *w*e and *y*ou) have been treated, when they occurred in initial positions of syllables in this sample, entirely as consonants (see Tables 2.2 and 2.3). An alternative method of classification would have been to treat them as parts of rising diphthongs and triphthongs. If this had been done the occurrences of w and j as single initial consonants, shown in Table 2.3 would have been replaced by a corresponding increase in the frequency of initial blanks. When w or j followed another consonant, the reallocation would have increased the numbers of those other consonants. The alternative method of classification would also have resulted in a redistribution of the frequencies of occurrence of the vowels and vowel combinations

and the various diphthongs and triphthongs containing w and j would have had to be included in Table 2.1.

Although it was not done in this sample, it would have been possible to treat the sounds w and j as final consonants; for example, the word 'true' could have been divided tr+u+w and the word 'boy' as b+ɔ+j instead of treating them, as here, as tr+u+— and b+ɔI+—. A consequence of this arrangement would be the designation of the vowel combinations, shown in Table 2.1 as eI, əU, aI, aU and ɔI respectively, by the following: ɛj, əw, aj, aəw and ɔj.

The other semivowels l and r cause less difficulty in their allocation in English but it is worth noting that r, which is not used as a vowel in Southern English, does occur in American English, for example in the word 'butt*er*'.

Note 2 Other diphthongs occur in English, for example the word 'boar' ought, strictly, to be pronounced bɔə although it is often spoken as bɔ. Triphthongs also exist, namely: eIə (key word pl*ayer*); aIə (key word f*ire*); aUə (key word p*ower*); oUə (key word s*ower*).

Note 3 Material published in the United States of America often includes key words which are confusing to a speaker of Southern English. The sources of confusion are mainly in the phonemes *a*, *ɒ* and æ. American English does not use *ɒ*, the words 'hot', 'dock' being pronounced as a Southern English speaker would pronounce 'heart' and 'dark', the distinction being made in American English by pronunciation of the r sound which, in Southern English, is silent in these final consonant positions. Note also that words such as 'cast' which, in Southern English use, the same vowel sound as 'cart', are pronounced in American English and in Northern English with the same vowel sound as 'back' and 'bang'. The words 'soul', 'boat' and 'moan' pronounced in Southern English with a diphthong variously sounding like oU, əU or *ɒ*U are sometimes shown in American English as using the pure vowel o which is not used in Southern English.

Note 4 In other publications, to avoid as much as possible the use of special printed or typewritten characters, the alternative sets of symbols shown in Table 2.1a have been used for the vowels; they are reproduced here only for convenience when referring to other published information and will not be mentioned again.

Table 2.1a

Symbol used here	*Alternative sets of symbols* 1	2	3
i	i:	ii	ē
I	i	i	i
ɛ	e	e	e
æ	a	a	à
ʌ	ʌ	ʌ	o
a	a:	aa	a
ɒ	o	—	—
ɔ	o:	oo	ò
U	u	u	u
u	u:	uu	ū
ɜ	ə:	əə	r̄
ə	ə	ə	—

Alernatives 1 and 2 can only, of course, be used when restricted to describing the vowels that have phonemic distinctions in English, the symbols properly belonging to the non-English vowels *e*, *a* and *o* having been used to replace ɜ, æ and *ɒ*.

Alternative 3 has been applied to American English where e and oU may be pronounced more nearly as the pure vowels *e* and *o*. The vowel in b*ir*d is articulated in American English with greater constriction of the vocal tract and can be denoted by she symbol ɝ to distinguish it from the Southern English version of the same word which sound is properly denoted by ɜ. r̄ is therefore perhaps better equated to ɝ rather than ɜ.

Table 2.2. SYMBOLS AND KEY WORDS FOR CONSONANTS

Symbol	*Class*	*Place of Formation*	*Voiced/ Unvoiced*	*Position in the syllable*	
				Initial	*Final*
p	Plosive	Lips	unvoiced	*p*at	u*p*
b	Plosive		voiced	*b*et	ro*b*
t	Plosive	Teeth or alveolus*	unvoiced	*t*en	se*t*
d	Plosive		voiced	*d*ay	ha*d*
k	Plosive	Velum*	unvoiced	*c*an	ba*ck*
g	Plosive		voiced	*g*one	bi*g*
f	Fricative	Lower lip and teeth	unvoiced	*f*at	i*f*
v	Fricative		voiced	*v*an	mo*v*e
θ	Fricative	Teeth*	unvoiced	*th*in	bo*th*
ð	Fricative		voiced	*th*en	ba*the*
s	Sibilant fricative	Alveolus*	unvoiced	*s*ee	thi*s*
z	Sibilant fricative		voiced	*z*one	choo*se*
ʃ	Sibilant fricative	Palate	unvoiced	*sh*ed	di*sh*
ʒ	Sibilant fricative		voiced	plea*s*ure†	rou*ge*†
h	Aspirate	Glottis	usually unvoiced	*h*ad	does not occur
m	Nasal	Lips closed	voiced	*m*ade	fro*m*
n	Nasal	Tongue to alveolus	voiced	*n*ot	i*n*
ŋ	Nasal	Back of tongue to velum—tip to lower teeth	voiced	does not occur	lo*ng*
w	Semi-vowel	Lips	voiced	*w*et	does not occur
l	Semi-vowel	Laterally at palate; tongue to alveolus	voiced**	*l*et	fa*ll*
r	Semi-vowel	Palate§	voiced	*r*ed	does not occur‡
j	Semi-vowel	Velum—tip of tongue to lower teeth	voiced**	*y*ou	does not occur
—	Blank	—	—	-at	go-

* For these sounds the tongue forms a closure or constriction at the place indicated.

† The consonant ʒ occurs very rarely as a separate sound in English; it occurs more commonly in the combination dʒ (key words: *j*ump; ri*dge*). Other related combinations are tʒ (key words: *ch*um: ri*ch*) and ts (rare in English except as final in plurals) which, as a group, are sometimes classed as single phonemes called affricates; the symbols ʃ͡, ʒ͡ and s͡ are sometimes used for these consonants.

‡ Final r is not sounded in Southern English but, in writing, the symbol denotes modification of the vowel; for example: b*i*d—b*ir*d; p*e*t—p*er*t; b*a*n—b*ar*n; b*u*n—b*ur*n; p*o*t—p*or*t.

§ Rolled r is formed by the tip of the tongue flapping against the alveolus.

** Unvoiced partners of l and j exist in other British languages, for example the Welsh ll and thc of Scottish English.

Table 2.3. FREQUENCES OF OCCURRENCE OF CONSONANTS AND THE MORE COMMON CONSONANT COMBINATIONS IN A SAMPLE OF ENGLISH CONVERSATIONAL SPEECH

Symbol	*Initial consonant of syllable*				*Final consonant of syllable*			
	a	b	c	f	a	b	c	f
Plosives								
p	350	23	184	557	205	37	54	296
pt	—	—	—	—	17	2	1	20
ps	—	—	—	—	25	1	2	28
pr	80	5	20	105	—	—	—	—
pl	68	6	26	100	—	—	—	—
b	418	156	536	1110	93	1	3	97
bz	—	—	—	—	7	—	1	8
br	37	2	5	44	—	—	—	—
bl	30	2	31	63	—	—	—	—
bj	7	—	1	8	—	—	—	—
t	501	87	1256	1844	1278	462	1304	3044
ts	—	—	—	—	90	65	214	369
tʃ	61	2	29	92	93	17	29	139
tj	9	1	17	27	—	—	—	—
tw	70	16	—	86	—	—	—	—
tr	121	6	54	181	—	—	—	—
d	598	117	472	1187	550	133	452	1135
dz	—	—	—	—	13	—	31	42
dʒ	155	47	52	254	34	2	40	76
dʒd	—	—	—	—	3	—	5	8
dr	21	1	56	78	—	—	—	—
dj	5	3	8	16	—	—	—	—
k	502	106	280	888	516	86	83	685
kt	—	—	—	—	46	5	19	70
ks	—	—	—	—	101	12	30	143
kst	—	—	—	—	37	4	2	43
ksθ	—	—	—	—	9	—	—	9
kw	73	6	14	93	—	—	—	—
kr	30	1	3	34	—	—	—	—
kl	54	3	1	58	—	—	—	—
g	502	154	132	788	—	—	—	—
gr	47	1	11	59	—	—	—	—
gl	7	—	7	14	—	—	—	—
Fricatives								
f	421	47	239	707	267	27	97	391
ft	—	—	—	—	11	1	—	12
fθ	—	—	—	—	8	1	—	9
fr	58	16	46	120	—	—	—	—
fj	25	3	—	28	—	—	—	—
v	115	22	110	247	438	115	455	1008
vd	—	—	—	—	7	1	—	8

Table 2.3. *(Contd.)*

Symbol	*Initial consonant of syllable*				*Final consonant of syllable*			
	a	b	c	f	a	b	c	f
Fricatives contd.								
vz	—	—	—	—	9	—	1	10
vj	8	2	1	11	—	—	—	—
θ	253	38	92	383	68	6	9	83
θr	55	12	1	68	—	—	—	—
ð	386	227	1526	2139	105	21	91	217
s	719	196	259	1174	723	263	477	1 463
sp	63	5	2	70	—	—	—	—
st	126	6	17	149	182	40	53	275
str	26	—	7	33	—	—	—	—
sk	9	—	—	9	14	2	—	16
skt	—	—	—	—	13	—	1	14
skr	10	—	—	10	—	—	—	—
sm	21	1	—	22	—	—	—	—
sw	8	—	—	8	—	—	—	—
sl	10	—	—	10	—	—	—	—
z	36	—	39	75	401	144	767	1 312
zd	—	—	—	—	22	3	2	27
ʃ	129	36	128	293	50	3	31	84
ʃt	—	—	—	—	4	1	6	11
ʒ	—	—	8	8	2	—	—	2
h	563	195	581	1339	—	—	—	—
Nasals								
m	645	134	354	1133	437	108	429	974
mp	—	—	—	—	7	—	2	9
mz	—	—	—	—	17	2	7	26
n	680	199	249	1128	1244	282	990	2 516
nt	—	—	—	—	260	99	183	542
nts	—	—	—	—	9	2	22	33
nd	—	—	—	—	174	21	469	664
ndz	—	—	—	—	21	—	2	23
ndʒ	—	—	—	—	15	1	1	17
nθ	—	—	—	—	8	—	—	8
ns	—	—	—	—	39	5	32	76
nz	—	—	—	—	48	5	46	99
nj	24	14	1	39	—	—	—	—
ŋ	—	—	—	—	108	14	670	792
ŋk	—	—	—	—	134	23	12	169
ŋks	—	—	—	—	22	—	—	22
ŋz	—	—	—	—	16	6	12	34

Table 2.3. *(Contd.)*

Symbol	Initial consonant of syllable a	b	c	f	Final consonant of syllable a	b	c	f
Semi-vowels								
w	734	394	1014	2142	—	—	—	—
l	581	62	247	890	650	207	577	1 434
ld	—	—	—	—	51	8	14	75
lf	—	—	—	—	29	6	6	41
lv	—	—	—	—	16	—	—	16
ls	—	—	—	—	12	—	—	12
lz	—	—	—	—	21	3	7	31
r	436	41	491	968	—	—	—	—
j	501	350	934	1785	—	—	—	—
Blank	1618	591	6606	8815	3153	1085	8398	12 636

Note Combinations that occurred less than 8 times in this sample have not been listed above

(various African languages). Certain languages distinguish aspirated from nonaspirated plosives (some Indian languages). The significance of stress in English has already been mentioned but other prosodic features are important in certain languages, examples being intonation (Chinese, Swedish and Serbian) and relative durations (Finnish and Hungarian).

Additional information is given later (in Section 2.1.3) on the frequency composition of particular sounds but, for more detailed information, reference should be made to the more specialised publications quoted in Section 2.1 of the bibliography.

It is possible to produce speech sounds artificially by means of a suitable model of the vocal tracts. Acoustical models have been constructed reproducing features of the speech production mechanism but these are difficult to control, especially if realistic successions of speech sounds are to be produced. Electrical analogues are more flexible and can be suitably controlled and the electrical form in which the speech is produced is quite suitable if it is merely desired to present the result to a listener. However, if a true representation of speech emitted from a mouth is required there are considerable difficulties in converting the electrical signals to sound pressures so that the radiation pattern of a real mouth is reproduced accurately. Such a converter is termed an artificial mouth and is required for testing close speaking microphones such as are used in telephone handsets. The fact that the source position varies from one sound to another is a serious complication.

Practical artificial mouths for testing the microphones of handset telephones consists of a loudspeaker pressure unit emitting sound through a small orifice, usually circular, to represent the mouth opening. It is difficult to provide great enough power handling capacity to represent real speech at natural levels below about 200 Hz but this is not a serious limitation when the microphone to be tested is linear and intended to transmit only the telephony frequency range. Measurement of carbon telephone microphones with artificial mouths poses certain additional problems; the performance of such microphones is much affected by the presence in real speech of very low frequency breath pressures* which are not easy to represent in an artificial mouth. Attempts to reproduce breath pressures have been made by use of a device which emits puffs of air to supplement the voice frequency sound pressure produced by a loudspeaker pressure unit.

Within the above mentioned limitations, artificial mouths are available which represent the sound pressure surrounding real mouths within the frequency range 200–4000 Hz so that sensitivities of linear handset microphones measured with such devices agree reasonably well with those measured by real voice methods (*see* Chapter 5). The location of the equivalent position of the lips is very important when close speaking microphones are being tested and this can be determined for a given artificial mouth as follows:

The decrease in sound pressure as a small measuring microphone is moved away from the human mouth is consistent with the assumption of an equivalent point source of radiation located 6 mm inside the mouth. The assumptions that the sound pressure varies inversely as the distance from this point gives a satisfactory description of the sound field required from the artificial mouth. Sound pressure measurements can be made in a similar manner on an artificial mouth and the equivalent point source located; this is usually some small distance *outside* the orifice of the artificial mouth, 5 mm outside in the case of the British Post Office equipment. Locating the equivalent lip position 6 mm further from the orifice than the equivalent point source (namely 11 mm from the orifice) then defines the equivalent lip position for the artificial mouth. The inverse

* These puffs are particularly prevalent with plosives, the amplitude rising some 20 dB above the accompanying speech sound pressure; they are very directional and are much attenuated beyond 40 mm from the mouth.

distance law for the decrease of sound pressure from the equivalent point source enables speech sound pressures measured at one distance to be referred to a point at a different distance; the law does not apply, of course, at very small distances from the lips, correction being needed for distances less than about 20 mm from the lips. Further information is given in Figure 2.5.

Use of a circular orifice does not, of course, attempt to reproduce the asymmetry of the real mouth but this would require considerable extra complication because sounds are not only radiated from the mouth but also from the nose and some radiation comes from the chest.

2.1.2 The specification of speech

When persons engage in conversation the streams of speech by which they communicate can be observed and analysed in many different ways according to the purpose of the analysis. When studying the performance of a speech communication system which is intended to facilitate such conversational activity, a physical description of the waveform as a function of time is required. Psychologists, linguists and phoneticians will be interested in certain rather detailed forms of description but even the telephone engineer must be able to take advantage of the fact that certain transformations of the waveforms concerned are more important than others; knowledge of these can be derived from study of the speech production and recognition mechanisms. A speech communication system can fulfil its purpose perfectly without necessarily providing faithful reproduction of all audible waveforms. (This aspect of speech transmission has been touched upon in Chapter 1.2.)

Knowledge of a sufficient specification of speech would permit considerable economy to be achieved in transmission by adopting highly sophisticated techniques of analysis and synthesis and attempts in this direction are mentioned in Section 6.5 of Chapter 6; the design of conventional transmission systems, however, requires knowledge only of relatively crude descriptions of the main features of the speech process.

In describing the stream of speech emitted by a given person when conversing, the following terms will be used:

Sentence; the least unit of speech capable of conveying intelligible information.

Utterance; a portion of the stream of speech that is spoken continuously without any noticeable pause.

Word; the smallest significant linguistic unit that can normally be spoken in isolation. (The smallest significant linguistic unit is termed a 'morpheme' but this may consist of part of a word and may contain more than one syllable.)

Syllable; the smallest unit of a word that is separately pronounceable.

Phoneme; one of a set of phonetic elements used to describe the speech sounds that distinguish meanings of words in a language.

Syllabic segment; that unit of speech during which the statistical properties of the waveform may be considered to be stationary.

The sentence is not a very useful subdivision for the present purpose except when constructing artificial test material as discussed in Chapter 3.4. Similar considerations apply to words (*see* Chapter 3.3).

The utterance is the first useful subdivision of a stream of speech because it enables periods of time when speech is present to be distinguished from those when speech is absent but possibly some other sound, eg noise, remains. In principle, a speech detecting instrument (speech detector) can be used to perform the necessary distinction.

Syllables can, fairly readily, be distinguished within utterances because they are characterised by consonant–vowel–consonant combinations (either or both consonants can be absent, *see* Tables 2.2 and 2.4). Most of the power of speech is associated with the vowels, which are, furthermore, generally of greater duration than the consonants so that fluctuation of power at syllabic rate is quite evident and can readily be detected.

Because the articulatory organs which are responsible for producing speech cannot move very quickly, the rates of change of the essential characteristics of the waveforms emitted while speaking are limited. It is therefore possible to segment the waveform into intervals of not less than 10 ms during which such features as the mean power, frequency spectrum and probability density distribution may be considered to remain constant; of course, some

Table 2.4. FREQUENCIES OF OCCURRENCE OF VARIOUS CLASSES OF CONSONANTS AND CONSONANT COMBINATIONS IN A SAMPLE OF ENGLISH CONVERSATIONAL SPEECH

Class of consonant or combination of consonants	*As initial consonant of syllable*				*As final consonant of syllable*			
	a	b	c	f	a	b	c	f
Plosives	2 871	643	2 860	6 374	2 676	723	1 905	5 304
Plosive combinations	881	102	337	1 320	483	108	375	966
Fricatives, excluding h	2 059	566	2 401*	5 026	2 054	579	1 927	4 560
h	563	195	581	1 339	—	—	—	—
Fricative combinations†	440	46	77	563	284	49	66	399
Nasals, m or n	1 325	333	603	2 261	1 681	390	1 419	3 490
ŋ	—	—	—	—	108	14	670	792
Nasal combinations, m or n	24	15	4	43	620	135	765	1 520
ŋ					174	29	24	227
Semivowels, w or r	1 170	435	1 505	3 110	—	—	—	—
l	581	62	247	890	650	207	577	1 434
j	501	350	934	1 785	—	—	—	—
Semivowel combinations‡	—	—	—	—	150	19	29	198
Blank (consonant absent)	1 618	591	6 606	8 815	3 153	1085	8 398	12 636
TOTAL	12 033	3338	16 155	31 526	12 033	3338	16 155	31 526

* of these, 1526 are ð.
† Fricative combinations do not contain h.
‡ Semivowel combinations contain only l (finals).

Note Combinations are classified according to the first consonant of the group; for example, *nd* from the word 'and' is classed as a nasal combination only and not as a plosive combination. *str*ict is a fricative combination, *ol*d is a semivowel combination and *pl*ease is a plosive combination. The frequencies shown for plosives, fricatives, nasals and semivowels are for those sounds occurring alone and do not include occurrences of the same sound when in combination with another.

sounds, particularly pure vowels, remain stationary much longer than this. (*See* definition, given above, of 'syllabic segment'.)

The actual waveforms associated with individual phonemes in conversation are difficult to isolate, identify and classify because their essential physical features are very complicated. In any case the physical features associated with given phonemes do not exist as separate successive waveforms independent of adjacent phonemes but tend to overlap and change gradually into one another and be considerably modified by their neighbours. The physical characteristics of the waveforms associated with certain phonemes when particular words are spoken carefully have been extensively studied in attempts, now partially successful, to recognise and synthesise speech automatically and these results are of general interest in considering conventional speech transmission systems. Their relevance must not, however, be overemphasised because the phonetic structure of speech during telephone conversations differs considerably from that of a person reading written material aloud in a standard form of the language. Results of an analysis of the words used in 60 conversations (120 persons) covering about 100 topics under telephone conditions are available and a brief description follows.

The speech used was transcribed into phonetic symbols and the words (being the largest units of the material that could conveniently be sorted, classified and counted) assembled in decreasing order of frequency of occurrence. Not only were different inflections treated as separate words but distinctions were made according to the amounts of stress actually employed*. Three degrees of stress were recognised, a. strong stress; b. secondary stress and, c. unstressed. The sample contained 24 782 words and these were classified as 2213 different words, ignoring stress. Of the whole sample, 48% of the words were strongly stressed, 13% had secondary stress and 39% were unstressed. The proportion of strongly stressed words in telephone conversations appears to be considerably higher than is usual in non-telephone, face to face conversation. The 24 782 words contained 31 526 syllables, an average of 1.27 syllables per word which is lower than the figure of 1.43 quoted by Dewey from his examination of written English; this confirms the impression that monosyllables are more common in telephone conversation than in writing.

Table 2.5 lists the 20 most commonly occurring words (taking all degrees of stress) and it will be seen that this includes a very high proportion of words having weak forms. The least commonly occurring words are represented by 880 different words that occurred once only in this sample; Table 2.7 gives a selection of 20 representatives and it will be seen that most are polysyllabic and none

* About 50 words very commonly used in conversation have more than one pronunciation and the weak forms are almost always unstressed but even the strong forms of the same words can also be used unstressed.

Table 2.5. LIST OF 20 WORDS THAT OCCURRED MOST FREQUENTLY IN THE SAMPLE

Word	*Pronunciation*	a	b	c	f
the	ði	11	—	—	836
	ðI	—	—	58	
	(w)	—	—	767	
you	ju	96	109	295	821
	(w)		—	321	
I	aI	105	112	510	727
yes	jɛs	283	198	152	633
to	tu	8	15	223	615
	tə	—	—	369	
and	ænd	8	5	105	455
	(w)	—	—	337	
it	It	5	32	355	392
a	eI	11	—	—	362
	(w)	—	—	351	
well	wɛl	71	82	182	335
that	ðæt	106	63	95	329
	(w)	—	—	65	
oh	əU	53	35	210	298
we	wi	18	44	202	264
in	In	38	11	189	238
of	ɒv	2	5	32	231
	(w)	—	—	192	
on	ɒn	61	32	134	227
no	nəU	121	60	36	217
have	hæv	74	44	36	211
	(w)	—	—	57	
is	Iz	50	18	136	204
are	*a*	59	15	43	199
	(w)	—	—	82	
see	si	123	45	19	187
Total for above words		1 303	925	5553	7781
Total for all words of sample		11 869	3186	9727	24 782
Percentage represented by above words		11%	29%	57%	31%

a indicates use with strong stress
b indicates use with secondary stress
c indicates use unstressed
f total number of occurrences
(w) weak pronunciation, not precisely identifiable

is unstressed. The occurrence of these words is much dependent on the topic of conversation and a sample of many more conversations than 60 would be needed to obtain a proper representation of all words likely to be used in telephone conversation. In this sample some words were over represented because they were particularly relevant to one of the 100 topics involved; for example the word 'mudguard' occurred 11 times in the sample, all in one conversation.

The commonly occurring words, particularly when unstressed, convey very little information and efficiency in transmitting them

Table 2.6. LIST OF 20 WORDS SPOKEN WITH STRONG STRESS THAT OCCURRED MOST FREQUENTLY

Word	*Pronunciation*	*Frequency*	*Rank ignoring stress*
yes	jɛs	283	4
see	si	123	20
no	nəU	121	16
that	ðæt	106	10
know	nəU	106	22
I	aI	105	3
you	ju	96	2
all	ɔl	94	40
got	gɒt	87	30
do	du	81	31
don't	dəUnt	80	24
good	gUd	79	54
down	daUn	76	62
time	taIm	76	65
have	hæv	74	17
what	wɒt	72	23
well	wɛl	71	9
think	θIŋk	70	55
there	ðeə	68	29
not	nɒt	68	39
Total for above words		1936	
Total for all words of sample		11 869	
Percentage represented by above words		16%	

faithfully is probably of little importance.* The rarer words are nearly all strongly stressed and are probably essential key words. The stressed words that occur frequently (*see* Table 2.6) are probably also very informative in their context or they would not have

Table 2.7 REPRESENTATIVE SELECTION OF 20 WORDS FROM THE 880 WORDS THAT OCCURRED ONCE ONLY IN THE SAMPLE

Word	*Stress*
altered	a
Belfast	a
carbonate	a
cloth	a
cricket	a
district	a
expected	a
frock	b
hanging	a
issues	a
lime	a
misreading	a
outside	a
ports	a
regards	a
salary	a
soul	a
Sunday	a
tinting	a
whatnot	a

a indicates use with strong stress
b indicates use with secondary stress

been uttered with stress. All stressed words contain a stressed syllable but very few contain more than one stressed syllable; there were 11 869 stressed words in the sample and these contained 12 033 stressed syllables. In studying the relative frequencies of occurrence of different phonemes and different combinations of phonemes it is therefore appropriate to confine attention to those

* Of course, distortion of an unstressed word must not make it seem to the listener to be stressed; other gross forms of distortion must also be avoided if they are likely to draw attention to certain sounds that would otherwise be unimportant to the listener.

Table 2.8. FREQUENCIES OF OCCURRENCE OF A SELECTION OF THE TRANSITIONS BETWEEN INITIAL CONSONANT AND VOWEL OF STRESSED SYLLABLES
Total: 12 033 stressed syllables

Vowel/Key word		*Total frequency*	*Initial consonant**						
			—	w	s	n	m	d	l
			1618	*734*	*719*	*680*	*645*	*598*	*581*
ɛ	bet	1644	224	145	169	60	24	21	93
I	bit	1091	167	57	71	8	79	63	66
əU	boat	1001	167	30	34	237	44	89	68
aI	bite	996	188	30	27	91	53	5	66
ɒ	bog	937	139	158	10	82	35	39	53
i	beat	934	41	68	176	13	78	14	47
ʌ	but	873	118	87	74	60	96	52	74
æ	bat	866	65	1	19	11	66	14	10
eI	bait	731	43	58	61	11	51	80	25
Total for above		9073	1152	634	641	573	526	377	502

Vowel/Key word		*Total frequency*	*Initial consonant**						
			h	k	g	t	j	r	*Total for these phonemes*
			563	*502*	*502*	*501*	*501*	*436*	*8580*
ɛ	bet	1644	30	3	81	109	318	55	1332
I	bit	1091	19	7	29	27	0	26	619
əU	boat	1001	51	9	186	16	5	12	948
aI	bite	996	8	0	2	84	0	157	711
ɒ	bog	937	19	66	121	13	0	18	753
i	beat	934	27	22	0	55	0	27	568
ʌ	but	873	52	110	0	17	5	7	752
æ	bat	866	178	46	6	3	0	11	430
eI	bait	731	1	67	29	47	0	49	522
Total for above		9073	385	330	454	371	328	362	6635

* Initial consonants comprise single phonemes and multiple combinations of two or more phonemes; multiple initial consonants are much less frequent than single phonemes, the most frequent multiples being dL (key word jump) (which occurred 155 times) and st (which occurred 126 times).

Table 2.9. FREQUENCIES OF OCCURRENCE OF A SELECTION OF THE TRANSITIONS BETWEEN VOWEL AND FINAL CONSONANT OF STRESSED SYLLABLES
Total: 1 2033 stressed syllables

*Final Conso-nant**	*Total frequ-ency*	Vowel									
		ε	I	əU	aI	ɒ	i	ʌ	æ	eI	*Total for these pho-nemes*
		1644	*1091*	*1001*	*996*	*937*	*934*	*873*	*866*	*731*	*9073*
—	3153	121	7	615	273	55	442	20	11	313	1857
t	1278	160	68	126	263	270	33	16	143	85	1164
n	1244	243	155	71	50	113	115	209	88	45	1089
s	723	350	89	4	58	35	15	10	26	39	626
l	650	188	63	26	32	27	26	2	17	18	399
d	550	81	53	14	22	17	33	13	69	41	343
k	516	37	44	5	47	66	75	22	102	53	451
v	438	110	56	1	58	3	42	32	107	11	420
m	437	40	26	19	90	33	5	147	14	29	403
z	401	15	84	42	17	38	61	17	8	14	296
Total for above	9390	1345	645	923	910	657	847	488	585	648	7048

* Final consonants comprise single phonemes and multiple combinations of two or more phonemes; combinations occur less frequently than single phonemes although the contrast is less marked than with initial consonants (*see* Table 2.8). The most frequent multiple final consonants were nt (which occurred 260 times), st (182 times), nd (174 times) and ŋk (key word thi*nk*) (174 times).

present in stressed syllables. Tables 2.1 and 2.3 show the frequencies of occurrence, of all the phonemes considered separately when taken from syllables having the three degrees of stress; considerations of transitions between phonemes within syllables has, however, been confined to syllables uttered with strong stress. An important use for this analysis is in the construction of artificial syllables (logatoms) for conducting articulation tests (*see* Section 3.3); unless these are based on strongly stressed syllables they are likely to contain many examples of unpronouncable combinations of sounds.† The pronunciations of weakly stressed and unstressed

† In articulation tests, logatoms are always spoken with strong stress.

syllables are often difficult to identify precisely because they have wide variations and their pronunciations are usually different from those of the same syllable spoken with strong stress.

Table 2.8 shows the thirteen most frequently occurring initial consonants and the frequency with which they combined (in the sample already referred to) with the nine most frequently occurring vowels. These combinations account for about 55% of all the stressed syllable combinations of initial consonant and vowel. It will be seen that the combinations are far from random and are strongly influenced by those in such common words as *yes*, *no*, *go*, *I*, *up*, *what*, *got*, *have*, *day*, *come* and *right*. A large proportion of those not included in the table is accounted for by other common words such as: *you*, *think*, *can't*, *be*, *back*, *more*, *for* and *all*.

Table 2.9 shows similar information for combinations of the same nine vowels with the ten most frequently occurring final consonants; these account for about 50% of all combinations of vowel and final consonant in stressed syllables. Again the distribution is far from random and certain very common endings of words can be identified.

2.1.3 Physical characteristics important in telephony

2.1.3.1 General

The foregoing description of speech in mainly phonemic terms is related to phonetic processes which can be described in terms of patterns which change continuously with time. These can be displayed as physical observations of patterns in the dimensions of amplitude and frequency which change with time and Figure 2.3 shows a short sentence displayed in this way. The illustration has been simplified very considerably. The information was obtained by use of 14 bandpass filters, the outputs of which were reproduced on graphic level recorder traces. These traces were examined and the amplitudes roughly quantised. The plain areas show where the output was not more than about 18 dB above the level of the weakest sounds; the dotted areas represent the range 18–23 dB above the weakest sounds; the hatched areas represent the range 23–28 dB and the black areas above 28 dB relative to this datum. Much more sophisticated methods are available for more detailed examination using the sound spectrograph but Figure 2.3 provides a useful introduction to speech patterns.

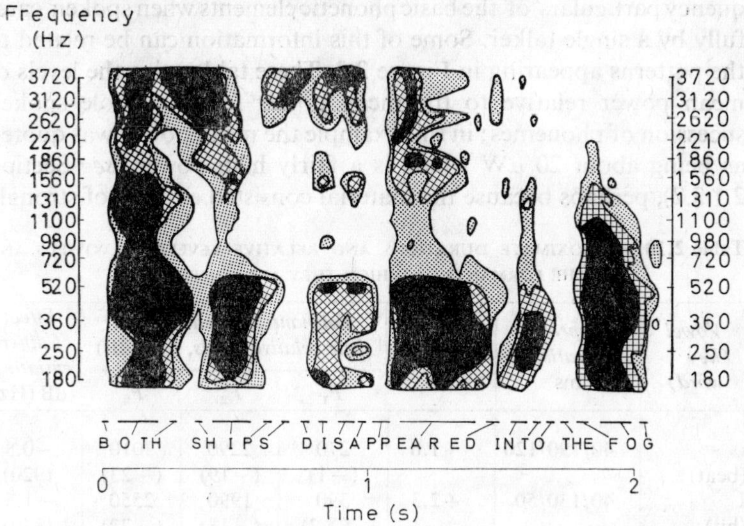

Figure 2.3. Speech intensities as functions of frequency and time

Figure 2.4 gives a broad picture of the relative amplitudes of different frequency components in the speech of a given talker and Tables 2.10 and 2.11 show the amplitude/duration/formant fre-

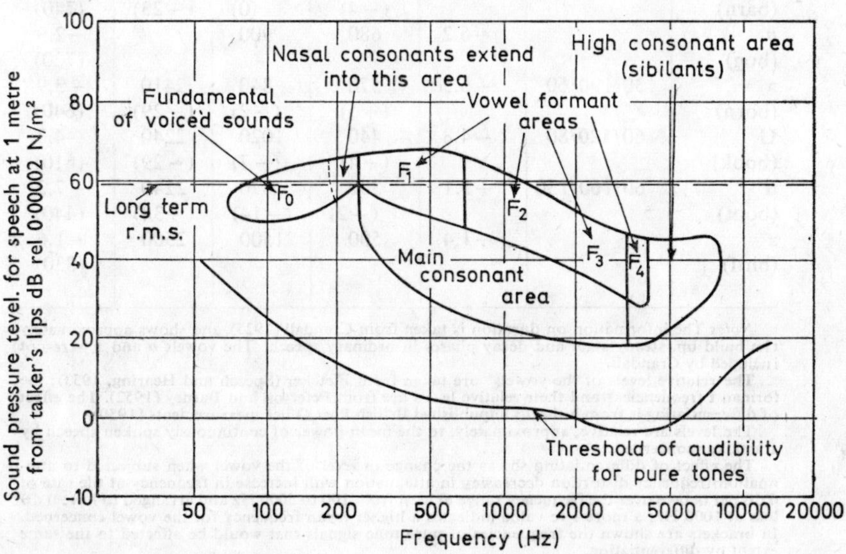

Figure 2.4. Intensity–frequency distribution of speech signals

quency particulars of the basic phonetic elements when spoken carefully by a single talker. Some of this information can be related to the patterns appearing in Figure 2.3. These tables give the levels of mean power relative to the mean power for the whole spoken sucession of phonemes; in this example the mean power was quoted as being about 20 μW which is a fairly high power (*see* Section 2.1.3.2), perhaps because the material consisted entirely of strongly

Table 2.10. APPROXIMATE DURATIONS AND RELATIVE LEVELS OF VOWELS AND THE FORMANTS OF WHICH THEY ARE COMPOSED

Vowel (key word)	*Approx. duration* ms	*Relative level* db	*Formant frequencies and their relative levels,* Hz (dB) F_1	F_2	F_3	*Effect of differentiating* dB (Hz)
i (beat)	40/190/120	+1.6	270 (+1)	2290 (−19)	3010 (−23)	−0.8 (920)
I (bit)	40/130/50	+2.3	390 (+2)	1990 (−18)	2550 (−22)	−1.5 (840)
ε (bet)	30/120/70	+3.6	530 (+3)	1840 (−12)	2480 (−19)	+1.0 (1130)
æ (bat)	40/180/80	+5.1	660 (+4)	1720 (−7)	2410 (−17)	−1.6 (830)
ʌ (but)	50/180/60	+5.3	640 (+4)	1190 (−5)	2390 (−22)	−1.9 (800)
a (barn)	30/200/80	+6.0	730 (+4)	1090 (0)	2440 (−23)	−2.6 (740)
ɒ (bog)		+6.2	680	900		−2.9 (720)
ɔ (born)	30/190/60	+6.5	570 (+5)	840 (−2)	2410 (−29)	−3.9 (640)
U (book)	60/120/80	+4.8	440 (+4)	1020 (−7)	2240 (−29)	−4.3 (610)
u (boot)	60/160/130	+3.1	300 (+2)	870 (−14)	2240 (−38)	−7.2 (440)
ɜ (bird)		+1.4	500	1500	2500	−1.6 (830)

Notes The information on duration is taken from Crandall (1925) and shows approximately the build up, steady state and decay phases in ordinary speech. The vowels *ɒ* and ɜ were not included by Crandall.

The relative levels of the vowels* are taken from Fletcher (Speech and Hearing, 1953); the forman t frequencies* and their relative levels are from Peterson and Barney (1952). The effect of differentiating is from hitherto unpublished British Post Office measurements (1950).

The levels are relative, approximately, to the mean power of continuously spoken speech by the talker concerned.

The effect of differentiating shows the change in level of the vowel when subjected to attenuation/frequency distortion decreasing in attenuation with increase in frequency at the rate of 6 dB per octave over the frequency range at least from 100 to 5000 Hz and arranged to have 0 dB loss at 1000 Hz; a more +ve value indicates a higher mean frequency for the vowel concerned. In brackets are shown the frequencies of pure tone signals that would be affected to the same extent by differentiating.

* Except *ɒ* and ɜ; the information on these vowels is less reliable.

Table 2.11. APPROXIMATE DURATIONS AND RELATIVE LEVELS OF CONSONANTS AND THEIR COMPONENTS

Con-sonant	*Class*	*Approx. dura-tion* ms	*Relative level* dB	*Spectral components and their relative levels;* Hz (dB)			
				F_1*	F_2	F_3	F_4
p	Plosives	30	−14	$= F_0$	800	1350	
b		160		(−15)			
t		60	to	$= F_0$	1700	2450	
d		120	−10	(−15)			
k		80		$= F_0$	rather indefinite		
g		110		(−12)			
f	Frica-	220	−15	$= F_0$	1150	2500	3650
v	tives	220	to −11	(−13)	(−32)		
θ		20	†	$= F_0$	1450	2550	
ð		190		(−13)	(−22)	(−40)	
s	Sibilant frica-	230	−10	$= F_0$	2000	2700	
z	tives	230	to − 9	(−11)	(−28)	(−37)	
ʃ		180	‡	$= F_0$	2150	2650	
ʒ		200		(−10)	(−19)	(−34)	
h	Aspirate						
m	Nasals	170	− 5	Close to F_0	1250	2250	2750
				(−6)	(−16)	(−18)	(−25)
n		160	− 6	(−8)	1450	2300	2750
					(−21)	(−23)	(−28)
ŋ		160	− 3	(−5)	2350	2750	
w	Semi-		+ 2				
l	vowels	160	− 2	450	1000	2550	2950
				(−3)	(−16)	(−15)	(−19)
r			+ 1	500	1350	1850	3500
				(0)	(−7)	(−16)	(−24)
j			− 2				

* Absent for unvoiced plosives and unvoiced fricatives
† Except θ which is particularly weak, about −22 dB
‡ Except ʃ which is particularly strong, about −3 dB

stressed syllables whereas only about 38% of syllables are strongly stressed in telephonic conversation.

Telephone speech paths are required to transmit all the essential features of patterns such as those illustrated in Figure 2.3 and, if great economy of transmission capacity is required, the transmission system can be matched to the task by limiting itself to transmission of a finite set of alternatives which provide a sufficient representation; further mention of the possibilities is made in Section 2.1.4 below and in Section 6.5 of Chapter 6.

Most telephone connections involve relatively unsophisticated speech paths and so only rather gross particulars of the speech process need usually be considered. The sections that follow are concerned with (2.1.3.2) amplitude structure of the total speech signal power, that is integrating the power over vertical slices of Figure 2.3 and considering probability density distribution in which the order of occurrence is neglected; (2.1.3.3) time structure, in which the vertical slices of Figure 2.3 are integrated but the pattern in time of the resulting power is considered; and (2.1.3.4) frequency structure, in which horizontal slices of Figure 2.3 are integrated within narrow frequency bands and the resulting long term mean power–density spectrum is considered. The resulting information is sufficient for the study of a very wide range of speech transmission systems at present in use and likely to remain important.

2.1.3.2. *Amplitude structure*

Introduction

Strictly speech is an acoustical phenomenon but, in telephony, it is mainly the electrical analogues of relevant acoustical quantities that must be considered. The power emitted by a talker is radiated in all directions from his vocal organs (mainly mouth and nose) as alternating sound pressure waves. Usually the conversion from sound pressure to electrical amplitude (voltage or current) is performed by a pressure sensitive microphone*. With a high quality microphone the acoustical/electrical conversion can be virtually free of distortion over the frequency range concerned (say 60–8000 Hz). The relationship between acoustical power and sound pressure is given by the following expressions:

$$\text{Sound intensity, } I = \frac{p^2}{\varrho c} \text{ W/m}^2$$

* Other types of microphone are sensitive to particle velocity, pressure gradient, etc. but they can still be calibrated in terms of sound pressure.

where p is the sound pressure in N/m², ϱ is the density of the medium in kg/m³ and c is the velocity of propagation in the medium concerned (m/s). When the medium is air, the product ϱc is equal to 400. Because of the analogy to the electrical formula for power, Watts = (Volts)²/Ohms, the quantity ϱc is sometimes treated as the 'characteristic impedance' of the medium, the units being 'acoustical ohms' or Rayls (after Lord Rayleigh).

It is convenient to use logarithmic measures for both sound pressure and sound intensity and the following reference values are employed; these correspond approximately to the threshold of audibility for sound at a frequency of 1000 Hz (*see* Section 2.2).

$$p_0 = 2\times10^{-5}\ \text{N/m}^2\ (0.0002\ \text{dyn/cm}^2)$$

$$I_0 = 10^{-12}\ \text{W/m}^2\ (10^{-16}\ \text{W/cm}^2)$$

The sound pressure level (SPL) = $20 \log_{10} (p/p_0)$ dB and the sound intensity level = $10 \log_{10} (I/I_0)$ dB. These two levels are, of course, numerically exactly the same when p_0 and I_0 are related by the expression given above.

The total power emitted by a sound source is obtained by integrating the intensity over the surface of a sphere which contains the source. An actual measurement is not very practical for speech and so the vocal power output of a talker uttering speech sounds is expressed as the level relative to 'reference vocal level' (RVL) defined in terms of the acoustical speech level measured at a point defined in position relative to the talker's lips. RVL is defined below but first it is necessary to describe the measuring equipment required.

Measuring equipment

Acoustical speech level is measured by means of a calibrated high quality microphone, equaliser (to render the sensitivity independent of frequency over the speech range), amplifier and speech voltmeter. The combination of microphone, equaliser and amplifier is conveniently provided by using the sending end of the ARAEN (*see* Section 3.1), the free field sensitivity of which is such that a free field sound pressure of 0.1 N/m² produces an output of 1 V across 600 Ω. The speech voltmeter employed must be specified together with any necessary rules for interpreting the readings. The following tables (Tables 2.12 and 2.13) show the characteristics of certain speech level meters, of the indicating instrument type, and the relative readings obtained with these and certain other, rather more sophisticated, speech level measuring equipment. Detailed descrip-

Table 2.12. CHARACTERISTICS OF SPEECH LEVEL METERS OF THE INDICATING INSTRUMENT TYPE

Item	*Name of instrument*	*Zero reference level*	*Integration time* ms	*Overswing* %	*99% rise time* ms
1	BPO speech voltmeter type 3 (SV3)	1 V	100	2.5	230
2	SFERT (CCITT) volume indicator	1.9 V (6 mW in 600 Ω)	200	0–5	410–650 depending on overswing
3	Vu-meter	0.775 V (1 mW in 600 Ω)	about 120	1–1.5	300

Notes The above instruments all employ nominally square law rectification. The rectification law for Items 1 and 2 is required to conform rather closely to the index 2 but the rectification law for Item 3 is better described as conforming to the index 1.4. The instruments are calibrated to read correctly with a continuously applied sinusoidal input at 1000 Hz.

The integration time is defined as the duration of a constant amplitude pulse of pure tone which will cause the needle just to rise to a value that is 63% of the angular deflection that would have been obtained if the same amplitude had been applied indefinitely.

The 99% rise time is the time that must elapse after application of a constant amplitude pure tone before the needle has risen to 99% of the final angular deflexion. The relationship between integration time and rise time is rather complicated and is a function of the amount of overswing.

tions of these and other speech level measuring devices are given elsewhere (*see* Bibliography).

According to CCITT definition the term 'volume'* denotes an electrical quantity analogous to 10 $\log_{10}$ (speech power) and is considered to be constant for a given talker while active. 'Volume indicators' and 'volume meters' are intended to measure such a quantity and so are scaled in dB relative to 1 mW or some other power. In fact, they are not power meters but are voltmeters and so read correctly only when applied to a circuit of given impedance (usually 600 Ω). The term 'speech voltmeter' is considered more logical and will be used here, although, for convenience, such meters are scaled in dB relative to 1 V (20 $\log_{10}$ V), the abbreviation for which is dBV.

To measure vocal level, the microphone must be located at a suitable and known distance from the lips of the talker. The microphone associated with the ARAEN is suitable for location in ordinary laboratory environments at distances from, say, 100–350 mm

* In some early material published in the United Kingdom the word 'volume' has been used in a wider, or more general, sense relating to the loudness of speaking or the loudness of the sounds heard; this practice is now deprecated.

Table 2.13. RELATIVE READINGS OF VARIOUS SPEECH LEVEL MEASURING INSTRUMENTS

Item	*Name of instrument or quantity measured*	*High quality speech*		*Speech from telephone sets; continuously spoken or conversational material*
		Continuously spoken material	*Conventional phrase or logatoms*	
1	BPO speech voltmeter type 3 (SV3)	0 dBV†	0 dBV	0 dBV†
2	SFERT (CCITT) VI*	−6.8 dB†	about −6.8 dB	−6.8 dB†
3	Vu-meter	−3.2 vu†	about +0.5 vu	−3.2 vu†
4	BPO speech voltmeter type 4A (SV4A)††	0 dBV	not suitable	0 dBV
5	BPO speech voltmeter type 5A or 5B (SV5A or SV5B)§	−5.5 dBV	−1.4 dBV	(a)‡ −7.0 dBV (b) −5.7 dBV (c) −6.0 to −6.5 dBV
6	Level of long term mean power while talker is active**	−3.3 dBm	—	(a)‡ −4.8 dBm (b) −3.5 dBm (c) −3.8 to −4.3 dBm

* Readings by the SFERT volume indicator are termed 'dB relative to reference volume' or 'dB relative to reference telephonic power (RTP)'.

† Certain rules apply for interpreting the fluctuations of the indicating instruments of these meters. The SV3 and SFERT VI are read in accordance with the CCITT method which is to observe that level which is exceeded, on the average, once every 3 s. The vu-meter is read differently, namely by taking the average of the peak deflections about every 10 s, after excluding the two or three highest; this method yields readings on the vu-meter about 3 dB lower than using the CCITT method on the same meter.

†† The SV4A contains an arrangement for interpreting the fluctuations in short term mean in a manner, resembling the CCITT method. It can be used equally well on fragmentary conversational speech and continuously spoken material; it cannot, however, be used on logatoms or formally repeated conventional phrases.

§ The SV5A yields the r.m.s. voltage (expressed in dBV) averaged over a fixed period of 1, 3 or 10 s at choice. The SV5B is identical in construction except that the timing is interrupted by a voice operated switch when the talker pauses so that the integrating period is 10 s of 'active' speech. For definition of 'active', see text. When measuring conversational speech with the SV5A, considerable skill on the part of the operator is required because inclusion of even a few of the normal conversational pauses will reduce the reading appreciably below that obtained with the SV5B. The SV5B is therefore more suitable than the SV5A for measuring fragmentary material.

** The values shown assume that the speech levels have been measured on a 600 Ω circuit and that 'activity' is defined in the manner used for the SV5B. It is, naturally, important to associate with this quantity the corresponding activity factor, i.e. the proportion of total time during which the talker is active.

‡ The relative readings depend to some extent upon the characteristics of the microphone. The values shown apply as follows: (a) Post Office transmitter inset No.13 obtained with SV5A; (b) Conversational speech from actual conversations originated at customers' telephone stations (probably mostly Post Office transmitters inset No. 16) obtained with the SV5B, comparison being made with readings by SV4A; (c) Linear microphones, mounted in handsets and having a sensitivity/frequency characteristics similar to that desirable for a modern commercial telephone (the values shown are estimates expected to apply for modern telephone microphones).

but a smaller probe microphone could be used down to distances of perhaps 25 mm for the lips. If the distance is too small, the microphone may distort the sound field being measured or, with the probe microphone, the simple inverse distance law relating sound pressure with distance from the equivalent point source may not apply sufficiently accurately. If too great a distance is used in an environment having surrounding reflecting surfaces, the reflections from these will interfere with the direct sound from the talker. The distance used by the United Kingdom Post Office with the ARAEN sending end is 337 mm but the distance of 140 mm specified to be used with the same microphone when it forms part of NOSFER (*see* Section 3.2) would also be suitable.

Reference vocal level was originally defined to enable loudness rating tests and articulation measurements to be conducted at controlled vocal levels. In such cases the speech material consists of a repeated conventional phrase or a sequence of logatoms, usually set in a carrier phrase. This enables a simple indicating instrument type of speech voltmeter to be used, such a meter being very convenient for the talkers to monitor their own vocal levels and to control them to a prescribed value. When ordinary meaningful speech is being uttered, especially in telephone conversations, the difficulty of measuring the electrical speech levels is greatly increased. The instrument used by the United Kingdom Post Office (and now by the CCITT Laboratory) for measuring conventional phrases is the speech voltmeter type 3 (*see* Item 1 in Tables 2.12 and 2.13) and, for such speech material, is read by noting the maximum excursions of the needle. The SV3 is identical to the 'ARAEN volume meter' described in Supplement No. 10, White Book, Volume V (1968).

Reference vocal level is defined, fundamentally, in terms of the reading on the SV3; Reference vocal level is identical, where conventional phrases or logatoms are concerned, with the level of acoustical speech power used by the CCITT Laboratory for articulation (AEN) measurements with SRAEN. The normal speech power specified to be used for measuring reference equivalents with NOSFER is also defined in terms of readings on the SV3 but corresponds to a level about 5.6 dB lower than RVL. (The normal speech power for measurement of reference equivalents was formerly defined in terms of the SFERT Volume Indicator.)

In principle, reference vocal level is defined for continuously spoken or fragmentary conversational speech in exactly the same way as for the conventional phrases, namely in terms of the reading on the SV3.

Although rules have been standardised by the CCITT for interpreting such readings, the results are rather unreliable and subject to considerable observer bias as well as rather large random errors. The SV4A was therefore developed to reproduce standardised readings on the SV3 when continuous or fragmentary conversational speech was being measured. The SV4A simulates, to a considerable extent, the principle underlying the SV3 and its associated rules for interpretation of the fluctuations of its pointer and so readings by SV4A are interchangeable with those of the SV3 for ordinary speech. The readings are not comparable, however, when conventional phrases or logatoms are being measured and the SV4A is then unsuitable.

Because it has been found very desirable to relate speech levels on telephone channels to the level of the long term mean power for system loading calculations, the SV5A (and later SV5B) was developed (*see* Table 2.13) and their success has rendered the SV4A unnecessary. The SV4A is no longer used but a great deal of published information quotes speech levels measured with it.

When the speech is spoken continuously with very little pausing, the readings on the SV5A and SV5B will not differ. When there are frequent pauses, as in telephone conversations, attempts have been made to use the SV5A by switching it on manually when the talker seemed likely to continue uttering for at least the integrating time used; this, however, seems to result in some bias and the results then differ to some extent from those obtained with the SV5B which will discard time intervals while no speech is present in accordance with the following definition of 'active' speech. The control signal to the speech switch is rectified and smoothed by a two section RC integrator of 30 ms time constant per section and the timer is activated whenever the smoothed signal exceeds −15 dB relative to the level of the final reading of the equipment. To avoid interrupting the timing for structural pauses, a hangover of 200 ms is provided in the switching logic so that pauses less than about 350 ms will not stop the timer (the balance of the hangover is provided by the decay time of the RC smoothing circuit, in association with the release threshold of the switch which is slightly lower than the operate threshold). When using the SV5B a few abortive readings have to be discarded until the sensitivity has been set properly.

The speech levels given below in defining reference vocal level apply when the measuring microphone is located at a distance of 337 mm from the talker's lips on the axis of his direction of speech. If the free field sound pressure level at any other distance along this axis is required, the allowances shown in Figure 2.5a can be made; these are calculated on the assumption that the equivalent point source of speech is 6 mm inside the mouth measured from the outer surface of the lips. Although the equivalent point source can be

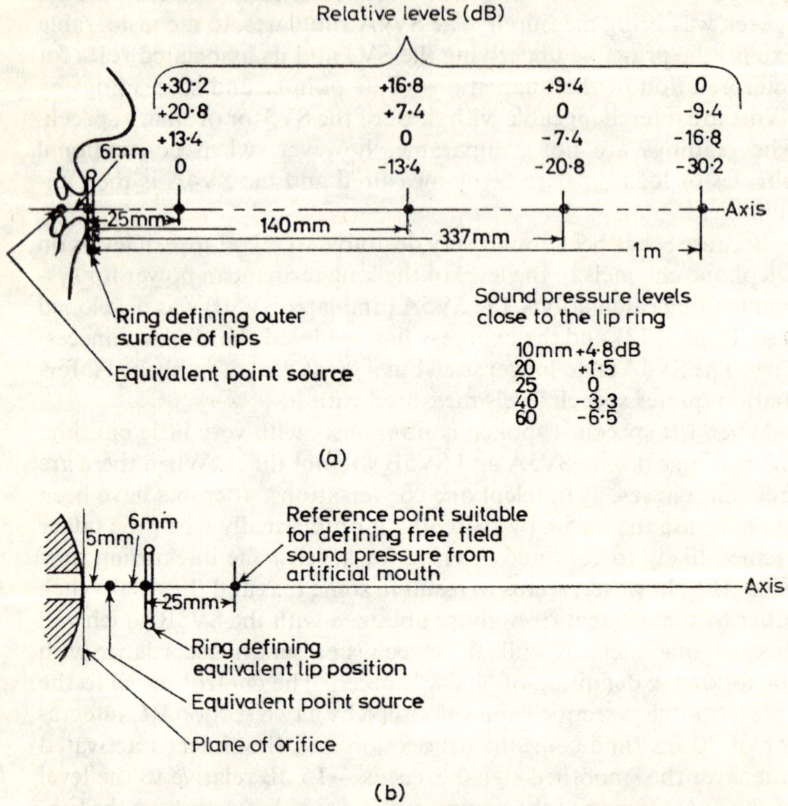

Figure 2.5. Reference points to define sound pressure levels from real and artificial mouths. (a) Real mouth. (b) Artificial mouth

taken for the present purposes at the point shown, this is an approximation, the different speech sounds being formed at different places in the mouth so varying the equivalent point source. For most types af artificial mouth, including that used by the British Post Office, the equivalent point source (so far as changes in free field sound pressure with distance are concerned) lies just outside the orifice as shown in Figure 2.5b. The equivalent lip position of the Post Office artificial mouth thus lies 11 mm in front of the plane of the orifice.

Definition of Reference Vocal Level (RVL)

A talker is said to be uttering speech at reference vocal level when the following free field r.m.s. sound pressures are obtained at a point 337 mm in front of the lips in a reasonably nonreflecting environment.

(a) Logatoms, spoken in a carrier phrase, or other selected conventional phrases. Using a SV3 and taking the maximum excursions of the pointer, the sound pressure should be 0.1 N/m^2 (-20 dB relative to 1 N/m^2) corresponding to a speech sound intensity level of $+74$ dB relative to 1 pW/m^2. Integrating the sound intensity in all directions from the talker, this corresponds to a total speech sound power of about 27 μW and would apply approximately to the vowel sounds which are mainly responsible for the reading of the SV3.

(b) Ordinary speech uttered continuously, such as reading from a book. Using a SV3 and reading it in the CCITT manner, the value obtained should be the same as for (a). Using a SV5A or SV5B, the speech sound pressure should be 0.053 N/m^2. (-25.5 dB relative to 1 N/m^2) corresponding to a speech sound intensity level of $+68.5$ dB relative to 1 pW/m^2. The total integrated speech sound power would then be approximately 7.6 μW.

(c) Telephone conversational speech*. Using a SV5B, the speech sound pressure and intensity level should be as (b).

Distributions of amplitude

The vocal levels of speech produced by different persons can, naturally, depart considerably from RVL. With very little ambient noise, under face to face conversational conditions, levels somewhat lower than RVL tend to be used, particularly by women; when addressing a large audience, however, levels are much higher, especially for men. Results for film actors, on the sound stage, averaged about $+7$ dB relative to RVL and film actresses averaged -1 dB. When conversing over telephone connections, the vocal level increases as conversa-

* It is very difficult to measure vocal level during a telephone conversation, even under laboratory conditions, because it is necessary to maintain a fixed distance between the lips of the participant and the measuring microphone. Furthermore, vocal level varies within any telephone conversation and so the 'vocal level for a conversation' must be averaged over the duration of the conversation.

tion becomes more difficult; on average, RVL applies under rather good telephonic conditions with little room noise, and levels perhaps 5 dB higher are found, on average, when the worst connections likely to be experienced in a public telephone network are encountered. Under such conditions users are likely also to speak more directly into their microphones.

When the physical conditions are constant, vocal levels are distributed approximately according to the Gaussian probability distribution function with a standard deviation of about 2.8 dB. When handset telephones are used, the distribution of speech levels observed on telephone lines is influenced also by variations in the manner of holding the handset and by the characteristics of the microphone so that the standard deviation becomes about 4.8 dB for a fixed commercial telephone connection.

Under field conditions, measurements at local exchanges or group switching centres are affected also by variation in the transmission loss from the talker's mouth to the point of measurement and then the standard deviation may be as great as 5–6 dB. Measurements on telephone connections containing carbon microphones are usually more dispersed than those on connections that contain linear microphones.

Telephony is mainly concerned with conversational speech produced by a population of talkers. In addition to the above mentioned variations from one talker to another and those attributable to variation in losses of the connection, it is necessary to have knowledge of the shorter term fluctuations in power of such speech, so that the load capacity of transmission equipment can be properly dimensioned both from the point of view of overloading and to ensure proper transmission of the lowest level speech components. It is convenient to consider first speech from a single talker or, perhaps more generally, that from different talkers after normalising their long term mean speech powers.

The amplitude distribution of speech can be expressed in various ways; using different integrating periods or none.

Different types of speech can also be considered. Figure 2.6 shows probability density functions for instantaneous amplitudes of speech from a given talker with the microphone at a fixed position in front of the lips. For comparison certain theoretical distributions are shown on the same scale.

The exponential distribution has been used for certain purposes as a theoretical model to represent the distribution of instantaneous speech amplitudes. As can be seen in Figure 2.6, the approximation is poor at very low levels and so it has been suggested that a more suitable model is to accept the exponential form for the vowel sounds which are taken to represent 40% of the total speech duration and to add to this a Gaussian distribution about 20 dB lower

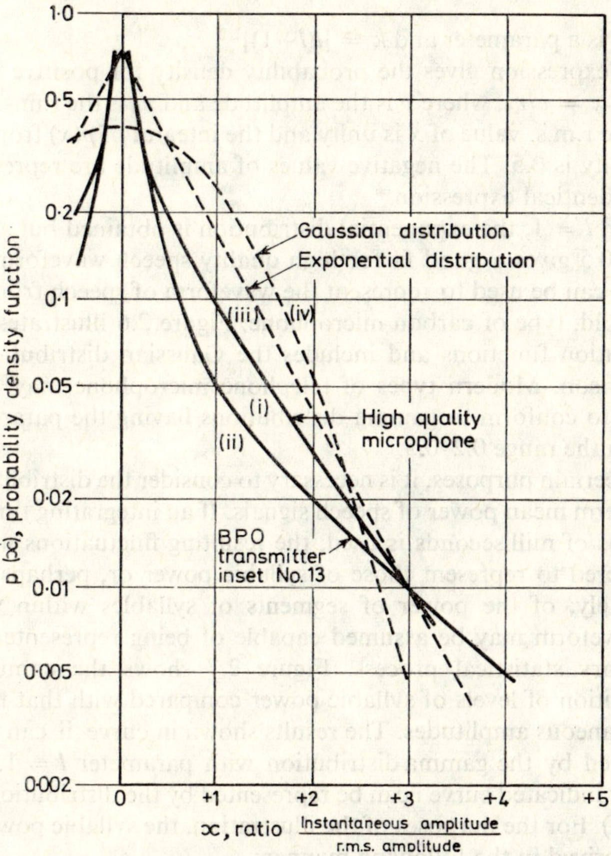

Figure 2.6. Probability density functions for instantaneous amplitudes of speech. Note. Negative values as for x positive

in power level to represent the consonants for the remaining 60%. Alternatively two Gaussian signals 23 dB different in level and equally probable (representing vowels and consonants) has been used.

A much better fit to the distributions shown in Figure 2.6 can be obtained by use of the gamma distribution with a suitable choice of value for the parameter. The gamma distribution can be represented by the expression:

$$p(x) = \frac{k}{2\Gamma(l)} \cdot (kx)^{l-1} \cdot e^{-kx}$$

where l is a parameter and $k = [l(l+1)]^{1/2}$

This expression gives the probability density for positive values of x. If $x = v/v_0$, where v is the amplitude and v_0 is the r.m.s. value of v, the r.m.s. value of x is unity and the integral of $p(x)$ from zero to infinity is 0.5. The negative values of amplitude are represented by an identical expression.*

When $l = 1$, the exponential distribution is obtained but a value of $l = 0.5$ gives a good fit for high quality speech waveforms and $l = 0.2$ can be used to represent the waveform of speech from one, rather old, type of carbon microphone. Figure 2.6 illustrates these distribution functions and includes the Gaussian distribution for comparison. Modern types of telephone microphone may be expected to conform to gamma distributions having the parameter l lying in the range 0.2–0.5.

For certain purposes, it is necessary to consider the distribution of short term mean power of speech signals. If an integrating time of a few tens of milliseconds is used, the resulting fluctuations may be considered to represent those of syllabic power or, perhaps more accurately, of the power of segments of syllables within which the waveform may be assumed capable of being represented as a stationary statistical process. Figure 2.7 shows the cumulative distribution of levels of syllabic power compared with that for the instantaneous amplitudes. The results shown in curve ii can be represented by the gamma distribution with parameter $l = 1.5$. (As already indicated curve i can be represented by the distribution with $l = 0.5$). For the purposes of the illustration, the syllabic power has been derived in the following manner:

The speech signals from a high quality microphone were rectified by a square law device and smoothed with a circuit having a fairly accurate rectangular impulse response of 50 ms duration. Speech was produced by the talker reading steadily from a book.

* The waveforms of speech are often found to be slightly unsymmetrical even when derived from high quality linear microphones, the extent of asymmetry being more marked for some talkers than others. The characteristics of non-linear microphones, like carbon types, may increase this asymmetry to some extent.

The distributions shown in Figure 2.7 are typical of the results for a single talker or for those of several talkers after standardising the values of long term mean power. The dispersion between talkers can be represented by that of the values of long term mean power for

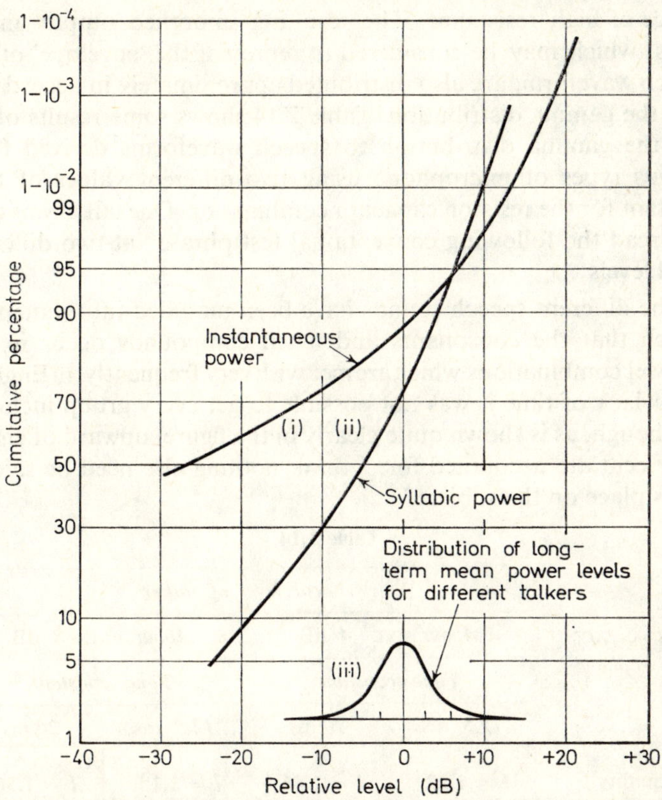

Figure 2.7. Cumulative percentages for distributions of levels of instantaneous and syllabic power of speech from a high quality microphone located at a fixed position relative to the mouth

different talkers and these are dispersed fairly widely. In a fixed environment, their levels (in dB) can be represented by a Gaussian distribution of about 2.8 dB standard deviation; such a distribution is shown by the inset curve iii included in Figure 2.7. The dispersion between talkers under conversational conditions, when observing speech signals from telephone handset microphones, can be considerably greater, especially when the ambient conditions and the efficiency of the telephone connection are also dispersed.

The fluctuation in short term mean power can be determined with somewhat simpler equipment than that mentioned above by rectifying the instantaneous amplitudes of the speech waveform, using an accurate 'linear' rectifier, and then smoothing the rectified amplitudes with a simple parallel resistor/capacitor combination fed from a source of high resistance. The resulting smoothed output amplitudes, which may be considered to represent the 'envelope' of the speech waveform, are also distributed approximately in accordance with the gamma distribution. Table 2.14 shows some results of fitting the gamma distribution to speech waveforms derived from various types of microphone using two different values of time constant for the resistor/capacitor combination. One talker was used who read the following conventional test phrase* at two different vocal levels:

> 'The different speech sounds have been moulded into sentences, such that the consonants and usual compounds occur in the vowel combinations which are met with very frequently in English. For lack of time it was not possible to get every group included although, as is shown quite clearly in the figure, upward of eighty per cent are accounted for. I think nothing else need be said in this place on the subject.'

Table 2.14

Type of microphone	*Vocal level of talker*			
	About RVL+*4* dB		*About* RVL+*8* dB	
	Time constant		*Time constant*	
	12.5 ms	*20* ms	*12.5* ms	*20* ms
High quality linear microphone in telephone handset. Frequency response as for telephone	$l = 1.22$	$l = 1.46$	$l = 1.15$	$l = 1.36$
	$l = 0.84$	$l = 0.96$	$l = 0.98$	$l = 1.12$
Commercial telephone (Type 706)	$l = 0.71$	$l = 0.78$	$l = 0.76$	$l = 0.83$

* This passage was originated by Dr. Collard and is quoted by Dunn and Farmsworth (1939).

The lower value of time constant included in Table 2.14 is, perhaps too small to represent properly the syllabic segment, resulting as it does in somewhat smaller values of the parameter *l*. Higher values of time constant from about 20 ms upward to about 50 ms yield values of *l* that are practically independent of the value of smoothing time constant.

It should be noted that all simple mathematical forms for representing the probability density distributions of speech amplitudes fail for very low amplitudes unless steps are taken to eliminate an appropriate proportion of time during which silences are present. Measurements at very low levels also encounter the difficulty of distinguishing genuine speech from noise, including noise from the breath striking the measuring microphone.

2.1.3.3 *Time structure*

The shortest sample of speech that can convey information to its hearer may be taken as the sentence although very simple forms of vocal emission are occasionally used for special purposes. In telephone conversations, sentences are usually shorter than in writing or in more formal speaking and normally last only a few seconds. For equipment design, it is more convenient to define the structural particulars of conversational speech in terms of physically measurable quantities rather than in terms of syntax and grammatical form and the first requirement is to be able to distinguish the presence of speech from its absence.

The simplest analysis of a speech waveform in the dimension of time is, perhaps, to determine the distribution of the intervals of time during which the amplitude exceeds, or fails to exceed, some assigned value. If the assigned amplitude is very close to zero and the waveform is free from any d.c. component and appreciable noise, the result is a distribution of time intervals between successive crossings of the zero axis; this distribution is closely related to the period of the predominant frequency component of the speech, usually the first formant. Some results are quoted by Davenport (*see* Bibliography) and 1 ms is a typical value for the most probable time interval.

If the assigned amplitude is well below the long term r.m.s. value but comfortably above the noise and the distribution of intervals of time during which the amplitude fails to exceed the assigned value is extended to time intervals of 100 ms and greater the distribution of the durations of silences can be obtained. In continuously spoken material, e.g. a passage read from a book, the probability density falls steadily over the range 100 to almost 300 ms, when it tends to level out up to about 1000 ms. The lower end of this plateau can be identified as approximately the minimum duration of nonstructural silences, i.e. those not related to the phonetic structure of individual phonemes.

When speech from one party in a telephone conversation is studied in this way, the result is rather similar except for silences greater than 1000 ms which are much more frequent in conversation because each party has to pause while the other is talking; in

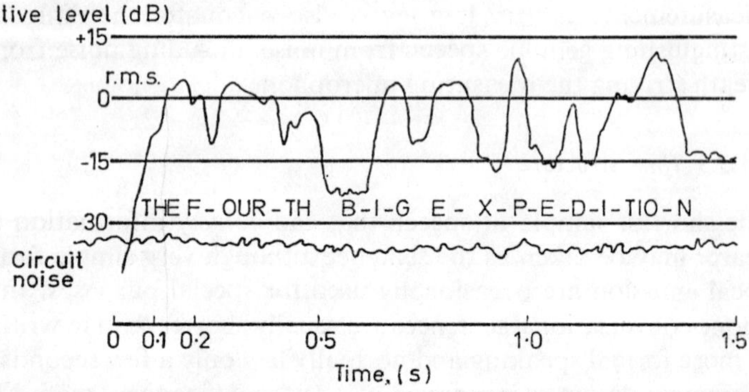

Figure 2.8. Envelope waveform of an utterance

continuously spoken passages, silences greater than 1000 ms are rather rare. If, the combined speech from both parties is examined, e.g. at a two wire point in a telephone connection, the relatively constant probability density in the range 300–1000 ms becomes less marked and, of course, the long (greater than 1000 ms) silences are much less frequent.

Silences can, alternatively, be defined in terms of the envelope of the speech waveform, or the short term mean (or r.m.s.). Figure 2.8 shows the envelope waveform of an utterance; the level has been standardised so that the r.m.s. value of the amplitude of the original speech waveform is unity (0 dB) when averaged over the duration of the utterance. For durations less than 100 ms, there are considerably fewer 'silences' when judged from the envelope than when the instantaneous amplitudes are considered. The silences that are left in the envelope within this range of durations are mostly attributable to the structure of certain speech sounds, such as plosives, when the vocal tract suffers momentary complete closure. For durations greater than about 100 ms, the two forms of detection will usually yield practically the same results.

From the foregoing it can be seen that momentary absence of signal or the signal falling momentarily below some defined amplitude does not, in itself, provide a useful means of registering the absence of speech. Additional difficulties are caused by the presence of noise which is usually associated with the speech under most practical situations. For the purpose of measuring the level of speech while the talker is active, it has proved convenient to define inactive time as that which elapses while the short-term speech power is below a certain value. This value corresponds in Figure 2.8 to a level of −15 dB provided each such period is at least 350 ms (*see* description of speech voltmeter type 5B given above in Section 2.1.3.2). As noted above, this critical duration is the minimum which a listener would recognise as a pause in the flow of speech from the talker.

The durations of definite features of active speech can also be studied and a few relevant particulars are given below. Some of the features are defined in linguistic or phonemic terms and, as explained in Section 2.1.1 above, it is not easy to associate these very exactly with physical features; the results should therefore be taken only as indicating orders of magnitude.

The rate at which persons speak varies from one individual to another but typical numbers of words per minute are from 80 to 200. The mean durations of words therefore range from 300 to 750 ms. On average words in conversational English contain 1.27 syllables per word and so the mean durations of syllables range from about 250–600 ms, depending on the talker; the corresponding frequencies appear fairly prominently when the spectrum of the envelope of a speech waveform is examined; syllables are characterised by the presence of a vowel which usually has a much greater intensity than that of the intervening consonants.

When speech is spoken fairly carefully, the components of each syllable of the consonant–vowel–consonant type can be identified and their separate durations measured and the following results are typical:

Initial consonants	70–250 ms; average about 120 ms
Vowels	150–500 ms; average about 250 ms
Final consonants	100–350 ms; average about 190 ms

These values include clusters of consonants which occur in nitial and final positions and the vowels include diphthongs. Some information is given in Table 2.10 on the build up, steady state and decay periods of vowel sounds and the approximate durations of certain specific consonants are shown in Table 2.11.

For certain purposes it may be desired to analyse speech in time intervals within which the waveform can be described by a stationary statistical process; a suitable time interval for this purpose lies in the range 20–40 ms.

Detailed consideration of actual instantaneous waveforms of speech is, in itself, rarely very informative; the same speech sound may be perceived by a listener although the instantaneous waveform of the sound pressure reaching his ear may assume many different forms when examined oscillographically. Only rather gross statistical descriptions of instantaneous speech waveforms in the time domain are therefore of interest. It is worth noting the following:

> The average number of crossings of the zero axis is about 500 per second; 1% peaks, i.e. those excursions of the instantaneous amplitude that exceed the 99 percentile, recur on average at about 5 per second.

Detection of the presence or absence of speech is a problem that requires different solutions according to the purpose. It will be seen from Figure 2.8 that a threshold detector will be restricted in its sensitivity by the presence of circuit and other noises accompanying the speech and so the problem will involve making some compromise between failing to detect (or being slow to detect) and giving false detections due to the noise. Various facets of the speech detection problems are discussed in Section 4.4.3.

2.1.3.4 Frequency structure

The short term frequency structure of speech within the duration of its quasi-stationary states has already been referred to in Section 2.1.2 and is illustrated in a broad manner by Figure 2.4; this will be mentioned again in Section 2.1.4 in relation to analysis/synthesis systems. The present section is concerned mainly with the long term mean frequency spectrum.

The spectrum density of speech from different individuals does vary, especially between males and females as shown in Figure 2.9. This figure also shows that speaking loudly or quietly also has a considerable effect on the spectrum shape. The actual speech material being emitted will also affect the results and particular care should be exercised in using information on average speech spectrum densities obtained from conversational speech for purposes where special forms of speaking, for example articulation tests, or special speech phrases, for example those in loudness rating measurements, are concerned. For convenience when making certain calculations, the average of the male and female results in Figure 2.9 have been assembled in tabular form in Tables 2.15 and 2.16 using a set of frequencies which will also be used throughout the other chapters. These tables also include certain other

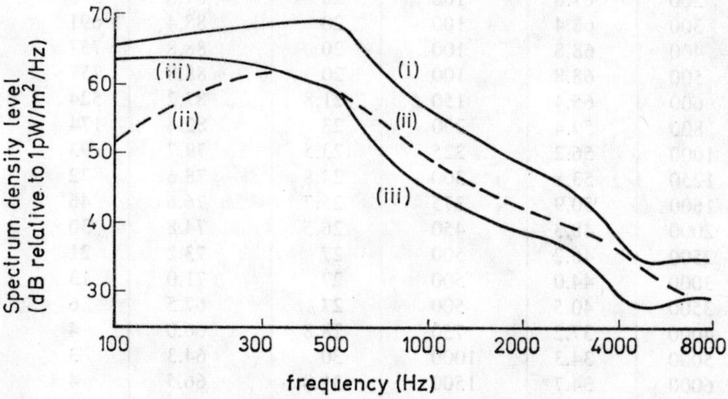

Figure 2.9. Spectrum density levels of speech referred to a point 25 mm in front of the lips. (i) Male, vocal level RVL*+6.9. (ii) Female, vocal level* RVL*−1.2. (iii) Male, vocal level reduced to* RVL

particulars, which will be used later, and shows the relative amounts of power attributable to the frequency bands chosen; it is of interest to note that most of the total power of speech is concentrated at the low frequencies, more than half lying below 450 Hz. Of course, telephone connections do not transmit the whole frequency range shown here and all parts of the range do not have the same importance; discussion of these matters forms important parts of Sections 3.2, 3.3 and 3.4. Many subjective tests are conducted at controlled vocal levels and the long term mean speech spectra appropriate to the levels used are given in the relevant sections of Chapter 3; information on the effects of changes in vocal level is given by Loye and Morgan (1939) and this has been used to derive the required spectrum shapes. Certain other spectrum shapes are used in Sections 3.2, 3.3 and 3.4 for spec

Table 2.15. LONG TERM, FREE FIELD, SPECTRUM DENSITY LEVELS FOR SPEECH MEASURED AT 25 mm FROM THE LIPS OF A MALE TALKER

Frequency Hz	*Spectrum density level* β'_s dB	*Frequency interval represented* Δf Hz	*10* $\log_{10}$ Δf dB	*Level of intensity represented by frequency interval* dB	*Intensity represented by frequency interval (See Notes)*
(1)	*(2)*	*(3)*	*(4)*	*(5)*	*(6)*
100	+65.3	80 (say)	19	+84.3	269×10^6
200	67.6	100	20	87.6	575
300	68.4	100	20	88.4	691
400	68.8	100	20	88.8	757
500	68.8	100	20	88.8	757
600	65.4	150	21.8	87.2	524
800	59.4	200	23	82.4	174
1000	56.2	225	23.5	79.7	93
1250	53.8	300	24.8	78.6	72
1600	50.9	375	25.7	76.6	46
2000	48.3	450	26.5	74.8	30
2500	46.2	500	27	73.2	21
3000	44.0	500	27	71.0	13
3500	40.5	500	27	67.5	6
4000	37.2	750	28.8	66.0	4
5000	34.3	1000	30	64.3	3
6000	34.7	1500	31.8	66.5	4
8000	34.8	2000 (say)	33	67.8	6
					Sum 4045×10^6

Total speech level +96.1 dB

Notes The frequency interval represented by the entry 100 Hz in col. 1 has been taken to extend from 70 to 150 Hz. The subsequent intervals given in col. 3 are between frequencies midway between the values entered in col. 1. The highest frequency considered is 9000 Hz which forms the upper limit of the last interval.

The units for intensity are pW/m².

An intensity of 1 pW/m² is equivalent to a sound pressure in air of 20 μN/m² which, by convention, is used as a datum close to the pure tone threshold of audibility at 1000 Hz.

The total speech level corresponding to reference vocal level is +89.3 and so the level given above is RVL+6.8 dB.

ial conditions and these have been derived from the information given here.

2.1.4 Acoustical descriptions relevant to recognition and synthesis

The desire to construct speech transmission systems capable of using information paths of limited capacity has stimulated a great deal of research having the objective of finding the necessary and

Table 2.16. LONG TERM, FREE FIELD, SPECTRUM DENSITY LEVELS FOR SPEECH MEASURED AT 25 mm FROM THE LIPS OF A FEMALE TALKER

Frequency Hz	*Spectrum density level* β'_s dB	*Frequency interval represented* Δf Hz	*10* $\log_{10}$ Δf dB	*Level of intensity represented by frequency interval* dB	*Intensity represented by frequency interval*
(1)	*(2)*	*(3)*	*(4)*	*(5)*	*(6)*
100	+51.4	80 (say)	19	+70.4	11.0×10^6
200	59.2	100	20	79.2	83.1
300	61.4	100	20	81.4	138.0
400	60.4	100	20	80.4	109.6
500	59.3	100	20	79.3	85.0
600	57.4	150	21.8	79.2	83.1
800	52.8	200	23	75.8	38.0
1000	49.4	225	23.5	72.9	19.5
1250	46.9	300	24.8	71.7	14.8
1600	44.0	375	25.7	69.7	9.3
2000	42.2	350	26.5	68.7	7.4
2500	40.6	500	27	67.6	5.8
3000	39.2	500	27	66.2	4.2
3500	37.5	500	27	64.5	2.8
4000	35.7	750	28.8	64.5	2.8
5000	32.4	1000	30	62.4	1.7
6000	30.6	1500	31.8	62.4	1.7
8000	29.1	2000 (say)	33	62.1	1.6
					Sum 619.4×10^6

Total speech level +87.9 dB

Notes See Table 2.15

The total speech level given above corresponds to RVL−1.4 dB

sufficient means of specifying the essential content of speech waveforms. These ideas stem from the work of Homer Dudley who showed that speech could be described in terms of relatively simple 'carriers', namely (a) larynx excitation and (b) a random waveform resulting from the turbulence which produces unvoiced consonants. These are modulated relatively slowly by changes in the short term spectrum. Only a very brief indication of this work can be given here and, for further information, reference should be made to the extensive material which has been published.

The earlier work was based on the channel vocoder principle in which the speech was synthesised with a set of fixed frequency bandpass filters fed with larynx or hiss excitation of amplitude derived from the analyser which used an identical set of filters. The frequency of the larynx excitation had to be measured by the analyser and this has proved a very formidable problem; the analyser has also to distinguish when the larynx is active and when not. Surprisingly intelligible speech was produced with the relatively crude equipment of the late 1940s using only 10 bands. The rate of change of the modulating signals could be kept quite low and, in fact, a 25 Hz low pass filter could be inserted in each channel information path. With this equipment a reduction in channel capacity to about one tenth could readily be achieved while retaining a surprising degree of intelligibility. Increasing the number of channels naturally improved the performance but reduced the economy.

The signals in the various channels of the fixed channel vocoder are considerably correlated and so, in principle, the efficiency ought to be capable of improvement by suitable extraction of more nearly independent signals to describe the speech. This can be achieved by use of the formant vocoder principle which seeks to determine, in the analyser, the frequency and, possibly, the bandwidth of each of a small number of regions of high spectrum density. The information concerning larynx excitation must also be obtained and considerably more effective means of doing this are now available using correlation techniques. Much improved performance is now possible as a result of studying how to make the most practical electrical analogues of the vocal tract.

The ultimate objective of studies of this kind would be to devise means for recognizing phonemic elements as they are emitted by the talker, to code them efficiently (not more than about 40 need be distinguished, see Tables 2.1 and 2.2), to transmit the coded information (the rate need not exceed about 30 bit/s), and to reconstruct replicas of the phonemic elements. Comparing this information rate with the theoretical information capacity of a typical telephone channel (3100 Hz bandwidth and 30 dB signal/noise ratio), which may be taken as about 30,000 bit/s, it will be seen that about 1000/1 improvement in transmission efficiency is to be sought. The improvement actually achieved by practical equipment so far is little more than 10/1.

2.2 HEARING

2.2.1 The reception mechanism (including artificial ears)

Figure 2.10 shows the mechanism of the human ear. It is convenient to divide the ear, for description, into: the outer ear, comprising the acoustical path coupling the telephone earphone to the eardrum (tympanum): the middle ear, comprising a mechanical system of small bones (ossicles) which convert movement of the eardrum into pressure applied to a membrane (oval window)

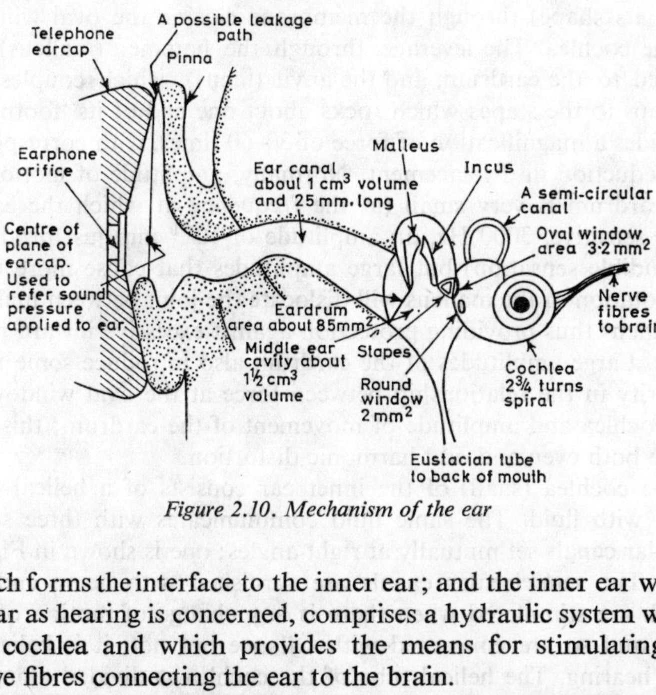

Figure 2.10. Mechanism of the ear

which forms the interface to the inner ear; and the inner ear which, so far as hearing is concerned, comprises a hydraulic system within the cochlea and which provides the means for stimulating the nerve fibres connecting the ear to the brain.

The outer ear provides the acoustical load impedance into which the telephone earphone works. The usual manner of placing the earphone against the ear does not normally provide a very good acoustical seal unless great care is taken and so some leakage, for example as shown in Figure 2.10, is liable to be present. When telephone earphones are tested by instrumental methods, for example to determine sensitivity/frequency characteristics, they are attached to an artificial ear, which provides an acoustical load intended to simulate that provided by a real ear, and which is equipped with a measuring microphone to determine the sound pressure produced. To relate the results obtained with artificial and real ears, it is necessary to take some point of reference that can be defined in common for the two situations; the point at the opening of the ear canal (meatus), which will coincide with the centre of the plane of the earcap, is convenient (*see* Figure 2.10).

The middle ear lies in a cavity which communicates through the Eustacian tube with the air at the back of the mouth. The three small bones form a system of levers that couple the 85 mm^2 area of the eardrum with the 3.2 mm^2 area of fluid in the cochlea actuated by the third small bone (called the stirrup or stapes, from its shape) through the membrane closing the oval window of the cochlea. The leverage through the hammer (malleus), attached to the eardrum, and the anvil (incus), which couples the malleus to the stapes which rocks about one end of its 'footplate' provides a magnification of force of 30–60 times, with corresponding reduction in displacement. Normally, amplitude of motion of the eardrum is very small (at the frequency at which the ear is most sensitive, 3000 Hz, an amplitude of 10^{-9} mm just produces an audible sensation) but large amplitudes that cause more than 36° rotation of the malleus will dislocate the joint between malleus and incus thus providing protection against excessive inward pressures. Large amplitudes of the eardrum also introduce some non-linearity in the relationship between force at the oval window of the cochlea and amplitude of movement of the eardrum; this can cause both even and odd harmonic distortion.

The cochlea (snail) of the inner ear consists of a helical tube filled with fluid. The same fluid communicates with three semi-circular canals set mutually at right angles; one is shown in Figure 2.10. The semicircular canals are furnished with nerve endings which are stimulated when there is any change of motion of the head and so are concerned with balance but not, it is believed, with hearing. The helical tube of the cochlea is divided along its entire length into two chambers, called scala vestibuli and scala tympani, which communicate with each other through a very small hole (0.25 mm^2 area), called the helicotrema, situated at the apex of the spiral farthest from the oval window. Figure 2.11 shows a diagram of the helical tube straightened out for illustration. The division between the two chambers is formed partly by a bony shelf and partly by a flexible membrane (the basilar membrane). Force applied to the oval window displaces the fluid along the scala vestibuli and through the helicotrema to the scala tympani along which the fluid is displaced to a second (the round) window, closed with a flexible and elastic membrane, facing the cavity of the middle ear. The basilar membrane carries the organ of Corti, which is equipped with four rows of hair cells; the nerve endings of these cells are connected through several thousand nerve fibres to the brain.

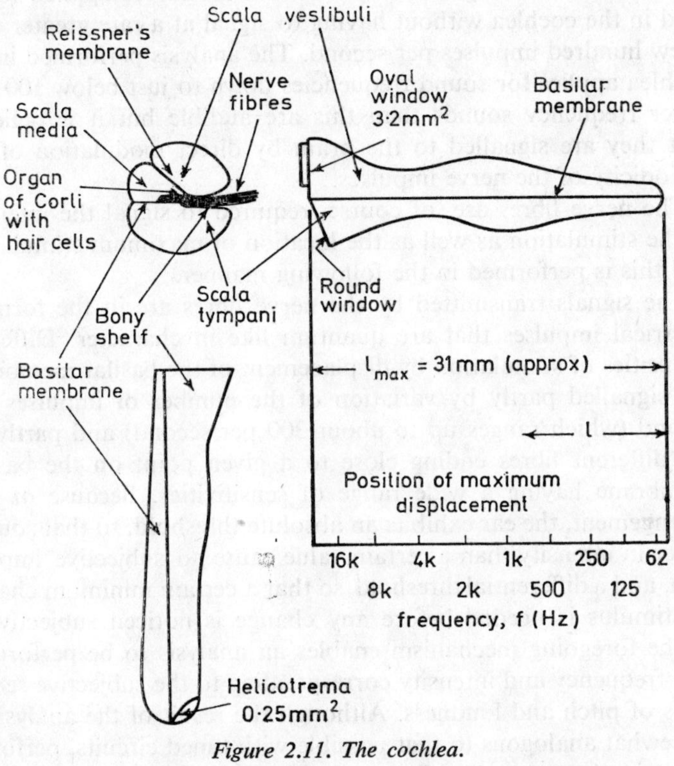

Figure 2.11. The cochlea.
Note. The $2\frac{3}{4}$ turns spiral of the cochlea is shown straightened out

When alternating forces are applied through the oval window to the fluid in the cochlea, a wave motion is set up in the fluid such that transverse displacement of the basilar membrane takes place. The velocity with which this wave is propagated along the basilar membrane is such that the wave takes 5 ms to reach the helicotrema which is about 30–35 mm from the oval window. The construction of the cochlea is such that the maximum displacement of the basilar membrane takes place at different points along its length according to the frequency of the applied forces; the location of the point of maximum displacement as a function of frequency is shown in Figure 2.11. It will be seen that this results in nerve endings near the oval window being stimulated to the maximum degree for the highest frequencies of the applied forces and those near the helicotrema being responsible for the lowest frequencies. In this way, the nerve fibres to the brain are enabled to transmit information concerning the frequencies of the forces applied to the fluid in the cochlea without having to signal at a rate greater than a few hundred impulses per second. The analysis performed in the cochlea applies for sound frequencies down to just below 100 Hz; lower frequency sounds than this are audible but it is believed that they are signalled to the brain by direct modulation of the periodicity of the nerve impulses.

The nerve fibres are, of course, required to signal the intensity of the stimulation as well as the location of maximum stimulation and this is performed in the following manner.

The signals transmitted by the nerve fibres are in the form of electrical impulses that are quantum like in character. Different intensities of stimulation by displacement of the basilar membrane are signalled partly by variation of the number of impulses per second (which ranges up to about 300 per second) and partly by the different fibres ending close to a given point on the basilar membrane having a wide range of sensitivities. Because of this arrangement, the ear exhibits an absolute threshold, so that sounds lower in intensity than a certain value cause no subjective impression, and a differential threshold, so that a certain minimum change in stimulus is needed before any change is noticed subjectively.

The foregoing mechanism enables an analysis to be performed into frequency and intensity corresponding to the subjective sensations of pitch and loudness. Although the result of the analysis is somewhat analogous to that possible with tuned circuits, performing it in that way would require the circuits to be very sharply tuned to account for the fine discrimination of frequency that is possible, but without the rather long build up and decay times which would be associated with such sharply tuned circuits; the response time of the analysing system would require the circuits to be about critically damped.

So far as mere audibility of sound is concerned, the two ears operate independently; in fact an attenuation of some 50 dB exists through the bones of the skull between the two ears. Binaural perception, by which different but to some extent related sounds reaching the two ears form a single image to the listener, involves correlation within the brain of the two sets of nerve impulse patterns (*see* Section 2.2.3.3).

The artificial ears in common use for telephony are those standardised by the American Standards Association (Ref.* 6) (often called the NBS–9A Coupler), the British Standards Association (Ref.* 12) (also called the Post Office Artificial Ear) and the International Electrotechnical Commission (referred to as the IEC Artificial Ear). The NBS–9A and the IEC types use capacitor microphones for measuring the sound pressure but the Post Office Artificial Ear uses a probe tube microphone (Ref.* 11) which can be located to measure the sound pressure at the ear reference point indicated in Figure 2.10. (In normal use, however, the tip of the probe is lowered to the base of the cavity to lessen the risk of accidental damage, especially when used for testing in the factory.) The physical design and the dimensions of these three artificial ears differ and so the sound pressures produced in them by a common earphone are not exactly the same. Each was designed to simulate the rather artificial condition of an earphone perfectly sealed to the ear and so it is not surprising to find that the sound pressure actually produced in a listener's ear from an earphone during actual telephone conversational conditions is somewhat (several decibels) lower than that indicated by measurements with these artificial ears. (This is discussed in Section 3.2.4.2 where suitable allowances are suggested.) Many other designs of artificial ear exist and these are listed in the references.

2.2.2 Perception

The significance of any sound is perceived in the brain by comparison of the patterns of nerve impulses, which correspond with acoustical features of the sound patterns of the waves entering the ears, with stored memories of previously interpreted patterns. The significant features of sound patterns are extremely difficult to identify because the theoretical information content of audible sounds vastly exceeds that which can be interpreted continuously by the hearer as a stream of informative events. Use is made of this redundancy when listening to speech to overcome interference by noise, distortion and loss introduced by the transmission medium between the mouth of the talker and the

* These references appear in Section 2.2.1 of the Bibliography.

ear of the listener. Conventional speech transmission systems do not at present exploit this redundancy to any great extent, although certain systems do make some use of it. Present speech transmission systems are designed mainly to reproduce at the ear a fairly faithful replica of the waveform of the sounds emitted by the talker. Some concessions are, of course, made to the economic limitations in transmission capacity by allowing telephone connections to introduce certain amounts of loss, noise and distortion. The information that follows illustrates those characteristics of the hearing mechanism that can be revealed by means of simple test signals. Descriptions of speech paths in such terms do not provide a very informative indication of the efficiency with which persons could communicate by speech although, with the aid of certain fundamental data derived from subjective assessment tests, some indication of the performance can be inferred when the transmission system is of a rather unsophisticated type.

Sophisticated transmission systems gain little from any description in terms of such simple test signals and require much more elaborate treatment when they are being assessed (*see* Chapter 6.5).

Speech, and many other sounds having significance to man, can be described to a limited, but useful, extent in terms of their patterns in time* as changes occur in quantities measured along the dimensions of frequency and intensity; indications of the discrimination possible by the ear in some of these quantities is given in the sections that follow.

2.2.3. Characteristics relevant to telephony

2.2.3.1 Steady state sounds

Pure tones

Much useful information concerning the characteristics of hearing can be derived from knowledge of the levels of pure tones that are just audible at different frequencies. Sound intensity is defined as the power transmitted per unit cross-section of the transmission medium (*see also* Section 2.1.3.2), the units usually employed being W/m^2: the symbol I will be used. A reference intensity, I_0, equal to 10^{-12}

* Speech displayed visually as a pattern in time is illustrated in Section 2.1 (see Figure. 2.3).

W/m² is used as the datum so that the intensity level, β, is expressed in dB relative to 10^{-12} W/m², so that $\beta = 10 \log_{10} (I/I_0)$.

In practice sound intensity cannot be measured directly, most measurements being made in terms of r.m.s. *sound pressure*, *p*, in N/m². Under free field, plane progressive wave conditions, the reference intensity corresponds approximately in air to an r.m.s. sound pressure, p_0, of 2×10^{-5} N/m² and so *sound pressure levels*,

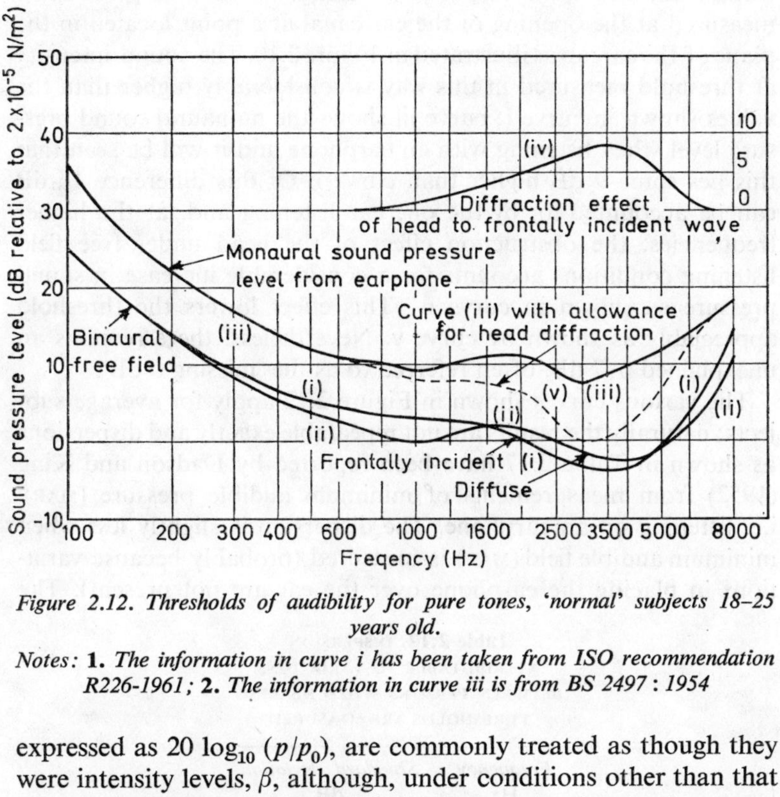

Figure 2.12. Thresholds of audibility for pure tones, 'normal' subjects 18–25 years old.

Notes: **1.** *The information in curve i has been taken from ISO recommendation R226-1961;* **2.** *The information in curve iii is from BS 2497 : 1954*

expressed as $20 \log_{10} (p/p_0)$, are commonly treated as though they were intensity levels, β, although, under conditions other than that stated, the identity does not strictly apply and only sound pressure or sound pressure level would be relevant.

Figure 2.12 shows the minimum intensity level (or sound pressure level) that is audible to an 'average' young person, aged 18–25 years, of normal hearing ability as a function of frequency. Curve i shows the result when using both ears and facing the source, the sound pressure being measured, with the head removed, at the point that was located midway between the centres of the openings of the ear canals; the results apply for a plane progressive wave.

When the sound field is diffuse, the waves arriving equally from all directions, the results are slightly different as shown in curve ii.

Listening with one ear only raises the threshold by about $1\frac{1}{2}$ dB at all frequencies up to about 3000 Hz and by a somewhat greater amount at higher frequencies (up to about 4–5 dB).

Under ordinary telephone conditions, the sound is applied to one ear only and is supplied by an earphone instead of from a rather distant source. With earphone listening, the sound pressure is measured at the opening of the ear canal at a point located in the plane of the earcap as illustrated in Figure 2.10. The sound intensity at threshold measured in this way is considerably higher than the values shown in curve i; curve iii shows the monaural sound pressure level when listening with an earphone and it will be seen that this lies some 9 dB higher than curve i. Of this difference $1\frac{1}{2}$ dB can be accounted for by the one ear listening and, at the higher frequencies, the obstruction effect of the head under free field listening conditions accounts for a considerable increase in sound pressure as shown in curve iv. This effect lowers the threshold appreciably as shown by curve v. Nevertheless, there remains an unexplained 6–7 dB, often referred to as the 'missing 6 dB'.

The various curves shown in Figure 2.12 apply for average subjects; naturally the results are not repeatable exactly and dispersions as shown in Table 2.17 have been reported by Dadson and King (1952) from measurements of minimum audible pressure (MAP), i.e. listening on an earphone. The dispersion is slightly less when minimum audible field (MAF) is measured (probably because variations in placing the earphone over the ear are not present). The

Table 2.17. DISPERSION ATTRIBUTABLE TO VARIATIONS BETWEEN PERSONS WHEN HEARING THRESHOLDS ARE MEASURED

Frequency Hz	*Standard deviation* dB
125	6.8
250	7.3
500	6.5
1000	5.7
2000	6.1
4000	6.9
6000	9.1
8000	8.7

dispersion can mainly be attributed to variations between different persons, including that of the volume of the ear canal.

Somewhat higher thresholds are found in older persons than the age group concerned in determining the results in Figure 2.12. In persons with otologically normal ears, the effects of aging are probably mainly in the middle ear and the effect is believed to be the same for women as for men. Normality of hearing for older persons is, however, difficult to establish because exposure to damaging sounds cannot readily be checked; the statistics show rather greater hearing loss with age for men than for women but this is believed to be due to the greater likelihood of damaging exposure of men in work and in leisure activities. Table 2.18 has been extracted from ISO Recommendation R 226 (1st Edition, 1961); strictly the information relates to minimum audible field but may be assumed to apply sufficiently well for most purposes also to minimum audible pressures.

The table gives the elevation of the threshold of audibility relative to the normal threshold for the age group 18–25 years as a function of age for otologically normal subjects.

Table 2.18. ELEVATION OF HEARING THRESHOLD AS A FUNCTION OF AGE OF SUBJECT

Frequency Hz	*Age, years*			
	30	*40*	*50*	*60*
1 000	0	0	0	0
1 250	0	0	0	0
1 600	0.1	0.4	0.6	0.8
2 000	0.4	1.2	2.0	2.3
2 500	0.8	2.8	4.6	5.4
3 000	1.2	4.0	6.5	7.7
3 500	1.5	5.2	8.5	9.9
4 000	1.9	6.4	10.5	12.3
5 000	2.4	8.0	13.1	15.3
6 000	2.4	8.2	13.4	15.7
8 000	2.5	8.6	14.0	16.5
10 000	3.5	11.8	19.3	22.6

Note. For frequencies of 1000 Hz and below, the corrections for age are zero.

When the intensity level (or sound pressure level), β, of a pure tone exceeds the relevant threshold level, the loudness will gradually increase as a function of the difference β–β_0, which is termed the 'sensation level' and denoted by the symbol Z, the units being dB. β_0 is the intensity level (or sound pressure level) at the threshold of audibility. Loudness, as a subjective intensity attribute, cannot be quantified directly, but two different sounds can be compared and one adjusted in intensity until they are equally loud as judged subjectively. The *loudness level* (Λ, dB) of a given 'unknown' sound is therefore defined as the intensity level (or sound pressure level) of a *reference tone* that has been adjusted in intensity to be equal

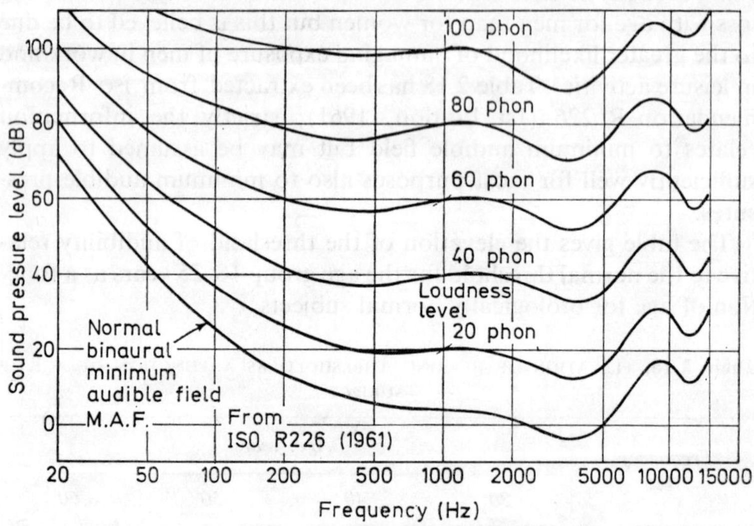

Figure 2.13. Equal loudness controus of sound pressure level and frequency

in loudness to the 'unknown'; a tone at the frequency of 1000 Hz is used as the reference tone and the units of loudness level defined in this way and applied to binaural listening in a free field are termed *phons.* In this way it is possible to display, on scales of intensity level or sound pressure level and frequency, contours of equal loudness which can be identified by their loudness level as the parameter, as shown in Figure 2.13. Naturally, at 1000 Hz, these curves pass through points where the loudness level is equal to the intensity level or sound pressure level.

The equal loudness contours of Figure 2.13 can, alternatively, be presented on the scales of sensation level, Z, and frequency by taking differences from the threshold values shown by the lowest curve in Figure 2.13; this form of presentation, shown in Figure 2.14, has some advantages because diffraction and other effects common to all curves of Figure 2.13 will be eliminated and the graphs therefore, in practice, become somewhat smoother.

For any particular frequency, the loudness level can be plotted as a function of the intensity level as is shown, for a few frequencies, in Figure 2.15. The nonlinearity of the relationship is apparent; it can also be seen that plotting loudness level as a function of sensation level, instead of against intensity level, has the effect of displacing the graphs horizontally relative to each other to pass through a common origin. In this form, the effect of frequency on the degree

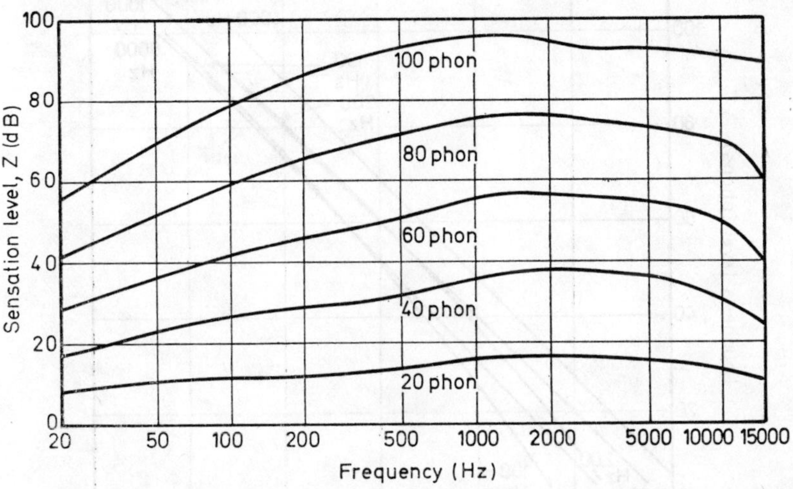

Figure 2.14. Sensation level as a function of frequency for various loudness levels

of nonlinearity becomes very clear but it will be found that, for the frequency range of greatest interest in telephony, say 500–3000 Hz, the nonlinearity is not very great.

The upper end of the range of sound pressure levels which the ear is capable of handling is bounded by regions where discomfort, pain and, eventually, permanent damage is suffered. Information on the precise locations of these boundaries is lacking (for obvious reasons) but it can usually be taken that discomfort will be felt when pure tones greater in sound pressure level than 110–120 dB are applied; sounds higher in level than about 140 dB will cause pain, and immediate damage can be caused by those of 150 dB and greater. Lesser sound pressures applied repeatedly can also cause permanent damage when these exceed about 130 dB, at which level temporary deafness is likely. In these regions of very high sound pressures, it is likely that the important criterion is the total r.m.s. sound pressure, irrespective of frequency content of the sound at least over the telephony range of frequencies.

Figures 2.13, 2.14 and 2.15 apply for pure tones present singly; the total loudness of a complex tone, composed of several pure tones applied simultaneously, can, according to Fletcher and Munson, be determined as follows:

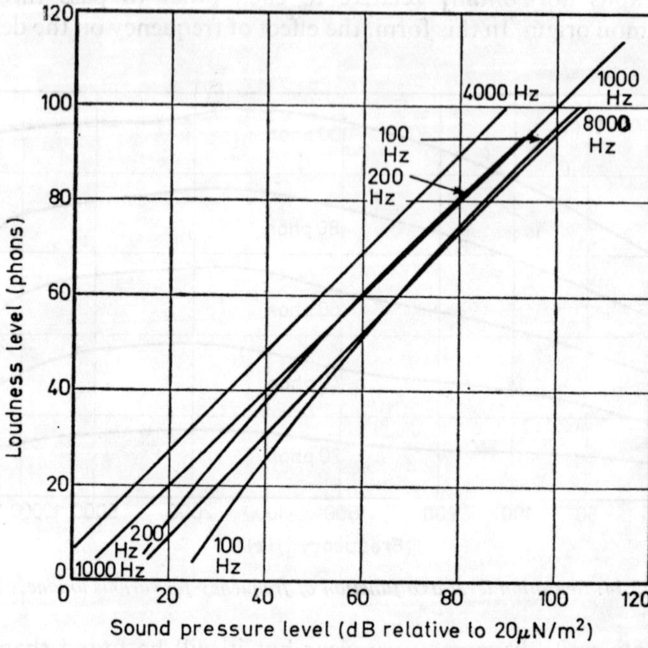

Figure 2.15. Loudness level as a function of sound pressure level for different frequencies

It is assumed that a measure of loudness, λ, can be defined such that λ is uniquely related to loudness level, Λ, by a monotonic function $\lambda = S(\Lambda)$ such that the total loudness, λ_r, is the sum of the loudness, λ_k, of each component at the different frequencies, f_k, determined in the absence of all other components. An allowance is included, by means of a multiplying factor, b_k, for intercomponent masking; b_k is less than, or equal to, 1. Algebraically, λ_r is given by the following expression:

$$\lambda_r = \sum_{k=1}^{k=n} b_k \lambda_k$$

where n is the number of components.

Unfortunately, except in rather simple cases, the determination of b_k is rather difficult. It may be stated, however, that a given tone is masked appreciably only by one of somewhat lower frequency and then only seriously when the masking tone is at a relatively high intensity level; a typical family of intertone masking curves is shown in Figure 2.16.

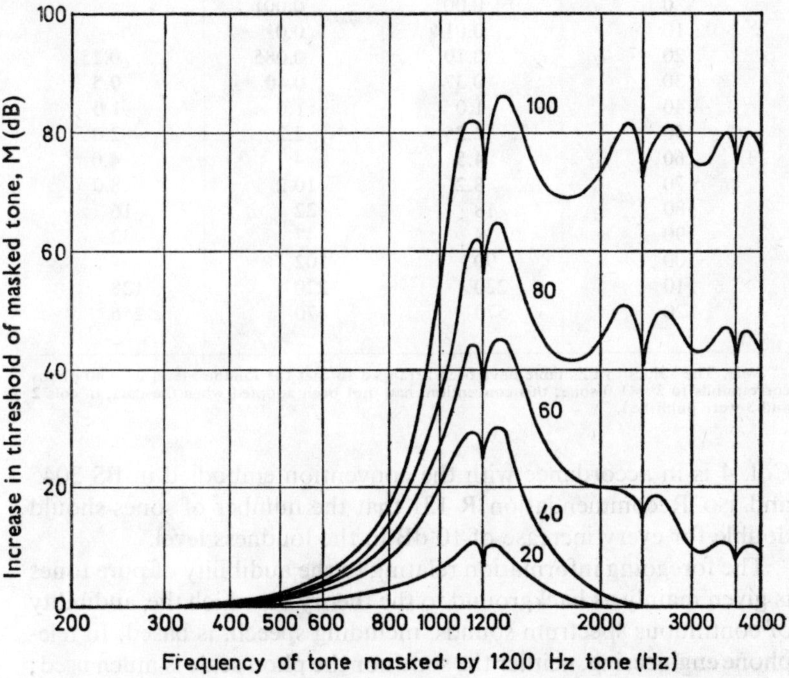

Figure 2.16. Masking effect of a tone at 1200 Hz on tones at other frequencies
Note: Parameter is level of masking tone relative to its threshold

Table 2.19 shows some different versions of the relationship between loudness, λ, expressed in sones, and loudness level, Λ, expressed in phons.

In Table 2.19, col. 2 gives the relationship originally published by Fletcher and Munson but with all the values of λ multiplied by a constant to make $\lambda = 1.0$ when $\Lambda = 40$. Col. 3 shows the results calculated from a formula suggested by Knauss which is of the form:

$$\text{sones} = \text{constant}\ [10^{\Lambda/10}\{10^{(\Lambda-25)/10}+1\}^{-2/3}]$$

Table 2.19. RELATIONSHIP BETWEEN LOUDNESS (SONES) AND LOUDNESS LEVEL (PHONS)

Loudness level, Λ phons	*Fletcher and Munson (1933) λ, sones*	*Knaus formula (1937) λ, sones*	*BS 3045 ISO/R131 λ, sones*
(1)	(2)	(3)	(4)
0	0.001	0.001	—
10	0.014	0.01	—
20	0.10	0.085	0.25
30	0.37	0.40	0.5
40	1.0	1.0	1.0
50	2.25	2.2	2.0
60	4.5	4.7	4.0
70	8.2	10.2	8.0
80	18	22	16
90	39	47	32
100	90	102	64
110	220	220	128
120	570	470	256

Note. The values in this table have been arranged so that the loudness level $\Lambda = 40$ phons corresponds to $\lambda = 1.0$ sone; this convention had not been adopted when the data, in cols 2 and 3 were published.

Col. 4 is in accordance with the convention embodied in BS 3045 and ISO Recommendation R 131 that the number of sones should double for every increase of 10 dB in the loudness level.

The foregoing information relating to the audibility of pure tones is given mainly as background to the theory on which the audibility of continuous spectrum sounds, including speech, is based. In telephone engineering, neither the sone nor the phon scale is much used; they would, in principle, be applicable when considering certain aspects of room noise but it is usually more convenient to express the sound level of room noise in terms of an objective sound level measuring instrument as explained in Section 4.2.1.1.

Continuous spectrum sounds

Certain rather easily applied rules have been evolved for determining the loudness level of relatively complex sounds, such as airborne noises from various sources of disturbance and these are embodied in ISO Recommendation R532. The principles which underlie these methods and those specially designed to treat the information bearing sounds of speech are closely related to Harvey Fletcher's development of the foregoing to treat continuous spectrum sounds.

When a sound can be considered to have a continuous spectrum, not necessarily uniform but free from single frequency components, the problem of masking can be treated very elegantly by use of the concept of 'critical bands' and this leads to several practical procedures for determining the threshold level and loudness of such types of complex sound.

According to the theory of hearing advanced by Fletcher and Munson, a sound at a single frequency f_k will cause maximum excitation of the nerve endings on the basilar membrane located at a distance l_k from the helicotrema (*see* Figure 2.11). This proposition has been convincingly supported by anatomical studies by Bekesy and by simulation with models by Peterson and Bogert. The excitation at this location due to the single frequency tone can be masked (i.e. the tone can be rendered inaudible) by the presence simultaneously of a continuous spectrum sound whose frequency bandwidth includes f_k; the masking will be marginally effective (i.e. the tone will be at its threshold of audibility) when the total excitation by the masking sound integrated over a critical length $(\Delta l)_c$* centred at l_k just exceeds that due to the tone at l_k. The audible frequency range mapped out along the length of the basilar membrane intercepts, in this critical length, a frequency band known as the 'critical frequency band' $(\Delta f)_c$ centred at f_k. Although the critical length $(\Delta l)_c$ is about the same for all locations along the basilar membrane, the corresponding critical frequency bands vary with f_k although they are nearly constant up to about 1000 Hz; the critical band is smallest at about 500 Hz (then being, according to Fletcher, 50 Hz), it is about 100 Hz at 2000 Hz, 200 Hz at 4000 Hz and 600 Hz at 8000 Hz. More complete information on the extent of the critical bands is given later.

The masking effect just described can be illustrated using Figure 2.17 as follows:

Consider a continuous spectrum sound S_1, in this example, for simplicity, of uniform spectrum density, I' W/m^2 per Hz, sufficiently

* According to Fletcher, the critical length extends over about 0.5 mm representing 1.56% (1/64) of the whole length of rather more than 30 mm; other authors report somewhat greater values, for example see Greenwood who quotes 1 mm. These disagreements are of little practical importance in the application of the theory described here because they do not affect the *relative* importance of different parts of the frequency range.

intense to be easily audible as shown in Figure 2.17. N′ is the spectrum density level, $N' = 10 \log_{10} I'/I_0$ dB. β in this figure is the minimum intensity level of the tone, S_2, for this to be just audible; thus, at the frequency f_k, the threshold intensity level β_k of the tone S_2 lies above the spectrum density level N_k of the sound S_1 by an amount, in dB,

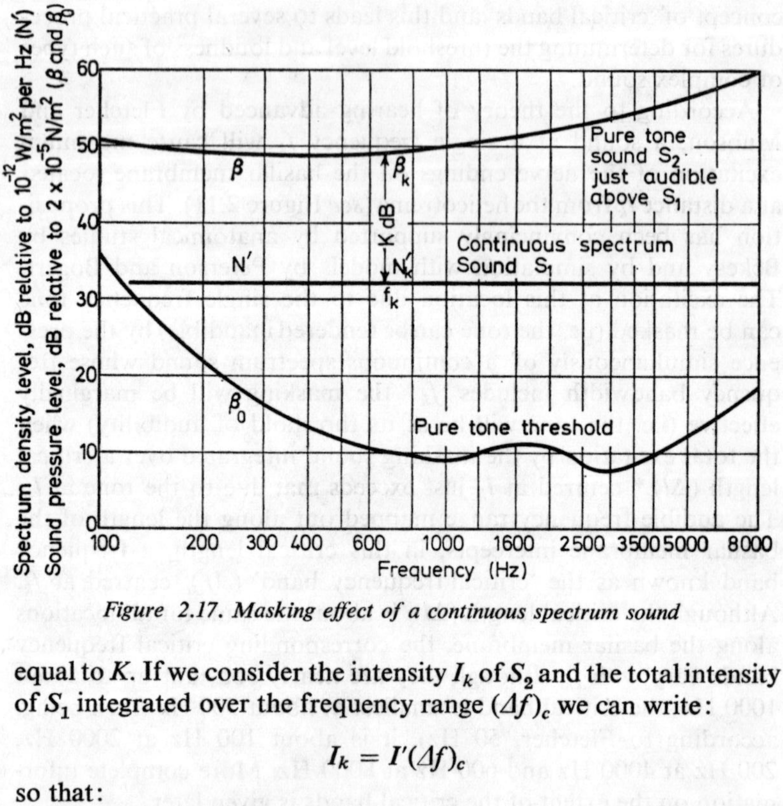

Figure 2.17. Masking effect of a continuous spectrum sound

equal to K. If we consider the intensity I_k of S_2 and the total intensity of S_1 integrated over the frequency range $(\Delta f)_c$ we can write:

$$I_k = I'(\Delta f)_c$$

so that:

$$\beta_k = 10 \log_{10} \frac{I'(\Delta f)_c}{I_0} = 10 \log_{10} \frac{I'}{I_0} + 10 \log_{10} (\Delta f)_c = N' + K \text{ dB}$$

where K is used to denote the critical bandwidth expressed in dB, i.e. $K = 10 \log_{10} (\Delta f)_c$; values of K as a function of frequency are shown in Table 2.20. A remarkable feature of the concept of critical bands is the stability of K (and of $(\Delta f)_c$); the values shown in Table 2.20 were derived from a wide variety of tests which yielded consistent results, the experiments being not only of the type described but also some based on anatomical information, physical models and subjective tests involving pitch discrimination, *see* references to publications by Bekesy, Peterson and Bogert and Stevens.

Table 2.20. VALUES OF K AND SOME DERIVED QUANTITIES FOR DIFFERENT FREQUENCIES

Frequency Hz	K dB	$(\Delta f)_c$ Hz	$B' = \frac{dl}{df}\frac{1}{l_{max}} = \frac{0.0156}{(\Delta f)}$	$10 \log_{10} B' = -K - 18.07$ dB	Frequency interval represented, Δf Hz	$10 \log_{10} \Delta f$ dB	$10 \log_{10} B' \Delta f$ dB
(1)	(2)	(3)	(4)	(5)	(6)	(7)	(8)
100	19.4	87	0.000179	−37.5	80(say)	19	−18.5
200	17.2	52	0.000297	−35.3	100	20	−15.3
300	17.0	50	0.000313	−35.1	100	20	−15.1
400	17.0	50	0.000313	−35.1	100	20	−15.1
500	17.0	50	0.000313	−35.1	100	20	−15.1
600	17.2	52	0.000297	−35.3	150	21.8	−13.5
800	17.6	57	0.000271	−35.7	200	23	−12.7
1 000	18.0	63	0.000247	−36.1	225	23.5	−12.6
1 250	18.5	71	0.000220	−36.6	300	24.8	−11.8
1 600	19.2	83	0.000187	−37.3	375	25.7	−11.6
2 000	19.9	98	0.000160	−38.0	450	26.5	−11.5
2 500	20.8	120	0.000130	−38.9	500	27	−11.9
3 000	21.5	141	0.000110	−39.6	500	27	−12.6
3 500	22.3	170	0.000092	−40.4	500	27	−13.4
4 000	23.1	204	0.000076	−41.2	750	28.8	−12.4
5 000	24.4	275	0.000057	−42.5	1000	30	−12.5
6000	25.7	371	0.000042	−43.8	1500	31.8	−12.0
8 000	27.7	588	0.000027	−45.8	2000 (say)	33	−12.8
10 000	29.2	832	0.000019	−47.3	—	—	—

Note. The total frequency range represented in this table (70–9000 Hz) accounts for only 89.7% of the total length l_{max} of the basilar membrane; the remaining 10.3% is concerned with the very high frequencies up to 20 kHz which are audible only to young persons of rather exceptional hearing.

If N' is not uniform as was considered in the example of Figure 2.17, but varies as a function of frequency, the same rules can still be applied to determine the threshold due to masking by a continuous spectrum sound. When the masking sound is not well above its threshold of audibility, as was the case in the example above, it is necessary to consider also the proximity of the threshold of audibility β_0 which applies in the absence of any noise. The values of β_0 shown in Figure 2.17 apply when one ear is used and the sound proceeds from an earphone (*see* Figure 2.12).

The principle can be explained by an example shown in Figure 2.18a in which is shown the spectrum density level N' of a sound

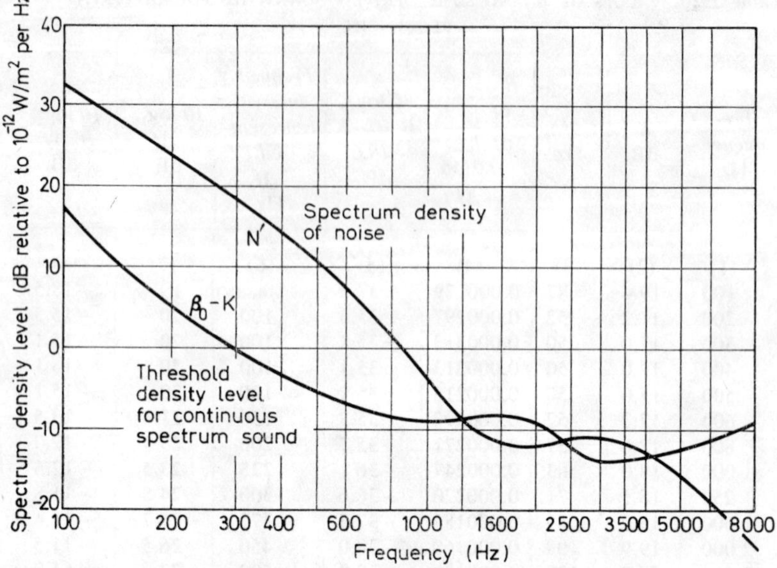

Figure 2.18. Determination of the audibility of a continuous spectrum sound. (a) Information expressed as spectrum density (levels of intensity per Hz)

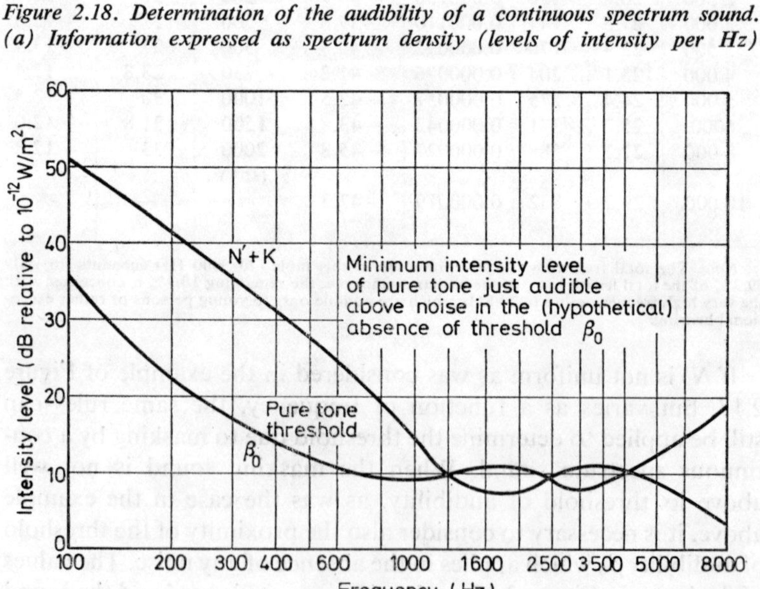

(b) Information expressed as intensity levels

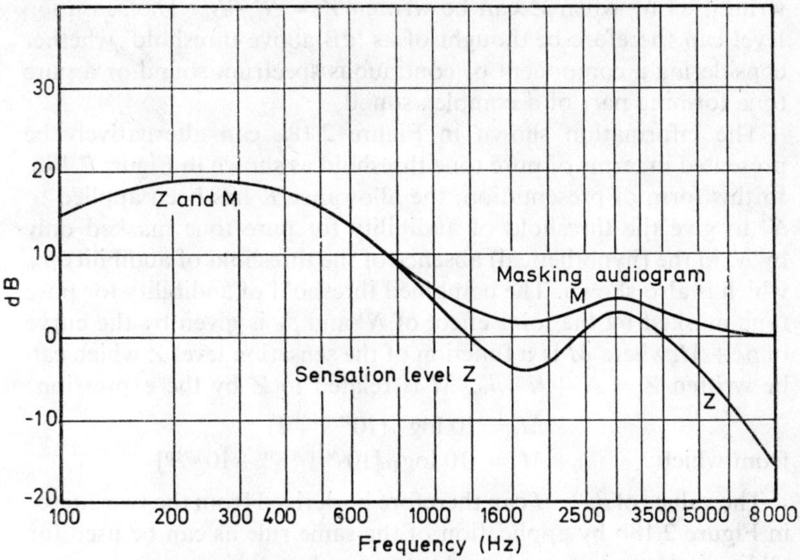

(c) Sensation levels and masking audiogram

whose degree of audibility is to be determined; the degree of audibility, or loudness level, will be expressed as the number of dB by which it must be uniformly attenuated to lower it to its threshold of audibility. The lower curve in the figure, β_0–K shows the threshold of audibility for pure tones (β_0) with allowance for the width of critical frequency bands; this curve shows the spectrum density level of a critical band of continuous spectrum sound that will, when present in isolation, be at its threshold of audibility. Subject to certain restrictions* concerning the maximum slope at any part of the spectrum of the sound under consideration the audibility of such bands is additive in the same way that the audibility of separate pure tones is additive: it should be noted that this applies even when each element would be below the threshold of audibility (i.e. inaudible) in the absence of the other components.

The interval N'–(β_0–K) is termed the sensation level and will be denoted by Z; because β_0–K can be regarded as analogous to a threshold for continuous spectrum sound, it is convenient to

* According to Fletcher, the maximum value of any positive slope must not exceed 80 dB per octave and the maximum value of any negative slope must not exceed a value of $75-Z/2$ dB per octave, where Z is the sensation level (defined later). If these values are exceeded some interband self-masking may be present.

write it as N'_0 when Z can be written $Z = N' - N'_0$. The sensation level can therefore be thought of as 'dB above threshold' whether considering a component of continuous spectrum sound or a pure tone forming part of a complex sound.

The information shown in Figure 2.18a can alternatively be presented in terms of pure tone threshold as shown in Figure 2.18b. In this form of presentation, the allowance K has been applied to N' to give the threshold of audibility for pure tone masked only by N' in the (hypothetical) absence of the threshold of audibility β_0, which is also shown. The combined threshold of audibility for pure tone masked by the joint effect of N' and β_0 is given by the curve of $\beta_0 + M$, where M is a function of the sensation level Z which can be written $Z = N' + K - \beta_0$. M is related to Z by the expression:

$$M = 10 \log_{10}(10^{Z/10} + 1)$$

from which

$$\beta_0 + M = 10 \log_{10}[10^{(N'+K)/10} + 10^{\beta_0/10}]$$

The values of $\beta_0 + M$ can therefore be derived from the two curves in Figure 2.18b by application of the same rule as can be used for adding independent random processes when their powers are expressed in dB.

A similar operation can be performed on the two curves given in Figure 2.18a to obtain $\beta_0 - K + M$.

In this way it is possible to obtain M from either Figure 2.18a or b and the resulting values are shown in Figure 2.18c; this figure also shows Z, which is simply the interval, in dB, between the pair of curves in Figure 2.18a or in Figure 2.18b.

According to Fletcher and Munson, the loudness of a sound having spectrum density level N' as a function of frequency can be determined by a rather simple operation performed on M (M as a function of frequency is termed the masking audiogram of the sound in question). The procedure is as follows:

The values of M are transformed to a quantity $F(M)$ which has additive properties and can be thought of as measuring elementary 'loudness units'. When $M = 0$ there is no contribution to loudness and so $F(M) = 0$. Different parts of the frequency range are of different degrees of importance and so are allocated weights B', which depend in a certain manner upon the frequency. According to the theory, B' is derived by assuming a given distance interval along the basilar membrane to have constant weight independently of its location; this leads to the result that:

$$B' = \frac{dl}{df} \frac{1}{l_{max}}$$

where l is the distance along the basilar membrane corresponding to the frequency f; l_{max} is the total length of the basilar membrane. From information given above:

$$B' = \frac{(\Delta l)_c}{(\Delta f)_c} \frac{1}{l_{max}} = \frac{0.0156}{10^{K/10}}$$

or $10 \log_{10} B' = -K - 18.1$ dB

The integral of B' with respect to frequency over the range 0 to ∞ is equal to 1 and so the weighted mean value of $F(M)$ is given by:

$$\overline{F(M)} = \int_0^\infty F(M)\, B'\, \mathrm{d}f$$

and the total 'loudness' is given by:

$$\lambda = 64\overline{F(M)}$$

the number 64 appearing because there are that number of independent critical bands of frequency to cover the complete range of audibility.

λ is a quantity, having arbitrary units, which has the property that all sounds producing the same number of units are equally loud*.

For better understanding, the theory has been presented in terms of M, it being intuitively reasonable to suppose that increments of loudness are closely related to the degree by which the sound in question would mask a pure tone at any given frequency. But Z is uniquely related to M and so the procedure can equally well (and more conveniently for calculation) be based on Z. In this case the function $F(M)$ is replaced by $Q(Z)$ which has an identical value; it turns out that $Q(Z)$ has convenient approximate forms as follows:

Near threshold (say, $Z \leqslant +10$ dB),

$$Q(Z) = C_1\,[10^{Z/10}] \tag{2.1}$$

and, for values of $Z \geqslant +30$ dB,

$$Q(Z) - C_4[10^{Z/10}]^m \tag{2.2}$$

where C_1 and C_4 are constants.

* λ as defined here is in different units from those used earlier to define the scale of sones; the units are, however, related by a constant multiplying factor.

The value of m has been taken variously from 0.25 to 0.33 but, for continuous spectrum sounds devoid of meaning (e.g. random noise), the value $m = 0.3$ is now generally accepted for the range of $Z \geqslant 20$ dB.

It should be noted that, whereas only positive values of M contribute to loudness, negative as well as positive values of Z contribute unless Z is lower than about -20 dB.

The foregoing procedure has, for explanation, been laid out in Table 2.21 which is concerned with the same spectrum density N' as in Figure 2.18a. The example actually relates to the sound reaching the ear by leakage past the earcap of a telephone earphone used in the presence of room noise of Hoth spectrum at a level of 50 dBA. (*See* Section 3.1.1, Section 4.2.1.1 and Section 6.1.4.)

Table 2.21. DETERMINATION OF AUDIBILITY OF A CONTINUOUS SPECTRUM SOUND

f Hz	N' dB	$\beta_0 - K$ dB	$Z = N' - \beta_0 + K$ dB	$Q(Z)$	$B'\Delta f$	$Q(Z)B'\Delta f$
(1)	(2)	(3)	(4)	(5)	(6)	(7)
100	32	17.5	14.5	28.84	0.0143	0.4124
200	24	5	19	79.43	0.0297	2.3591
300	18.5	0	18.5	70.79	0.0313	2.2157
400	14	− 3	17	50.12	0.0313	1.5688
500	10.5	− 5	15.5	35.48	0.0313	1.1105
600	7.5	− 6	13.5	22.39	0.0446	0.9986
800	1	− 8	9	7.94	0.0542	0.4303
1 000	− 4	− 9	5	3.16	0.0556	0.1757
1 250	− 8.5	− 8.5	0	1.00	0.0660	0.0660
1 600	−11.5	− 8	− 3.5	0.45	0.0701	0.0315
2 000	−12	− 9	− 3	0.50	0.0720	0.0360
2 500	−11	−11.5	0.5	1.12	0.0650	0.0728
3 000	−11	−14	3	2.00	0.0550	0.1100
3 500	−11.5	−13.5	2	1.58	0.0460	0.0727
4 000	−13	−13	0	1.00	0.0570	0.0570
5 000	−16	−12.5	− 3.5	0.45	0.0570	0.0256
6 000	−19	−11.5	− 7.5	0.18	0.0630	0.0113
8 000	−25	− 9	−16	0.03	0.0540	0.0016
10 000		−10			—	

Sum 0.8974 9.7556

$= -\overline{Q(Z)}$

Therefore $\lambda = 624$

and $\bar{Z} = +9.9$ dB

The frequencies used in col. 1 of Table 2.21 are the same as in Table 2.20 which indicates the intervals of frequency, Δf, represented by each entry. Col. 2 shows the spectrum density level of the continuous spectrum sound in question, the units being dB relative to the reference intensity per unit bandwidth; intensity is expressed in pW/m^2, and the reference intensity is 1 in these units. The values of β_0 are taken from curve iii of Figure 2.12, which applies for one ear listening with an earphone; K is taken from Table 2.20. Because Z in this example does not greatly exceeded 10 dB, $Q(Z)$ will, as an approximation, be taken from Eqn. 2.1 and, for simplicity C_1 has been taken as 1; thus $Q(Z) = 10^{Z/10}$. B' and Δf are given in Table 2.20 and their products are entered in col. 6. Col. 7 is the product of cols 5 and 6. The total of Δf accounts for the frequency range 70–9000 Hz; this does not, as explained above, represent the whole audible range and so the sum of $B' \Delta f$ does not quite reach 1. The sum of the last column gives the weighted mean of $Q(Z)$ and this times 64 gives λ. The value of $\overline{Q(Z)}$ can be restored to the units of Z to give $\overline{Z}$ the equivalent value of Z if it were the same at all frequencies.

λ is the total of contributions by all the critical bands to the audibility of the sound in question; there is a critical value of λ (and of $\overline{Z}$) below which the sound will be inaudible and this is determined below. The present sound produces values of λ and $\overline{Z}$ much greater than this and so the sound will be easily audible.

The multiplications required in the above table can be avoided by expressing all the necessary quantities in logarithmic units; this has been done in the following example and the practice will be continued in all subsequent calculations of this nature.

Hawkins and Stevens present information which enables a value to be put on λ for the threshold of audibility for a sound of the type considered above. (The threshold of audibility of speech is considered in Section 4.5.2.) The value of N', for a continuous spectrum sound that is just audible, and β_0 are given together with evidence that supports the values of K used here (Table 2.20). Table 2.22 follows the same arrangement as Table 2.21 for the first four columns: col. 5, however, uses the logarithmic form of Eqn 2.1, namely:

$$10 \log_{10} Q(Z) = 10 \log_{10} C_1 + Z \qquad (2.3)$$

If all the values of Z exceed 30 dB, the logarithmic form of Eqn 2.2 would be used, namely:

$$10 \log_{10} Q(Z) = 10 \log_{10} C_4 + m.Z \qquad (2.4)$$

Table 2.22. DETERMINATION OF λ AND $\overline{Z}$ FOR A CONTINUOUS SPECTRUM SOUND AT THE THRESHOLD OF AUDIBILITY

Frequency Hz	N' dB	$\beta_0 - K$ dB	$Z = N' - \beta_0 + K$ dB	$10 \log_{10} Q(Z)$ dB	$10 \log_{10} B' \Delta f$ dB	$10 \log_{10} Q(Z) B' \Delta f$ dB	$Q(Z) B' \Delta f$
(1)	(2)	(3)	(4)	(5)	(6)	(7)	(8)
100	−42	24	−66	−66	−18.5	−84.5	0.00000
200	−32.5	12.5	−44.5	−44.5	−15.3	−59.8	0.00000
300	−29	4.5	−33.5	−33.5	−15.1	−48.6	0.00001
400	−26	0	−26	−26	−15.1	−41.1	0.00008
500	−26.5	− 2.5	−24	−24	−15.1	−39.1	0.00012
600	−26.5	− 5.5	−21	−21	−13.5	−34.5	0.00035
800	−26.5	− 8.5	−18	−18	−12.7	−30.7	0.00085
1000	−27.5	−10.5	−17	−17	−12.6	−29.6	0.00110
1250	−28	−12	−16	−16	−11.8	−27.8	0.00166
1600	−29	−13	−16	−16	−11.6	−27.6	0.00174
2000	−31.5	−14.5	−17	−17	−11.5	−28.5	0.00142
2500	−33	−17.5	−15.5	−15.5	−11.9	27.4	0.00182
3000	−29.5	−20.5	− 9	− 9	−12.6	−21.6	0.00692
3500	−30.5	−22.5	− 8	− 8	−13.4	−21.4	0.00725
4000	−33.5	−23.5	−10	−10	−12.4	−22.4	0.00576
5000	−35.5	−22	−13.5	−13.5	−12.5	−26.0	0.00252
6000	−34	−16.5	−17.5	−17.5	−12.0	−29.5	0.00112
8000	−36	−13	−23	−23	−12.8	−35.8	0.00026

$\overline{Q(Z)}$ = Sum = 0.03298
$\therefore \lambda = 2.1$
and $\overline{Z} = -14.8$ dB

Note. The values of β_0 used here are those quoted by Hawkins and Stevens which differ somewhat from those used in Table 2.21.

In this example, using $C_1 = 1$ eliminates the first term of Eqn 2.3 and so col. 5 of Table 2.22 is the same as col. 4; this would not, of course, be the case if the values of Z had been much greater and use of Eqn 2.4 had been necessary. (That range of loudness is treated in Section 3.2.4.1 when applying the method to speech sounds.) Col. 6 is taken from Table 2.20 and col. 7 is the sum of cols 5 and 6. The final column is obtained by taking antilogarithms of col. 7, and the sum of this column yields, as before, $\overline{Q(Z)}$ from which λ is 2.1 and $\overline{Z}$ is −14.8 dB; these quantities are the required critical values for threshold of audibility.

The method can be applied to speech sounds to determine the threshold of audibility but suitable values of the parameters must be used and this matter is treated is Section 4.5.2 which deals with crosstalk; the gradually increasing loudness of speech as its level rises above the threshold of audibility is also considered in Section 4.2.1.5. The loudness, in itself, of sounds other than speech is rarely of interest in telephony; it is more common to consider the extent to

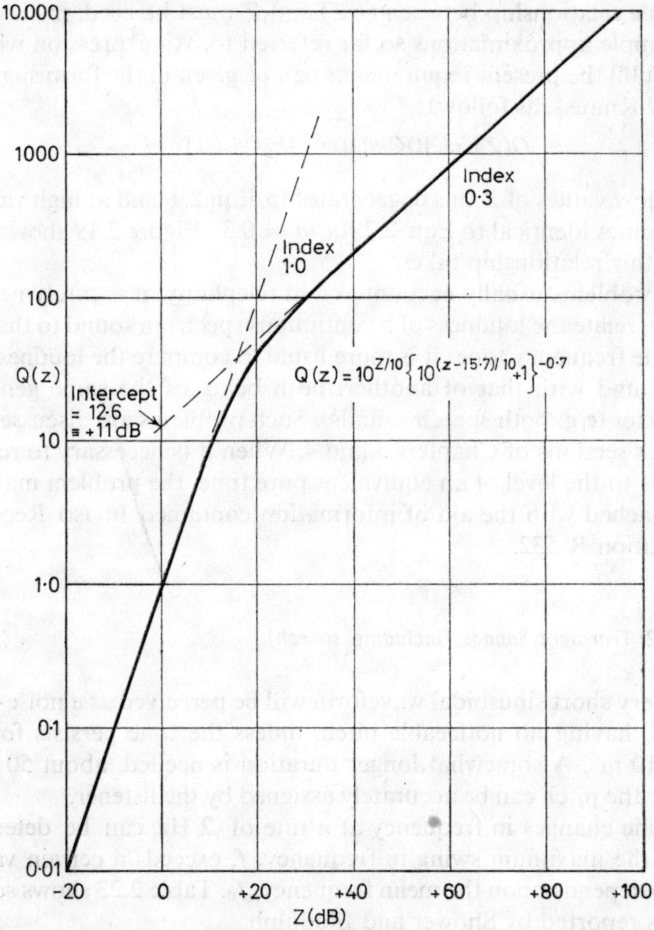

Figure 2.19. Loudness growth function for continuous spectrum, steady, noise sounds

which unwanted sounds (i.e. noises) interfere with the reception of speech and this matter is considered in detail in Section 4.2.1.

When the sensation level, Z, exceeds about 10 dB in at least some part of the frequency range, the simple expressions of Eqns 2.1 and 2.3 become inaccurate. If Z exceeds about 30 dB for most of the range, it is usually possible to use Eqn 2.2 or 2.4 with a suitable choice of value for the parameter m, the constant C_k being chosen arbitrarily; clearly, this leads to rather simple computations. When, however, Z ranges widely, especially in the interval 10–30 dB, a suitable relationship between $Q(Z)$ and Z must be used, instead of the simple approximations so far referred to. An expression which may fulfil the present requirements can be given in the form suggested by Knauss, as follows:

$$Q(Z) = 10^{Z/10}\{10^{(Z-15\cdot7)/10}+1\}^{-0\cdot7}$$

At low values of Z this degenerates to Eqn 2.1 and at high values it becomes identical to Eqn 2.2 for $m = 0.3$; Figure 2.19 shows the form this relationship takes.

In problems usually encountered in telephony, it is rarely necessary to relate the loudness of a continuous spectrum sound to that of a single frequency tone; it is more usual to compare the loudness of one sound with that of another, both being of the same general character (e.g. both speech sounds). Such problems are discussed in various sections of Chapters 3 and 4. When it is necessary to relate sounds to the level of an equivalent pure tone, the problem may be approached with the aid of information contained in ISO Recommendation R 532.

2.2.3.2 *Transient Sounds (including speech)*

A very short sinusoidal waveform will be perceived as a noise-like sound, having no noticeable pitch, unless the tone persists for at least 10 ms. A somewhat longer duration is needed, about 50 ms, before the pitch can be accurately assigned by the listener.

Cyclic changes in frequency at a rate of 2 Hz can be detected when the maximum swing in frequency, f, exceeds a certain value which depends upon the mean frequency, f_0. Table 2.23 shows some results reported by Shower and Biddulph.

It will be noticed that the minimum detectable cyclic variation in

Table 2.23. MINIMUM CHANGE IN FREQUENCY THAT CAN BE DETECTED

f_0, Hz	62.5	125	250	500	1000	2000	4000	8000
f, Hz	2.7	3.1	2.9	2.5	3.0	4.6	9.0	29.5
f/f_0	0.043	0.025	0.012	0.005	0.003	0.0023	0.00225	0.0037

frequency is approximately constant up to 1000 Hz and is approximately proportional to frequency above this value.

When the cyclic rate of change of frequency is greater than about 3 Hz, discrimination deteriorates. It may therefore be argued that the ear requires a minimum duration of almost 200 ms to perceive pitch to the full extent.

If the transient characteristics of hearing are thought of in terms of damped resonators having a certain time constant, the foregoing can be explained only by postulating a rather short time constant, of the order 20 ms, effective for very short duration sounds, which increases for longer duration sounds to about 250 ms. There is no physical part of the ear mechanism to account for resonances with the longer time constant mentioned and so pitch discrimination must be explained differently. It is supposed that this takes place in two stages; the first by resonators that are well-damped and the second by location of the middle of the resonance frequency region with a precision that increases (up to a certain degree) with increase in duration of the stimulus. A useful discussion of this matter is given by Gabor.

Additional evidence supports the proposition that only sounds having durations of at least a few tens of ms are fully perceived and that events separated by less than this order of time interval are not distinguishable as separate events. Delayed speech superposed on an undelayed version is not distinguishable as an echo unless the delay is at least 30 ms; this fact can partly be accounted for by the characteristics of speech but it also illustrates how well the characteristics of speech* and bearing are adapted to each other.

2.2.3.3 Use of both ears

A steady state sound that produces a given free field sound pressure will be perceived as louder when heard with both ears than with only one. If the hearing acuity is the same for the two ears, the sound will appear twice as loud so that the threshold sound pressure level using both ears would be 3 dB lower than with one only. If, as is usual even with normal hearing, the two ears differ slightly in acuity, the reduction in threshold will, perhaps, be only $1\frac{1}{2}$ dB.

When a human head is placed in a sound field, it will cause some disturbance to that field. The effect of this will be to render the

* The time interval over which the waveforms of most speech sounds may be regarded as statistically stationary is of the order 20–40 ms.

waveforms reaching the two ears slightly different; the difference may be in sound pressure level, in the time of arrival (or phase) or in both these features. With periodic sounds, phase will be important at frequencies not exceeding about 700 Hz (when the ear to ear spacing is about one half wavelength) and, at higher frequencies than this, intensity differences, caused by the shadowing effect of the head, become important to resolve the ambiguities in interpreting the relative phase information. Transient sounds, including speech, can be identified in their time of arrival by means of recognisable features of their envelopes; in such cases differences in time of arrival of even less than 1 ms are important clues to the direction from which the sound has arrived. The mechanism by which the time and intensity information is compared by the two ears certainly lies within the nerve and brain system and it is believed that the impulses arriving at the brain from the two ears undergo some form of cross-correlation process. The ability to separate speech arriving from different talkers at about the same level has been termed the 'cocktail party' effect.

Experiments in binaural hearing are usually conducted by supplying the two ears with suitably controlled signals through separate earphones and finding out where the listener localises the sound in his head and discovering the circumstances when he hears a single, fused, sound and not two separate sounds. In stereophony, the problem is rather different; a desirable object is then to present a listener in a small room with a sound field that is a replica of that in which he would find himself if present at the real performance. This can only be done accurately by providing a rather large number of separate channels from suitably spaced microphones; it is possible, however, to produce acceptable results with only two channels arranged to reproduce at the listener's two ears the same sound pressures and phase differences between them as he would experience in the real location. For further details reference should be made to published information indicated in the bibliography.

2.3 CONVERSATION

2.3.1 Introduction

The two previous sections (2.1 and 2.2) are concerned with descriptions of speech, the means by which it is produced and the method by which sounds (including speech) are heard and understood. Certain speech transmission problems can be studied by considering a system comprising a talker who emits message bearing signals and a listener who receives those signals and extracts the meanings of the messages. Such a model is a considerable oversimplification; it is incomplete for general study of telephone transmission problems because it omits consideration of the effects of the participants alternating their talking and listening roles, thereby introducing considerable interactions between their behaviours in both roles.

Accordingly, conversation (direct or over telephone connections) must be treated as a process in which the participants cooperate with each other and neither can be regarded exclusively as a talker or a listener; instead, they must be regarded as adopting, as occasion demands, talking and/or listening roles. It must not be supposed, either that these roles are always mutually exclusive; in fact an important fraction of conversation time is usually devoted to simultaneously talking and receiving speech from the other participant. The talking participant may, under these circumstances, be merely *hearing* the other participant's speech and not actually be paying attention to the content by *listening*; nevertheless, hearing the other person's speech does affect subsequent talking behaviour.

The importance to the telephone engineer of studying *conversation* as well as *speech* and *hearing* is underlined by noting that several possible sources of degradation in telephone connections cannot affect the participants at all if one exclusively adopts the role of talker and the other says nothing but listens to the talker's speech. An important example of such a factor is the presence of an abnormally long propagation time which will manifest itself to the participants only when they alternate their talking and listening roles.

Conversation, as a mode of human behaviour, has received little attention by psychologists; publications concerned with conversation are very few indeed compared with the enormous literature that has been built up on speech and on hearing as separate processes. Telephone engineers have recognised the problem of considering the coordination of talking and listening activities in telephone conversations for a long time and information has had to be gathered for the proper study of requirements for echo suppressors (Ref. 1)* and other devices† that rely, for successful

* This and other references in this section appear in Section 2.3 of the Bibliography.

† Another example of a problem for which information on conversation structure has been assembled is TASI (Time Assignment Speech Interpolation)

performance, upon proper detection of the presence or absence of speech on one or more speech paths.

During the early 1960s, the prospect of introducing synchronous orbit communication satellite circuits into telephone connections stimulated further study of the structure of conversation (Refs 3 and 4) and this has been followed by proposals for various statistical models to simulate the on/off sequences of speech activity in conversation (Refs 6 and 7).

2.3.2 Description of conversation

As explained above, conversation between two persons proceeds by alternation of talking and listening roles. The simplest alternation occurs when one participant stops talking and this event is followed by the other starting to talk. Any reasonably fluent conversation, however, contains other types of alternation. The second participant may anticipate the end of the first's utterance (by his tone of voice or by the content) and start talking before the first has finished, i.e. a period of 'double talk' intervenes. Another situation resulting in double talk occurs when the second participant wishes to interrupt the flow of speech from the other; this can take place when the first momentarily pauses intending to resume or, as is frequently the case, without waiting for an actual gap but judging his timing from the content.

In general, separate descriptions are needed for the conversation as observed by the two participants from their different locations. The two descriptions will, of course, be exactly complementary if no delay or mutilation occurs in transmission from one end to the other of the connection. Figure 2.20 shows a flow diagram illustrating the states of conversation, as perceived by one participant, and the events that change one state into another. The four states can be distinguished by combinations of the roles of talking and hearing as follows:

00 The participant is neither talking nor hearing (i.e. not receiving any speech from his partner).

(see Section 6.5 of Chapter 6) which is a channel economy device that exploits the silent intervals in speech from a given talker (while he needs no speech path to his partner) by offering the channel he has been using to another person who is actively speaking and so does need a speech path to his partner. TASI enables about $2N$ conversations to be served by N channels in each direction.

01 The participant is hearing speech from his partner but is not talking.

10 The participant is talking but not hearing.

11 The participant is talking while receiving speech from his partner.

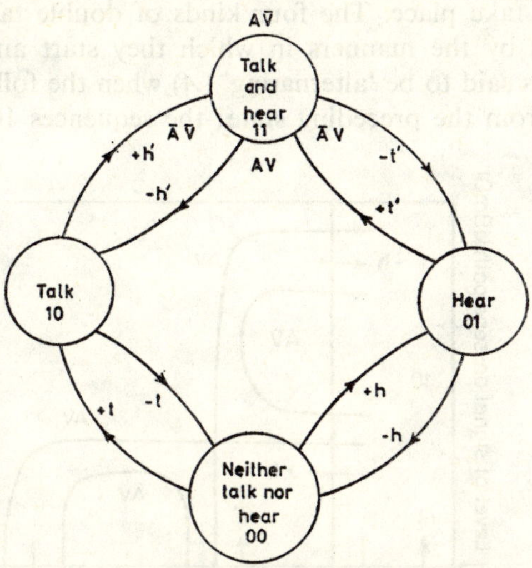

Figure 2.20. Diagram summarising the states and events in conversation as perceived by one participant

The four states are changed from one to another by the occurrence of the following conversational events which are assumed never to take place simultaneously:

- $+t$ The participant commences to talk while not hearing speech from his partner.
- $+t'$ The participant commences to talk while hearing speech from his partner.
- $-t$ The participant ceases from talking while not hearing from his partner.
- $-t'$ The participant ceases from talking while he is hearing.
- $+h$ The participant commences to hear speech from his participant while not talking himself.
- $+h'$ The participant commences to hear while talking himself.
- $-h$ The participant ceases from hearing while not talking himself.
- $-h'$ The participant ceases from hearing while he is himself talking.

Naturally, a participant has no control over the last four of these events but he is responsible for the first four. Double talks are of particular interest in the design of voice operated devices like loudspeaking telephones, echo suppressors and antisinging devices and it is useful to distinguish the various circumstances under which double talks take place. The four kinds of double talk state are distinguished by the manners in which they start and finish. A double talk is said to be 'alternating' (*A*) when the following state is different from the preceding state; the sequences 10-00-01 and

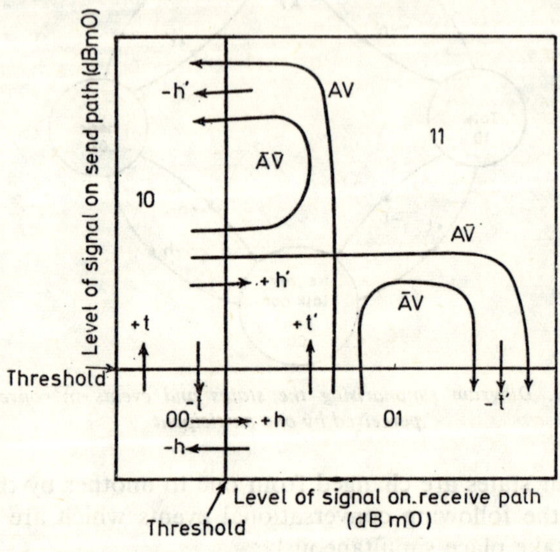

Figure 2.21. Diagram summarising conversational events as perceived by one participant

01-00-10 involve alternation but 10-00-10 and 01-00-01 are non-alternating ($\bar{A}$). In addition, those double talks for which the participant is responsible himself are termed 'voluntary' (*V*) and those over the onset of which he can have no control are non-voluntary ($\bar{V}$). The four different combinations, AV, $\bar{A}V$, $A\bar{V}$ and $\overline{AV}$ are indicated in Figure 2.20.

Detection of the presence of speech signals on a telephone circuit involves the discrimination between amplitudes greater and less than some threshold amplitude. When the speech is to be distinguished from noise, the problem is fairly simple unless the speech to noise ratio is poor; when speech is to be distinguished from other speech signals and these are of comparable or greater amplitude than the wanted speech, the problem becomes very difficult. Considerable ingenuity has been exercised to achieve a reasonably satisfactory performance of echo suppressors by taking into account the joint amplitudes of the speech that is wanted and the interfering speech. For such study, it is convenient to convert the flow diagram of Figure 2.20 into the more practical form shown in Figure 2.21 where the simultaneously present amplitudes

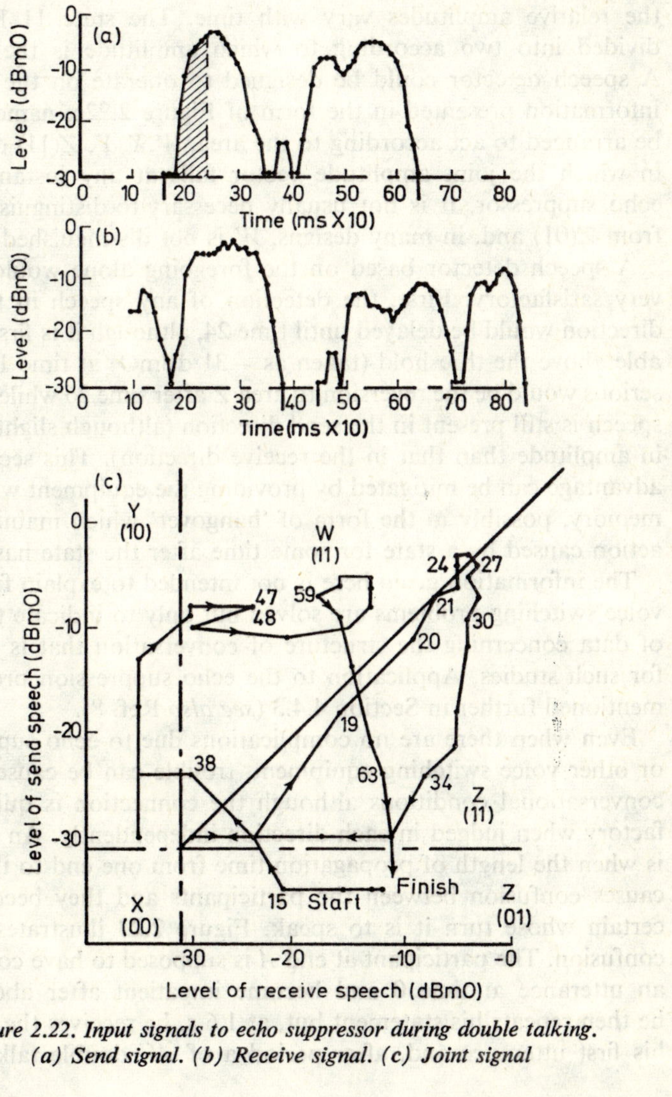

Figure 2.22. Input signals to echo suppressor during double talking.
(a) Send signal. (b) Receive signal. (c) Joint signal

are plotted against each other; the four states, the eight conversational events and the four kinds of double talk already explained are shown in the figure.

Figure 2.22a and b shows the envelopes of two speech waveforms during a period of double talk. The envelopes have been smoothed and sampled at intervals of 10 ms and quantised in amplitude into 32 values. Figure 2.22c shows the waveforms plotted jointly, sample by sample, on the same basis as Figure 2.21. This shows how the relative amplitudes vary with time. The state 11 has been divided into two according to which amplitude is the greater. A speech detector could be designed to operate on the basis of information presented in the form of Figure 2.22c, namely could be arranged to act according to the area, *W*,*X*, *Y*, *Z*(11) or *Z*(01), in which the joint amplitude vector falls at any instant. In an echo suppressor, it is not usually necessary to distinguish *Z*(11) from *Z*(01) and, in many designs, *W* is not distinguished from *Y*.

A speech detector based on the foregoing alone would not be very satisfactory. First, the detection of any speech in the send direction would be delayed until time 24, although it is first noticeable above the threshold (taken as −31 dBmO) at time 16. More serious would be the reversion to area *Z* after time 26 while genuine speech is still present in the send direction (although slightly lower in amplitude than that in the receive direction). This second disadvantage can be mitigated by providing the equipment with some memory, possibly in the form of 'hangover' which maintains the action caused by a state for some time after the state has ceased.

The information given here is not intended to explain fully how voice switching problems are solved but only to indicate the form of data concerning the structure of conversation that is required for such studies. Application to the echo suppression problem is mentioned further in Section 4.4.3 (*see also* Ref. 8).

Even when there are no complications due to echo suppressors or other voice switching equipment, trouble can be caused under conversational conditions although the connection is quite satisfactory when judged in each direction independently. An example is when the length of propagation time from one end to the other causes confusion between the participants and they become uncertain whose turn it is to speak. Figure 2.23 illustrates such a confusion. The participant at end *A* is supposed to have completed an utterance at time 0 and become impatient after about 1 s; he then repeats his statement but, at 1.6 s, he receives the reply to his first utterance and, after period a of $A\bar{V}$ double talk, ceases talking. Meanwhile, *B* is interrupted in his reply by the arrival of *A*' s repetition causing another $A\bar{V}$ double talk. *A* then resumes, only to be interrupted again by the arrival of an unexpected utterance from *B*, which causes another nonvoluntary double talk.

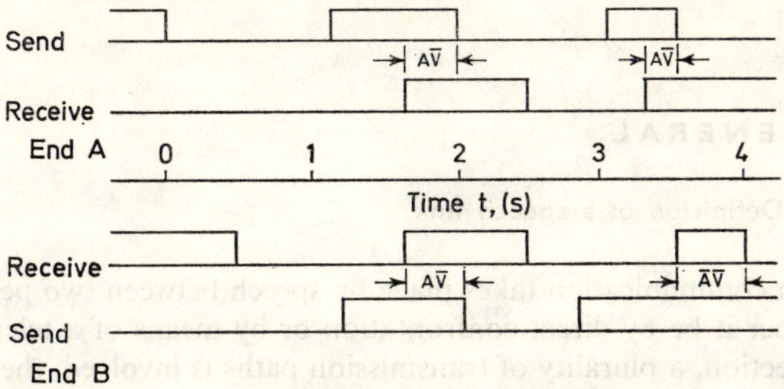

Figure 2.23. Confused situation in conversation

The succession of resumptions and interruptions punctuated by nonvoluntary double talks is characteristic of this state of confusion and can continue for several cycles until one party waits long enough before resuming. The incidence of such confusions as a function of the propagation time of the connection is discussed later in Section 4.4.2.

3. METHODS OF MEASURING THE COMMUNICATION EFFICIENCY OF A SPEECH LINK

3.1 GENERAL

3.1.1 Definition of a speech link

When communication takes place by speech between two persons, whether it be by direct confrontation or by means of a telephone connection, a plurality of transmission paths is involved, the more important of which are indicated in Figures 3.1 and 3.2. The term 'speech link' is used to denote the complex of transmission paths involved when such conversation is in progress and it is the efficiency of this complete system that must be measured. When the conversation is over air paths (*see* Figure 3.1), the characteristics of the following must be considered: *a* and *b* are the paths by which the speech emitted from the participants' mouths reaches the ears of their respective respondents, and *c* and *d* are the sidetone paths by which the participants hear their own voices.

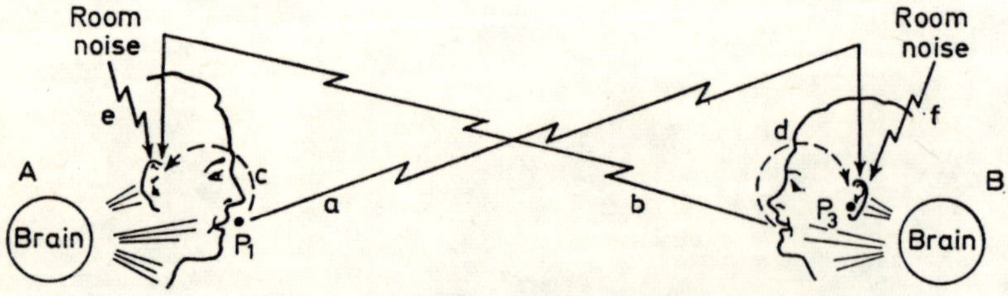

Figure 3.1. Basic paths involved in face to face conversation

Paths *a* and *b* are clearly the principal speech paths but the others cannot be disregarded; to a considerable extent, the characteristics of the sidetone paths govern the manner and level of speaking.

All these paths are affected by any room noise which may be present and would then reach the ears of the participants via paths *e* and *f*. Room noise that masks a participant's own speech, heard via *c* or *d* will usually cause the participant to talk louder. Room noise that masks speech from the other station will make conversation more difficult.

When the conversation takes place over a telephone connection (*see* Figure 3.2) the above mentioned paths are still relevant but will have become more complicated because of the telephone

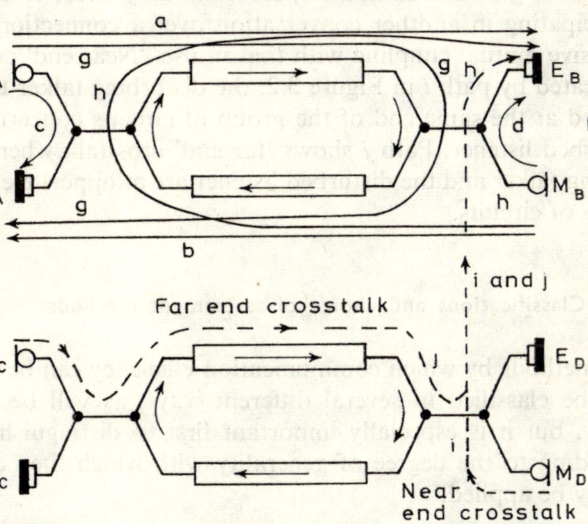

Figure 3.2. A selection of paths present in conversation over a telephone connection

apparatus and lines through which they pass. For example, any room noise will not only reach the ear being used for telephoning* by leakage under the earcap of the earphone but also through the microphone of the telephone and via the electrical sidetone path to the telephone earphone. Two kinds of sidetone must therefore be distinguished, namely 'speech' sidetone and 'room noise' sidetone.

In some telephone connections, the sidetone path through the electrical components of the telephone set will often be supplemented by one through a more distant point of electrical mismatch as illustrated by paths *g* and *h* in Figure 3.2; this will have no

* The other, open, ear will also receive the room noise but interaural masking can be neglected; noise heard via the nontelephoning ear may, of course, be distracting if the sounds heard have significance to the person concerned.

significant influence where the transmission circuit is short or is of a 'high velocity' type but when it is very long appreciable transmission delay may be present and, if so, the speech returned to the talker will appear to him as an echo, which can be troublesome.

Certain other paths provide unwanted couplings that affect telephony. Instability may be provoked by inappropriate combinations of gain and mismatch; crosstalk may occur from a talker participating in another conversation over a connection that has excessive mutual coupling with that in use. 'Near end' crosstalk is illustrated by path *i* in Figure 3.2, the disturbing talker then being located at the same end of the group of circuits concerned as the disturbed listener. Path *j* shows 'far end' crosstalk where the disturbing talker and the disturbed listener are at opposite ends of the group of circuits.

3.1.2 Classifications and choice of assessment methods

The methods by which communication efficiency can be measured may be classified in several different ways, as will be described below, but it is especially important first to distinguish methods according to the degree of generality with which they can legitimately be applied.

Class I comprises those methods that can be applied for assessing the widest possible range of types of speech link, degraded in any manner.

Such methods must, of course, treat the speech link as a complete system in which all the normal conditions of use are maintained, or closely reproduced, and all parts perform, as far as possible, their normal functions. These requirements lead to methods in which users, or subjects using the speech link in question, are interrogated and express opinions, their behaviour examined or the assessment based on features observed in the structure of conversations that have taken place over the speech link.

Class II comprises methods that are less general than those in Class I but are very commonly applied because they are correspondingly less laborious and time consuming to conduct. Most, but not necessarily all, Class II methods require the use of subjects but a special team trained for the work will sometimes suffice.

An example of a Class II method would be the treatment of one particular path, especially one of the principal paths *a* or *b* in Figure 3.2, as an information bearing channel and expressing its information capacity in some significant manner. Articulation tests offer a well known example of a method of this kind; similar examples are other one way methods, utilising combinations of talker and listener, in which sentence material is transmitted. The class also includes methods in which subjects perform rather specialised, nonconversational tasks related to the study of the effects of variations in the characteristics of particular paths of a speech link; for example, threshold determinations to obtain the lowest level of some undesirable feature which can just be detected or which just becomes objectionable or troublesome.

Class III consists of methods which seek to describe relevant fundamental characteristics of particular speech paths.

This class embraces all the instrumental methods for measuring such characteristics as the sensitivity/frequency response, noise level and various specific distortions. In principle, all these ought to be treated as acoustical measurements because the input and output ports of each path are interfaces with an acoustical medium; in practice, electrical measurements usually suffice with appropriate allowances for the electrical to acoustical conversions. Loudness rating methods, particularly determinations of reference equivalent, are very important members of this class because they seek to define loudness weighted, overall, mouth to ear transmission losses of speech paths.

Specifications for individual items of telephone transmission equipment are usually framed in terms of measurements appropriate to Class III. It must be remembered, however, that the foregoing discussion is concerned with complete connections and the quantities under consideration are those applicable for the whole; for example, noise limits or tolerances in attenuation/frequency distortion assigned to any individual item in the chain will usually have to be much less than that tolerable for the whole connection of which the item may form a part. This aspect of the problem is discussed later, for example in Chapters 5, 6 and 7, where hypothetical reference connections are formulated for the purpose.

When choosing an assessment method for investigating the performance of a particular speech link, it is necessary to consider the nature of the physical causes of loss of efficiency that will be present. Factors causing loss of communication efficiency can be classified according to the following three types:

Type A Factors that cause difficulty in listening when the transmission is unidirectional and no assistance is given to the listener by the talker.

Type B Factors that cause difficulty in talking

Type C Factors that cause difficulty specifically in conversing, that is, factors associated with the alternation of talking and listening roles of the participants.

Although not concerned directly with classification of assessment methods, mention must be made of a classification of factors causing loss of communication efficiency which is described more fully in Section 4.1.1. The distinction is made between X factors, associated with the physical plant provided by the telephone operating organization, Y factors concerned with the environment in which telephone conversations take place and Z factors which relate to the characteristics of the participants.

An essential feature of factors of Type A is that listening to speech from the other participant is involved but, in many cases, appropriate tests can be conducted with the aid of recordings, so avoiding use of a live talker while the listening is taking place. Very many of the practical causes of loss of communication efficiency are of this type and so one way, listening only, test methods falling into Class II are very commonly employed.

Methods suitable for studying factors of Type B will, clearly, require subjects to talk; the relevant observations may, however, consist merely of measurements of speech volume. Alternatively, or additionally, the subjects may be required to express opinions after they have spoken or their behaviour may be examined in other ways. Assessment methods suitable for such studies can often be found in Class II.

Factors of Type C almost always require use of a Class I assessment method to ensure that the proper conversational structure of the speech will be present. Naturally, a Class I method can also be used for studying factors of Types A or B but this is frequently unnecessary and such tests can often be limited to special circumstances such as the treatment of some novel factor.

This classification of types of factor causing loss of communication efficiency in speech links will be used later (e.g. in Chapter 4) when the effects of specific factors are considered in detail.

The possible methods for measuring the communication efficiency of a speech link can be considered as falling into four basic categories of test; the following subdivision will be used when describing assessment methods in detail:

Category 1. Examination and description of the speech link itself, particularly the individual paths of which it is composed,

Category 2. Measurement of the information capacity of the particular paths of the speech link,

Category 3. Examination of conversations that take place over the speech link in question,

Category 4. Examination of the participants in such conversations.

It is worth noting, too, that a useful distinction can be made between measurements specifically intended to assign a quantity directly relating to the speech link itself and those to determine a degree of 'satisfactoriness' for one particular conversation. Of course, a 'poor' speech link will result in a relatively larger proportion of 'unsatisfactory' conversations than a 'good' speech link but it must be remembered that even a 'poor' speech link will permit a substantial proportion of quite 'satisfactory' conversations to be conducted over it. It is also important to remember that the aim for a public telephone service should be to ensure that only a very small proportion of conversations over the speech links offered are permitted to fall below an acceptable degree of 'satisfactoriness'. It may be permissible, and indeed appropriate economically, to employ considerably lower grades of transmission performance for certain specialised services such as military communications, air traffic control or police networks than for a public telephone service; in such cases, the users exercise some discipline in communicating and usually undergo some training in the use of the facilities.

3.1.3 Reference systems, transmission standards and representative connections

The results of an assessment measurement apply, strictly, only to the speech link actually included in the experiment but it is usually desired to express the information gathered in some more generally applicable form so that it can be used to form a basis for planning the installation of telephone equipment. To this end, use is made of reference systems, transmission standards and representative connections.

A reference system is a speech link composed of stable and specifiable components and intended for laboratory use under conditions that enable all essential particulars to be defined precisely in simple physical terms. One important form of reference system is based on some very simple fundamental principle so that it can be realised without restriction to a particular design of the essential components (mainly the microphones and earphones). Historically, the first reference system to be defined internationally was the SFERT (Système Fondamental Européen de Référence pour la transmission Téléphonique: European Master Telephone Reference System) which was officially adopted by the CCIF (Comité Consultatif International Téléphonique*; International Telephone Consultative Committee) in 1928 (*see* Section 3.2 for further information). The basis for the current internationally used telephone reference system is one that represents a one metre airpath and was developed by the United Kingdom Post Office in 1948 and then adopted by the CCITT as the basis for the international reference systems SRAEN (Système de Référence pour la détermination de l'Affablissement équivalent pour la netteté; Reference System for determining Articulation Ratings†) and NOSFER (Nouveau Système Fondamental pour la détermination de l'Equivalent de Référence; New Master System for determining Reference Equivalents). SRAEN and NOSFER are described in detail in Sections 3.3 and 3.2 respectively but the basis of both is the ARAEN (Appareil de Référence pour la détermination de l'Affablissement Equivalent pour la Netteté; Reference Equipment for determining Articulation Ratings) which will now be described. (The official description of ARAEN is given in CCITT Recommendation P.41).

The principle of a one metre airpath is illustrated in Figure 3.3. It is supposed that a talker, A, faces a listener, B, who also faces A in anechoic conditions with the distance between A's lips and the opening of B's ear equal to one metre. The reference system is normally used for rating speech links containing handset tele-

* The letter F was used (German — Fernsprecher) to distinguish the organisation from the CCIT, the T of which stands for Télégraphique. Both organisations have now been merged in the CCITT (Comité Consultatif International Télégraphique et Téléphonique; International Telegraph and Telephone Consultative Committee).

† Although 'l'affablissement équivalent pour la netteté' can be translated into English as: 'articulation ratings', it is quite usual to use the form 'AEN values'.

phone sets in which one ear only is used for listening to speech from the other end and so it is supposed that only one ear is used by the listener, B, in Figure 3.3. In principle, the system shown in Figure 3.3 could be realised by use of a microphone

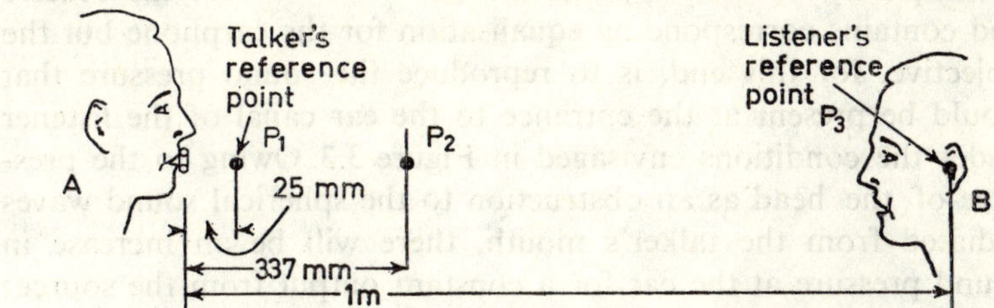

Figure 3.3. Fundamental principle of one metre airpath reference system

mounted in a dummy head situated in place of the real head; this would, however, limit use of the equipment to anechoic conditions. By placing the microphone of the reference system closer to the talker, for example at P_1 (approximately the position occupied by the centre of the mouthpiece opening of the microphone of a typical telephone handset), the influence of the acous-

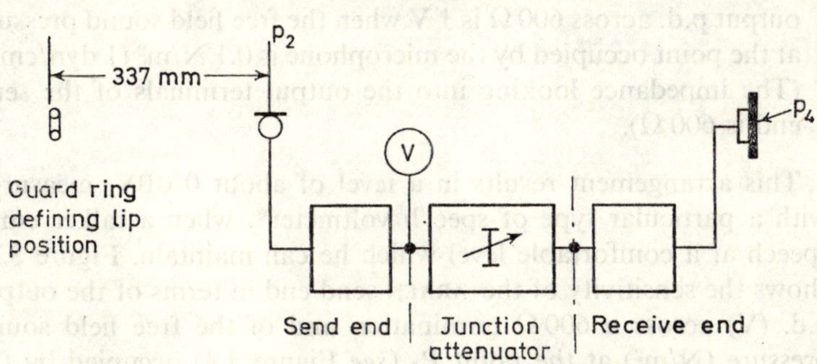

Figure 3.4. Reference system representing a one metre airpath

tical surroundings in an ordinary room would be eliminated. Such a close spacing between talker's lips and microphone of the reference system may, however, be difficult to maintain accurately and so a somewhat larger talking distance is more convenient for a fundamental reference system. The ARAEN uses a distance between the lips of the talker and the centre of the microphone of 337 mm. (The NOSFER uses a different distance, *see* Section 3.2).

Figure 3.4 shows the realisation of the one metre airpath principle as used in ARAEN and consisting of a send end (microphone, equaliser and amplifier), a receive end (amplifier, equaliser and earphone) and an adjustable junction attenuator. The send end contains equalisation to render the free field sensitivity independent of frequency at least over the range 100–6000 Hz; the receive end contains corresponding equalisation for the earphone but the objective, for this end, is to reproduce the sound pressure that would be present at the entrance to the ear canal of the listener under the conditions envisaged in Figure 3.3. Owing to the presence of the head as an obstruction to the spherical sound waves radiated from the talker's mouth, there will be an increase in sound pressure at the ear for a constant output from the source; this increase is negligible at frequencies below about 700 Hz but rises to some 8.3 dB at 3600 Hz. Equalisation appropriate to the simulation of this effect is therefore included in the receive end as well as that needed to compensate for imperfections causing variations in sensitivity with frequency in the earphone itself

The absolute values of sensitivity for the send and receive ends separately are arbitrary but, for convenience, the following arrangement was adopted:

1. The absolute sensitivity of the send end is such that the output p.d. across 600 Ω is 1 V when the free field sound pressure at the point occupied by the microphone is 0.1 N/m² (1 dyn/cm²). (The impedance looking into the output terminals of the send end is 600 Ω).

This arrangement results in a level of about 0 dBV, measured with a particular type of speech voltmeter*, when a talker emits speech at a comfortable level which he can maintain. Figure 3.5a shows the sensitivity of the ARAEN send end in terms of the output p.d. (V) across a 600 Ω termination and of the free field sound pressure (N/m²) at the point P_2 (*see* Figure 3.4) occupied by the microphone. The sensitivity referred to the point P_1 as shown in Figure 3.3 is given in Table 3.1 for a selection of frequencies that will be used later. The relationship between the sound pressure levels at P_1 and P_2 is explained in Section 2.1 (*see especially* Figure 2.5).

* The speech voltmeter referred to here is the British Post Office speech voltmeter type 3 which is of the nominally square law indicating instrument type reading approximately the r.m.s. voltage of each syllable. Detailed information about this instrument and the relationships between its readings and those of other speech voltage measuring devices is given in Section 2.1.3.2.

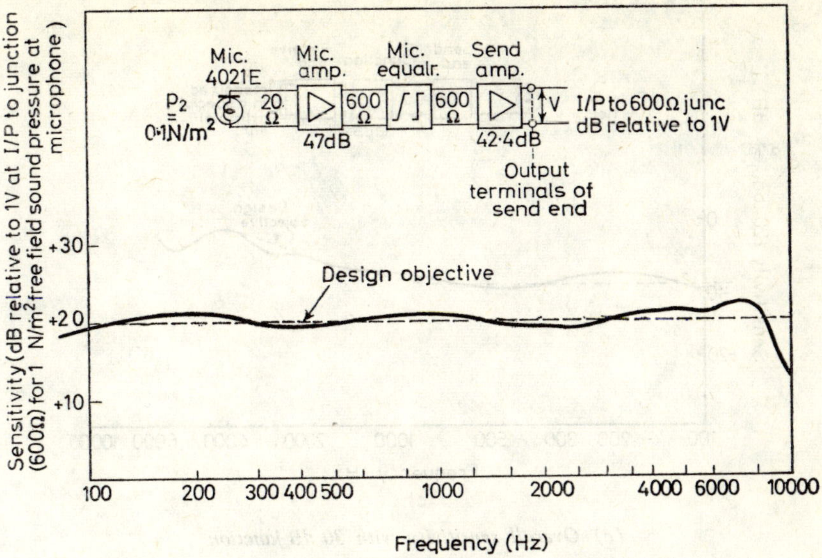

(a)

Figure 3.5. Reference system based on one metre airpath.

(a) Send end sensitivity

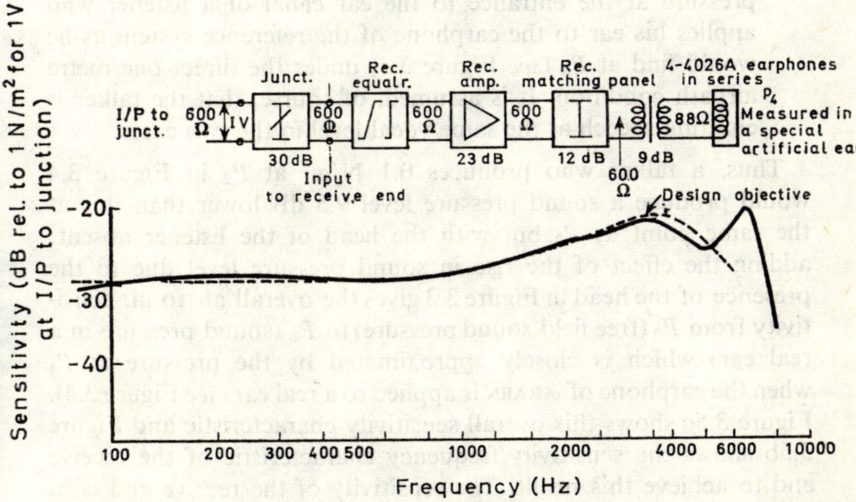

(b)

(b) Receive end sensitivity with 30 dB junction.

Note. Receive equaliser introduces the 'head effect' as well as correcting distortion of the earphones

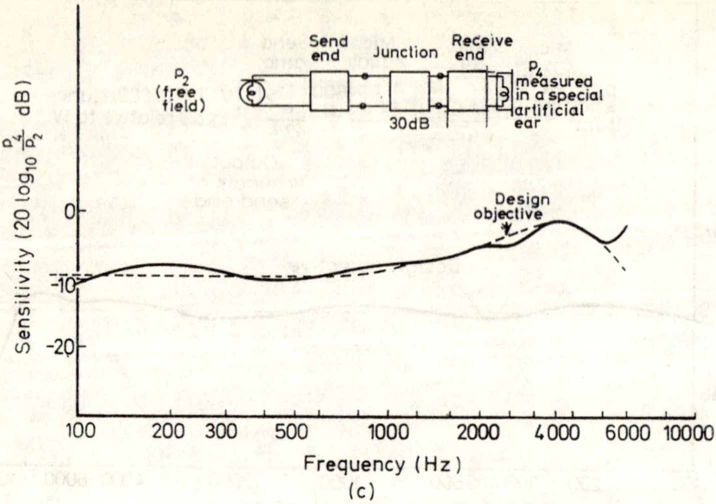

(c) Overall sensitivity with 30 dB junction

2. The combination of send end, of sensitivity as 1, 600 Ω junction attenuator set at 30 dB and receive end should result in the production at P_4 (*see* Figure 3.4) of the same sound pressure at the entrance to the ear canal of a listener who applies his ear to the earphone of the reference system as he would find at P_3 (*see* Figure 3.3) under the direct one metre airpath condition. It is assumed, of course, that the talker is emitting speech at the same vocal level in the two cases.

Thus, a talker who produces 0.1 N/m² at P_2 in Figure 3.4 would produce a sound pressure level 9.5 dB lower than this at the same point as P_3 but with the head of the listener absent; adding the effect of the rise in sound pressure level due to the presence of the head in Figure 3.3 gives the overall air to air sensitivity from P_2 (free field sound pressure) to P_3 (sound pressure in a real ear) which is closely approximated by the pressure at P_4 when the earphone of ARAEN is applied to a real ear (*see* Figure 3.4). Figure 3.5c shows this overall sensitivity characteristic and Figure 3.5b shows the sensitivity/frequency characteristic of the receive end to achieve this result. The sensitivity of the receive end is in terms of sound pressure measured in a special artificial ear which closely simulates the performance with a representative sample of real ears and the p.d. (V) across the input terminals of the receive end, the input impedance of which is 600 Ω.

Table 3.1. SENSITIVITIES OF SEND AND RECEIVE ENDS OF REFERENCE SYSTEM ARAEN, BASED ON A ONE METRE AIRPATH

Frequency Hz	*Send end sensitivity* dB	*Receive end sensitivity* dB	*Overall mouth to ear loss,* L_{ME}, dB
100	−1.4	+0.2	+1.2
200	+0.1	+1.0	−1.1
300	−0.9	+1.0	−0.1
400	−1.3	+0.5	+0.8
500	−1.0	+0.5	+0.5
600	−0.5	+0.5	0
800	−0.2	+0.9	−0.7
1000	−0.1	+1.7	−1.6
1250	−0.5	+2.4	−1.9
1600	−1.4	+3.8	−2.4
2000	−1.5	+5.5	−4.0
2500	−1.8	+6.0	−4.2
3000	−0.8	+7.4	−6.6
3500	−0.2	+8.0	−7.8
4000	+0.1	+7.0	−7.1
5000	+0.6	+4.2	−4.8
6000	+0.5	+7.2	−7.7
8000	+0.5	−5.6	+5.1

Notes 1. Sending sensitivities are expressed in dB relative to 1 V per N/m², the input sound pressure being referred to a point 25 mm in front of the lips with the sound field unobstructed. The output voltage was measured across a 600 Ω load.

2. Receiving sensitivities are expressed in dB relative to 1 N/m² per V, the input voltage being measured across a 600 Ω load terminating the 600 Ω source. Because the input impedance of the receive end is also 600 Ω, this is identical to the p.d. across the input terminals. The output sound pressure is referred to a point located at the centre of the opening of the ear canal under the earphone. Actually a special artificial ear was used and a small correction applied to allow for the difference in sound pressure in a median real ear from that in the artificial ear. The special artificial ear is referred to in CCITT Recommendation P.41 and a detailed description is included in Annex 5 to CCITT Red Book, V, 197–214 (1962)

3. The overall mouth to ear loss is given by minus the sum of send and receive end sensitivities.

Table 3.1 shows the sensitivity of the ARAEN receive end at a selection of frequencies and the mouth to ear loss referred to the points P_1 and P_3 is also included, assuming the junction loss is set at 0 dB.

3. The setting of 30 dB for the junction attenuator mentioned in 2 was chosen for convenience. It is a coincidence that this renders the mouth to ear loss approximately (within a few dB) the same as the setting of the junction attenuator of ARAEN; the mouth to ear loss specified in the manner explained above (i.e. referred to the points P_1 and P_3) also happens (within a

few dB) to be approximately the same as the reference equivalent.

The preferred value of overall loss when the reference system is spoken into at reference vocal level (*see* definition of RVL in Section 2.1.3.2) corresponds to about a 20 dB setting of its junction attenuator. The overall mouth to ear losses of practical telephone connections correspond roughly to a range of settings of from 0 to 40 dB.

Other important reference systems serving similar purposes to the ARAEN will be referred to in Sections 3.2 and 3.3 where specific assessment methods are described.

Reference systems of the type described so far differ in transmission quality a great deal from that of ordinary handset telephones and so they provide a useful datum for comparison in only certain respects. For example, loudness ratings for commercial telephone speech paths can be determined by making direct subjective comparisons with a high quality reference system (as described in Section 3.2) and the results are useful indications of the effective overall mouth to ear losses; more sophisticated comparisons, where the effects of noise and distortions are to be included, are, however, not practicable with such a reference system if the results are to be applied in the planning of telephone networks. For such studies, it is usual to make use of a so-called 'transmission standard' which comprises working type telephone apparatus representative of equipment in service. Such a transmission standard consists of two complete local telephone circuits* connected together by a fixed junction attenuator, the whole being chosen to represent the most adverse connection that can occur in a national telephone network. Because the meaning of the word 'standard' can be misinterpreted in this context, it is perhaps more appropriate to refer to such an assembly as a 'representative limiting connection', RLC for short.

When the same, or equivalent, local telephone circuits as form part of the 'transmission standard' (RLC) are used in tests with an adjustable junction attenuator, the equipment may be termed a 'standard speech link' or just a 'representative telephone connection'. Such equipment is required as a means for studying the effects of all the practical factors that can cause loss of communication

* The term 'local telephone circuit', LTC for short, is used to denote the combination of a subscriber's set, a subscriber's line and a feeding bridge.

efficiency in real telephone connections. A particularly important feature of such a 'standard speech link' is that its 'ends' can be used separately from each other to enable ratings (especially, but not exclusively, loudness ratings) to be allocated separately to the sending and receiving directions of any local telephone circuit under consideration. A 'standard speech link' may also be arranged to include circuit noise, additional attenuation/frequency distortion and other distortions to represent in a significant manner those that could occur in real telephone connections. This procedure will be explained in Section 3.2.

Improvements in modern telephone sets now make it feasible to represent practical telephone connections containing them by an 'intermediate reference system' (IRS) which consists of telephone type handsets having linear microphones and earphones, which are quite stable and can be calibrated quite accurately; if necessary a desired sensitivity/frequency characteristic can be incorporated by additional equalisation. Such a handset in association with suitable circuit components, including amplification, can then represent, and be used as an equivalent to, a 'standard speech link' as defined above. The greater stability avoids many difficulties encountered in assembling basic telephone planning data such as those attributable to manufacturing tolerances and other variations in commercial telephone microphones and earphones.

In the more detailed descriptions of assessment methods given below, the appropriate reference systems, 'standard speech links' (or representative telephone connections) and intermediate reference systems will be described together with the manner in which they should be used.

3.1.4 Rating scales and the impairment principle

It would be convenient if the telephone transmission performance of any given speech link could be expressed as a unique value along a unidimensional scale graduated in clearly definable units. Certain rather simple relevant subjective criteria do appear to be unidimensional, such as 'comparative loudness' and 'comparative fitness for a defined purpose' and much advantage can be taken of these properties. Unfortunately, however, these only enable a particular set of speech links to be ranked according to certain criteria and they do not yield an unequivocal absolute scale which would have universal meaning. It is true that certain assessment methods yield a 'score' that is absolute in the sense that use of a particular speech link produces a number which can be repeated fairly precisely in a given laboratory (particularly if the same subjects are used again) when all the essential conditions are reproduced; the articulation method provides an example. In practice, it is usually not possible to specify sufficiently accurately all the essential conditions: reproduction of even articulation scores is not possible with any great precision in different laboratories.

For the above reasons, much use is made of 'ratings' related to equivalent settings of some adjustable feature of a reference or 'standard' speech link. Thus, loudness ratings are expressed as the values of loss setting of the attenuator in a specified reference system to produce the equivalent sensations of loudness of speech under defined conditions of comparison. When criteria of equivalence other than loudness are used, it is no longer necessary to restrict the variable feature to loss; for example, the bandwidth or the amount of nonlinear distortion or the noise can be used. Direct comparisons between 'unknown' and 'reference' can be made using listening effort as the criterion of equivalence (*see* Section 3.4). The bandwidth, the amount of nonlinear distortion or the level of the noise (each expressed in convenient and appropriate physical units) would then provide the rating scale instead of loss when this is used and loudness is used as the criterion.

Three different procedures for making the comparisons are possible and these are illustrated in Figure 3.6. In the explanation above, the 'variable reference' method was considered, i.e. the reference speech path was adjusted to secure equivalence to the 'test' or 'unknown' speech path. (*See* Figure 3.6a). This clearly results in the balance being made at a value of transmission performance that depends on the performance of the test item; poor test items will be balanced at worse values of transmission performance than good test items. This may be appropriate but it must not also be expected that removal or addition of the same amount of degradation from or to both reference and test speech paths will maintain equivalence in all cases; for example, a loudness balance that results in equivalence at 40 dB loss in the reference speech path will not necessarily yield equivalence at 15 dB loss in the reference speech path if 25 dB loss were removed from the test speech path. For some purposes the 'variable reference' method may not be appropriate and it may be desirable to ensure that all balances are made at a constant value of transmission performance defined by a fixed setting of the reference speech path;

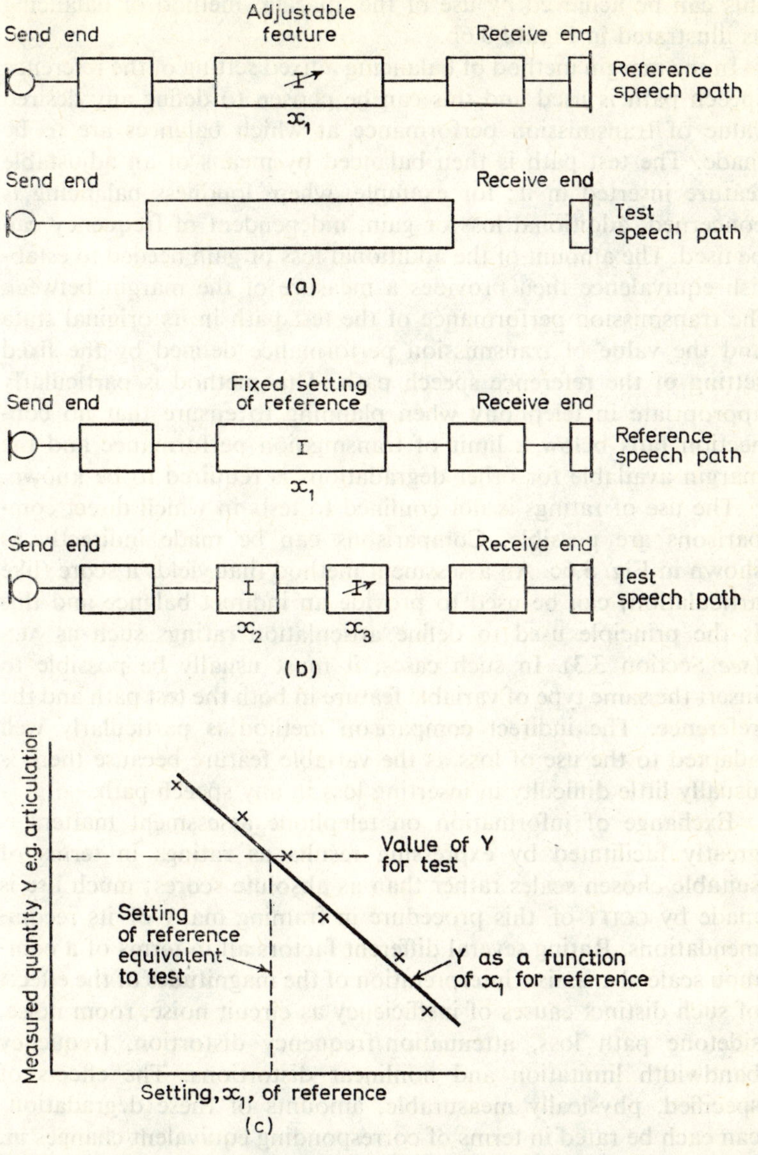

Figure 3.6. Methods of rating a test speech path relative to a reference speech path. (a) Measurement of a rating by the 'variable reference' method. (b) Measurement of a rating by the 'margin' method. Note: x_2 represents the 'normal' setting of the test speech path, x_3 represents the 'margin' and if necessary, can be a gain. (c) Measurement of a rating by the 'indirect balancing' method

this can be achieved by use of the 'margin' method of balancing as illustrated in Figure 3.6b.

In the margin method of balancing a fixed setting of the reference speech path is used and this can be chosen to define any desired value of transmission performance at which balances are to be made. The test path is then balanced by means of an adjustable feature inserted in it; for example, where loudness balancing is concerned, additional loss or gain, independent of frequency can be used. The amount of the additional loss or gain needed to establish equivalence then provides a measure of the margin between the transmission performance of the test path in its original state and the value of transmission performance defined by the fixed setting of the reference speech path. This method is particularly appropriate in telephony when planning to ensure that no connection falls below a limit of transmission performance and the margin available for other degradations is required to be known.

The use of ratings is not confined to tests in which direct comparisons are possible. Comparisons can be made indirectly as shown in Fig. 3.6c. An assessment method that yields a score (like articulation) can be used to provide an indirect balance and this is the principle used to define articulation ratings such as AEN (*see* Section 3.3). In such cases, it must usually be possible to insert the same type of variable feature in both the test path and the reference. The indirect comparison method is particularly well adapted to the use of loss as the variable feature because there is usually little difficulty in inserting loss in any speech path.

Exchange of information on telephone assessment matters is greatly facilitated by expressing results as ratings in terms of suitable chosen scales rather than as absolute scores; much use is made by CCITT of this procedure in framing many of its recommendations. Rating several different factors all in terms of a common scale also assists interpretation of the magnitudes of the effects of such distinct causes of inefficiency as circuit noise, room noise, sidetone path loss, attenuation/frequency distortion, frequency bandwidth limitation and nonlinear distortions. The effects of specified, physically measurable, amounts of these degradations can each be rated in terms of corresponding equivalent changes in, for example, overall transmission loss. Such values of equivalent changes in loss are termed 'impairments'. (The term 'impairment' is usually applied to equivalent changes in loss but it may also be used to denote a value obtained by rating in terms of any other scale.)

Expressing the results of assessment tests as impairments does not confer on them any special property of 'additivity'. That is, an impairment due to one factor determined with another factor also present will not necessarily be the same as that determined in the absence of the second factor. For this reason, when impairments in terms of the scale of loss are determined, it is convenient to use the 'margin' method of balancing rather than the 'variable reference' method; this may not improve 'additivity' but it does serve to standardise the conditions and avoids variations attributable to changes in listening level. The problem of estimating the performance of a speech link containing several sources of loss of communication efficiency, given the results when each source is operating in isolation, is a very difficult one to solve and, in fact there is no generally valid solution. This problem will be referred to later (in Chapter 4) when the various individual factors are being considered.

Assessment in terms of absolute scores are, of course, desired when decisions are to be made on the suitability of a particular arrangement of transmission plant in a new telephone network. These can be based on such statistics as the proportion of calls made over connections in the network concerned that will be acceptable to the users. The fundamental information for making these predictions will need to be obtained from conversational assessment tests of a particularly realistic nature so that the closest possible approach can be made to a determination of absolute quantities such as the proportion of customers likely to experience difficulty in conversing over the telephone connections provided. Data from observations in actual service may be of some assistance but these usually suffer from lack of sufficient knowledge of the characteristics of the actual speech links in use.

3.2 LOUDNESS RATING

3.2.1 Aim

The aim of a loudness rating method is to provide a standardised measure of the transmission loss of a given speech path from the mouth of a talker to the ear of a listener. The measure is single valued and related to the loudness with which the listener perceives speech emitted by the talker at a specified constant level. The following factors are intended to be taken into account so far as they affect loudness of speech:

a. The coupling between mouth and microphone

b. The coupling between earphone and ear

c. Any attenuation/frequency distortion of the path being measured

d. Any nonlinear distortions introduced by the path

The following are, however, in principle, specifically *excluded:*

e. Variations between talkers

f. Variations between listeners

g. Any influence the characteristics of the path being measured may have on the behaviour of talker or listener

h. Any influences on the behaviour of talker or listener of any other characteristic of the speech link of which the path being measured forms part

i. The effects of any circuit noise or room noise associated with the path being measured

The effects of e, f, g, h and i are excluded from measurements of loudness ratings by taking the following precautions:

1. The position of the talker's lips relative to the microphone is standardised in a manner intended to be representative of actual conditions of use
2. The level at which the speech is emitted by the talker is controlled
3. The coupling between earphone and ear is arranged to represent actual conditions, assuming reasonable care in placing the earphone to the ear but without a perfect seal
4. Any circuit noise or room noise normally associated with the path being measured is removed

The loudness rating method therefore provides a means whereby certain particulars of a speech link can be described in a useful manner; a complete description of a speech link would, however, require a set of loudness ratings describing the paths of, say, Figure 3.2 to be supplemented by information derived from other, suitable, kinds of measure, e.g. levels of circuit noise and of room noise, etc.

3.2.2 Principle and method

Loudness ratings involve comparisons and so it is necessary to employ a reference system. The most commonly used loudness ratings are termed 'reference equivalents' which were originally defined with the SFERT as the basis. The SFERT was a reference system composed of high grade laboratory components supplied in 1928 to the SFERT Laboratory, in Paris, by the American Telephone and Telegraph Company. It was a replica of the fundamental reference system then used in the United States of America as the basis for loudness rating determinations. The SFERT was officially adopted by the CCIF on 16th June 1928 as the basis for defining loudness ratings for international telephony and rules were established for making the measurements. These rules included standardisation of the means of calibrating the system, the talking distances to be used when testing handset telephones, and the vocal level to be used with the conventional phrase 'Paris–Bordeau–Le Mans–Saint-Leu–Leon–Loudun', which was adopted. These rules have, on various occasions, been revised by the CCIF/CCITT but they still include certain features which the British Post Office considers undesirable and does not now apply in its own laboratories; these are discussed later.

The ARAEN, based on the principle of a one metre airpath, was developed by the United Kingdom Post Office and a model was supplied to the CCIF in 1948 to form a basis for defining ratings by articulation (AEN) as will be described in Section 3.3. In 1960, the SFERT had become impossible to maintain because the essential components were no longer manufactured and replacements had become unavailable. A series of very careful tests was therefore made to determine what changes could be made to the ARAEN to render it suitable as a direct replacement for SFERT. The new system, NOSFER, was set up by making certain changes to ARAEN and those finally standardised are shown in Figure 3.7. The changes, in detail, are as follows:

Send end

1. Move the microphone of ARAEN from its normal position, centred at 337 mm from the lip position, to 140 mm. This is equivalent to increasing the sensitivity of the send end by 7.4 dB.
2. Insert a 600 Ω network which introduces attenuation/frequency

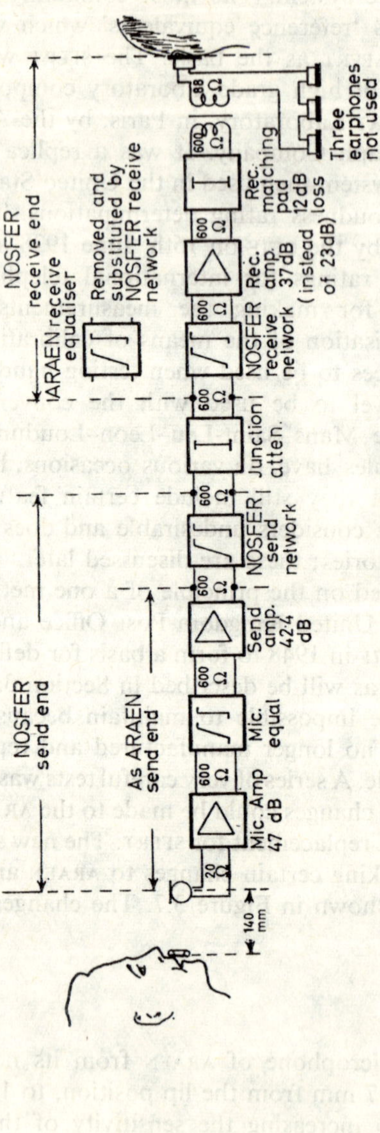

Figure 3.7. Physical basis for reference equivalents (NOSFER)

distortion corresponding to that present in the send end of SFERT. The loss/frequency characteristic of this network is given in Table 3.2 which follows.

Table 3.2. INSERTION LOSSES OF NOSFER NETWORKS AND RECEIVE NETWORK OF ARAEN

Frequency *(1)*	NOSFER *send network* *(2)* dB	NOSFER *receive network* *(3)* dB	ARAEN *receive network* *(4)* dB	NOSFER *receive* — ARAEN *receive* *(5)=(3)−(4)* dB	*Overall loss* NOSFER—ARAEN *(networks only)* *(6)=(2)+(5)* dB
100	12.6	28.7	15.8	12.9	25.5
200	12.3	27.3	15.6	11.7	24.0
300	12.2	25.8	15.4	10.4	22.6
400	11.5	23.8	15.0	8.8	20.3
500	11.0	21.4	14.2	7.2	18.2
600	10.5	21.2	13.5	7.7	18.2
800	9.6	20.2	11.8	8.4	18.0
1000	8.7	19.0	9.9	9.1	17.8
1250	8.0	18.2	8.2	10.0	18.0
1600	7.3	17.8	6.7	11.1	18.4
2000	6.8	18.0	6.0	12.0	18.8
2500	6.5	19.8	7.5	12.3	18.8
3000	6.2	23.3	8.8	14.5	20.7
3500	6.4	27.6	9.8	17.8	24.2
4000	7.6	27.0	10.1	16.9	24.5
5000	12.2	20.2	9.4	10.8	23.0
6000	19.0	16.4	8.2	8.2	27.2
8000	23.8	22.2	14.0	8.2	32.0

Notes: NOSFER send and receive networks are taken from CCITT Recommendation P.42
ARAEN receive network is from measurements made by the United Kingdom Post Office
The total difference between the mouth to ear losses of NOSFER and ARAEN must include the effects of the different microphone distance and a change in the gain of the receive amplifier (see text).

Receive end

3. The gain of the receive amplifier of the ARAEN is increased by 14 dB.
4. The ARAEN receive equaliser is removed and replaced by a 600 Ω network which changes the sensitivity to include a representation of the attenuation/frequency distortion present in the SFERT receive end. The ARAEN receive equaliser fulfils

two functions: (a) It corrects the distortion of the ARAEN earphones and (b) It provides the equivalent of the obstruction effect of the human head placed one metre from the lips of a talker in anechoic conditions. The NOSFER receive network therefore includes allowance for the distortion of the earphones. Table 3.2 gives the losses of the ARAEN receive equaliser as well as that of the NOSFER receive network.

The effects of these changes have been combined with the sensitivities of the ARAEN send and receive end, taken from Table 3.1, to give, in Table 3.3 the send and receive end sensitivities of NOSFER. These have been presented in a uniform manner with those given for ARAEN and so the sensitivities and overall losses of the two systems can be directly compared; Table 3.3 includes the differences.

Table 3.3. SENSITIVITIES OF SEND AND RECEIVE ENDS OF NOSFER

Frequency Hz	*Send End*	*Receive End*	*Overall mouth ear loss*
	NOSFER	NOSFER	NOSFER
100	−6.6(−5.9)	+1.3(+1.2)	+5.3(+4.7)
200	−4.8(−4.4)	+3.3(+3.5)	+1.5(+0.9)
300	−5.7(−5.2)	+4.6(+4.2)	+1.1(+1.0)
400	−5.4(−5.7)	+5.7(+5.9)	−0.3(−0.2)
500	−4.6(−4.3)	+7.3(+7.3)	−2.7(−3.0)
600	−3.6(−3.1)	+6.8(+7.0)	−3.2(−3.9)
800	−2.4(−2.5)	+6.5(+6.6)	−4.1(−4.1)
1000	−1.4(−1.4)	+6.6(+6.9)	−5.2(−5.5)
1250	−1.1(−1.3)	+6.4(+7.0)	−5.3(−5.7)
1600	−1.3(−1.2)	+6.7(+7.1)	−5.4(−5.9)
2000	−0.9(−1.1)	+7.5(+7.6)	−6.6(−6.5)
2500	−0.9(−1.5)	+7.7(+7.6)	−6.8(−6.1)
3000	+0.4(0)	+6.9(+6.9)	−7.3(−6.9)
3500	+0.8(+1.3)	+4.2(+5.5)	−5.0(−6.8)
4000	−0.1(+0.5)	+4.1(+6.5)	−4.0(−7.0)
5000	−4.2(−3.8)	+7.4(+10.7)	−3.2(−6.9)
6000	−11.1(−10.9)	+13.0(+18.0)	−1.9(−7.1)
8000	−15.9	+0.2	+15.7

Notes Sending sensitivity is in dB relative to 1 V per N/m²; sound pressure is referred to a point 25 mm in front of the lips.

Receiving sensitivity is in dB relative to 1 N/m² per V; sound pressure is measured in the special ARAEN artificial ear with a correction applied for the slight difference found between sound pressure measured in real ears and that in the artificial ear. The values in brackets are taken from CCITT Red Book, V, 45 and 49 (1962); allowances have been included for change of units and of sound pressure reference point.

The rules for adjusting the vocal level of the talkers when making reference equivalent measurements require the speech voltmeter, connected to the output terminals of the NOSFER send end to read 10.0 dB lower than when the same speech voltmeter is connected to the output terminals of the ARAEN send end and the ARAEN is spoken into at reference vocal level. This defines the vocal level to be used for measuring reference equivalents and, when the sensitivities given in Table 3.3 are taken into account in conjunction with the spectrum density levels for speech (*see* Section 2.1.3.4), the vocal level for reference equivalent measurements proves by calculation to be 5.6 dB lower than reference vocal level. A direct measurement made by the United Kingdom Post Office, using the conventional phrase 'can-con-by-dace-also' yielded about 6 dB.

The reference equivalent of a 'test' speech path is defined as the setting of NOSFER (i.e. the loss of a 600 Ω attenuator inserted between send and receive ends) which produces to the listener the same sensation of loudness as that produced to the same listener when he uses the 'test' speech path. The vocal level of the talker is the same in both cases and, when a handset telephone is being measured, the manner in which he talks (including the distance between the microphone of the 'test' item and his lips) is specified as described in Chapter 5. This definition, strictly, requires that the 'variable reference' method of balancing (*see* Section 3.1) must be used, with certain attendant disadvantages. In practice reference equivalents of complete 'test' speech paths are rarely measured, except for sidetone paths; more frequently some separation into send and receive ratings is required and, according to CCITT Recommendation P.72, this is performed as shown in Figure 3.8.

To measure the send reference equivalent of a 'test' telephone set associated with its feeding bridge, and if desired, subscriber's line (this combination of 'test' items is termed a 'local telephone circuit' (LTC)), the arrangement shown in Figure 3.8a is used. The send and receive ends of the reference system (NOSFER) are used separately, facilities for changing from the reference system send end to that of the 'test' LTC are provided under the control of the talker as he talks alternately into the reference system and the 'test' microphone. The listener has control of the attenuator x_1, which he adjusts so that the loudness he experiences of the speech sounds from the talker are the same from both reference and 'test' send ends. To avoid the possibility of bias on the part of the listener, a 'hidden loss', x_2 is inserted in the 'test' send end and which is set by the talker for each balance at a value chosen at

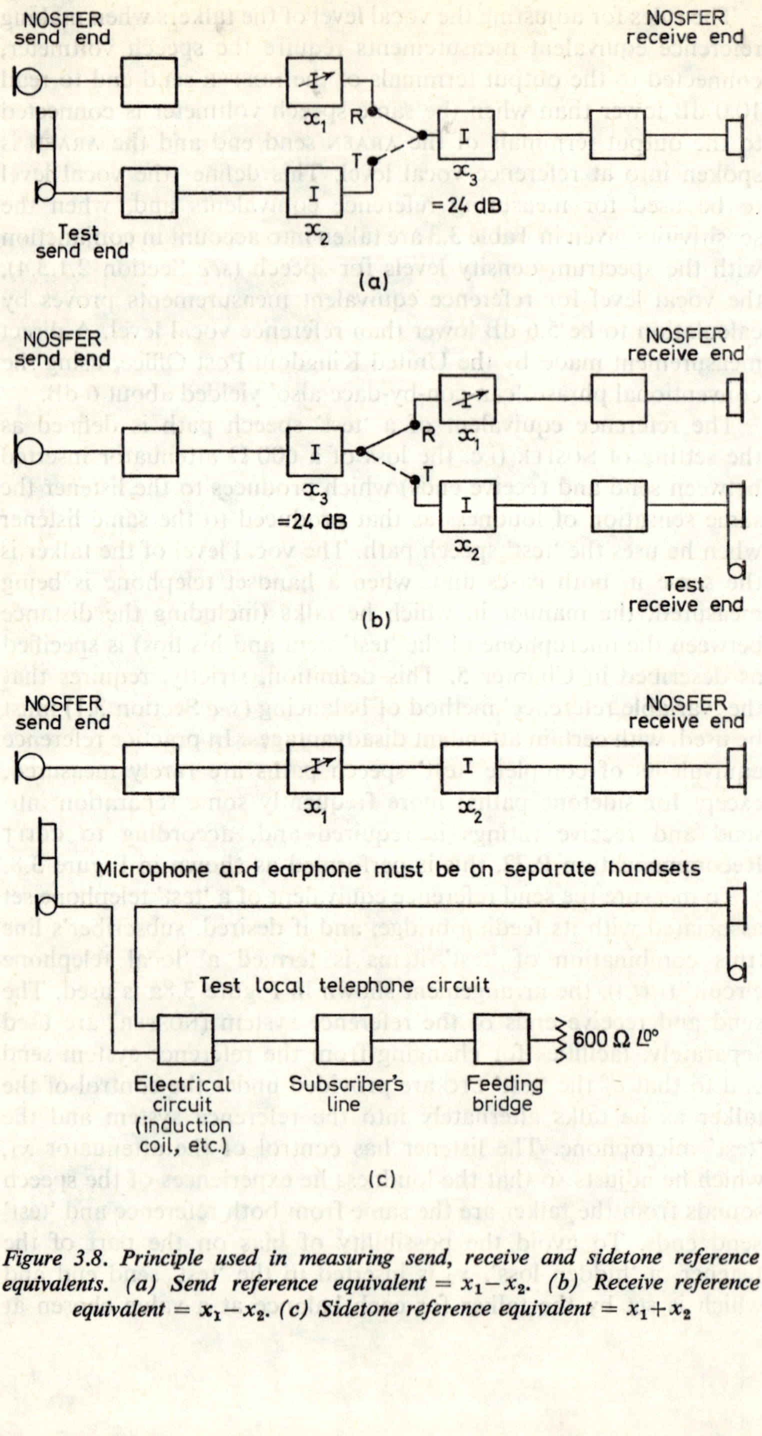

Figure 3.8. Principle used in measuring send, receive and sidetone reference equivalents. (a) Send reference equivalent = $x_1 - x_2$. (b) Receive reference equivalent = $x_1 - x_2$. (c) Sidetone reference equivalent = $x_1 + x_2$

random in the range 0–10 dB (1 dB intervals) and unknown to the listener. An additional attenuator, x_3, is inserted in the common path to the reference system receive end; this is to avoid having to make balances at unrealistic listening levels but it also serves to stabilise the impedances to which the send and receive ends are connected. A setting of 24 dB is used for this attenuator so that during balancing the total loss in the speech path 'test' send end to reference system receive end will range from 24 to 34 dB.

Measurement of receive reference equivalents follows a corresponding procedure as illustrated in Figure 3.8b. In this case the talker speaks continuously into the reference system microphone and the talker controls the switch which changes the receive ends; at the same time the listener changes the earphone to which he is listening, the circuit change being indicated to him by a lamp. The listener adjusts the attenuator x_1, in the path to the reference receive end until he judges the loudnesses to be the same. The attenuator x_2 provides the hidden loss (unknown to the listener) which again is selected at random in the range 0–10 dB. The common attenuator x_3 is again set at 24 dB.

When the balance for sending or receiving has been made the value of the respective reference equivalent is equal to $x_1 - x_2$, dB.

The reference equivalent for a sidetone path* can be measured either with speech or using room noise; the method for speech will be described. A sidetone path cannot usually be interrupted for insertion of any additional loss and so it must be measured by the variable reference method and this often results in the need to perform the balancing at a rather high listening level. The hidden loss attenuator x_2 must therefore be inserted in the reference system as shown in Figure 3.8c. The procedure is similar to that for a sending or a receiving reference equivalent measurement except that the talker and listener must synchronise their changes. The sidetone reference equivalent is given by $x_1 + x_2$. When very small sidetone reference equivalents are being measured it may be necessary to insert some fixed, but known, amplification in the reference path so that the full range of hidden losses can be used.

The overall reference equivalent of a complete 'test' path would be measured in the manner described for sidetone except the hidden loss x_2 will be inserted in the 'test' path together with an additional attenuator x_3 (24 dB). The value of overall reference equivalent is then equal to $x_1 - x_2 - x_3$ dB.

The method for arriving at the settings of x_1 for balance described

* See CCITT Recommendation P.73.

above, the two operator hidden loss method, is not, of course the only one that could be used. It is possible to use a third operator to adjust x_1 in response to signals from the listener indicating which path he finds louder; this method is open to criticism on grounds that the third operator could introduce bias. The method preferred and used for all such balancing operations by the United Kingdom Post Office is the 'quantal response' method which will be described below.

Although measurement of reference equivalents must continue in the manner described for the sake of continuity until another method has been agreed upon internationally, it is possible for individual laboratories to use other methods for measuring loudness ratings for their own purposes, although the results must not be called 'reference equivalents'. The following description is of the method now used in the United Kingdom Post Office for measuring loudness ratings of LTCS containing handset telephones. The method incorporates the following differences in certain features which are considered undesirable in the reference equivalent method as described above:

1. The fundamental reference system used is the ARAEN without any of the distortion incorporated in NOSFER merely to simulate SFERT.
2. The send and receive ends of the fundamental reference system are never used separately.
3. Sending and receiving ratings of 'test' LTCS are determined by use of an 'intermediate reference system' (IRS). This is used to define the separate sending and receive ratings in such a way that reasonably accurate additivity will be achieved and this requires that the characteristics chosen for the IRS shall approximate more closely to those of commercial LTCS than does either of the high quality reference systems, ARAEN or NOSFER.
4. The vocal level specified for measuring reference equivalent is considered to be too low; reference vocal level (5.6 dB higher) is used by the United Kingdom Post Office and this seems convenient for subjective measurements because it can be maintained for long periods. It ia also believed to be much nearer to the median value used by actual telephone customers.
5. The talking distance, when a handset is used as the 'test' item, ought to be more realistic than that specified for the

measurement of reference equivalents according to CCITT Recommendation P.72. This is discussed in detail in Chapter 5.*

6. The position in which the handset is held while being tested needs to be specified to ensure that an appropriate attitude is used for carbon microphones (*see* Chapter 5). Clamping in a fixed stand is used by the United Kingdom Post Office because it ensures precise control; holding in the hand may be suitable and would obviate, or at least reduce, the need for preconditioning procedures.
7. The 'margin' method of balancing is preferable to the 'variable reference' method (*see* Section 3.1) because it enables the listening level to be set at a suitably chosen constant value.

The procedure used by the United Kingdom Post Office is shown in Figure 3.9. The reference system used is the ARAEN with x_1 set at 30 dB (the metre airpath setting). Each of the four paths 2, 3, 4 and 5 is balanced in turn† against the reference speech path, 1, the attenuators x_2, x_3, x_4 and x_5 respectively being adjusted. The method of adjustment (quantal response method) consists of choosing, before each presentation but unknown to the listener, at random but

* The talking distance used by the United Kingdom Post Office for conducting loudness rating measurements is the same as that recommended since 1949 by the CCITT in Recommendation P.45 for articulation measurements (*see* Section 3.3). The talking distance specified in CCITT Recommendation P.72 for measuring reference equivalents brings the mouthpiece of the handset microphone about 12 mm closer to the lips and moves it across the mouth by about 10 mm; telephone handsets designed with rather short handles (which are advantageous in actual service) cannot be tested properly at the reference equivalent position (*see* Section 5.1.1). This somewhat unrealistic mouth location seems to have been chosen originally from erroneous interpretation of a large body of measurements of male and female heads made in the United States of America before 1931, in which year the mouth location was provisionally adopted at the Paris meeting of CCI (8th Plenary Assembly); the provisional adoption was confirmed in 1934 at the Budapest meeting (10th Plenary Assembly of CCIF). The location appearing in CCITT Recommendation P.45 was derived from very extensive and well diversified tests sponsored by the CCIF during 1931–35; this information and the better interpretation of the original data resulted in the recommendation at the Paris meeting in 1949 (15th Plenary Assembly of CCIF) that this more realistic location should be used for articulation measurements (determinations of AEN), which, at that time, were expected to supersede use of reference equivalents in planning telephone networks. (This expectation was not, however, realised and the less realistic location has remained.)

† The complete procedure must be conducted in accordance with a properly designed experimental arrangement so that unwanted variations will not bias the results but will be assessable by suitable statistical analysis. Ref. 12 explains in detail the procedure used by the United Kingdom Post Office.

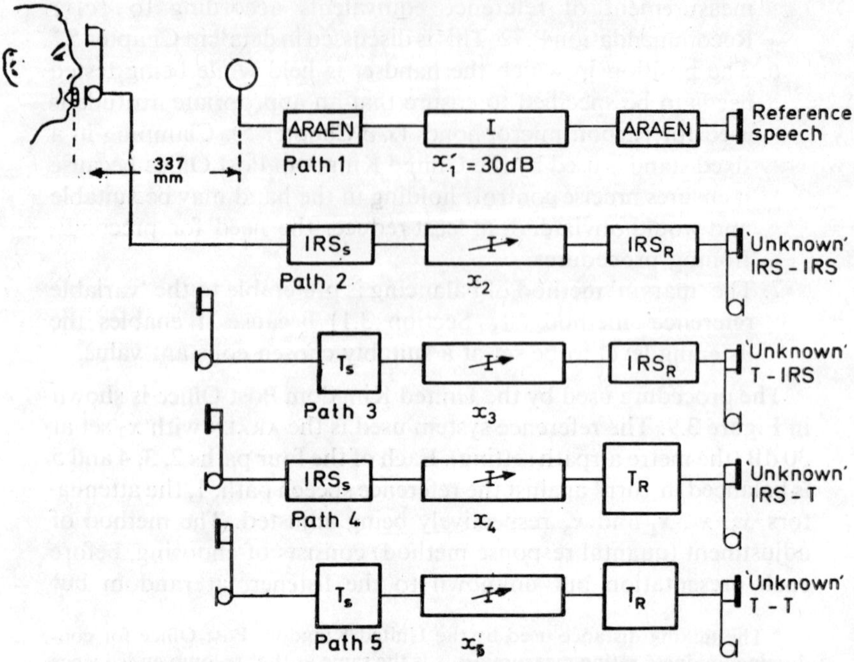

Figure 3.9. Separation send of and receive ratings using an intermediate reference system.

Note: Each 'unknown' is balanced in turn against the reference speech path. IRS_S and IRS_R represent send and receive ends of the intermediate reference system. T_S and T_R represent send and receive ends of the test local telephone circuit which is being measured

without replacement, a setting for the variable attenuator; on presentation of this condition to the listener, he decides whether the 'unknown' (2, 3, 4 or 5, as the case may be) or the 'reference' speech path produces the louder speech. His decision is noted but does not influence the choice of the next setting to be presented. When ten settings having 1 dB intervals have been presented and the decisions taken, the balance point is estimated as shown in Table 3.4.

Table 3.4. QUANTAL RESPONSE METHOD OF BALANCING; SUBJECT'S DECISIONS WITHOUT ANY 'CROSSOVER'

Setting	20	21	22	23	24	25	26	27	28	29	dB
Decision	*U*	*U*	*U*	*U*	*R*	*R*	*R*	*R*	*R*	*R*	

In this example, the highest setting producing '*U* is louder than *R*' is 23 dB and the lowest that produces '*R* is louder than *U*' is 24 dB; the estimated balance point is then taken as 23.5 dB, namely the mean of these settings. The balance may not, however, be so well defined as in the above example; suppose that the results had been as shown in Table 3.5.

Table 3.5. QUANTAL RESPONSE METHOD OF BALANCING; SUBJECT'S DECISIONS PRODUCE A 'CROSSOVER'

Setting	20	21	22	23	24	25	26	27	28	29	dB
Decision	*U*	*U*	*R*	*U*	*U*	*R*	*R*	*R*	*R*	*R*	*R*

In this example the highest setting producing *U* is 24 dB and the lowest producing *R* is 22 dB; the estimated balance point is still the mean of these, namely 23 dB. The rule is always to take the mean of the highest *U* and the lowest *R*. The occurrence of an anomalous decision as illustrated in Table 3.5 is termed a 'crossover' and justification for use of the rule indicated is explained elsewhere (Ref. 12)* (*see also* Section 1.3).

Different members of the crew may well produce consistently different balance points from each other and so the best group of settings from which to select, so that the balance points will be within their range, must depend upon some preliminary crude balances using much wider intervals, say 5 or 10 dB.

The four 'unknown' paths, 2, 3, 4 and 5, consist respectively of:

Path 2; the complete IRS
Path 3; the 'test' send end connected to the receive end of the IRS
Path 4; the send end of the IRS connected to the 'test' receive end
Path 5; the complete 'test' speech path.

The respective balance settings of their attenuators are x_2 x_3, x_4 and x_5 when each is balanced against the fixed setting, x_1, of the reference speech path. The rating of each 'unknown' can then be denoted as follows:

Path 2, relative to the reference path 1,	(1)	$= x_1 - x_2$
Path 3,	*A*	$= x_1 - x_3$
Path 4,	*B*	$= x_1 - x_4$
Path 5	*AB*	$= x_1 - x_5$

* This and other references in this section appear in Section 3.2 of the Bibliography.

By adopting the commonly used statistical conventions for separating main effects and interactions, the following four quantities express the same information in a more convenient form:

$$a = \tfrac{1}{2}(A+AB-(1)-B)$$
$$b = \tfrac{1}{2}(B+AB-(1)-A)$$
$$ab = \tfrac{1}{2}(AB+(1)-A-B)$$
$$\bar{G} = \tfrac{1}{4}(A+B+(1)+AB)$$

These quantities are defined as follows:

a is the send rating of the 'test' LTC relative to the IRS
b is the receive rating of the 'test' LTC relative to the IRS
ab is the interaction between send and receive ratings
$\bar{G}$ is the grand mean and is not usually of any interest.

If the interaction *ab* should be negligible, the send and receive atings simplify to:

$$a = A-(1) \text{ and } b = B-(1)$$

If this procedure is compared with that followed in the measurement of reference equivalent, it will be seen that the reference equivalent method corresponds to replacing both paths 1 and 2 by NOSFER and usually omitting path 5. Reference equivalents are then defined in the form of the expressions for *a* and *b* applicable when the assumption is made that no interaction exists. In fact, a considerable interaction can exist as is illustrated later (see Section 3.2.4) and this can be either positive or negative.

The IRS now in use by the United Kingdom Post Office has evolved from the previous practice of using, for the purpose of separating sending and receiving ratings, selected commercial telephone sets associated with certain simulated subscribers' lines and a type of feeding bridge. These LTCS have been termed 'transmission standards' but are perhaps better described as physical representations of the limiting (worst) transmission performance permitted according to the network plan in force. The principles underlying such use are described later in Chapters 4 and 7. The work of selecting components from supplies of the commercial items manufactured by different companies has proved very onerous and the results are somewhat unreliable; it is largely for this reason and because future designs of commercial items are liable to change, that the IRS principle has been adopted. Besides forming the means for separating loudness ratings into sending and receiving components, the transmission standards and modified versions to represent better than limiting transmission performance have been used for conversation experiments and assessments by various other criteria than loudness. It is therefore desirable that the IRS which replaces these transmission standards shall also be capable of the same wide variety of use. Figure 3.10 therefore shows only the basic and essential parts and their most important characteristics; later, in Chapter 4, the arrangement used when a complete multipath speech link is required (for example, for use in laboratory conversation tests) is explained. Additional information on the use of a 'theoretical IRS' for calculating loudness ratings is given elsewhere, *see* Richards (1971).

The IRS must have certain similarities to commercial telephone sets but it must be constructed in such a way that it can be effectively reproduced and calibrated in successive tests and in different laboratories. The essential characteristics must be capable of being checked with the aid only of commonly available electro-acoustical measuring equipment. The most important features are as follows:

1. The type of earphone should be like those of commercial telephone sets, particularly so far as acoustical impedance and form of earcap are concerned.
2. A close speaking microphone mounted in a telephone handset should be used. A stable type of linear microphone is necessary.
3. The sensitivities and frequency bandwidths of the send and receive ends should be typical of those of actual commercial telephone sets. Those used in the UKPO IRS are shown in Figure 3.10.

The shapes of the sensitivity/frequency characteristics shown in Figure 3.10 were chosen to correspond to the design objectives used in the United Kingdom Post Office since 1947 (*see* Chapter 5) for new microphones and earphones for telephone sets. (They are now (1970) being reviewed but it is not likely to result in very drastic changes.)

The handset used at present is the type used in the current UKPO 700 type telephone sets and the earphone is also of the type used in those sets but with some attenuation equalisation to attain the

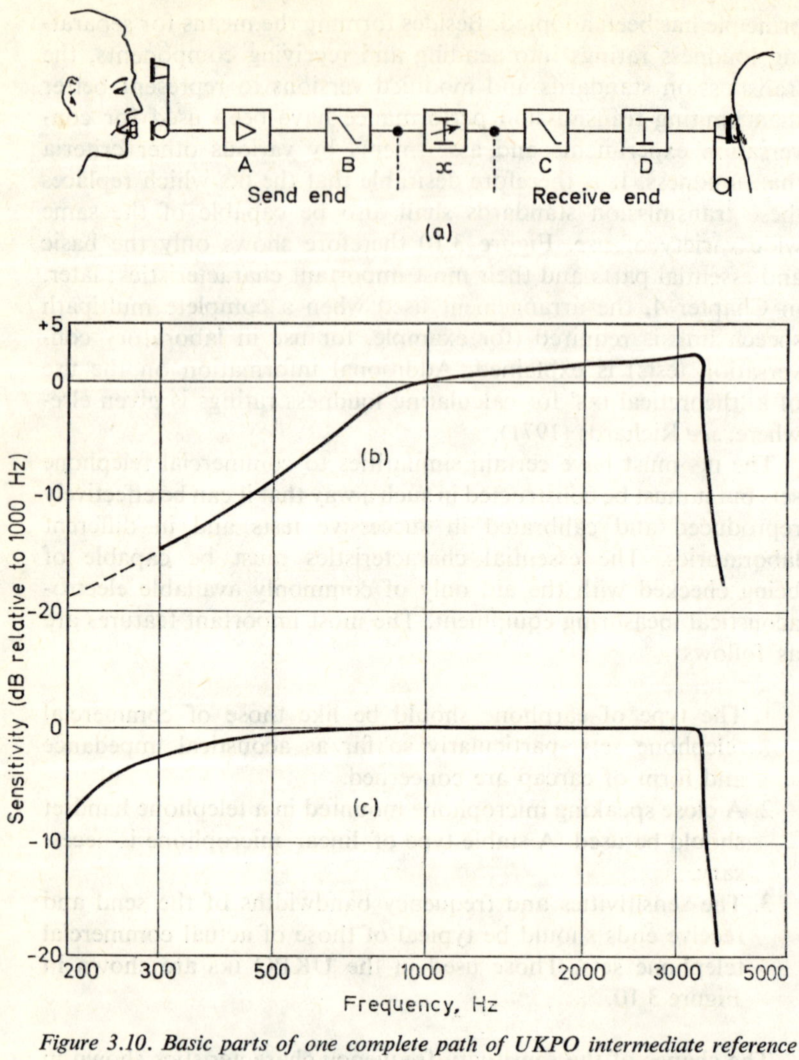

Figure 3.10. Basic parts of one complete path of UKPO intermediate reference system

(a) Schematic diagram. (A) Microphone amplifier. (B) Send equaliser. (C) Receive equaliser. (D) Earphone matching pad or transformer. (b) Relative sensitivity of send end. (c) Relative sensitivity of receive end

required sensitivity/frequency characteristics when measured on real ears. The microphone is a rocking armature type of capsule which can be fitted with slight modification into the handset. The desired real voice sensitivity/frequency characteristic requires provision of some equalisation; amplification is also provided to achieve comparable sensitivity with that of typical commercial LTCS.

The sensitivities shown in Figure 3.10 correspond approximately to SRE = 0 dB and RRE = +10 dB when the sensitivities at 1000 Hz are 1 V per N/m^2 and 1 N/m^2 per V, respectively for sending and receiving; this assumes that the sending sensitivity relates to speaking at BMSP. Table 3.7 gives the sensitivities required for an IRS as discussed above to yield sending and receiving loudness ratings (SLR and RLR) numerically equal, respectively to SRE and RRE.

3.2.3 Use

Loudness ratings are laborious, rather time consuming and fairly expensive to measure and so they are rarely performed on complete paths present in an actual speech link. Measurements are usually confined to those of local telephone circuits (combinations of a subscriber's telephone set and a feeding bridge together with, if desired, a particular subscriber's line). The send and receive loudness ratings, e.g. reference equivalents, determined for this combination are used to estimate the overall loudness rating or reference equivalent of any complete path containing the same local telephone circuit; this is done by adding the appropriate send and receive ratings together with suitable values for the losses of all the intervening elements. Some simple rules are used to determine the losses of these intervening elements such as those recommended by the CCITT for determining 'The nominal reference equivalent of a complete connection' (CCITT Recommendation P.11). The matter is also referred to in CCITT Recommendation P.21. These rules can be interpreted as follows:

Local telephone circuits, i.e. between customer and local exchange including the feeding bridge at the local exchange: Use the send and receive reference equivalents as determined according to Recommendation P.72.

Junction circuits, i.e. between local exchange and group switching centre: No specific rule is recommended by the CCITT but the loss at 800 Hz is suitable for amplified circuits and loaded unamplified cables provided the nominal upper frequency cutoff of 3400 Hz is respected. For unloaded cables the United Kingdom Post

Office uses the loss at 1600 Hz but another frequency or the mean of the losses over the entire frequency range could be used.

Trunk circuits, which are now invariably provided by some form of modulation that provides a fairly uniform response over the range 300–3400 Hz: Use the loss at 800 or 1000 Hz (the United Kingdom Post Office uses 800 Hz).

Clearly, such rules rely on the assumption that the separate quantities are additive, i.e. their sum gives the same result as making a direct measurement of the complete mouth to ear path. In general this is not true as is shown in Section 3.2.4.3. For particular circumstances, however, positive and negative discrepancies can compensate each other, particularly for connections containing a good deal of attenuation/frequency distortion. A fairly extensive series of tests was carried out by the CCITT Laboratory on connections designed to simulate conditions likely to arise in international connections and these show reasonably accurate additivity.* Nevertheless, non-additivity of reference equivalents is a severe problem when studies of local connections are made and it becomes necessary to use loudness ratings defined in terms of a suitable intermediate reference system.

The most important uses of loudness ratings are for defining transmission losses of speech paths for network planning as explained in Chapters 4 and 7.

3.2.4 Relationships between loudness ratings of speech paths and their absolute sensitivities

3.2.4.1 Theoretical basis for calculating loudness of speech

The theory of hearing described in Section 2.2 has been extended by Braun, Fletcher and Munson to include the special treatment of speech. The theory can be recapitulated in terms of the funda-

* The discrepancies were generally less than ±1 dB when the sum of send and receive reference equivalents measured separately for the local telephone circuits was compared with the overall reference equivalent of the combination of the two local telephone circuits connected together through bandpass filters representing the attenuation/frequency distortion of various numbers of international circuits, allowance being made in each case for the passband loss of these filters at 800 or 1000 Hz. (CCITT Tech. Rep. 305 (1964) refers).

mental expression:

$$\lambda = C \int_0^\infty Q(Z)B' \, df \qquad (3.1)$$

or

$$\lambda = C\overline{Q(Z)} \qquad (3.1a)$$

where Z is the *sensation level* (dB) and is a function of frequency.

$$Z = \beta'_S - L_{ME} - (\beta_0 - K). \qquad (3.2)$$

β'_S is the long term spectrum density level of speech referred to an arbitrary reference point close to the talker's lips; this point will be taken here as located 25 mm in front of the lips (*see* Section 2.1.3).

β_0 is the pure tone threshold of audibility (*see* Section 2.2.3.1); in most cases discussed here, the threshold for monaural listening with an earphone will apply and the sound pressures will be referred to the centre of the opening of the ear canal (*see* Section 2.2.1).

$K = 10 \log_{10} (\Delta f)_c$, where $(\Delta f)_c$ is the width of a critical band of frequencies (*see* Section 2.2.3.1).

$(\beta_0 - K)$ can be treated as though it were the threshold of audibility for sounds having continuous spectra (*see* Section 2.2.3.1).

L_{ME} is the mouth to ear transmission loss between the two reference points used for defining β'_S and β_0.

Z can be regarded as the level of the speech in a particular frequency region above its threshold of audibility in that region*.

$Q(Z)$ is a function of Z, the form of which will be described below. B' is a frequency weighting factor.

C is a constant inserted to enable a convenient scale to be used for λ.

$\overline{Q(Z)}$ is the weighted mean of $Q(Z)$.

The forms of $Q(Z)$ and B' are chosen so that, so far as possible, λ will have the property that speech sounds yielding the same value of λ will appear to a listener as equally audible. In the original theory, developed to deal with simple, noise-like, sounds, B' is directly related to $(\Delta f)_c$ which is supposed (on very convincing

* It must be noted that components in different frequency regions all contribute to total audibility and so a component that is below its threshold of audibility when considered in isolation (i.e. when $Z < 0$ db) can make a definite contribution and so must be included.

grounds, *see* Section 2.2.3.1) to represent equal lengths of patches of nerve endings along the basilar membrane of the ear.

$$\text{That is:} \qquad (\Delta f)_c = \frac{\mathrm{d}f}{\mathrm{d}l}(\Delta l)_c$$

$$\text{and} \qquad B' = \frac{\mathrm{d}l}{\mathrm{d}f}\frac{1}{l_{\max}}$$

$$\text{therefore} \qquad B' = \frac{(\Delta l)_c}{l_{\max}}\frac{1}{(\Delta f)_c} = \frac{(\Delta l)_c}{l_{\max}}10^{-K/10}$$

It is stated that each $(\Delta l)_c$ occupies 1/64 of the total length, $l_{\max}$, of the basilar membrane and so:

$$B' = \frac{1}{64}10^{-K/10} \quad \text{or} \quad 10\log_{10}B' = -18.07-K$$

Figure 3.11 shows the values of $10\log_{10}B'$ determined as above. B' can also be determined experimentally direct from loudness balancing experiments using high pass and low pass filters inserted in an otherwise distortionless speech path and these yield results that differ in some cases from those in Figure 3.11 (*see* Braun, Fletcher and Galt, Japanese Telephone Company, etc) but the above theoretical expression gives values within the range of those determined experimentally and so will be used here.

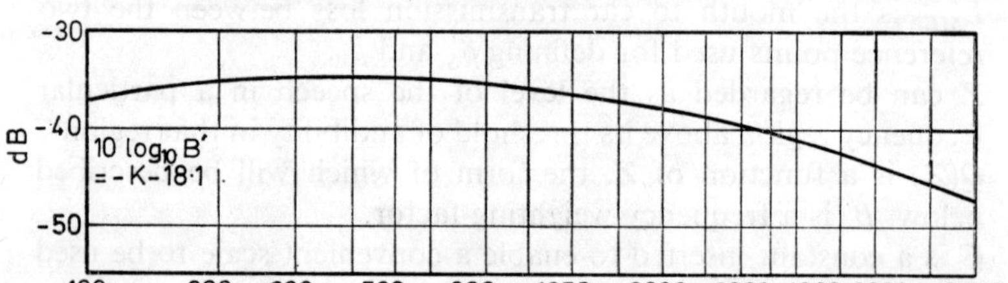

Figure 3.11. Frequency weighting factor for audibility and loudness

It is convenient and appropriate to approximate the function $Q(Z)$ by the following:

$$Q(Z) = C_k(10^{Z/10})^m \tag{3.3}$$

where C_k and m are parameters.

Although m does not remain constant for the whole range of Z (*see* Figure 3.12), it is sufficient for many purposes when telephony speech is being considered to distinguish two regions as follows:

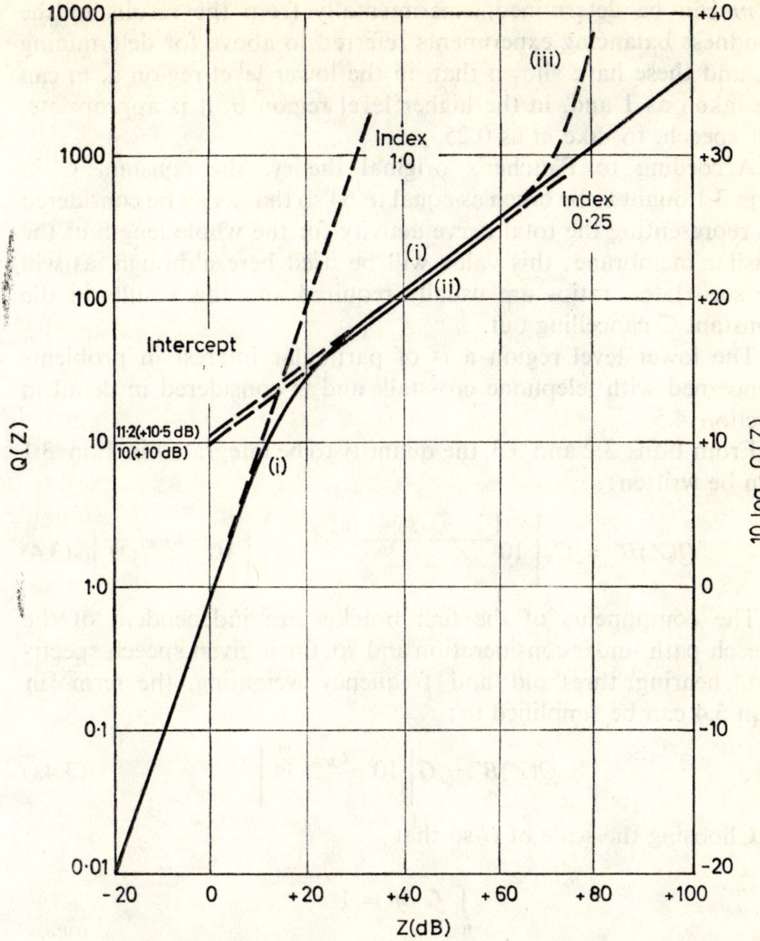

Figure 3.12. Loudness growth function, Q(A) for speech.
(i) An approximation to Fletcher's Q_S $Q(Z) = 10^{Z/10}\{10^{(Z-14)/10}+1\}^{-0.75}$
(ii) An approximation to (i) useful over the range $13 \leqslant Z \leqslant 60$ *dB*
$10 \log_{10} Q(Z) = 0.25Z+10$
(iii) Probable form of Q(Z) for high speech levels

(a) A region concerned with thresholds of audibility where the values of Z range from about -20 dB to perhaps $+13$ dB.

(b) A region concerned with the loudness of fairly easily audible speech where Z exceeds about $+13$ dB over at least the major portion of the audible speech frequency range.

m can be determined experimentally from the results of the loudness balancing experiments referred to above for determining B' and these have shown that, in the lower level region a, m can be taken as 1 and, in the higher level region b, it is appropriate, for speech, to take m as 0.25.

According to Fletcher's original theory, the constant C in Eqn 3.1 ought to be taken as equal to 64 so that λ can be considered as representing the total nerve activity for the whole length of the basilar membrane; this value will be used here although, as will be seen later, ratios are usually required and this results in the constant C cancelling out.

The lower level region a is of particular interest in problems concerned with telephone crosstalk and is considered in detail in Section 4.5.

From Eqns 3.2 and 3.3, the quantity to be integrated in Eqn 3.1 can be written:

$$Q(Z)B' = C_k\left[10^{\frac{\{\beta'_S-(\beta_0-K)\}m+10\log_{10}B'}{10}}\right]\left[10^{\{-L_{ME}\}\frac{m}{10}}\right] \quad (3.4)$$

The components of the first bracket are independent of the speech path under consideration and so, for a given speech spectrum, hearing threshold and frequency weighting, the terms in Eqn 3.4 can be simplified to:

$$Q(Z)B' = G\left[10^{(-L_{ME})\frac{m}{10}}\right] \quad (3.4a)$$

Choosing the scale of G so that

$$\int_0^\infty G\,df = 1$$

enables G to be treated as a weighting factor (called a reciprocal loss frequency weighting factor) in determining the weighted mean of the function of L_{ME} represented by the term dependent on the speech path under consideration. The results of various determinations of G are shown for comparison in Figure 3.13.

λ and C, which appear in Eqn 3.1, can be dispensed with by using the concept of a weighted mean value of L_{ME} defined as follows:

$$\overline{L_{ME}} = -\frac{1}{m}10\log_{10}\int_0^\infty Q(Z)B'\,df \quad (3.5)$$

The integral in Eqn 3.5 can be replaced by a summation using a suitable choice of contiguous frequency bands each Δf Hz wide. Δf need not be independent of frequency but can be chosen arbitrarily; it has proved convenient to standardise a set of midband frequencies as shown in Tables 3.6–3.11.

$$\overline{L_{ME}} = -\frac{1}{m} 10 \log_{10} \Sigma 10^{\frac{1}{10}(10 \log_{10} G\, \Delta f - mL_{ME})} \quad (3.6)$$

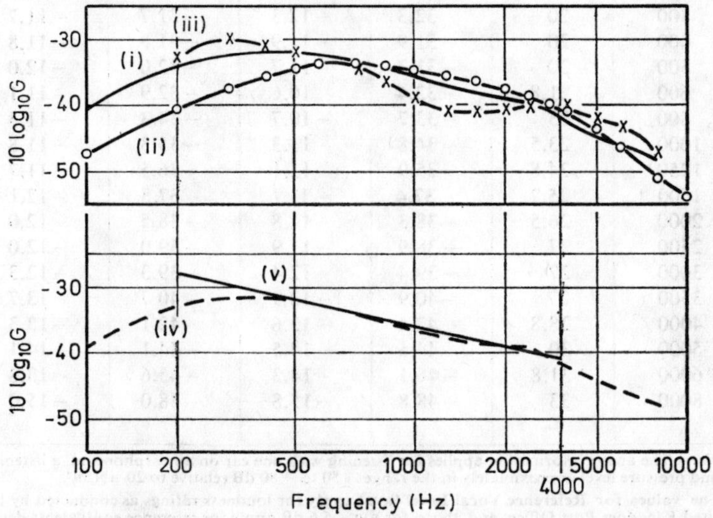

Figure 3.13. Reciprocal loss frequency weighting factor for loudness, G (i) From Table 3.6 (RVL) (m=0.25). (ii) Fletcher and Galt (m=0.25). (iii) Japan (m=0.27). (iv) From Table 3.6 (RVL−5.6 dB). (v) As used in Braun's reference equivalent meter

Appropriate values of $10 \log_{10} G$ and $10 \log_{10} G\Delta f$ are shown in Table 3.6 for $m = 0.25$; slightly different values are needed when the speech is spoken at the level appropriate to the determination of reference equivalents, in accordance with CCITT Recommendation P.72, from those required for speech at reference vocal level, because the spectrum shape of β'_S is affected,

The loudness rating of a given speech path relative to a specified reference system can be determined by calculating $\overline{L_{ME}}$ for the two speech paths concerned and taking their difference.

Let $\overline{L_{RME}}$ be the value for the reference system and $\overline{L_{ME}}$ be that for the 'unknown' speech path; the loudness rating of the 'unknown' is then:

$$\text{Loudness rating} = \overline{L_{ME}} - \overline{L_{RME}} \quad (3.7)$$

Table 3.6. RECIPROCAL LOSS WEIGHTING FACTORS, G, FOR TWO VOCAL LEVELS

Frequency f Hz	10 $\log_{10}$ Δf	RVL		RVL—5.6 dB	
		10 $\log_{10}$ G	10 $\log_{10}$ $G\,\Delta f$	10 $\log_{10}$ G	10 $\log_{10}$ $G\,\Delta f$
(1)	(2)	(3)	(4)	(5)	(6)
100	19	−40.2	−21.2	−38.8	−19.8
200	20	−33.9	−13.9	−32.9	−12.9
300	20	−32.3	−12.3	−31.7	−11.7
400	20	−31.9	−11.9	−31.8	−11.8
500	20	−31.7	−11.7	−32.0	−12.0
600	21.8	−32.4	−10.6	−32.9	−11.1
800	23	−33.7	−10.7	−34.3	−11.3
1000	23.5	−34.8	−11.3	−35.3	−11.8
1250	24.8	−35.9	−11.1	−36.5	−11.7
1600	25.7	−37.4	−11.7	−37.8	−12.1
2000	26.5	−38.3	−11.8	−38.5	−12.0
2500	27	−38.9	−11.9	−39.0	−12.0
3000	27	−39.4	−12.4	−39.3	−12.3
3500	27	−40.9	−13.9	−40.7	−13.7
4000	28.8	−42.4	−13.6	−42.1	−13.3
5000	30	−44.5	−14.5	−44.1	−14.1
6000	31.8	−46.1	−14.3	−45.6	−13.8
8000	33	−48.8	−15.8	−48.0	−15.0

Notes The above information applies for listening with one ear on an earphone at a listening sound pressure level approximately in the range +50 to +80 dB relative to 20 $\mu N/m^2$.

The values for Reference Vocal Level (RVL) apply for loudness ratings as conducted by the United Kingdom Post Office and those for RVL−5.6 dB apply for reference equivalents determined in accordance with CCITT Recommendation P.72,

If the reference system used is NOSFER and the values of G appropriate to a vocal level of RVL−5.6 dB are taken, Eqn 3.7 gives the reference equivalent of the 'unknown' speech path. If, however, the values of G corresponding to RVL are used another reference system, such as ARAEN, etc. is used the results must be termed 'relative equivalents' or, simply, loudness ratings; naturally the reference system used must be stated.

Satisfactory agreement can be achieved between loudness ratings calculated as above and direct subjective determinations provided that the values of L_{ME} used in the calculations have been based on suitable measurements of sensitivity of the microphones and earphones concerned.

3.2.4.2 *Determination of proper sensitivity/frequency characteristics*

The sensitivity/frequency characteristics included in Tables3.1 and 3.3 for the reference systems ARAEN and NOSFER are well defined. When practical telephone sets are being treated, however, serious problems are encountered in defining their sending and receiving sensitivities for the present purpose. At present, there is no generally accepted method for measuring these quantities in a manner suitable for calculating loudness ratings. To provide a basis for study, the following rules have been adopted provisionally by the United Kingdom Post Office and the sensitivity data used in the examples given here have been obtained in this way, unless otherwise stated.

Sending sensitivity

A suitable artificial mouth is set up with the 'direction of speech' horizontal. The United Kingdom Post Office normally uses the 'Post Office' artificial mouth mentioned in Section 2.1.1. When a handset with a linear microphone is being tested, the sound pressure level used is of little importance provided that it is known and referred to a well defined point in front of the artificial mouth; with carbon microphones, the sound pressure level is of considerable importance. A person speaking at RVL produces a long term r.m.s. speech pressure at the 'mouth reference point', located on the axis of the mouth and situated 25 mm from the mouth side of the lip ring (which defines the equivalent lip position of an artificial mouth) in the absence of any obstruction of 0.34 N/m^2 (−4.7 dB relative to 1 N/m^2 or +89.3 dB relative to 20 $\mu N/m^2$). It is convenient to use this sound pressure for linear microphones and to use several values of sound pressure for carbon microphones, including this value, extending at least 10 dB above and 10 dB below this value. This procedure enables the 'upper envelope' of the family of sensitivity/frequency characteristics to be determined as explained in Section 5.1.2.

The handset is placed at the British modal speaking position (BMSP) located as described in Section 5.1.2; this corresponds, in talking distance and in orientation of the handset, to the manner that average customers may reasonably the expected to hold their handsets.

The electrical output level from the LTC is measured in dBV across a 600 Ω termination connected to the 'junction' side of the feeding bridge.

The sensitivity, expressed in dB relative to 1 V per N/m^2 is given by:

$$20 \log_{10} \frac{\text{output voltage}}{\text{input sound pressure in N/m}^2}$$

Naturally, with carbon microphones, appropriate conditioning treatment must be given to stabilise the measurements and prevent packing.

When sending reference equivalent is to be calculated from the measurements, special separate measurements at the reference equivalent speaking distance can be avoided by increasing the sending sensitivities determined at BMSP by 3 dB, at all frequencies.

Receiving sensitivity

A suitable type of artificial ear must be used, for example the IEC/CCITT artificial ear mentioned in Section 2.2.1.

The electrical source must be of 600 Ω resistance, angle 0°, and be set to produce a level comparable with that of speech signals likely to enter the LTC; −18 dBV, when the source is terminated in 600 Ω, is suitable (i.e. e.m.f. −12 dBV). The source is connected to the junction terminals of the feeding bridge and the sound pressure measured in the artificial ear.

The sensitivity, expressed in dB relative to 1 N/m² per V is given by:

$$20 \log_{10} \frac{\text{output sound pressure in N/m}^2}{\text{input voltage}}$$

The value of input voltage used is the terminated p.d. with 600 Ω load, i.e. −18 dBV when the above suggestion is adopted.

The receiving sensitivities measured in this manner are somewhat higher than those applicable when the handset is held in the normal conversational manner and so a correction must be applied (*see* Section 5.1.3). A simple reduction of 3 dB independent of frequency seems to yield appropriate sensitivity values for calculating loudness and other ratings for modern types of telephone earphone but corrections based on measurements of sound pressure at the ear reference point (described in Section 2.2.1) in real ears may be used if desired.

Table 3.7 shows the sensitivities, sending and receiving, of two types of commercial telephone sets on approximately average subscribers' lines. The results for the 13-2P-27 set were obtained as follows:

Sending; by real voice method, using half and quarter octave bandpass filters;

Table 3.7. SENDING AND RECEIVING SENSITIVITIES OF UNITED KINGDOM TELEPHONE SETS ON 1 MILE $6\frac{1}{2}$ lb SUBSCRIBER'S LINE AND 50 V 200+200 Ω FEEDING BRIDGE TOGETHER WITH THEORETICAL SENSITIVITIES OF AN INTERMEDIATE REFERENCE SYSTEM (IRS)

Frequency Hz	*13-2P-27*			*16-4T-31 (with regulator)*			*Theoretical* IRS	
	Sending		*Receiving* dB	*Sending*		*Receiving* dB	*Sending* BMSP dB	*Receiving* dB
	BMSP dB	RE dB		BMSP dB	RE dB			
(1)	(2)	(3)	(4)	(5)	(6)	(7)	(8)	(9)
200	−26.1	−23.1	+ 5.2	−16.8	− 3.8	+ 1.3	−18.8	+ 2.5
300	−24.0	−21.0	+ 8.5	−15.3	−12.3	+11.7	−14.6	+ 7.7
400	−18.8	−15.8	+ 9.9	−14.9	−11.9	+15.3	−11.6	+ 9.1
500	−17.3	−14.3	+10.7	−14.5	−11.5	+17.1	− 8.9	+ 9.7
600	−18.9	−15.9	+11.5	−13.7	−10.7	+17.8	− 6.2	+10.0
800	−18.3	−15.3	+11.7	−11.1	− 8.1	+18.5	− 1.2	+10.2
1000	−13.8	−10.8	+11.9	− 7.4	− 4.4	+17.7	+ 0.1	+10.2
1250	− 6.4	− 3.4	+13.0	− 4.4	− 1.4	+17.5	+ 0.7	+10.2
1600	− 1.4	+ 2.6	+15.8	− 6.3	− 3.3	+17.7	+ 1.1	+10.2
2000	− 3.4	+ 0.4	+11.5	− 7.8	− 4.8	+12.4	+ 1.6	+10.2
2500	− 4.8	− 1.8	+ 7.2	− 9.5	− 6.5	+ 5.4	+ 1.9	+10.2
3000	− 7.9	− 4.9	+11.5	− 7.2	− 4.2	+ 9.6	+ 2.2	+10.2
3500	−11.2	− 8.2	− 0.2	−11.7	− 8.7	+12.0	−11.4	− 2.3
4000	−13.7	−10.7	<-15	−18.9	−15.9	−1.6	<-28	<-20

Notes 1 Sending sensitivities are expressed in dB relative to 1 V per N/m^2, the input sound pressure being referred to a point 25 mm in front of the lips with the sound field unobstructed.

2 The sending sensitivities of the real telephone sets were measured at BMSP and the values at RE speaking position obtained by adding 3 dB.

3 Receiving sensitivities are expressed in dB relative to 1 N/m^2 per V, the input voltage being measured across a 600 Ω load terminating the 600 Ω source. The output sound pressure is referred to a point located at the centre of the opening of the ear canal under the earcap.

Receiving; using the BS2042 artificial ear and applying a uniform correction of −6 dB.

The results for the 16-4T-31 (with regulator) set were obtained as follows:

Sending, using the Post Office artificial mouth and taking the 'envelope' as described above;
Receiving, using the BS2042 artificial ear and applying corrections derived from the information given in Figure 5.8a which appears in Section 5.1.3.

When direct measurements of sensitivity of complete LTCs are not possible (or are inconvenient), the results required can be

Table 3.8. DETERMINATION OF SENDING AND RECEIVING REFERENCE EQUIVALENTS OF A LOCAL TELEPHONE CIRCUIT; for speech emitted by a talker at RVL −5.6 dB and allowing for RE speaking position

Frequency f Hz	*10* $\log_{10}$ $G\,\Delta f$	*16-4T-31 (reg.) on 1* m/$6\frac{1}{2}$ *to* NOSFER				NOSFER *to 16-4T-31 (reg.) on 1* m/$6\frac{1}{2}$			
		L_{ME} dB	$-0.25\ L_{ME}$	$10 \log_{10} G\Delta f$ $-0.25L_{ME}$	antilog *1/10* *Col (5)*	L_{ME} dB	$-0.25\ L_{ME}$	$10 \log_{10} G\,\Delta f$ $-0.25L_{ME}$	antilog *1/10* *Col (9)*
(1)	(2)	(3)	(4)	(5)	(6)	(7)	(8)	(9)	(10)
200	−12.9	+10.5	−2.6	−15.5	0.0282	+3.5	−0.9	−13.8	0.0416
300	−11.7	+7.7	−1.9	−13.6	0.0437	−6.0	+1.5	−10.2	0.0955
400	−11.8	+6.2	−1.6	−13.4	0.0458	−9.9	+2.5	−9.3	0.1176
500	−12.0	+4.2	−1.0	−13.0	0.0501	−12.5	+3.1	−8.9	0.1290
600	−11.1	+3.9	−1.0	−12.1	0.0617	−14.2	+3.6	−7.5	0.1778
800	−11.3	+1.6	−0.4	−11.7	0.0677	−16.1	+4.0	−7.3	0.1865
1000	−11.8	−2.2	+0.6	−11.2	0.0759	−16.3	+4.1	−7.7	0.1699
1250	−11.7	−5.0	+1.2	−10.5	0.0892	−16.4	+4.1	−7.6	0.1739
1600	−12.1	−3.4	+0.8	−11.3	0.0742	−1.64	+4.1	−8.0	0.1588
2000	−12.0	−2.7	+0.7	−11.3	0.0742	−11.5	+2.9	−9.1	0.1231
2500	−12.0	−1.2	+0.3	−11.7	0.0677	−4.5	+1.1	−10.9	0.0813
3000	−12.3	−2.7	+0.7	−11.6	0.0693	−10.0	+2.5	−9.8	0.1049
3500	−13.7	+4.5	−1.1	−14.8	0.0331	−12.8	+3.2	−10.5	0.0892
4000	−13.3	+11.8	−3.0	−16.3	0.0235	+1.7	−0.4	−13.7	0.0427
				Sum	0.8043			Sum	1.6918

$\overline{L_{ME}} = \frac{1}{0.25}(10 \log_{10} \text{Sum}) = +3.8$ dB

$\overline{L_{RME}}$ (NOSFER–NOSFER) $= -3.3$ dB

∴ SRE $= +7.1$ dB

$\overline{L_{ME}} = -9.1$ dB

$\overline{L_{RME}} = -3.3$ dB

∴ RRE $= -5.8$ dB

obtained by combining results of measurements on the microphone and earphone with measured or calculated values of losses associated with the electrical components. This procedure is described with an example in Section 6.1.2. When hypothetical LTCs are being studied, all the sensitivity values will be theoretical but must be defined in a manner compatible with actual measurement methods.

The sensitivities, S_M and S_E, of microphone and earphone and values of electrical loss required to establish appropriate acoustical (air to air) losses of sidetone paths are described in Section 6.1.4.

3.2.4.3 Calculation of reference equivalents

The theoretical basis described in Section 3.2.4.1 can be applied using sensitivity/frequency characteristics for telephone sets and associated line plant determined as described in Section 3.2.4.2. Calculation of sending and receiving reference equivalent is shown in Table 3.8, the table is arranged to evaluate $\overline{L_{ME}}$ in accordance with Eqn 3.6, which appears in Section 3.2.4.1. The values of mouth to ear loss at each frequency are obtained by adding together the appropriate sensitivity of the LTC being rated (*see* Table 3.7) and the sensitivity of the complementary end of NOSFER (*see* Table 3.3). The sending sensitivities used for the LTC are those applying for the RE speaking position. The remaining columns in Table 3.8 are arranged in a manner that needs no further explanation. The value of $\overline{L_{RME}}$ for NOSFER is calculated in the same manner using the overall mouth to ear losses of NOSFER derived from Table 3.3; the values of sending and receiving reference equivalent (SRE and RRE) follow from Eqn 3.7 (Section 3.2.4.1).

The values of sending and receiving reference equivalent (SRE and RRE), together with the appropriate values of loss for the intervening electrical circuits, enable the nominal overall reference equivalent of any given connection to be determined. Although not usually required for planning studies, the overall reference equivalent (ORE) can also be calculated by assembling the set of values of L_{ME} from the sending and receiving sensitivities of the LTC and performing corresponding calculations as shown in Table 3.9. It will be observed that the value of ORE (−0.8 dB) differs from the nominal ORE, namely the sum SRE+RRE (+1.3 dB), and this is caused by the use of the wideband receive and send ends of NOSFER used to define SRE and RRE, respectively. As explained in

Table 3.9. DETERMINATION OF OVERALL REFERENCE EQUIVALENT OF A SPEECH PATH
For speech emitted by a talker at RVL −5.6 dB
and allowing for RE speaking position

Frequency f Hz	*16-4T-31 (reg.) on 1* m/$6\frac{1}{2}$ *to ditto*				
	10 $\log_{10}$ $G\,\Delta f$	L_{ME} dB	-0.25 L_{ME}	*10* $\log_{10}$ $G\,\Delta f$ $-0.25L_{ME}$	antilog *1/10* *Col* *(5)*
(1)	(2)	(3)	(4)	(5)	(6)
200	−12.9	+12.5	−3.1	−16.0	0.0252
300	−11.7	+ 0.6	−0.2	−11.9	0.0647
400	−11.8	− 3.4	+0.8	−11.0	0.0795
500	−12.0	− 5.6	+1.4	−10.6	0.0872
600	−11.1	− 7.1	+1.8	− 9.3	0.1177
800	−11.3	−10.4	+2.6	− 8.7	0.1350
1000	−11.8	−13.3	+3.3	− 8.5	0.1414
1250	−11.7	−16.1	+4.0	− 7.7	0.1700
1600	−12.1	−14.4	+3.6	− 8.5	0.1414
2000	−12.0	− 7.6	+1.9	−10.1	0.0978
2500	−12.0	+ 1.1	−0.3	−12.3	0.0590
3000	−12.3	− 5.4	+1.4	−10.9	0.0814
3500	−13.7	− 3.3	+0.8	−12.9	0.0514
4000	−13.3	+20.5	−5.1	−18.4	0.0145
				Sum	1.2262

$$L_{ME} = \frac{1}{0.25}\,(10 \log_{10} \text{Sum}) \qquad = -4.1 \text{ dB}$$

L_{RME} (NOSFER–NOSFER) $= -3.3$ dB

∴ ORE $= -0.8$ dB

Section 3.2.4.4, the trouble can be much alleviated by defining sending and receiving loudness ratings in terms of receive and send ends of an intermediate reference system (IRS) having characteristics more like those of the commercial LTCs that are to be rated.

According to CCITT Recommendation P.73, sidetone reference equivalent (for speech) is determined in the same manner as ORE and so the appropriate values of L_{ME} must be made available for the speech sidetone mouth to ear path. Table 6.13 in Section 6.1.4 shows how the appropriate information can be derived using suitable sensitivities for microphone and earphone.

It must be remembered that the values of L_{ME} for sidetone reference equivalent must (like SRE) include appropriate sensitivity allowances for the reference equivalent speaking position. Table 3.10

Table 3.10. DETERMINATION OF REFERENCE EQUIVALENT OF A SPEECH SIDETONE PATH;
For speech emitted by a talker at RVL −5.6 dB and allowing for RE speaking position

Frequency f Hz	*16-4T-31 (reg.) on 1 m/6½* $10 \log_{10} G\Delta f$	L_{ME} dB	$-0.25 L_{ME}$	$10 \log_{10} G\Delta f - 0.25 L_{ME}$	antilog *1/10 Col (5)*
(1)	(2)	(3)	(4)	(5)	(6)
200	−12.9	+ 9.3	−2.3	−15.2	0.0302
300	−11.7	1.6	+0.4	−11.3	0,0741
400	−11.8	− 4.7	+1.2	−10.6	0.0871
500	−12.0	− 6.4	+1.6	−10.4	0.0912
600	−11.1	− 6.5	+1.6	− 9.5	0.1125
800	−11.3	− 8.3	+2.1	− 9.2	0.1202
1000	−11.8	− 9.6	+2.4	− 9.4	0.1148
1250	−11.7	−10.5	+2.6	− 9.1	0.1230
1600	−12.1	5.4	+1.4	−10.7	0.0851
2000	−12.0	+ 5.7	−1.4	−13.4	0.0457
2500	−12.0	+24.5	−6.1	−18.1	0.0155
3000	−12.3	+10.7	−2.7	−15.0	0.0316
3500	−13.7	+ 6.7	−1.7	−15.4	0.0288
4000	−13.3	+24.0	−6.0	−19.3	0.0118
				Sum	0.9716

$L_{ME} = \frac{1}{0.25}$ (10 $\log_{10}$ Sum) $= +0.5$ dB

L_{RME} (NOSFER–NOSFER) $= -3.3$ dB

∴ STRE $= +3.8$ dB

shows the calculation of sidetone reference equivalent (STRE) for a 16-4T-31 (with regulator) set on 1 mile 6½ lb cable with 50 V 200+200 Ω feeding bridge and 600 Ω junction. Note that, as explained in Section 5.1.5, the sidetone performance of a telephone set is critically dependent upon the impedance of the lines to which it is connected.

3.2.4.4. Alternative reference systems

The foregoing examples are based on the present CCITT definitions of reference equivalent (Recommendations P.72 and P.73) which require use of the NOSFER which is, at present (1970), the only internationally standardised fundamental reference system for determination of loudness ratings of telephone speech paths. NOSFER is defined in CCITT Recommendation P.42 (*see* Section 3.2.2) and the internationally accepted method for using this reference system for this purpose is that described in CCITT Recommendations P.72 (SRE, RRE and ORE) and P.73 (STRE). Certain alternative reference systems are used nationally by individual telephone organisations, including the United Kingdom Post Office and the American Telephone and Telegraph Company (Bell System).

The United Kingdom Post Office has, for many years, used the ARAEN set at 30 dB loss in its junction as a convenient representation of a one metre airpath. Overall loudness ratings and sidetone loudness ratings have been determined relative to this reference speech path in a manner similar to that recommended for reference equivalent determinations except:

(a) The vocal level used is RVL (*see* Section 2.1.3.2) instead of the reference equivalent vocal level which is 5.6 dB lower.
(b) BMSP (see Section 5.1.2) has been used for setting up the handset in front of the talker's mouth instead of the reference equivalent speaking position which is about 10 mm closer to the mouth (*see* Section 5.1.1).
(c) The telephone handset is clamped in position instead of being held in the hand.
(d) English speech sounds are used instead of French.

a, c and d probably have only secondary effects but b has a direct effect amounting to 2–4 dB decrease in sensitivity of the handset under test. The calculated overall reference equivalent of ARAEN with 30 dB loss is +30.8 dB* (measured values obtained by the CCITT Laboratory are +33.2 dB by comparison with SFERT when it existed and +29.1 dB in a direct comparison with NOSFER, using common microphone and earphone for the two systems). A telephone handset speech path having a rating of +30 dB relative to ARAEN (using BMSP) is therefore equivalent in loudness rating to a one metre airpath and may be expected to have an overall reference equivalent of about 30.8–3 dB (say, 28 dB) where the allowance of 3 dB is to account for the different handset locations in front of the mouth. This conversion has been used for overall reference equivalent (ORE) and sidetone reference

* Calculated values of SRE and RRE of the sending and receiving ends of ARAEN (relative to NOSFER) are −2.5 dB and +2.8 dB respectively; these are in reasonably good agreement with values measured in the CCITT Laboratory.

equivalent (STRE) of United Kingdom Post Office telephone sets, note, however that nominal overall reference equivalents are usually employed for complete telephone connections and, because of the manner used to separate SRE and RRE, these are likely to exceed the directly determined ORE by about 3 dB. An overall telephone connection 'equivalent in loudness rating to a one metre airpath' is likely to have a nominal overall reference equivalent of about 31 dB (a value of 32 dB is now being used).

Table 3.11. DETERMINATION OF LOUDNESS RATINGS OF A SPEECH SIDETONE PATH RELATIVE TO A ONE METRE AIRPATH AND RELATIVE TO AN IRS; For speech emitted by a talker at RVL

Frequency f Hz	16-4T-31 (reg.) on 1 m/6½				
	$10 \log_{10} G\,\Delta f$	L_{ME} dB	$-0.25\,L_{ME}$	$10 \log_{10} G\,\Delta f - 0.25 L_{ME}$	antilog 1/10 Col (5)
(1)	(2)	(3)	(4)	(5)	(6)
100	−21.2				
200	−13.9	+12.3	−3.1	−17.0	0.0200
300	−12.3	+ 1.4	−0.4	−12.7	0.0537
400	−11.9	− 1.7	+0.4	−11.5	0.0708
500	−11.7	− 3.4	+0.8	−10.9	0.0813
600	−10.6	− 3.5	+0.9	− 9.7	0.1072
800	−10.7	− 5.3	+1.3	− 9.4	0.1148
1000	−11.3	− 6.6	+1.6	− 9.7	0.1072
1250	−11.1	− 7.5	+1.9	− 9.2	0.1202
1600	−11.7	− 2.4	+0.6	−11.1	0.0776
2000	−11.8	+ 8.7	−2.2	−14.0	0.0398
2500	−11.9	+27.5	−6.9	−18.8	0.0132
3000	−12.4	+13.7	−3.4	−15.8	0.0263
3500	−13.9	+ 9.7	−2.4	−16.3	0.0234
4000	−13.6	+27.0	−6.8	−20.4	0.0091
5000	−14.5				
6000	−14.3				
8000	−15.8				
				Sum	0.8646

$\overline{L_{ME}} = \frac{1}{0.25}(10 \log_{10} \text{Sum}) = +2.5$ dB

$\overline{L_{RME}}$ (One metre Airpath; ARAEN with 30 dB loss) $= +27.3$ dB

∴ Sidetone rating rel. MAP $= -24.8$ dB

$\overline{L_{ME}} = +2.5$ dB

$\overline{L_{RME}}$ (IRS) $= -3.7$ dB

∴ Sidetone rating rel. IRS $= +6.2$ dB

As a check on the foregoing relationship between reference equivalent and the overall loudness rating relative to a one metre airpath, represented by ARAEN with 30 dB loss and no filter, Table 3.11 shows calculation of the loudness rating of the speech sidetone path treated in Table 3.10. In Table 3.11, the values of L_{ME} relate to use of BMSP and RVL and the values of $10 \log_{10} G\, \Delta f$ apply for the speech spectrum appropriate to RVL (*see* Table 6.13 in Section 6.1.4 and Table 3.6 in Section 3.2.4.1). This yields a sidetone loudness rating (STLR) relative to a one metre airpath of −24.8 dB as compared with the STRE shown in Table 3.10 of +3.8 dB, a difference of 28.6 dB.

When properly additive sending and receiving ratings are required, such as when studying relatively good quality telephone connections, an IRS may conveniently be employed, as explained in Section 3.2.2. For theoretical studies, the frequency responses shown in Figure 3.10 are used in the United Kingdom and the absolute values of sensitivity are chosen in such a manner that sending and receiving ratings using this system are in very good agreement in value with SRE and RRE, respectively. With this IRS, overall loudness ratings of commercial telephone sets agree fairly well with the sum of sending and receiving ratings, and therefore with *nominal* ORE but not *true* ORE. Sidetone loudness ratings, using this IRS are about 2–3 dB greater in numerical value than STRE. With this IRS, sending and sidetone sensitivities measured at BMSP must be used. The sensitivity values are shown in Table 3.7.

Table 3.12. COMPARISON OF SUBJECTIVELY DETERMINED REFERENCE EQUIVALENTS,

Reference equivalent *or* *Loudness rating* (LR)	*Send*		
	332	*706*	*Opt.*
(1)	(2)	(3)	(4)
Measured RE	+7	+7	—
Calculated RE	+8.4	+7.1	+6.3
Calculated LR(IRS)	+8.5	+7.2	+5.9

Notes 1. The measured values of reference equivalent (RE) are the best estimates from available tests in CCITT Laboratory; Overall RE has not been measured.
2. The calculated values use sensitivity data from Table 3.7
3. 332 denotes Telephone No 332 (13–2P–27)
706 denotes Telephone No 706 (16–4T–31, with regulator) Both types are associated with

Table 3.12 shows the results of some calculations of sending, receiving and overall loudness ratings using this IRS (SLR, RLR and OLR); for comparison, measured and calculated reference equivalents are also included. Noting that $\overline{L_{RME}}$ for this IRS is equal to -3.7 dB and that $\overline{L_{ME}}$ for the sidetone path treated in Table 3.11 is $+2.5$ dB, the sidetone loudness rating relative to the IRS is $+6.2$ dB, which is 2.4 dB greater than STRE determined in Table 3.10. In Table 3.12, the small discrepancies between measured and calculated RES can be ascribed to measurement error in the subjective tests and uncertainty concerning the most appropriate methods for determining the sensitivity/frequency response characteristics.

Practice in the Bell System is analogous and use is made of the Electro-Acoustic Rating System (EARS) which is employed in a similar manner to the United Kingdom IRS. Unfortunately direct comparisons between the results obtained with EARS and IRS are not available and only approximate estimates are at present possible because the two systems are calibrated differently.

An alternative reference system also based on the one metre airpath is described in Annex 2 to CCITT Recommendation P.42; this is termed SETED (Système Etalon de Travail avec microphone et recepteur Electrodynamique) and was originally intended as an alternative for SRAEN which is used for determinations of articulation ratings (*see* Section 3.3). SETED can, however, be used also for determining, indirectly via its calibration relative to NOSFER, reference equivalents.

CALCULATED REFERENCE EQUIVALENTS AND CALCULATED LOUDNESS RATINGS USING IRS

Receive			*Overall*		
332	*706*	*Opt.*	*332*	*706*	*Opt.*
(5)	(6)	(7)	(8)	(9)	(10)
−1	−4	—	—	—	—
−1.4	−5.8	+0.6	+4.3	−1.5	+4.1
			(+7.0)	(+1.3)	(+6.9)
−1.6	−5.4	+0.6	+6.7	+1.5	+6.5
			(+6.9)	(+1.8)	(+6.5)

subscribers' lines of 1.6 km 0.5 mm (1 mile 6½ lb) cable on 50 V 200+200 Ω feeding bridges

Opt. denotes a hypothetical telephone set conforming with the 'optimum' frequency characteristics proposed in Section 4.2.1.5

4. Values in brackets in cols. 8, 9 and 10 are nominal ORES obtained by adding together SRE. and RRE, and nominal OLRS obtained by adding together SLR and RLR.

3.3 ARTICULATION

3.3.1 Aim

Articulation measurements provide an example of a method intended to determine the information capacity of a path when it is transmitting information in the form of speech signals (*see* Section 1.2). The acceptability of a complete speech link cannot, of course, in general be expressed in terms of the performance of one particular path only, but it is sometimes possible to attribute a key role to one path of a speech link. In such a case, the information capacity of that key path may give a useful indication of the performance of the whole speech link. It must be remembered, however, that the aim of an articulation measurement is limited and the effects of the following features which are important in telephony are specifically excluded in an articulation test.

1. Variations between talkers
2. Variations between listeners
3. Any influence the characteristics of the path being measured may have on the behaviour of talker or listener
4. Any influence on the behaviour of talker or listener of any other characteristic of the speech link of which the path being measured forms part.

Like loudness ratings, the results of an articulation measurement apply to one direction of transmission in isolation, so that conversational factors (*see* Section 3.1.2) are totally excluded. In fact, also like loudness ratings, articulation can deal only with listening factors (Type A as defined in Section 3.1.2).

Unlike loudness ratings, articulation ratings can claim to deal with the effects of circuit noise and room noise; distortions are taken into account, furthermore, in a more sophisticated manner by articulation than by loudness rating.

3.3.2 Principle

In an articulation test the ensemble of symbols is defined by the set of phonemes used in the language concerned. These must be built into syllables that can be pronounced and this may be done according to various statistical and linguistic principles. For example, taking only single phoneme initial and final consonants and the six 'short' vowels, it is possible to generate nonsense syllables (logatoms), pronounceable by English speaking persons, by combining independently at random all the initial consonants, with uniform frequency of occurrence, with each vowel, also occuring with uniform frequency, and with each final consonant, uniformly frequently occurring. An example of a list generated in this manner is given in Table 3.13. This type of logatom is very unlike syllables found in the English language (and probably unlike any real language) but it has the advantage that analysis of the scores for individual sounds is greatly facilitated and confusion matrices can readily be constructed from the resulting errors. A better approximation to real language can be achieved by introducing redundancy and arranging the various parts of the logatom (initial–vowel–final, including 'blank' initial and final sounds) to occur with frequencies like those found in syllables of the language; a further step (with more redundancy) is achieved by matching the transition frequencies of the initial–vowel and vowel–final digrams. Table 3.13 includes a list of such logatoms generated from the frequencies and digram frequencies of English telephonic conversational speech (*see* Section 2.1.2 for details of the statistical structure). Table 3.13 shows also a list of English single syllable words, which have been constructed by selecting from the English logatoms just described those that were actual English words (rather more than half the logatoms); naturally, words are more redundant than logatoms. Other types of speech material have been used for articulation testing, including so-called phonetically balanced (PB) word lists (phonetic balancing does not, in general, result in uniform difficulty as judged by the incidence of errors with various kinds of degradation). Words of two syllables have also been employed, especially those having the metrical form of two long syllables (spondee words). An advantage of using words is that relatively untrained listeners can, more easily, report what they hear.

An articulation test therefore consists, in principle, of determining the percentage of information bearing elements, emitted by the talker, that are recognised correctly by the listener. As more redundant material is used for transmission over a given communication system, the performance, in terms of the percentage of elements correctly recognised, will improve (as communication theory predicts) and this is illustrated in Figure 1.7 which appears in Section 1.2.

Table 3.13. EXAMPLES OF SPEECH MATERIAL USED IN ARTICULATION TESTS BY THE UNITED KINGDOM POST OFFICE

Equal probabilities Random transitions Restricted set of 'short' vowels		*Probabilities and digram frequencies as in stressed syllables of conversational English when spoken*		*English single syllable words*
Logatom	*Pronounced as*	*Logatom*	*Word in English (if any exists)*	*Word*
vʌm	VUM	θ aU t	—	SHOES
fɒt	FOT	f ɜ n	FERN	ME
lɒk	LOCK	l ʌ st	LUST	AS
næm	NAM	g ɒ nts	—	AND
mIl	MILL	w ɒ p	—	IT
gæv	GAV	w aI —	WHY	FORCE
kʌs	CUSS	g ɒ t	GOT	BIT
dɛʃ	DESH	k eI dʒd	CAGED	THAN
pɛŋ	PENG	d ɒ s	DOSS	IS
zIz	ZIZZ	r ɛ s	—	HIT
θʌθ	THUTH	– *a* –	ARE	SAND
vʌv	VUV	r aU z	ROUSE	RING
nɒθ	NOTH	k u s	—	NEW
mIs	MISS	g U ʃ	—	HAT
mɒk	MOCK	h æ t	HAT	MAN
nam	NOM	b aI –	BY	MORE
lUl	LOOL (to rhyme with	s eI z	SAYS	SICK
	PULL)	p aU n	—	HIS
lId	LID	d aU t	DOUBT	SPEAK
jɛt	YET	l ɒ t	LOT	IT
–æb	–ABB	– ɔ –	OR	OF
–ap	–OP	j ɛ s	YES	SAY
pUʃ	PUSH	j ɛ m	—	CHURN
bʌŋ	BUNG			CHAT
tIz	TIZZ			AM
væv	VAV			

3.3.3 Method

Articulation tests are usually conducted with a trained testing team. The talking conditions are carefully controlled by providing the talker with means of monitoring the level of his utterances, for example a speech voltmeter connected to a high quality microphone located at a controlled distance from his lips. Control of talking is facilitated by emitting each logatom or word in a carrier phrase, which is repeated with only the logatom changed. The logatoms illustrated in Table 3.13 are used in the carrier phrase: CAN CON BY logatom ALSO. When words have been used by the United Kingdom Post Office, the carrier phrase has been: THE NEXT WORD WILL BE word ALSO.

The listening conditions are also controlled including provision of suitable listening cabins and furnishing them with definite levels of room noise. It is convenient to use a team of five persons of which, at any time, four listen simultaneously on parallel connected receiving ends, while the fifth does the talking. The members are rotated systematically with each change in the test condition according to a properly designed and administered experimental design. (Some details of the methods of design and analysis developed by the United Kingdom Post Office and used by the CCITT Laboratory are included in CCITT Recommendation P.45).

Articulation tests can be scored in different ways, even when a given type of logatom or word material is used. Strictly, the score ought to be made in terms of the number of complete logatoms or words recognised completely correctly; this is termed the logatom, syllable or word articulation score. However, scoring is often done in terms of the fractions of the complete logatom, that is, a score of three is given for a complete consonant–vowel–consonant combination recognised correctly with partially correct combinations allocated scores of two or one if one or two errors are present in the combination written down by the listener. Scoring in this way, out of a total equal to three times the number of logatoms emitted by the talker, yields the sound articulation score.

The scores obtained from an articulation test have very little absolute significance and can vary widely when a different crew is used or changes made in the procedure. For this reason the scores are used to obtain ratings, relative to a defined reference system, by the 'indirect balancing' method described in Section 3.1.4. The CCITT has defined a reference system for this purpose derived from the ARAEN which was described in Section 3.1.3; the system is termed SRAEN and the definition is found in CCITT Recommendation P.44. Ratings in terms of equivalent settings of SRAEN are called 'AEN values' and these are determined at a value of articulation, scored in sounds, of 80%.

The SRAEN is set up by introducing into the junction of the ARAEN a defined 300–3400 Hz bandpass filter (see CCITT Recommendation

Table 3.14. INSERTION LOSS, BETWEEN 600 Ω TERMINATIONS, OF SRAEN BANDPASS FILTER

Frequency Hz	*Insertion loss* dB	*Frequency* Hz	*Insertion loss* dB
100	12.5	1600	0.1
200	8.3	2000	0
300	2.8	2500	— 0.5
400	1.4	3000	0.4
500	0.7	3500	8.2
600	0.4	4000	>30
800	0.1	5000	>30
1000	0	6000	>30
1250	0	8000	>30

P.44, Figure 2) and arranging to inject a certain level of circuit noise, having a defined frequency spectrum, into the receive end. Table 3.14 shows the loss of this filter at a number of frequencies. The SRAEN is used without any noise at the listening location but the receive ends of all commercial local telephone circuits under test are used with a room noise of 60 dBA (Hoth spectrum) at the listening location.

An alternative reference system was developed for conducting AEN determinations and this is termed SETED (*see* Section 3.2.4.4); this system is arranged with its microphone somewhat closer to the talker's mouth than is the case with SRAEN. The method of calibration is also different and so, when SETED is used, allowance must be made to the results in accordance with information included in its description (*see* Annex 2 to CCITT Recommendation P.42).

When determining the AEN values of local telephone circuits, the loss setting corresponding to 80% sound articulation is determined by transforming the scores for four simultaneous listeners, according to the logistic transformation function (*see* Section 3.7) and fitting linear regression lines to the response metameter. Analysis in terms of such transformed units also permits confidence intervals to be estimated and tests of significance to be made to check that the results do not exhibit any unusual features. Details of the procedure are included in the version of CCITT Recommendation P.45 that appears in the Red Book, Volume V (1962).

Other methods are in use for articulation testing; for example, the procedure can be simplified so that relatively untrained and inexperienced persons can be used, e.g. volunteer students for work conducted in universities. Closed response sets can be constructed which require the listener merely to select one of a small number of responses offered on his score sheet. Words that are difficult to distinguish from other common words, e.g. by their initial consonant only (rhyming words), have the merit that they can readily be administered to persons without any appreciable special training. Other special sets of words have been compiled for diagnostic testing, especially when studying the problems surrounding development of channel economy transmission systems. These methods will be found reported in the literature but some caution is needed when they are used with subjects having native languages different from those for which the test material was developed. Section 3.7 includes some information concerning use of articulation measurement procedures immediately after a pair of persons have conversed with each other.

3.3.4 Use

Use of articulation scores as a measure of the communication efficiency of a speech link is limited to links of rather poor performance (by public telephone standards). The practical reasons for this restriction are explained in Section 4.1.5. It was hoped in the 1950s that articulation ratings (especially AEN values) would supersede reference equivalents and other loudness ratings for international planning purposes; the method has, however, proved impracticable for this purpose. Although noise and distortions do reduce articulation scores, it has been established that telephone speech paths having different impairments but producing the same AEN values can be far from equally acceptable so far as their use for ordinary telephone conversations are concerned. The AEN method is still mentioned in the current version of CCITT Recommendation P.45 (White Book, 1968, Volume V) but it is now very little used except for checking the quality of prototype designs of telephone set. Reference should be made to Recommendation P.45 in CCITT Red Book, 1962, Volume V for the complete description.

The United Kingdom Post Office does not use articulation testing procedures to obtain planning data but occasionally for diagnostic purposes use is made of one of the specialised articulation techniques of which the results can be presented as a set of confusion matrices. It must be noted that analysis of speech into phonemic elements (such as are used to construct lists of logatoms) is related only in a very complicated manner to phonetic and acoustical analysis and these relationships are not completely understood (*see* Section 2.1.1, where these distinctions are mentioned). Measurement of articulation scores can therefore be expected to provide only a very limited assessment of speech communication efficiency; it is not suitable for dealing with circumstances where long 'runs' of speech material are being interpreted, e.g. in ordinary conversation.

3.3.5 Relationships between articulation scores and certain physical particulars of speech paths

Although articulation methods have not proved directly useful for planning telephone networks, they have inspired valuable theoretical models which have been used with advantage to reproduce the scores of actual articulation tests by calculation based on knowledge of the physical particulars of the speech transmission system concerned. The particulars required are the mouth to ear loss (L_{ME}) of the principal path from the mouth of the talker to the ear of the listener, together with information on the spectrum of noise (from the circuit and from the room) reaching the listener's ear.

The first theoretical model was that proposed by Collard in 1928 and, although this had been later elaborated by others to extend its range of application, most of the essential fundamental principles remain unaltered. The later adaptations are reported by French and Steinberg, Fletcher and Galt, Beranek, and Kryter. As reported elsewhere (Richards and Archbold (1956)), the Collard principle can be used to reproduce articulation scores according to the AEN method of measurement. It has also been employed as a basis for predicting the best frequency responses of hearing aids and telephone microphones. (This matter is referred to again in Section 4.2.1.5.).

The term 'sensation level' (Z) has been used in Section 3.2.4.1 to represent the interval, in dB, between the level of speech in a particular frequency region and the threshold of audibility for continuous spectrum sounds in that region. This definition was sufficient for the loudness criterion because use of this criterion for rating telephone speech paths is in the absence of any noise. The definition must now be extended to denote the interval between the level of speech and the hearing threshold when it is masked by the presence of noise; this corresponds to an upward shift of the noise free threshold by a certain amount which, as explained in Section 2.2.3.1, depends on the spectrum of the noise at the ear and the unmasked threshold of audibility in each frequency region.

According to the Collard model, a quantity termed 'band articulation' is defined in accordance with the following expression:

$$\text{Band articulation} = \int_0^\infty P(Z)B'\,\mathrm{d}f \tag{3.8}$$

$$= \overline{P(Z)} \tag{3.8a}$$

$P(Z)$ is a defined function of Z, analogous to $Q(Z)$ used for loudness; B' is a frequency weighting factor, analogous to B' used for loudness; $\overline{P(Z)}$ is the weighted mean of $P(Z)$.

Band articulation is not the same as any articulation score that can actually be measured in a real articulation test but is related to each such score by a unique function.

The complete expression may therefore be written:

$$\text{Articulation score (in a real test)} = \phi\{\overline{P(Z)}\} \tag{3.9}$$

Unlike $Q(Z)$, which increases indefinitely as Z increases, $P(Z)$ is a sigmoid which is asymptotic to 1; the actual form of $P(Z)$ is shown in Figure 3.19 which appears in Section 3.4.4.

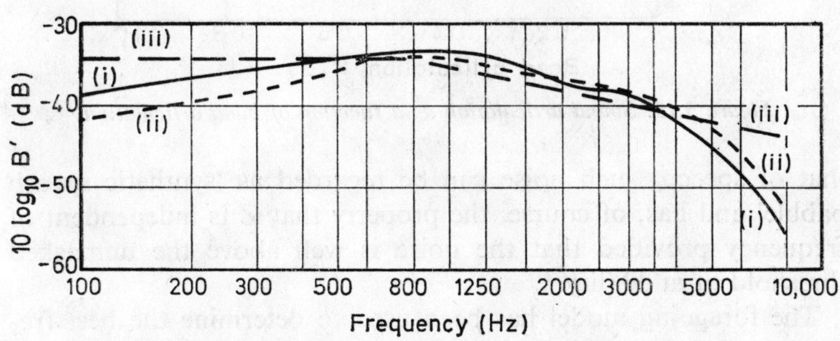

Figure 3.14. Frequency weighting factors for articulation. (i) Ackroyd (1948) male subjects. (ii) Ackroyd (1948) female subjects. (iii) Richards and Archbold (1956).

Note: Results similar to curves i and ii have been published by Pocock (1939)

B' differs slightly in values from the analogous function appropriate for loudness; Figure 3.14 shows results used by different authors. It will be noticed that the articulation frequency weighting factor places rather more emphasis on the middle frequencies and less on the lower and upper ranges than the frequency weighting factor for loudness (compare Figure 3.11 in Section 3.2.4.1).

Figure 3.15 shows a relationship (ϕ) obtained between sound articulation scores for English logatoms used in a carrier phrase by a very well trained crew and band articulation as defined in Eqn 3.8 above. The results were obtained from tests using the ARAEN with injected circuit noise of spectrum designed to match

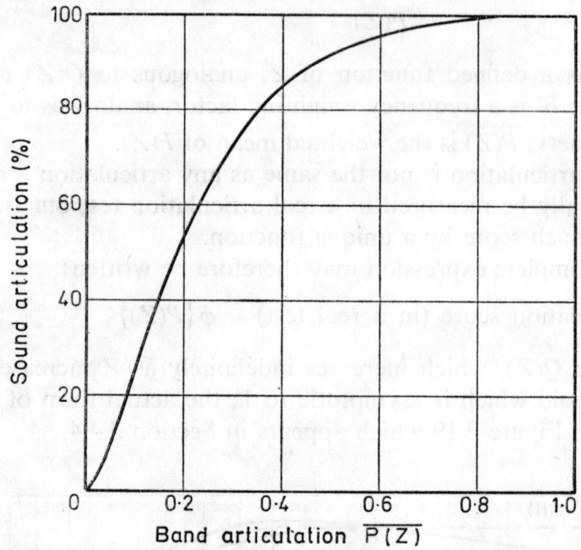

Figure 3.15. Sound articulation as a function of band articulation

that of speech; such noise can be regarded as 'synthetic speech babble' and has, of course, the property that Z is independent of frequency provided that the noise is well above the unmasked threshold of audibility.

The foregoing model has been used to determine the best frequency response characteristic for a local telephone circuit in its sending direction; this process is explained in Section 4.2.1.5 and a practical application is referred to in Section 5.1.2.

More elaborate models have been evolved which claim to deal with a broader range of transmission quality of speech paths than can be treated by formal articulation tests and the Collard method which reproduces their results. Some further reference is made to these in Section 3.4.4.

3.4 LISTENING METHODS USING SENTENCE MATERIAL

3.4.1 Aims and principles

Sentence material can be used to determine an 'information capacity' for a path of a speech link in a similar manner to that used with phonemes, syllables or words in an articulation test. Scoring in terms of amount of material received correctly is, however, rather difficult; writing down what the listener hears is excessively laborious and so scoring by verbal repeat or in terms of 'key' words is often resorted to.

To avoid excessive writing down but to retain the idea of scoring in terms of amount of material 'satisfactorily' received, it is possible to use the 'immediate appreciation' method, devised by Grinsted in 1937. A fixed, one way speech path is used with talking and listening conditions held constant, as for loudness rating and articulation measurement methods and the scoring is done by the listener noting whether or not he understood the meaning of each sentence without mental effort. The method is, however, rather too insensitive for assessing ordinary telephone speech paths; a score of about 95% is given for a speech path of 33 dB nominal overall reference equivalent without appreciable circuit noise. Such a path represents approximately the greatest transmission loss permitted in a modern telephone network (*see* Figure 4.8 for comparisons with other assessment criteria; Section 4.1.5).

The principle of scoring in terms of the amount of effort needed to understand the meanings of simple sentences having no semantic connections is very attractive for telephony and valuable results can be obtained by use of a multigrade rating scale with suitably defined categories. Such a scale can alternatively be designed to assess loudness preferences or to determine thresholds of detectability or objectionableness.

It is important to distinguish two different aims. First, a rather general one of obtaining assessment scores that will enable predictions to be made of the degree of satisfaction likely to be expressed by subjects conversing over symmetrical speech links of which the principal paths are identical to the path in question. For such purposes it is necessary to confirm that the listening assessment method used does indeed properly rate at least the effects of loss, noise and distortion. To ensure this, it is important that the listener should have sufficient opportunity to appreciate the nature of the degradation he is being asked to assess. It is therefore undesirable to expose the listener to a succession of short examples of speech each degraded in an entirely different way; for this reason, the intensity of only one factor should be changed within a test sequence lasting, say, a few minutes. Other precautions are necessary as described below.

The second aim is a more restricted one of rating correctly the relative effects of different varieties of one type of degradation. For example, different kinds of circuit noise may need to be compared for potency in their interference with speech, attenuation/frequency distortions of different shapes may need to be compared or different varieties of nonlinear distortions. For such purposes, sufficient exposure is obtained during each example of speech because they possess considerable common features. This more limited aim can often be satisfied by using pair comparisons; further information on this method is given below.

3.4.2 Methods

The sentence material used by the United Kingdom Post Office is obtained by selecting, say, one sentence from each page of a number of novels or other light literature. The sentences chosen are then placed in random sequence and arranged as lists of ten groups of five sentences. Examples from part of such a list are as follows:

1. This may lead to another man's ruin
2. At least he is an improvement on our last tenant
3. It took him many miles out of his way
4. Is there no way in which I can help you?
5. His story is different from the first one we heard

6. They ought to be ashamed of themselves
7. They sat together under the canvas roof
8. There had never been so much humiliation before in his life
9. Someone is knocking on the bedroom door
10. Do you mind if I listen to what you have to say?

11. These words naturally caused me some pain
12. Is that the cousin you mentioned?
13. You would not like to spend the night in this room
14. This was not a usual relationship
15. She had not spoken to him for a whole week

16. You can see what a precarious footing we are on
17. A frost is coming, and the trees are bare
18. You can scarcely call him a parasite
19. The woman felt in need of sympathy
20. Do girls always speak like that?

21. He knocked gently on the side door
22. Money makes very little difference to me
23. He stared at the wall, thinking
24. There was nothing to eat for breakfast
25. You would not easily forget.

When making recordings (*see* below), ten groups are usually recorded by one talker and the next ten by another so that a variety of voices is used. Naturally, no sentence may be repeated, even by a different talker, so that each sentence is unknown to the listener before hearing it in the test.

Recordings made in this way can be used for immediate appreciation tests, listening opinion tests (preferred loudness or listening effort) or for pair comparisons.

When sentence material is used, the subjects taking part in listening tests should preferably be naive, although it is sometimes possible to conduct certain tests with an experienced crew, e.g. immediate appreciation tests on rather poor speech paths.

Because the effects of talking factors (like sidetone and echo) and conversational factors (presence of a return path which permits the talker to adapt his speech to the listener) are specifically excluded, it is usually possible to record the sentence lists using the type of microphone appropriate to the various speech paths to be included in the experiment. One set of recordings can, therefore, serve for several experiments in which the same types of microphone feature. (Naturally the subjects who listen must be different so that they have not previously heard any of the sentences.

The recordings are prepared in three stages as shown in Figure 3.16.

Stage 1 The talkers read the sentences in a fluent but not unduly precise manner with only very small pauses between successive sentences of each group; they should adopt a level that is comfortable to them as individuals and which they can maintain fairly constant. A calibration tone is recorded immediately before the start of each talker's recording; the level is arbitrary but should be the same for all the recordings.

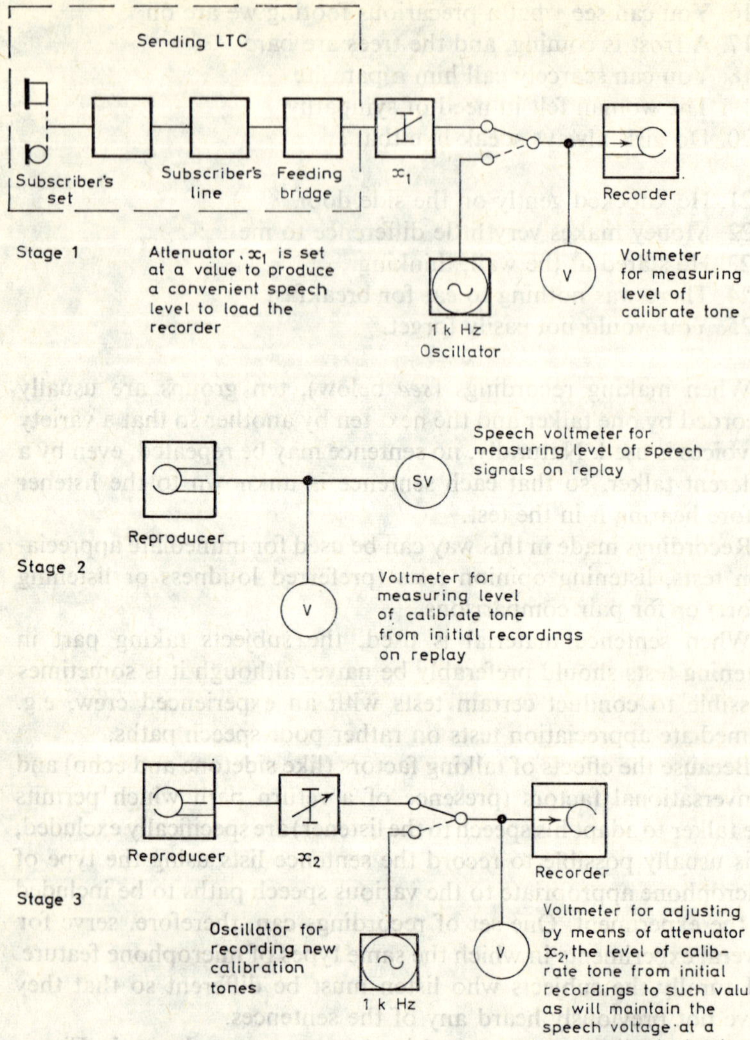

Figure 3.16. Preparation of magnetic tape recordings

Stage 2 The recordings are then replayed to calibrate them for speech level relative to the calibration tone; this operation yields corrections needed to render the levels of all groups of five sentences exactly the same.

Stage 3 The recordings are then replayed and rerecorded with the appropriate corrections in level. A new calibration tone is then associated with the final recordings at a convenient level relative to the constant speech level of these recordings; the original calibration tones are not rerecorded.

On replay during the listening tests, the speech must be reproduced through an attenuator and appropriate listening arrangements; the level at the input to the attenuator can be set by means of the calibration tone and it is necessary that the speech level should have a known relationship to that produced at the same point in a telephone connection used for conversation. An example of the application of this technique is given in Section 4.2.1.2.

Immediate appreciation tests consist of the transmission of a list of 25 or 50 sentences and asking the listener to indicate, for each sentence, whether or not he has understood the meaning immediately, without mental effort. The experiment should be arranged so that observations are collected for the various speech paths under study, usually when each is associated with several settings of added loss or other form of adjustable degradation. The results obtained from such measurements can then be plotted as a function of the amount of added loss, etc., as shown in Figure 3.17.

Listening opinion tests are conducted by arranging for the listener to hear a succession of groups of five sentences, each group being reproduced over a different setting of a speech path. After each group, the listener expresses his opinion on an appropriate scale. Each opinion is based on exposure to only five sentences and the replay machine is stopped between groups to give the listener time to vote. Voting can be by marking a form, indicating by a key or verbally. It is particularly important, when the results are to be used for predicting conversational performance, to ensure that the listeners are given sufficient opportunity to appreciate the nature of the degradation to which they are being exposed; for this reason it has been found desirable to vary one feature only within the sequence of ten groups of sentences, e.g. the loss, the level of circuit noise or the cutoff frequency of a high or low pass filter, and to make other changes in test condition only between such sequences. Another necessary precaution is to ensure standardised performances for the extreme conditions within each sequence. Consistent results have been obtained by ensuring that the best condition within each sequence corresponds to a fairly noiseless and distortionless telephone speech path, that the worst setting within each sequence yields a listening mean opinion score of about 1 (see below) and that the intervening settings of the single variable feature (usually loss but could be injected circuit noise or some al-

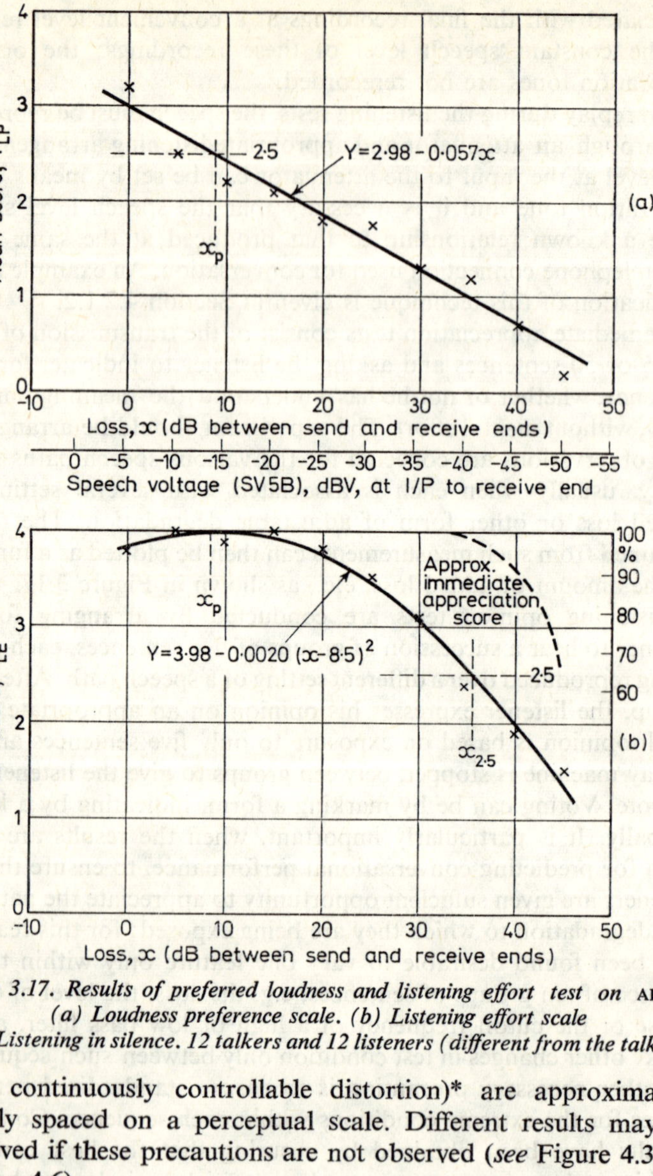

Figure 3.17. Results of preferred loudness and listening effort test on ARAEN *(a) Loudness preference scale. (b) Listening effort scale*
Note. Listening in silence. 12 talkers and 12 listeners (different from the talkers)

most continuously controllable distortion)* are approximately equally spaced on a perceptual scale. Different results may be observed if these precautions are not observed (*see* Figure 4.39 in Section 4.6).

The following opinion scales have been found most useful for listening tests:

* When certain opinions scale are used (e.g. Nos 4 or 4A), the variable feature must be loss.

Loudness preference scales

Opinion scale No. 4		*Opinion scale No. 4A*	
A	Much too loud	A	Much louder than preferred
B	Too loud	B	Louder than preferred
C	OK	C	Preferred
D	Too quiet	D	Quieter than preferred
E	Much too quiet	E	Much quieter than preferred

Note: The wording of Opinion scale No. 4A is preferable to that of No. 4.

Listening effort scale
Opinion scale No. 7

A Complete relaxation possible; no effort required
B Attention necessary; no appreciable effort required
C Moderate effort required
D Considerable effort required
E No meaning understood with any feasible effort

The categories A to E with any of these scales are scored 4, 3, 2, 1 and 0 respectively; typical results are shown in Figure 3.17. It will be noticed that a mean score of 2.5 is obtained with Opinion scale No. 4A at the loss setting that yields the maximum value of mean score with Opinion scale No. 7; this loss setting is termed the 'preferred value of loss' (x_p). Further comments on this relationship are made in Section 4.2.1.2 and, for loudspeaking telephone sets, in Section 4.4.4. The setting of loss that yields a score of 2.5 with Opinion scale No. 7 can be said to mark the threshold of listening effort (see category descriptions above) and this intercategory boundary is convenient for use when determining impairments (*see* Section 4.2.1); this loss setting is denoted by the symbol $x_{2.5}$. Determination of a set of values of $x_{2.5}$, each applying to a particular speech path, enables ratings of one in terms of another to be determined indirectly (*see* Section 3.1.4). Direct comparison is also possible by making use of recordings prepared as described above but replayed into an arrangement that permits the listener to select one path or another and to listen alternately to decide which enables him for example, to understand the meanings of the sentences with less mental effort. It is usually sufficient to allow only five sentences before requiring the listener to decide. This method is termed that of pair comparisons.

Pair comparisons of this type can be conducted in a manner similar to that described in Section 3.2.2 for loudness rating, namely making all comparisons against a reference speech path that can be adjusted in respect of a single feature, with the object of determining settings of subjective equality. The variable feature need not be loss but can be level of injected circuit noise, amount of simulated nonlinear distortion or amount of bandwidth limitation or attenuation/frequency distortion; some examples of the application of this method are given elsewhere (*see* Section 4.2.2.1). The method has some important limitations, for example it is suitable only when all the test paths are appreciably worse than the best performance setting of the reference path. Other methods of performing pair comparisons are possible. For example, when a number n of test conditions are to be ranked, it is possible to determine the rankings by performing $n(n-1)$ comparisons in pairs. It may be possible to reduce this number by half if it can be assumed that A versus B yields the complement of B versus A.

3.4.3 Use

The methods described above have very extensive use and several examples appear in Section 4.2 (*see* Section 4.2.1.2 for applications in the study of circuit noise and attenuation/frequency distortion and 4.2.2 for applications to nonlinear distortion). It is important to relate the results of listening only tests to the conversational situation in a proper manner and this is treated in Section 4.2.1.3. The use of other forms of opinion scales designed for determining thresholds of audibility, detectability, objectionableness and disturbance under listening conditions is illustrated in Sections 4.2.1.6 and 4.5.

3.4.4 Relationships between listening opinion scores and certain physical particulars of speech paths

The Collard model, explained in Section 3.3.5, has proved sucessful in predicting the articulation scores obtained using a trained team of testers; application of the same model in the attempt to predict the scores given by ordinary, untrained, persons taking part in listening opinion tests has proved more difficult. The first difficulty encountered can be illustrated using Figure 3.18. The fundamental particulars of the listening opinion situation are the same physically as for articulation testing, namely:

1. Speech reaches the listener's ear with a particular mean spectrum density:
2. Noise (room and circuit) enters the listener's ear with a particular mean spectrum density and, if audible, will produce a certain masked threshold of audibility.

This information may reasonably be expected to be sufficient to form a basis for calculation. In Figure 3.18, β'_S represents the mean spectrum of speech leaving the talker's mouth and curve ii shows

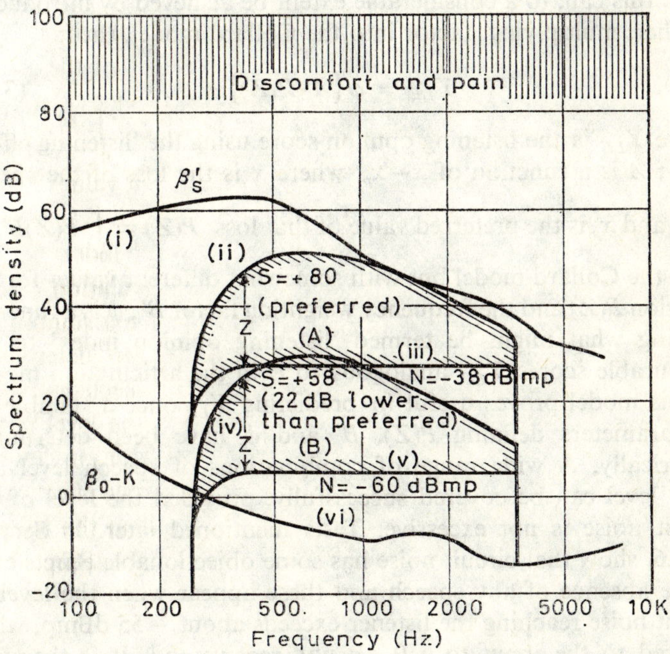

Figure 3.18. Conditions where the sensation areas are equal but listening levels are different.

Note: Areas (A) and (B) will yield the same articulation but the listening opinion score for (B) will be much lower than for (A)

the spectrum on arrival at the listener's ear over a path of preferred overall loss. Curve iii shows the spectrum of noise sufficiently above the threshold of audibility, curve vi, to represent the masked threshold. The vertical interval between curves ii and iii is the sensation level, Z. Now suppose that the levels of both speech and masked threshold are reduced by the same amount (22 dB in the example shown in the figure); the sensation level, as a function of frequency will remain unaltered and therefore the articulation score represented by the area A will be the same as that represented by area B. The listening opinion score, however, will be different in these two cases; although the values of Z are the same, the area B will correspond to a lower listening opinion score because the fainter speech will be judged more of a handicap and this will not, in the opinion of the listener, be offset by the quieter noise.

In listening opinion tests, a theoretical model must be used that distinguishes between faintness of the speech and the sensation level; this can, to a considerable extent be achieved by introducing another, multiplying, factor into the Collard model, thus:

$$Y_{LE} = \phi\{A \cdot \overline{P(Z)}\} \tag{3.10}$$

where Y_{LE} is the listening opinion score using the 'listening effort' scale; A is a function of $x - x_p$, where x is the loss of the speech path and x_p is the preferred value of that loss; $\overline{P(Z)} = \int_0^\infty P(Z)B'\, df$, as in the Collard model but with somewhat different values for the function $P(Z)$ and the frequency weighting factor B'; ϕ is a function relating what might be termed 'listening opinion index' to the measurable scores Y_{LE} analogous to ϕ in the articulation model.

This model proves useful in predicting Y_{LE} once a suitable set of parameters defining $P(Z)$, B' and ϕ have been determined empirically. A wide range of combinations of speech level and noise level can be covered successfully, provided the level of the circuit noise is not excessive. Tests mentioned later, in Section 4.2.1.6, show that circuit noise has some objectionable effects even in the absence of any speech and these appear when the level of circuit noise reaching the listener exceeds about -55 dBmp, when referred to the input to a 0 dB RRE receive end. It is therefore necessary to insert another multiplying factor, D, which is a function of the level of the circuit noise; the complete model can then be written:

$$Y_{LE} = \phi\{A \cdot D \cdot \overline{P(Z)}\} \tag{3.11}$$

$A = 1$ when $x - x_p$ is within the range 0 to $+7$ dB and $D = 1$ when the level of circuit noise is not greater than -55 dBmp (referred to 0 dB RRE).

The function $P(Z)$ is illustrated in Figure 3.19 which includes, shown broken, the corresponding function for articulation* using

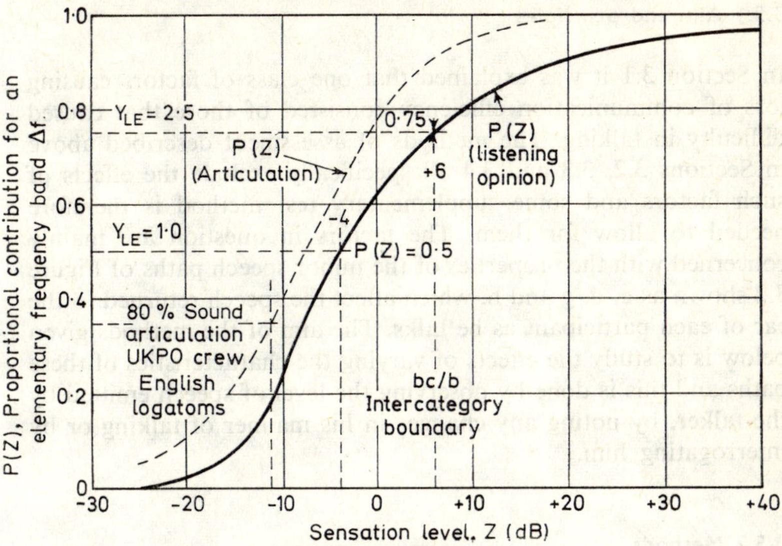

Figure 3.19. P(Z) as a function of sensation level

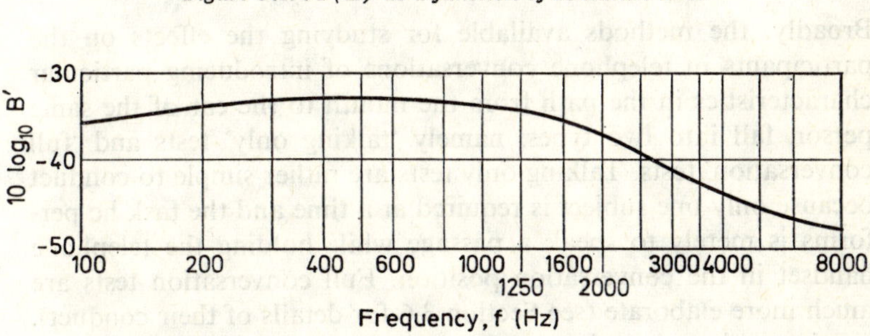

Figure 3.20. Frequency weighting factors for listening effort scores

a highly trained crew of operators. Figure 3.20 shows the frequency weighting factor B'; *see* Figure 3.14 for its counterpart for the articulation testing conditions.

* Comparison of this function with the corresponding function for articulation is rather complicated because the speech material used has different statistical properties and so the abscissa scales can be mutually located only by assuming a particular type of speech level measuring device. For the purposes of drawing the graph shown broken in Fig. 3.19, the SV5B has been assumed.

3.5 TALKING METHODS

3.5.1 Aim and principles

In Section 3.1 it was explained that one class of factors causing loss of communication efficiency consisted of those that caused difficulty in talking. The methods of assessment described above in Sections 3.2, 3.3 and 3.4 all specifically exclude the effects of such factors and some supplementary test method is therefore needed to allow for them. The factors in question are mainly concerned with the properties of the minor speech paths of Figure 3.2 shown as c, d, g and h, which affect the speech returned to the ear of each participant as he talks. The aim of the methods given below is to study the effects of varying the characteristics of these paths and this is done by observing the level of speech emitted by the talker, by noting any changes in his manner of talking or by interrogating him.

3.5.2 Methods

Broadly, the methods available for studying the effects on the participants in telephone conversations of introducing particular characteristics in the path from the mouth to the ear of the same person fall into two types, namely 'talking only' tests and 'full conversation' tests. Talking only tests are rather simple to conduct because only one subject is required at a time and the task he performs is merely to speak a passage while holding the telephone handset in the conversation position. Full conversation tests are much more elaborate (see Section 3.6 for details of their conduct), require subjects in pairs and involve much more time on the part of the subjects. However, talking only tests are rather artificial and the results they yield need to be verified by at least some conversation testing. With either type of test, the behaviour of the subjects or their opinions can be examined.

The paths concerned in this work, so far as public telephone connections are concerned, are those usually described as speech *sidetone* and *talker echo*. In principle, information is needed not only for a wide range of loudness of the speech heard over such paths but also for certain practical variations in loss of these paths as functions of frequency; it should be noted that speech sidetone and talker echo paths each contain an element associated with a return loss at some point (or points) in the connection where reflections occur and return losses are notoriously variable in frequency response (*see* Section 6.1.4). With sidetone, the reflection takes place mainly at the interface between telephone set and subscriber's line and so the speech returned to the talker is not noticeably delayed; with echo, however, the reflection is at the remote end of the connection and suffers twice the mean one way propagation time of the connection from the talker to the point of reflection.

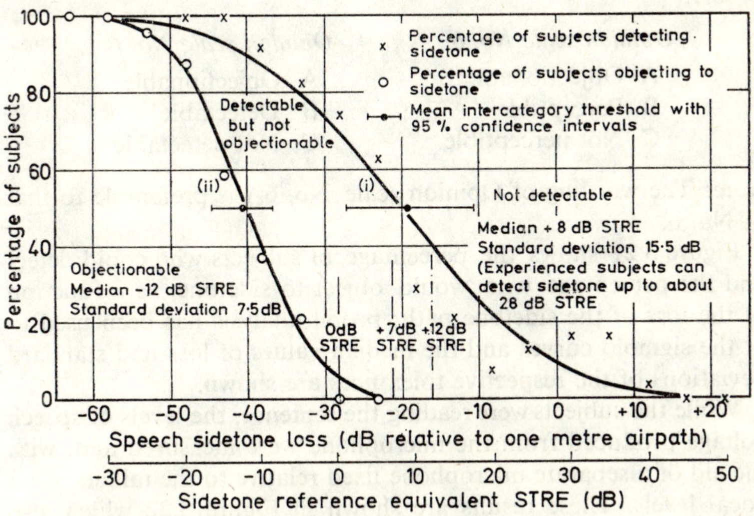

Figure 3.21. Talkers' opinions of sidetone. Talking only test, naive subjects.

As examples, some results from both types of test will be described. Figure 3.21 shows the results of a talking only test in which the subjects spoke a 15 s passage* into a handset furnished with a linear microphone connected to a circuit which reproduced the talker's speech in his ear with uniform frequency response from about 200–3500 Hz. No room noise was present and the circuit noise was negligible. The subject read the passage repeatedly into the telephone, the sidetone loss being changed between readings. The losses used ranged, randomly in 5 dB steps, from the equivalent of 12 dB loss to 48 dB gain relative to a one metre airpath (represented by ARAEN with 30 dB loss). This range is approximately equivalent to the range 40 dB to −20 dB sidetone

* The passage used was that reproduced in Section 2.1.3.2: 'The different speech . . .'

reference equivalent (STRE) determined in accordance with CCITT Recommendation P.73, allowing for the fact that the one metre airpath loudness ratings were made using the BMSP and the CCITT STRE determinations require the reference equivalent speaking position to be used. The range covered extends from sidetone being practically inaudible to it being painfully loud; exceptionably, values outside this range were administered to obtain unequivocal opinions. After speaking the passage, the subject expressed his opinion of the sidetone in accordance with Opinion scale No. 6 or 6A.

Opinion scale No. 6	*Opinion scale No. 6A*
A Objectionable	A Objectionable
B Perceptible	B Detectable
C Not perceptible	C Not detectable

Note: The wording of Opinion scale No. 6A is preferable to that of No. 6.

Figure 3.21 shows the percentage of subjects who could detect and the percentage who would object to sidetone, as a function of the loss of the sidetone path; probit analysis has been used to fit the sigmoid curves and the median values of loss and standard deviations of the respective tolerances are shown.

While the subjects were reading the sentence, the levels of speech voltage produced from the microphone were measured and, with the aid of a separate microphone fixed relative to the talkers, their vocal levels. These results are shown in Figure 3.22 which also reproduces (curves iv and v) information from Figure 3.21). The amplification following the microphone was chosen so that, when reference vocal level (RVL) was used and the microphone placed at the British modal speaking position (BMSP), the speech voltage would be unity (0 dBV). The measurements of vocal level made use of the sending end of ARAEN but the microphone of this equipment was located in the test cabinet so that the sound pressure at this microphone due to the subject's speech was affected as little as possible by his normal movements. To facilitate the measurements, the speech signals from the handset microphone and the vocal level microphone were recorded simultaneously on a two channel recorder and the speech levels measured subsequently on replay.

Curve i in Figure 3.22 shows how the vocal level changes as the sidetone loss increases; the range of vocal level is hardly more than 1 dB. The speech voltage from the handset microphone, however, (shown in curve ii) changes by more than 3 dB. Clearly the subjects are moving the mouthpiece from their lips when the sidetone loss is less than about 7 dB STRE. This is made clearer in curve iii which shows the handset microphone output corrected for vocal level. Comparison of the speech voltage results with the subjects opin-

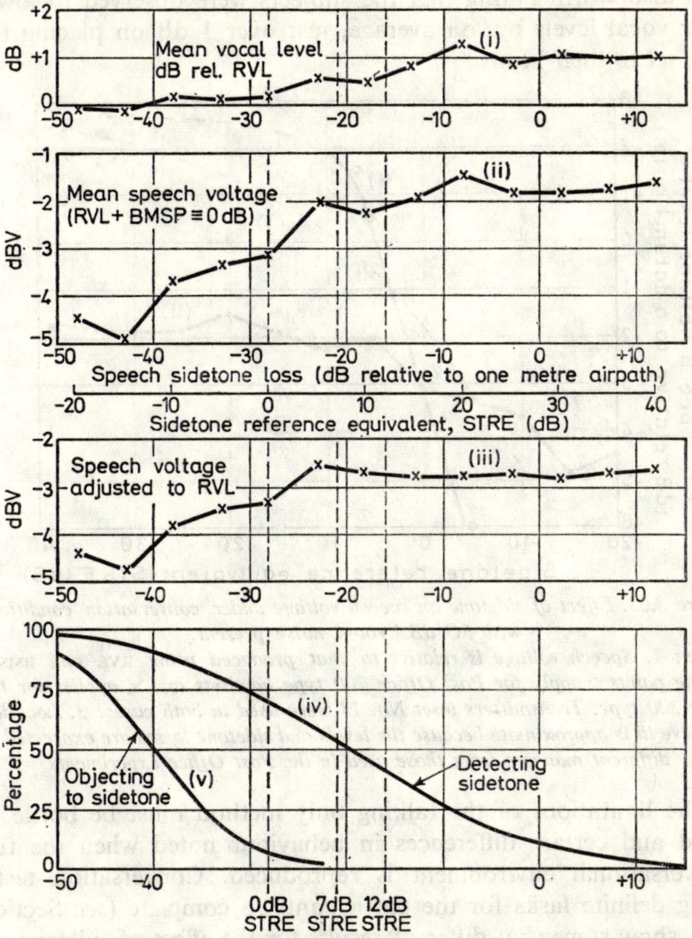

Figure 3.22. Effect of sidetone on vocal level and speech voltage. Talking only test, naive subjects; no room noise

ions shows that the median threshold for detection is approximately that sidetone loss at which the handset is no longer moved away from the lips.

Other interesting information shown in Figure 3.22 is that, with good sidetone suppression (greater than 12 dB STRE) the vocal level was about +1 dB relative to RVL: if room noise had been present, the level would have been even higher. Also, from curve iii, it is clear that a considerably less favourable position than BMSP is adopted by the subjects even when sidetone is very low. It is also worth noting that the subjects were observed to lower their vocal levels by, on average, just over 1 dB on placing the handset to their heads.

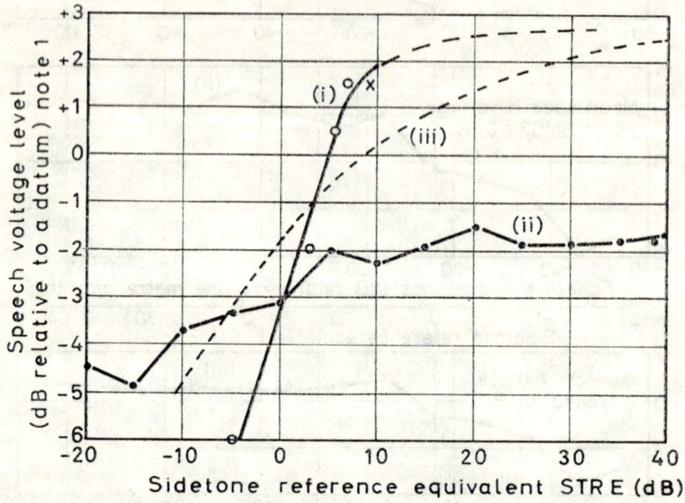

Figure 3.23. Effect of sidetone on speech voltage under conversation conditions with 50 dBA room noise present.
Notes: 1. Speech voltage is relative to that produced using RVL *and* BMSP*: 2. The points* o *apply for Post Office 700 type handsets and* × *applies for the older 300 type. Transmitters inset No. 13 were used in both cases: 3. Location of curve iii is approximate because the levels and sidetone losses are expressed in different manners from those used in the Post Office experiments*

The limitations of the talking only method must be borne in mind and certain differences in behaviour noted when the full conversational environment is reproduced. Conversation tests, using definite tasks for the participants to complete (*see* Section 3.6), show somewhat different results for the effect of sidetone on speech level. Figure 3.23 shows some conversation test results (curve i), using handset telephones, in which the sidetone loss was changed simultaneously at the two ends. Curve ii shows, for comparison, the talking only test results reproduced from curve ii of Figure 3.22; it will be seen that the rate of change of speech voltage from the telephone set with sidetone loss is much greater under conversation conditions than under the talking only conditions. With reasonably good sidetone suppression (STRE greater than 7 dB), the speech voltage from the handset telephone is much higher during conversation than when the subjects were merely talking; this is partly due to the presence of some room noise, partly to the need by the subjects to hold the handset more carefully to hear the speech from the other end, and partly to the tendency to talk louder when using a telephone than in direct conversation over an airpath. Reliable information on the relative magnitudes of these components is not available but there is some confirmation that the tendency, revealed in the results shown in Figure 3.22, to move the microphone relative to the mouth rather than to change the vocal level, is also present during telephone conversations when the characteristics of the speech paths are changed. Curve iii in Figure 3.23 reproduces some results of tests published by Noll, approximately located to the scales used in the figure, in which the subjects talked and did not converse but were specifically instructed to hold the earphone of the handset against the ear; the slope of the resulting relationship is seen to lie between those of the talking only test without instructions concerning holding the handset and the conversation tests.

Although the speech level output from a handset microphone under conversational conditions is affected by the level of room noise*, the relationship between speech voltage output and sidetone loss is unaffected by the level of the room noise, at least over the range 40 to 60 dBA. Furthermore, the speech voltage output versus sidetone loss relationship does not interact with the values of transmission loss in either direction, talker to listener or listener to talker.

Information on the effect of sidetone on the opinions of subjects while conversing is rather scanty; a telephone connection having STRE equal to −5 dB has a severe effect on conversational opinions but, when the STRE exceeds about +3 dB, conversational opinions seem to be unaffected in spite of the substantial effect that this amount of sidetone has on the speech voltage output from the microphones and on the opinions obtained from talking only tests.

The methods described above can be applied to the study of echo as well as of sidetone. Sigmoids showing the percentage of subjects detecting and objecting can be determined and these have

* Increasing the level of room noise from 40 to 60 dBA caused the participants using a connection of about 30 dB nominal overall reference equivalent to raise the speech voltage from their telephone sets by about 4 dB.

rather similar forms to those shown in Figure 3.21 for sidetone although they may be displaced and can have slightly different standard deviations. Some information is given in Section 4.3 which enables these items to be compared and attention here will be confined to differences in method.

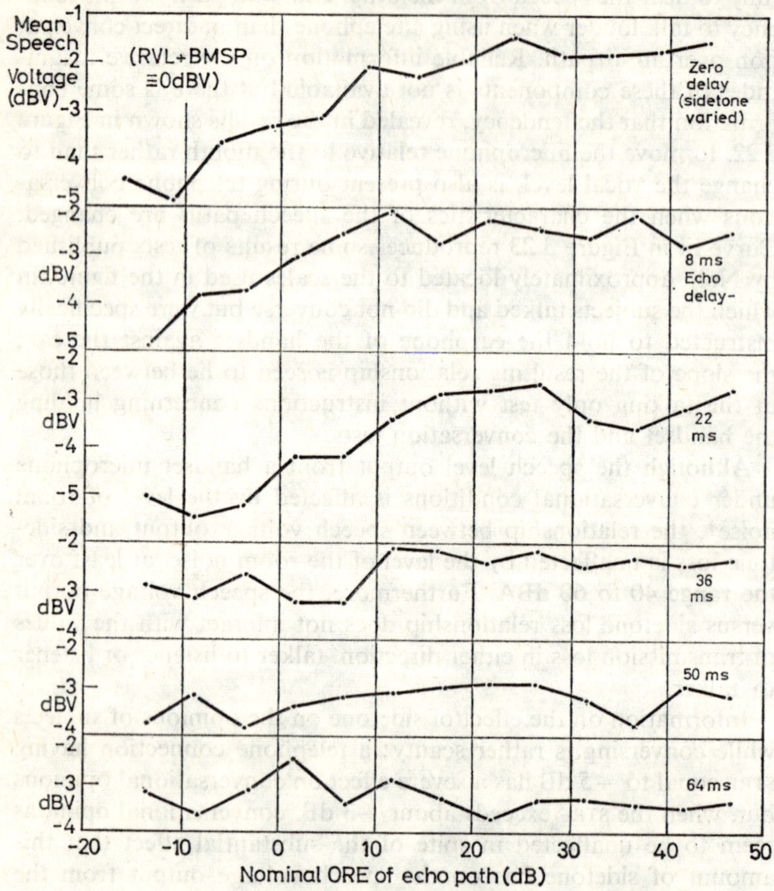

Figure 3.24. Speech level as a function of nominal overall reference equivalent of echo path and its round trip delay

Note: Sidetone reference equivalent = 15 dB

Echo paths that affect the participants while they are talking differ in form from sidetone paths only by the presence of appreciable delay in the round trip route taken by the speech signals. In fact, echo can only be distinguished subjectively from sidetone when the round trip delay exceeds about 30 ms. This is illustrated by the two families of graphs shown in Figures 3.24 and 3.25. Figure 3.24 shows the changes in speech voltage output from the

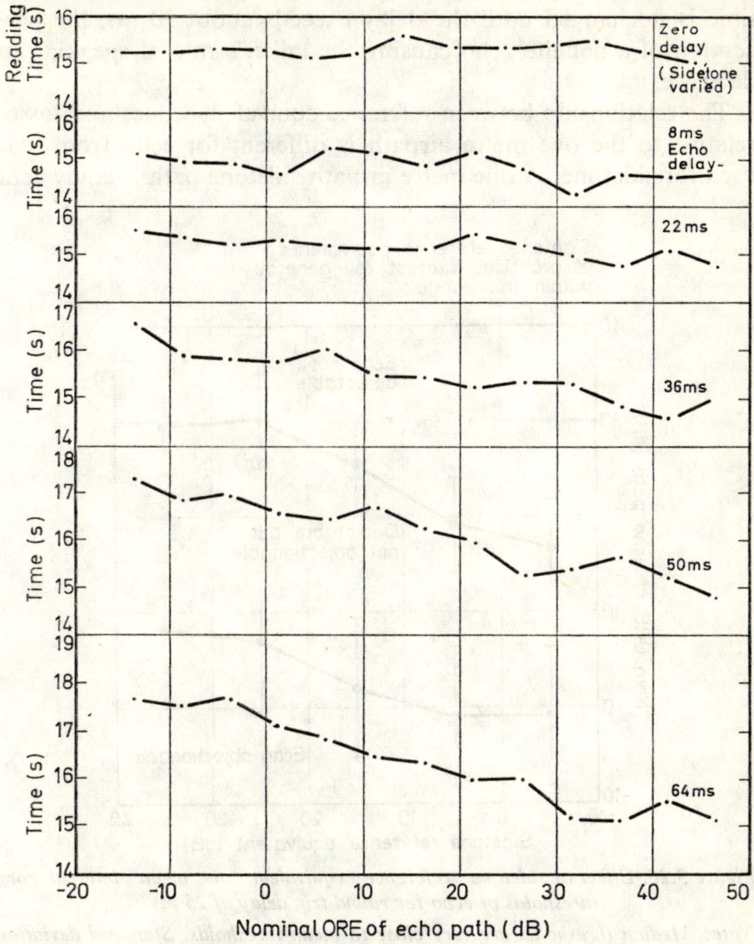

Figure 3.25. Mean reading time as a function of nominal overall reference equivalent of echo path and its round trip delay
Note: Sidetone reference equivalent = 15 dB

handset microphone when tests are conducted in the same manner as those that yielded the results shown in Figure 3.22 but with various amounts of delay introduced into the mouth to ear path, as indicated against the graphs. It will be seen that there is negligible change in speech voltage when the round trip delay exceeds about 30 ms, i.e. when a definite echo can be perceived. Figure 3.25 shows the results of timing the speaking of the passages and plotting the mean times against the loss of the echo path; in this case, the time is unchanged until the delay exceeds about 30 ms, the perception of a definite echo causing the talker's rate of speaking to decrease.

The relationship between reference equivalent values and losses relative to the one metre airpath is different for echo from that used for sidetone; A one metre airpath sidetone path is equivalent

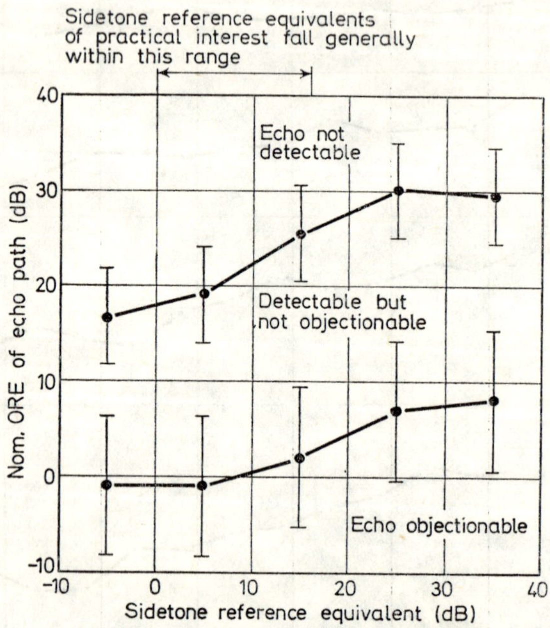

Figure 3.26. Effect of sidetone reference equivalent on mean intercategory thresholds of echo for round trip delay of 25 ms

Note: Median thresholds are very close to mean thresholds. Standard deviation of distributions of subjects' tolerances is approx. 9 dB

to about 28 dB STRE but a one metre echo path is equivalent to about 32 dB nominal overall reference equivalent. The difference arises from peculiarities in specifying the values of the quantities employed at present by telephone engineers for expressing such losses (*see* Section 3.2.4 and Section 6.4).

Although echo and sidetone are separately distinguishable by a subject (provided the echo delay exceeds 30 ms round trip), there can be some interaction between their effects and this is illustrated in Figure 3.26 which shows the thresholds of detectability and objectionableness of echo under conditions where the subjects could, at will, connect or disconnect the echo path while the side-tone remained fixed. These tests were administered by presenting a prearranged set of echo path losses in random sequence at each

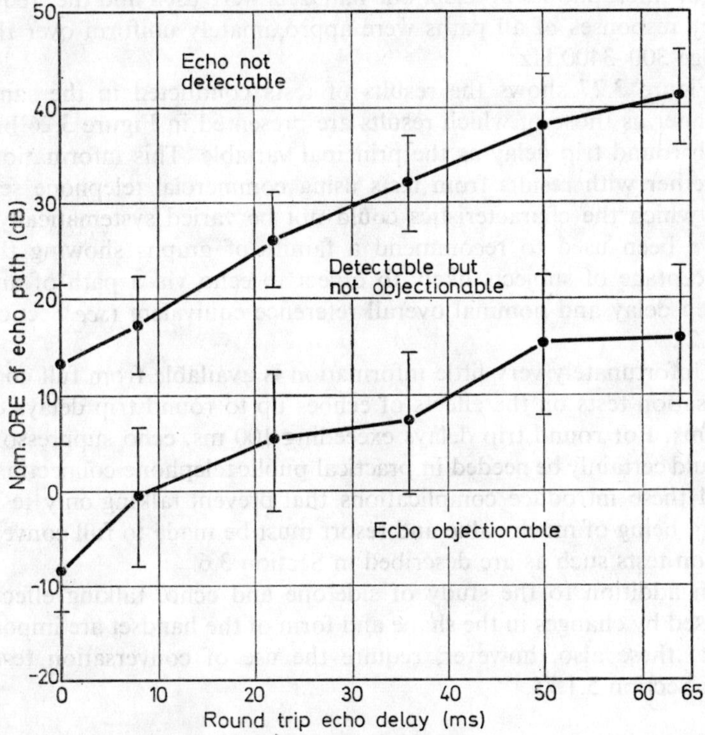

Figure 3.27. Effect of echo delay on mean intercategory thresholds of echo for sidetone reference equivalent of 15 dB

Note: Median thresholds are very close to mean thresholds. Standard deviation of distributions of subjects' tolerances is about 13 dB

setting of the abscissa. The method does not yield exactly the same values of threshold for the median subject as obtained by presenting a random sequence of abscissa values to a large number of subjects (as was done in obtaining the results given in Figure 3.21) but the method has the advantage that individual thresholds are obtained for each subject and so the dispersion attributable to the subjects can be separated from that attributable to repeated application to a given subject. The results shown in Figure 3.26 were obtained using naive subjects; use of trained 'speech testers' yields somewhat higher absolute thresholds but the interval between the two thresholds remains about the same. The tests were conducted with the aid of a laboratory simulation (*see* Section 4.1.4) in which the various quantities could readily be varied independently of each other; linear microphones in telephone handsets were used and the frequency responses of all paths were approximately uniform over the range 300–3400 Hz.

Figure 3.27 shows the results of tests conducted in the same manner as those of which results are presented in Figure 3.26 but with round trip delay as the principal variable. This information, together with results from tests using commercial telephone sets (of which the characteristics could not be varied systematically), have been used to recommend a family of graphs showing the percentage of subjects likely to object to echo via a path of any given delay and nominal overall reference equivalent (*see* Section 4.3.2).

Unfortunately very little information is available from full conversation tests on the effects of echoes up to round trip delays of 60 ms. For round trip delays exceeding 100 ms, echo suppressors would certainly be needed in practical public telephone connections and these introduce complications that prevent talking only tests from being of much value and resort must be made to full conversation tests such as are described in Section 3.6.

In addition to the study of sidetone and echo, talking effects caused by changes in the shape and form of the handset are important; these also, however, require the use of conversation tests (*see* Section 5.1.7).

3.5.3 Use

The above methods can be used to derive allowances in terms of changes in speech volume produced on lines when, for example, the sidetone reference equivalent is changed. Such an allowance can then be combined with a listening rating to give the total effect of making some change in a telephone connection which affected talking factors as well as listening factors (*see* Section 4.3.1).

Reference has already been made above to the derivation of echo tolerance curves which are presented in Section 4.3.2.

In some circumstance requiring study, changes in talker behaviour are not governed only by the characteristics of one or more direct mouth to ear paths terminated at the talker but also by interaction with the participant at the other end of the connection. The talking only methods are then unsuitable and a complete conversation type method of assessment must be used such as those described in Sections 3.6 and 3.7.

3.6 EXAMINATION OF CONVERSATIONS

3.6.1 Aim and principle

Conversation is the normal mode of employment of telephone connections by members of the public. The connections provided by the telephone operating organisation ought to enable the conversation to be conducted with minimal difficulty and no special skills ought to be expected of the participants. When some difficulty is encountered, it will, in most cases, be surmounted by the participants making some slight adjustment in their behaviour, such as talking louder or closer to the mouthpiece, listening more carefully and holding the earphone more firmly to the ear; they may also respond by requesting some material not understood to be repeated or otherwise indicating trouble by saying, for example, 'I can't hear you'; 'This connection is poor', etc. Such changes in behaviour and other responses can, in principle, be observed and their frequencies of occurrence used as an indication of failure to provide completely adequate performance.

The aim of conversational methods of assessing speech links and telephone connections is therefore; (1) to reproduce situations that result in pairs of subjects conversing with each other over the connections to be assessed; and (2) to observe relevant features of their behaviour so that abnormalities attributable to deficiencies in the connections can be identified.

The observations possible can be classed as (a) those made upon the conversations themselves and (b) those involving examination of the participants, either during or after their conversations. The present section is concerned with methods for producing conversations in appropriate environments and with observation methods falling into (a).

3.6.2 Methods

In principle, use of actual conversations over working telephone connections seems attractive because the environment is undisturbed; not only is this not permissible on grounds of infringing privacy but it also proves extremely difficult to make reliable observations of conversational events, even under good laboratory conditions. The results also become rather uninformative because they would apply for connections whose particulars can often be known only imperfectly. Any field observation method would therefore provide only very generalised information averaged over many unknowns and therefore unsuited to the purpose of predicting performance under any different conditions. Conversation tests conducted within laboratories enable much more reliable observations to be made but particular care is necessary to ensure that any effects caused by the artificiality of the situation are minimised. The following precautions are important.

1. The various paths of the telephone connection being simulated must be arranged to present the speech link to the participants as they would experience an actual telephone connection. Not only must the sidetone, room noise, attenuation/frequency distortion, circuit noise and other physical features be realistic but also the equipment used by the participants must be like a telephone.
2. Each conversation should be initiated as would a real telephone call, that is one participant should dial a number which causes the other's bell to ring and, on answering, the connection must be found by the participants to have been established so that the conversation will commence with the usual introductions.
3. The topic of the conversation should form a definite task which has a clear objective which, when achieved, allows the call to be terminated in a normal manner by replacing the handset.
4. The participants in the experiment should be untrained and as representative as possible of normal members of the public. They should take part in such tests not more frequently than, say, once in six months and, on each occasion, the number of conversations should be limited so that the tasks do not become too familiar to them.
5. The range of test conditions included in any experiment should be wide enough to ensure that the participants experience conditions likely to be considered 'excellent' down to those likely to be considered 'poor' by average subjects.

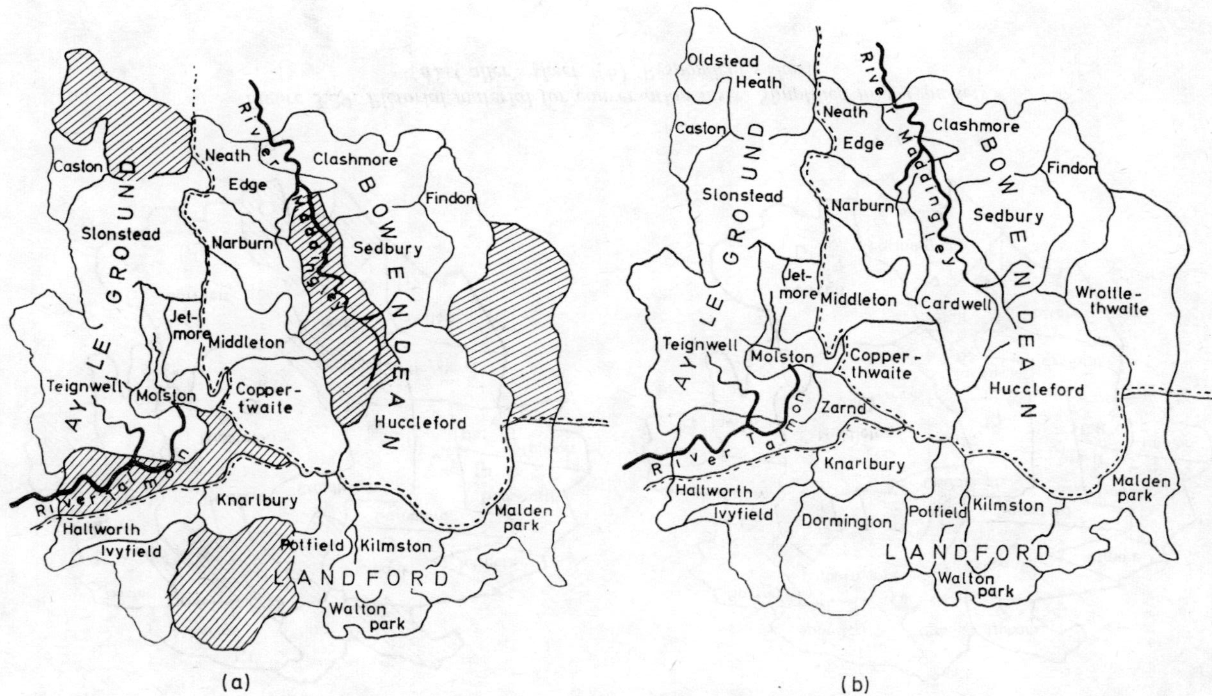

Figure 3.28. Pictorial material for conversation tests. First map type sets.
(a) Caller's sheet. (b) Respondent's sheet

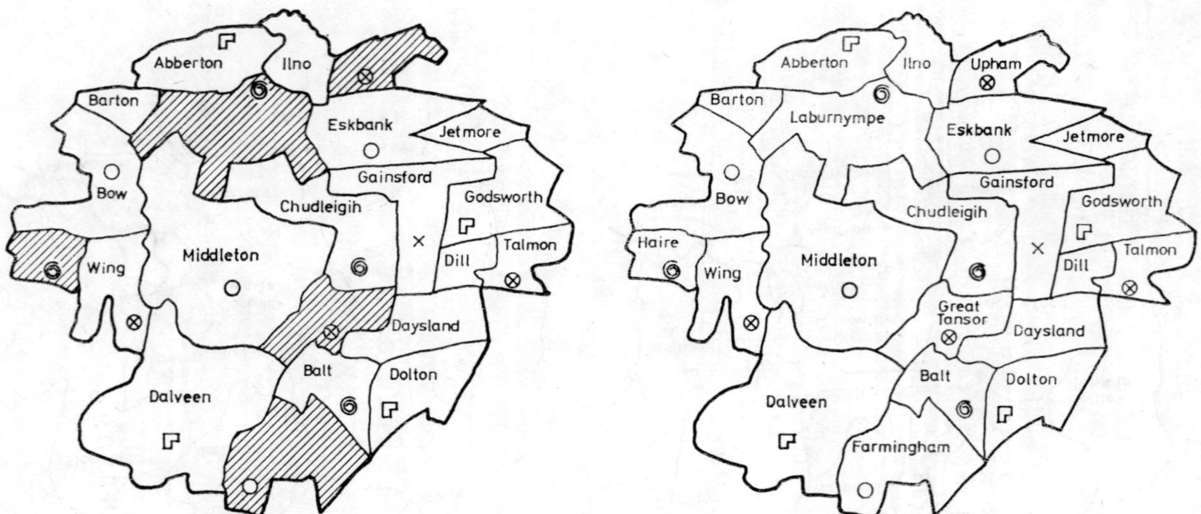

(a) (b)

Figure 3.29. Pictorial material for conversation tests. Simplified map type sets. (a) Caller's sheet. (b) Respondent's sheet

Item 3, has been the subject of a great deal of development and various tasks have been used; these have been presented to the subjects in mainly pictorial form to avoid dictating the words likely to be employed. Clearly one of the participants must originate the call and he is termed the *caller*; the scene must be set to provide him with a reason for making the call and many of the tasks used have required the caller to find some information he is lacking and which is in the hands of the other person, called the *respondent*. An early type of task is that using the map-like diagrams shown in Figure 3.28; the caller's sheet is identical to that of the respondent except that certain areas shown shaded lack names which appear without shading in the respondent's sheet. The caller is required to find out the missing names. This task proved excessively difficult for the purpose and the maps were simplified as shown in Figure 3.29; these were used in the same manner, proved somewhat easier as a task but suffered from the disadvantage that few of the subjects were familiar with maps and found the whole exercise strange. Those accustomed to use of maps found the task almost trivial. The next stage is shown in Figure 3.30 where the pictorial material is much more diversified and of about the required order of difficulty; titles are provided for both participants so that little spelling is necessary. The caller's task is to find out the titles for the five pictures given on his sheet which also appear, with 15 others, on the respondent's sheet.

It is worth noting that the differing difficulty experienced in solving the puzzles shown in Figures 3.28, 3.29, and 3.30 was reflected in changes of the level of speech sent to line by the participants; information on this matter is given in Section 4.6.

The material shown in Figure 3.30 was quite satisfactory and was used for a large series of experiments in which a substantial number of different subjects each made a small number of calls. A different, but equivalent, pictorial set was necessary for each call made by any participant. Later requirements were for a very large supply of different pictorial material so that a smaller number of subject pairs could be used in any experiment and make several calls without being given the same picture set again. The material in Figure 3.30 does not lend itself to such use and a simpler means of generating different but approximately equivalent designs was required. The method illustrated in Figure 3.31 was therefore developed. A polar coordinate graticule as shown in Figure 3.31a is used having 8 radii and 5 concentric circles. From a source of random decimal digits, a sequence of eight is selected, for example, 58235185. Starting from the 12 o'clock radius, the first random digit locates a point on the circle marked 5/6 and, being an odd digit is marked with a cross. The second digit defined a point on the 7/8 circle and, being even, is marked with a dot. All eight radii are treated in the same manner so that a sequence of crosses and dots is built up as shown in Figure 3.31b. A line is then drawn through the points passing smoothly through the dots and with discontinuities at the crosses; Figure 3.31c shows the final shape. The dots and crosses are erased before use. Random shapes produced in this way are assembled into call-

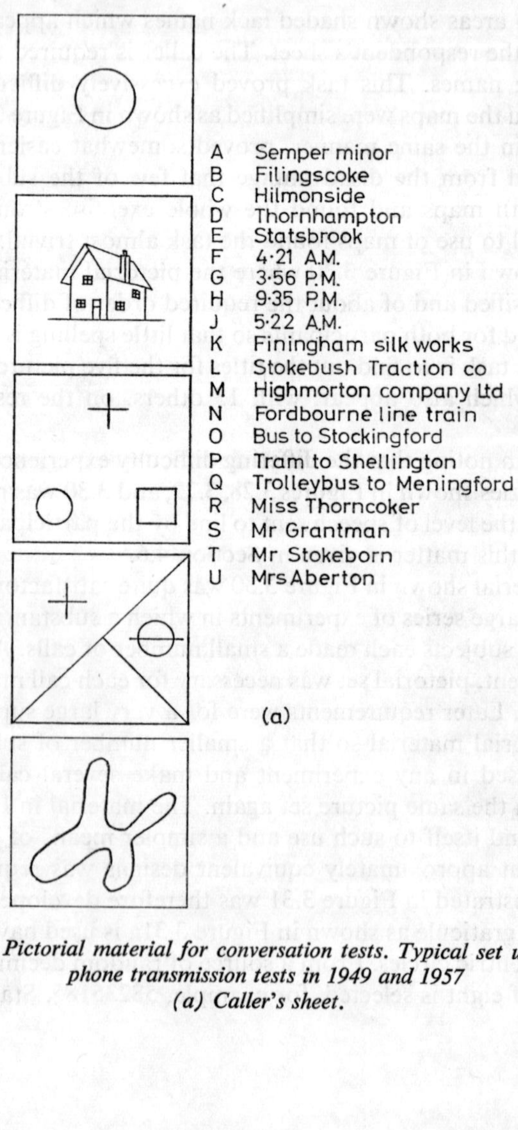

Figure 3.30. Pictorial material for conversation tests. Typical set used for telephone transmission tests in 1949 and 1957
(a) Caller's sheet.

ers' and respondents' sheets as shown in Figure 3.31d and e together with word titles selected from the first word in a newspaper column, excluding proper names, particles, pronouns, prepositions and other unsuitable words. In theory 10^8 different shapes can be generated but the number of separately distinguishable shapes may be rather fewer. About 25% of the shapes generated were stated to

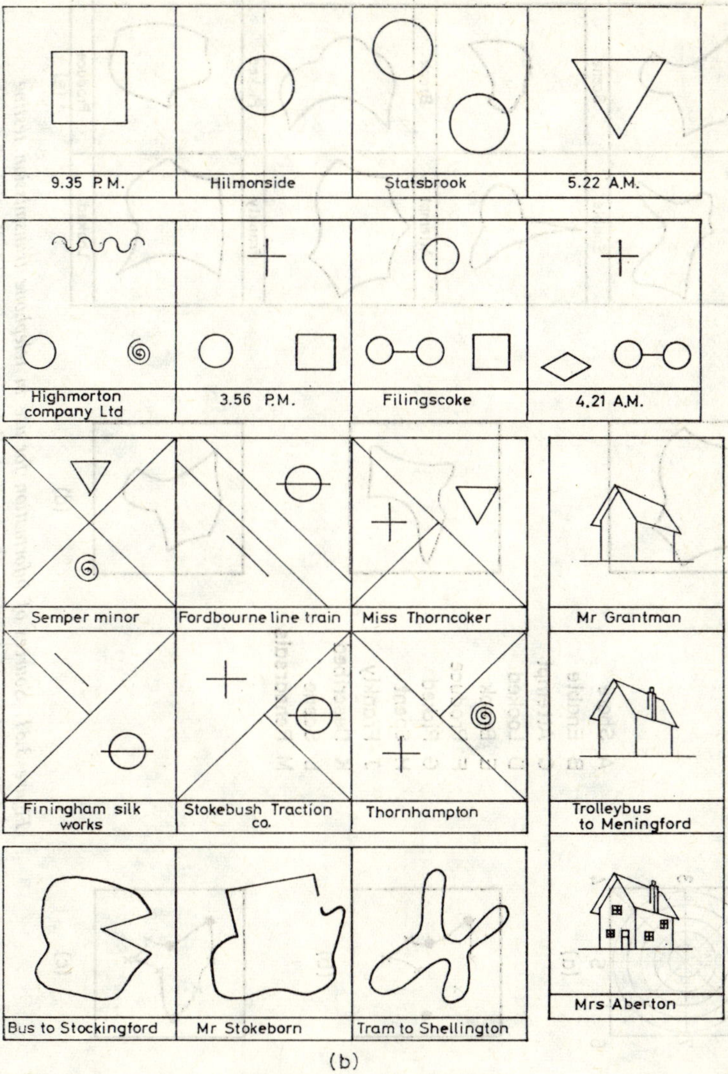

(b) Respondent's sheet.

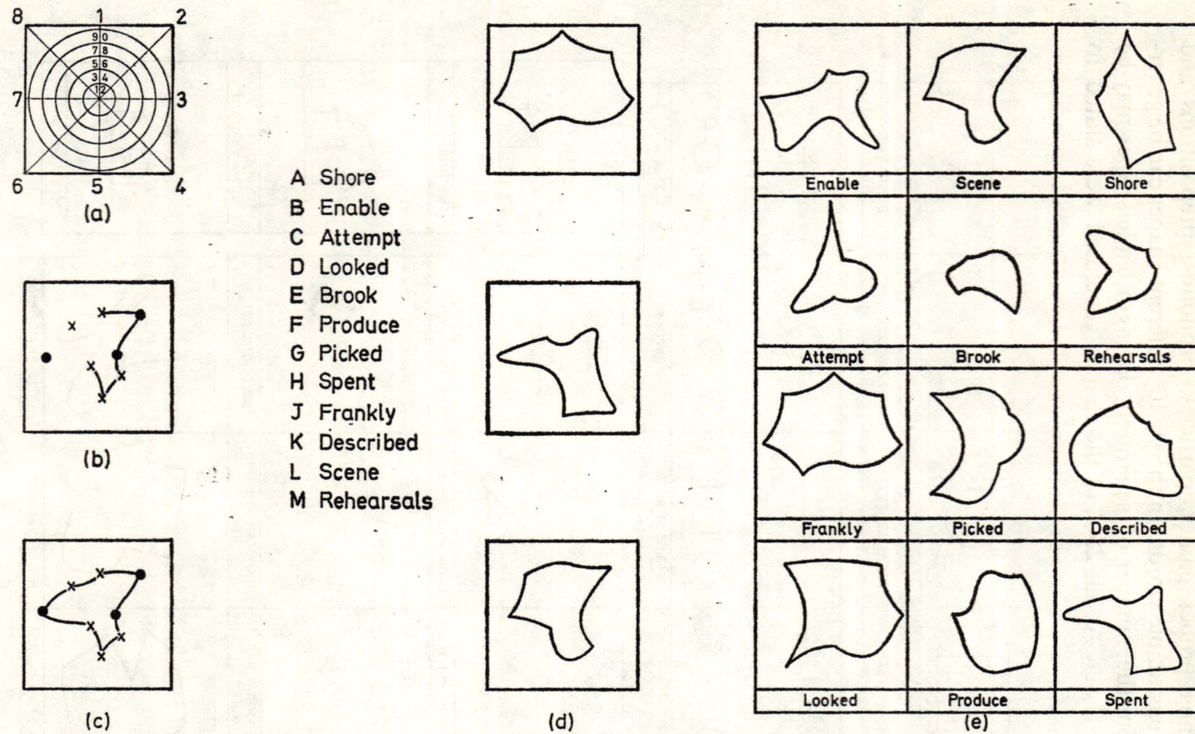

Figure 3.31. Sources of information for use in telephone transmission testing Construction of random sheets. (a) Polar graticule. (b) Partial shape. (c) Completed shape. (d) Caller's sheet. (e) Respondent's sheet

represent something (e.g. the bottom right example in Figure 3.31e) but only about 6% were consistently recognised as being like the same thing by two or more persons. These 6% were therefore not used.

The next stage in development was reached when studies were to be made on telephone connections containing long propagation times. The performance was expected to be affected by the structure of the conversation, and especially by the frequency of occurrence

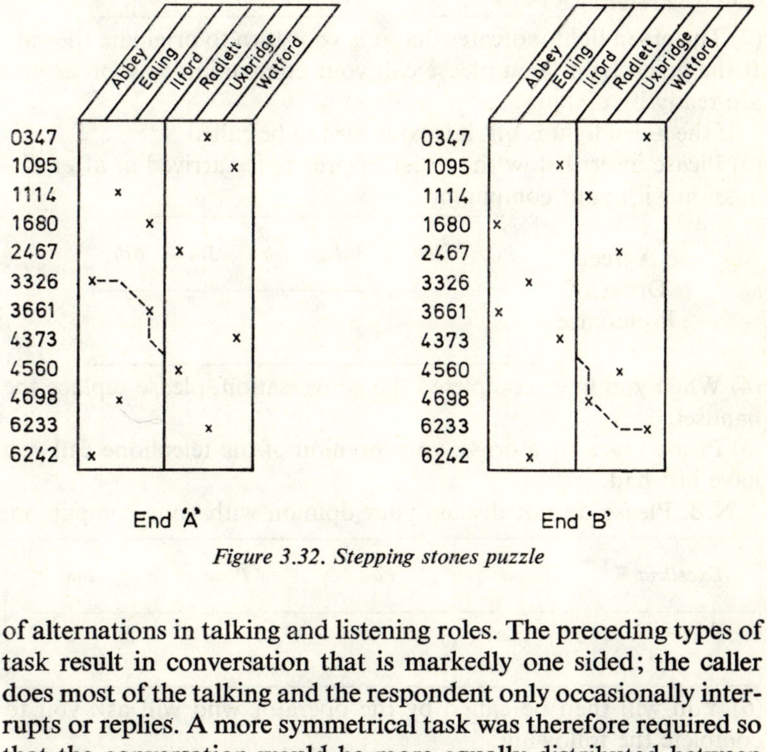

Figure 3.32. Stepping stones puzzle

of alternations in talking and listening roles. The preceding types of task result in conversation that is markedly one sided; the caller does most of the talking and the respondent only occasionally interrupts or replies. A more symmetrical task was therefore required so that the conversation would be more equally distributed between the participants. Figure 3.32 shows the design of puzzle used. It is supposed that the diagram represents a river, running up or down the figure and separating the two participants; *A* is supposed to be on the left bank and *B* on the right. The river is dotted with stepping stones located as shown by the crosses; half the number of stepping stone are visible to *A* and the other half to *B*. The task of the participants is to cooperate to find a route using adjacent stepping stones to the centre where they are supposed to shake hands. One

Opinion form No. 12

For conversation tests in which the participants use picture cards

(1) Before you commence your call, would you please insert below your order of preference for the picture cards. For identification, please use the serial numbers on the backs.

	1st	*2nd*	*3rd*	*4th*	*5th*	*6th*
Your Order of Preference						

(2) The green light indicates that it is your turn to originate the call. If the green light is on please call your companion as soon as you are ready by dialling...

If the green light is off, it is your turn to be called.

(3) Please insert below the order of preference arrived at after discussion with your companion:

	1st	*2nd*	*3rd*	*4th*	*5th*	*6th*
Agreed Order of Preference						

(4) When you have completed the conversation, please replace the handset.

(5) Please mark by a cross, your opinion of the telephone call you have just had.

N.B. Please do not discuss your opinion with your companion.

Excellent	*Good*	*Fair*	*Poor*	*Bad*

(6) You will then be called by the operator who will ask you to complete the following:

Did you, or the person who spoke to you, have any difficulty in talking or hearing over the connection?

YES	
NO	

If YES, please explain the nature of your difficulty to the operator.

solution is indicated by the broken line; the designs actually used were larger (16×16) with five stepping stones in each row and each column. Telephone exchange names and four-digit numbers are used to identify the locations of the stepping stones.

Although a large number of experiments were conducted using the stepping stone puzzles, it was decided that the conversation was rather formalised and the vocabulary limited. In addition it was desired to stimulate rather more vigorous conversation than results from puzzles where a certain amount of intellectual effort is required which slows down the responses. The next type of task to be developed was therefore intended to include an element of competitive negotiation between the participants. This was achieved by requiring the participants to form personal opinions about a set of pictures before establishing any call; after setting up the call, they are required to agree upon a common, compromise, opinion concerning the set. The pictures used are postcard size reproductions of a wide variety of paintings hanging in the Tate Gallery, London. These are arranged in duplicate sets of six and the members of each pair, who are to converse, first decide their personal order of preference and note it down on the form shown opposite.

Other tasks, based on use of pictorially presented information, have been used. The United Kingdom Signals Research and Development Establishment (SRDE) has used pairs of photographs which are almost identical but were taken with a slight separation in time of scenes containing some movement; typical subjects are street scenes, groups of persons walking along a railway platform. One participant is given the earlier photograph and the other the later; they are not told which was taken first but their conversational task consists of finding this out by discussion.

Examination of conversations can be facilitated by recording the spoken material in a suitable manner to preserve the time relationship between what a participant hears and what he says. Such recorded material can, provided that suitable calibration tones have been included to enable the correct levels to be reproduced on replay, be used for the following kinds of examination:

1. Speech voltages at some definitely located points in the connection
2. Durations of the conversation, i.e. time to complete the task
3. Monitoring by independent observers to report significant events and, possibly, to obtain the opinions of the observers on the quality of the conversation.

Speech voltages can be measured on conversational speech by many different instruments but the present method used by the United Kingdom Post Office is the speech voltmeter type 5B (see Section 2.1.3.2); this yields the long term r.m.s. voltage of speech while the talker is active. Information on speech voltages during conversations over laboratory telephone connections (where all the physical characteristics are known) is included in Section 4.2.1.3 and results of measurements on real connections are presented in Section 6.3. Considerable caution is necessary if any attempt is made to use speech level as an assessment criterion; some relevant information is given in Section 4.6.

The time required to complete a defined task does increase as the telephone connection becomes more degraded as shown in Figure 3.33, which illustrates the relationship between time taken and the subjects' opinions of the connection. The relationship is such, however, that measurement of time taken is a very insensitive indication of transmission performance except when the connection is so bad that it would be practically unusable for ordinary telephony. The ratio of time required to complete a given task over a good connection to that required over the given connection is termed *message rate efficiency*; some information relating this quantity to the overall transmission loss of telephone connections is given in Section 4.1.5. The durations of prepaid telephone calls do not seem to bear any well defined relationship to their transmission performance; the duration might be expected to increase for the reasons given above or to decrease because the customers might abandon the call prematurely if the transmission performance were inadequate.

At one time observations of simulated telephone calls to count the incidence of requests for repetitions of material not understood was highly thought of (*see* Pocock, 1948) and study of this method showed that the mean incidence increased exponentially with the amount of transmission loss inserted in a connection. Observations of adverse comments on the transmission made during the conversation also showed a similar relationship. The method has proved useful is studying the nature of difficulties experienced in conversing, but it has not proved possible to base any quantitative transmission planning information on such observations. Clearly, the method cannot be used in the field because privacy of telephone conversations over the public network would be infringed. The method has, in any case, proved very troublesome to implement; even when recordings of laboratory conversations are available (so that repeated examination is possible) the training required for the observers is considerable and disagreement between members of an observation team are not uncommon. Some detailed information on conversational behaviour based on such observations is given in Section 4.6 (*see also* References).

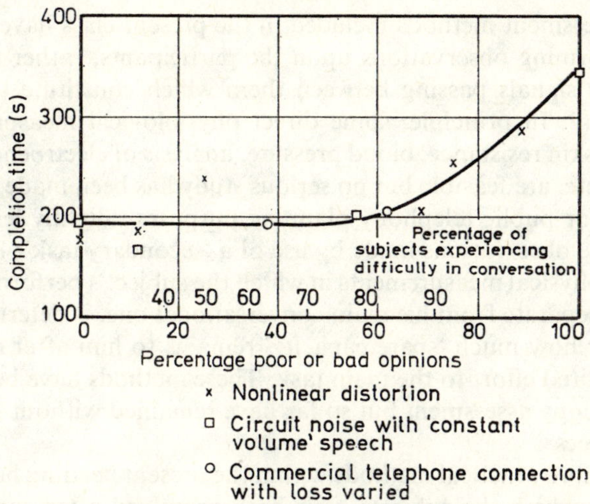

Figure 3.33. Time to complete a task involving conversation as a function of percentage 'difficult' and percentage 'poor or bad' opinions

3.6.3 Use

Assessment tests involving conversations stimulated by suitable tasks for the participants are of direct use in providing examples of conversational material which can be recorded and examined in detail. These examinations, however do not provide information directly useful for planning. The participants themselves, however, can be examined in various ways and these examinations are particularly fruitful; the methods are described in Section 3.7.

3.7 EXAMINATION OF PARTICIPANTS DURING OR AFTER CONVERSATION

3.7.1 Aim and principle

The assessment methods included in the present class have the aim of performing observations upon the participants, rather than the physical signals passing between them which constitute the 'conversation'. In principle, some direct physiological measurements, such as skin resistance, blood pressure, analysis of electroencephalograms, etc. are feasible but no serious study has been made by these means for public telephony planning purposes. Slightly less direct are those observations made by use of a 'secondary task'; these are psychophysical measurements in which the subject's performance of a task, separate from his main conversational one, is determined to discover how much 'spare capacity' remains to him after devoting the required effort to the main task. These methods have been tried in telephony assessment but so far have remained without appreciable success.

The more practical methods within the present heading have been those in which the subjects have been examined after completing the conversational task without any disturbance. The purely subjective method of interrogating the participants has proved the most fruitful and various forms of questionaire have been used, administered verbally, by writing on a form or by indicating by keys or pushbuttons. Use can also be made of certain forms of the talker/listener (as opposed to conversational) test methods mentioned in Section 3.3 and 3.4; the idea of this technique is to examine the participants after they have been conversing by requiring them to perform certain definite talker/listener tasks, like articulation, which lend themselves very readily to scoring. These supplementary tasks are performed over exactly the same speech paths as have been used for conversation and follow almost immediately after the conversation so that the 'mental contact' formed during the conversation is maintained. Examples of applying these principles are given below.

3.7.2 Methods

Methods for generating conversations between pairs of subjects were described in Section 3.6.2; on completion of the test call, e.g.

by replacing the handset, the participants are interrogated independently. That is they are not permitted to discuss their opinions of the connection. The criteria on which the interrogation is based can make use of various opinion scales which have been constructed for the purpose. The earliest such scale used in laboratory tests by the United Kingdom Post Office employed the two categories 'usable' and 'not usable' and the participants were invited to judge the connection (rather than the particular conversation) accordingly. (Some results using this scale are given in Section 4.6.)

This scale was considered too coarse and a four category scale devised as follows:

Opinion scale No. 1

G Good circuit; no appreciable effort required

F Fair circuit; some effort required in conversing

P Poor circuit; conversation possible but much difficulty on unfamiliar words

B Bad circuit; usable only with extreme difficulty

Several series of tests were performed using this scale and the votes were quantified by allocating scores of 4 for *G*, 3 for *F*, 1 for *P* and 0 for *B*; mean scores were calculated and, when divided by the maximum score (4) and converted to percentages, were termed 'Opinion Index'. The tests in which this procedure was used were intended to explore the usefulness of a multicriterion method of assessment and the conversation phase of the test was followed, almost immediately after the interrogation, by one of the participants reading a list of 25 short sentences (*see* Section 3.4) and the other replying 'yes' or 'no' according to whether he understood the meaning without mental effort. Any hesitation in replying was marked as 'no'. The score was kept by the subject who read the list of sentences and he marked a ticksheet. After one subject had read a list the other read another list so that two separate 'no mental effort' scores (expressed as percentage of 'yes' responses) were obtained.

The same pair of subjects then read a list of 25 single syllable words to each other (*see* Section 3.3) using the carrier phrase 'The next word will be . . . also'. The listening subject wrote down what he heard and the results were subsequently compared with the list read by the talker and percentage correct recognition scores obtained. Partially correct words scored nothing.

Finally the pair of participants read to each other a list of 25 meaningless monosyllables (logatoms). Because these tests were concerned with naive subjects who could not read or write in any unequivocal symbols, special scoring was necessary. The logatoms were of a special type using a vocabulary limited to single phoneme initial and final consonants and only five 'short' vowels (*see* Section 3.3). Initial/vowel and vowel/final transitions were random. Such combinations are not only all pronouncable but can be transcribed fairly readily into ordinary alphabetical characters so that they can be read correctly by ordinary, untrained persons (an example of this transcription is given in Section 3.3). By this means the person at one end could read a list of meaningless monosyllables, but the problem remained of determining whether the sound groups are correctly recognised by the listener. This problem was solved by requiring the listener to repeat audibly what he heard; his repetition was recorded through a separate high quality microphone in his cabinet together with a separate recording made simultaneously of what the talker had emitted. The two recordings were then compared by a team of well trained speech testing operators experienced in articulation testing and percentage correct recognitions determined. By this method occasional wrongly read utterances by the talker were unimportant.

Several experiments were conducted using this method and the speech links used included the following:

1. Simulated telephone connection with various values of overall loss and used in the presence of 60 dBA room noise (Hoth spectrum),
2. High quality reference system degraded by injected noise having a spectrum shape that matched that of speech,
3. High quality reference system degraded by a 'reference distortion circuit (RDC)' that interrupted the paths in a random manner at a mean rate of about 500 times per second and with variable 'speech/time' fraction,
4. High quality reference system degraded by a delta modulator, with the quantising rate varied. This delta modulator had constant step size.

The results for 1 and 3 are reproduced in Figures 3.34 and 3.35; those for 2 show similar features to 1 and those for 4 are similar to 3. If the various scores, expressed as percentages, are plotted against some convenient scale of increasing degradation, sigmoid graphs would be produced, rather like those shown in Figure 1.9 which

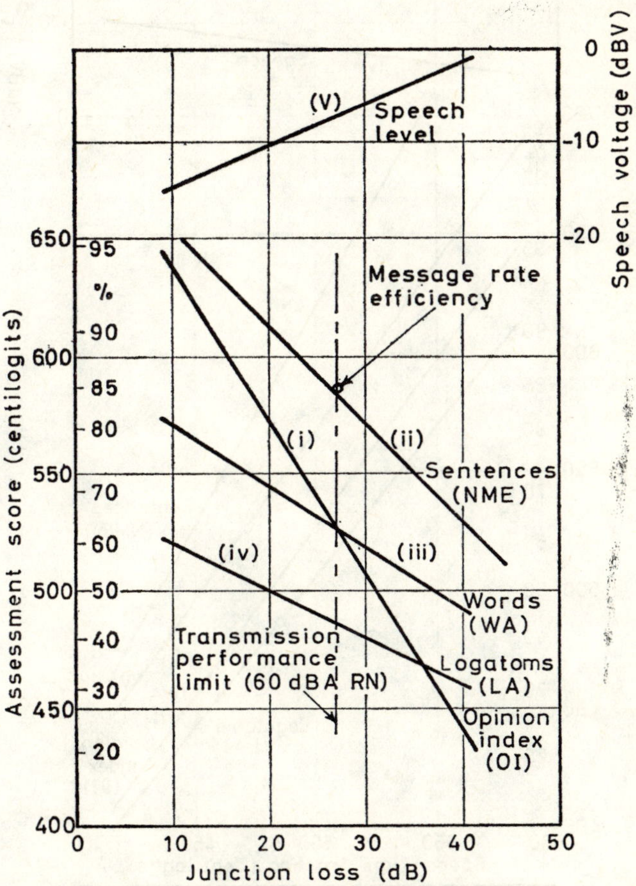

Figure 3.34. Assessment scores, after conversation, according to various criteria. Limiting LTCs *in 60 dBA room noise*

appears in Section 1.3. To simplify the curve fitting, the percentages have been transformed to 'centilogits' according to the expression:

$$y = 500 + 100 \text{ arg tanh } (2p - 1)$$

where y is the transformed value of score, in centilogits, and $100\,p$ is the score as a percentage.

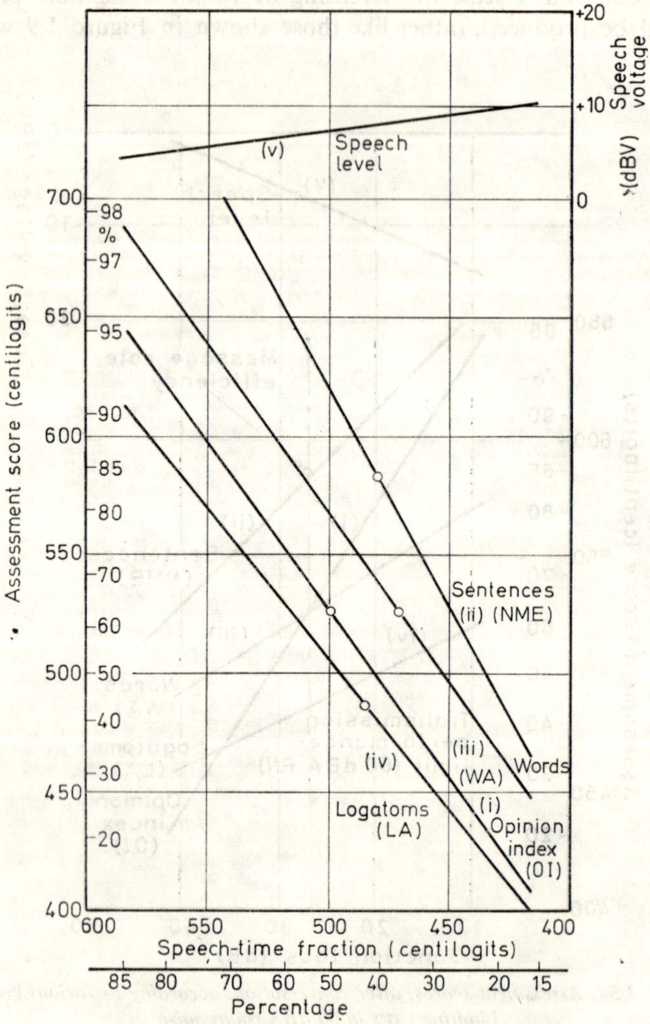

Figure 3.35. Assessment scores, after conversation, according to various criteria. High quality speech link degraded by distortion

Table 3.15 gives this transformation from p to y and Table 3.16 gives the inverse.

Table 3.15. TRANSFORMATION OF PERCENTAGES TO CENTILOGITS
(a) FROM 1 TO 99% IN STEPS OF 1%

Percentage	*0*	*1*	*2*	*3*	*4*	*5*	*6*	*7*	*8*	*9*
0	—	270	305	326	341	353	362	371	378	384
10	390	396	400	405	409	413	417	421	424	428
20	431	434	437	440	442	445	448	450	453	455
30	458	460	462	465	467	469	471	473	476	478
40	480	482	484	486	488	490	492	494	496	498
50	500	502	504	506	508	510	512	514	516	518
60	520	522	525	527	529	531	533	535	538	540
70	542	545	547	550	552	555	558	560	563	566
80	569	573	576	579	583	587	591	595	600	605
90	610	616	622	630	638	647	659	674	695	730

(b) FROM 97.0 TO 99.9% IN STEPS OF 0.1%

Percentage	*0.0*	*0.1*	*0.2*	*0.3*	*0.4*	*0.5*	*0.6*	*0.7*	*0.8*	*0.9*
97	674	676	677	679	681	683	685	687	690	692
98	695	697	700	703	706	709	713	717	721	725
99	730	735	741	748	756	765	776	790	811	845

In Figure 3.35, the speech/time fractions for the distortion setting of the RDC have also been transformed according to the same table; this renders the settings analogous to signal/noise ratio in decibels.

In Figure 3.34, the setting of the transmission performance limit of 27 dB (with the local telephone circuits used in the experiments) with 60 dBA room noise present is marked and the following scores apply:

Opinion Index (OI);	527 centilogits (63%)
No mental effort sentences (NME);	583 centilogits (84%)
Word articulation (WA);	527 centilogits (63%)
Logatom articulation (LA);	486 centilogits (43%)

These values of score have been marked on the graphs in Figure 3.35 and it will be noticed that they no longer correspond to a single setting of the speech link; the discrepancy is slight for NME, WA and

Table 3.16. TRANSFORMATION OF CENTILOGITS TO PERCENTAGES

(a) From 400 to 599 centilogits in steps of 1 centilogit

Centilogits	0	1	2	3	4	5	6	7	8	9
400	11.9	12.1	12.4	12.6	12.8	13.0	13.2	13.5	13.7	13.9
410	14.2	14.4	14.7	14.9	15.2	15.4	15.7	16.0	16.2	16.5
420	16.8	17.1	17.4	17.6	18.0	18.2	18.5	18.8	19.2	19.5
430	19.8	20.1	20.4	20.8	21.1	21.4	21.8	22.1	22.4	22.8
440	23.2	23.5	23.9	24.2	24.6	25.0	25.4	25.7	26.1	26.5
450	26.9	27.3	27.7	28.1	28.5	28.9	29.3	29.7	30.2	30.6
460	31.0	31.4	31.9	32.3	32.7	33.2	33.6	34.1	34.5	35.0
470	35.4	35.9	36.4	36.8	37.3	37.8	38.2	38.7	39.2	39.6
480	40.1	40.6	41.1	41.6	42.1	42.6	43.0	43.5	44.0	44.5
490	45.0	45.5	46.0	46.5	47.0	47.5	48.0	48.5	49.0	49.5
500	50.0	50.5	51.0	51.5	52.0	52.5	53.0	53.5	54.0	54.5
510	55.0	55.5	56.0	56.5	57.0	57.4	57.9	58.4	58.9	59.4
520	59.9	60.4	60.8	61.3	61.8	62.2	62.7	63.2	63.6	64.1
530	64.6	65.0	65.5	65.9	66.4	66.8	67.3	67.7	68.1	68.6
540	69.0	69.4	69.8	70.3	70.7	71.1	71.5	71.9	72.3	72.7
550	73.1	73.5	73.9	74.3	74.6	75.0	75.4	75.8	76.1	76.5
560	76.8	77.2	77.6	77.9	78.2	78.6	78.9	79.2	79.6	79.9
570	80.2	80.5	80.8	81.2	81.5	81.8	82.0	82.6	82.4	82.9
580	83.2	83.5	83.8	84.0	84.3	84.6	84.8	85.1	85.3	85.6
590	85.8	86.1	86.3	86.5	86.8	87.0	87.2	87.4	87.6	87.9

(b) From 600 to 990 centilogits in steps of 10 centilogits

Centilogits	0	10	20	30	40	50	60	70	80	90
600	88.08	90.02	91.68	93.09	94.27	95.26	96.08	96.77	97.34	97.81
700	98.20	98.52	98.79	99.00	99.18	99.33	99.45	99.55	99.63	99.70
800	99.75	99.80	99.83	99.86	99.89	99.91	99.93	99.94	99.95	99.96
900	99.97	99.97	99.98	99.98	99.98	99.99	99.99	99.99	99.99	99.99

LA, suggesting that these measures are closely related with each other, but the OI setting for 527 centilogits is appreciably different. The discrepancy becomes even more marked if the scores are taken for a lower loss setting in Figure 3.34 but the results for NME, WA and LA still retain about their same relationships.

For information, the two figures also include mean values of speech voltage output from the telephone sets and the sending end of the reference system. Speech voltage increases with increased degradation even in the case of the RDC where the signal reaching the listener is in no way improved by any increase in level from the talker because a constant volume amplifier was included.

Two conclusions of considerable importance were drawn from the results of this series of tests. First, that NME, WA and LA scores are definitely related to each other but not to OI; it follows that only one of the former need be used and can be chosen as convenient (sentences are probably easiest to administer). Second, OI is a much more discriminating measure for telephone type degradations, perhaps because features such as faintness are taken into consideration as well as the extent by which the speech is masked by noise or destroyed by distortion.

More recently (the tests just described were conducted in the early 1950s) tests intended for assembling planning data have concentrated on use of the subjective rating scales and not the more objective measures, like articulation. Furthermore, scales like Opinion scale No. 1, which include rather wordy descriptions of the categories, seem less useful than those in which each category is defined by a single word. The scale most commonly used now in the United Kingdom is that included, as item (5)*, in Opinion form No. 12 which appears in Section 3.6.2 together with the response to the question concerning 'any difficulty', shown as item (6) in Opinion form No. 12. Naturally, other opinion scales may be required for specialised purposes, for example that based on conversational effort (Opinion scale No. 8A) mentioned in Section 4.1.5. The scoring of the five categories of these scales is now in terms of equal weights, namely 4, 3, 2, 1 and 0 respectively, and not as used in the early tests.

It is convenient, when analysing the opinions obtained from the subjects in a conversation test, to use mean opinion scores (MOS) obtained by taking averages of scores allocated as above for each test condition. More detailed analysis of variance enables confidence intervals to be estimated and the effects of the various factors included in the experimental design to be separated. For example it is usual to treat subjects as 'blocks' and to exclude the dispersion

* This scale, excellent–good–fair–poor–bad, is sometimes used without the remainder of Opinion form No. 12; such a scale is referred to as Opinion scale No. 9.

attributable to subjects from the experimental error. Examples of information presented as MOS are given in Sections 4.2.1.3 and 4.4.

The final results must then be converted to a form suitable for network transmission performance calculations (*see* Section 8.4). This form may be as the proportion of poor or bad opinions but it is now the practice in the United Kingdom to present final results as percentages of persons who would say, if asked, that they had experienced some difficulty in the conversation. That is in terms of the responses to item (6) of Opinion form No. 12.

A quite definite relationship has been found between the percentage of poor or bad responses and the mean opinion score and this has been tabulated using the following model.

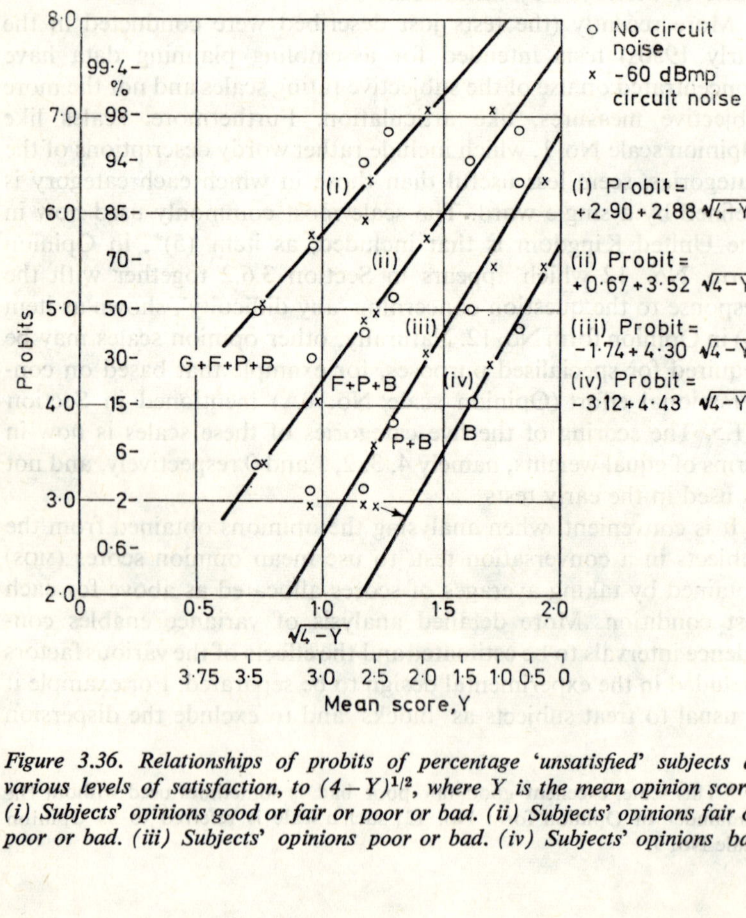

Figure 3.36. Relationships of probits of percentage 'unsatisfied' subjects at various levels of satisfaction, to $(4-Y)^{1/2}$, where Y is the mean opinion score. (i) Subjects' opinions good or fair or poor or bad. (ii) Subjects' opinions fair or poor or bad. (iii) Subjects' opinions poor or bad. (iv) Subjects' opinions bad

It has been established empirically that mean opinion score (MOS) as a function of overall loss can be represented over the practical ranges of loss by a parabola, $Y = \alpha+\beta(x-\gamma)^2$. For a high quality system, the parameter $\alpha = 4$ and so $(4-Y)^{1/2}$ will then be linearly related to x. It is also an empirically observable fact that the percentage $P+B$ as a function of x is a sigmoid definable as a Gaussian probability integral. The probit transformation (see Section 1.4) of such percentages will therefore be linearly related to x. From these two empirical features, it is possible to establish a linear relationship between the probit of % $P+B$ and $(4-Y)^{1/2}$ the parameters of which can be determined from the results of a

Table 3.17. PERCENTAGE 'POOR' OR 'BAD' OPINIONS AS A FUNCTION OF CONVERSATIONAL MEAN OPINION SCORE, Y_C

Y_C	0.00	0.01	0.02	0.03	0.04	0.05	0.06	0.07	0.08	0.09
1.0	76.4	76.0	75.6	75.2	74.8	74.4	74.0	73.6	73.2	72.7
1.1	72.3	71.9	71.5	71.0	70.6	70.2	69.7	69.3	68.8	68.3
1.2	67.9	67.4	67.0	66.5	66.0	65.6	65.1	64.6	64.1	63.6
1.3	63.1	62.6	62.1	61.6	61.1	60.6	60.1	59.5	59.0	58.5
1.4	58.0	57.5	56.9	56.4	55.9	55.4	54.8	54.3	53.8	53.2
1.5	52.7	52.2	51.6	51.0	50.5	50.0	49.4	48.9	48.3	47.8
1.6	47.2	46.6	46.1	45.6	45.0	44.4	43.9	43.4	42.8	42.2
1.7	41.7	41.2	40.6	40.0	39.5	39.0	38.4	37.8	37.3	36.8
1.8	36.2	35.7	35.1	34.6	34.1	33.6	33.0	32.5	32.0	31.4
1.9	30.9	30.4	29.9	29.4	28.9	28.4	27.8	27.3	26.8	26.3
2.0	25.8	25.3	24.8	24.4	23.9	23.4	22.9	22.4	22.0	21.5
2.1	21.0	20.6	20.2	19.7	19.3	18.9	18.5	18.1	17.6	17.2
2.2	16.8	16.4	16.1	15.7	15.3	15.0	14.6	14.2	13.9	13.5
2.3	13.1	12.8	12.4	12.1	11.8	11.4	11.1	10.8	10.5	10.1
2.4	9.9	9.6	9.4	9.1	8.8	8.6	8.3	8.0	7.7	7.5
2.5	7.2	7.0	6.8	6.5	6.3	6.1	5.9	5.7	5.4	5.2
2.6	5.0	4.8	4.7	4.5	4.4	4.2	4.0	3.9	3.7	3.6
2.7	3.4	3.3	3.2	3.0	2.9	2.8	2.7	2.6	2.4	2.3
2.8	2.2	2.11	2.02	1.94	1.85	1.76	1.67	1.58	1.50	1.41
2.9	1.32	1.24	1.17	1.10	1.04	0.99	0.94	0.89	0.84	0.79
3.0	0.75	0.71	0.67	0.63	0.59	0.55	0.52	0.49	0.46	0.43
3.1	0.40	0.37	0.35	0.32	0.30	0.28	0.26	0.24	0.22	0.20

Note: Percentages are given to three significant figures to assist interpolation; final values should be rounded to two significant figures.

suitable series of tests. Figure 3.36 shows the results obtained from a very extensive experiment and includes not only $P+B$ but also the three other dichotomies possible with a five category scale. The percentages $P+B$ have been tabulated in Table 3.17 according to this model.

More recent information has led to the conclusion that the percentage of 'difficulty' responses can be related, empirically, to MOS by a second order binomial model. This is based on the expansion of $(p+q)^u$ where $q = 1-p$ and $u = km-1$. m is the number of categories in the scale and k is the order of the binomial model.

Table 3.18. PERCENTAGE 'DIFFICULT' AS A FUNCTION OF CONVERSATIONAL MEAN OPINION SCORE, Y_C

Y_C	0.00	0.01	0.02	0.03	0.04	0.05	0.06	0.07	0.08	0.09
1.0	92.6	92.4	92.2	91.9	91.7	91.4	91.2	90.9	90.7	90.4
1.1	90.1	89.8	89.6	89.3	89.0	88.7	88.4	88.1	87.8	87.5
1.2	87.2	86.9	86.5	86.2	85.9	85.5	85.2	84.8	84.5	84.1
1.3	83.8	83.4	83.0	82.7	82.3	81.9	81.5	81.1	80.7	80.3
1.4	79.9	79.5	79.1	78.7	78.2	77.8	77.4	76.9	76.5	76.1
1.5	75.6	75.2	74.7	74.3	73.8	73.3	72.9	72.4	71.9	71.5
1.6	71.0	70.5	70.0	69.5	69.0	68.5	68.0	67.5	67.0	66.5
1.7	66.0	65.5	65.0	64.5	64.0	63.4	62.9	62.4	61.9	61.3
1.8	60.8	60.3	59.8	59.2	58.7	58.1	57.6	57.1	56.5	56.0
1.9	55.4	54.9	54.4	53.8	53.3	52.7	52.2	51.7	51.1	50.5
2.0	50.0	49.4	48.9	48.3	47.8	47.2	46.7	46.2	45.6	45.1
2.1	44.4	43.9	43.4	42.8	42.3	41.8	41.2	40.7	40.2	39.6
2.2	39.2	38.6	38.2	37.8	37.3	36.8	36.3	35.7	35.1	34.6
2.3	34.0	33.6	33.0	32.4	31.9	31.4	30.9	30.5	30.0	29.5
2.4	29.0	28.5	28.0	27.5	27.0	26.5	26.0	25.6	25.1	24.7
2.5	24.2	23.8	23.4	23.0	22.5	22.1	21.8	21.4	20.9	20.5
2.6	20.1	19.6	19.3	19.0	18.6	18.2	17.9	17.5	17.2	16.8
2.7	16.4	16.0	15.7	15.3	14.9	14.6	14.2	13.9	13.5	13.2
2.8	12.8	12.4	12.1	11.8	11.5	11.2	11.0	10.7	10.4	10.2
2.9	10.0	9.6	9.4	9.1	8.8	8.6	8.4	8.1	7.9	7.6
3.0	7.4	7.2	7.0	6.8	6.6	6.4	6.2	6.0	5.9	5.7
3.1	5.5	5.3	5.1	5.0	4.8	4.6	4.4	4.2	4.0	3.9
3.2	3.7	3.6	3.4	3.3	3.2	3.0	2.9	2.8	2.7	2.5
3.3	2.4	2.3	2.2	2.1	2.0	1.9	1.8	1.8	1.7	1.6
3.4	1.5	1.4	1.4	1.3	1.2	1.2	1.1	1.0	1.0	0.9
3.5	0.8	0.8	0.8	0.7	0.7	0.6	0.6	0.5	0.5	0.5

For $m = 5$ and $k = 2$:

$$p_E = p^9 + 9p^8q$$

$$p_G = 36p^7q^2 + 84p^6q^3 \qquad p_P = 84p^3q^6 + 36p^2q^7$$

$$p_F = 126p^5q^4 + 126p^4q^5 \qquad p_B = 9pq^8 + q^9$$

p_E, p_G, etc. are the proportions of votes for E, G, etc.

The mean opinion score is equal to:

$$4p_E + 3p_G + 2p_F + p_P$$

and the proportion of 'difficulty' responses is given by:

$$126p^4q^5 + 84p^3q^6 + 36p^2q^7 + 9pq^8 + q^9$$

Table 3.18 has been computed according to this model and has been found to provide a useful general relationship for deriving 'percentage difficult' from mean opinion score, which can usually be determined with more precision from actual experiments. This relationship has been used to present information in Section 4.2 in the form of percentage difficult.

3.7.3 Use

Assessment by examination of participants, especially by interrogation after conversation, now forms the fundamental basis on which telephone network planning data are assembled in the United Kingdom. Not all data can be determined directly by such a method and use must also be made in suitable circumstances of simpler and less laborious methods (such as listening opinion tests using sentences and scoring with a scale based on listening effort); these less general techniques are used to interpolate between data points determined by conversational methods, as explained in Section 4.2.1.3. Sections 4.2 and 4.4 include examples of planning data obtained in this way; in principle, the information on sidetone and echo (Section 4.3) ought also to be determined in the same way but such test results are rather scarce at present.

Customers can also be interrogated after completing conversations over telephone connections that can be identified, for example by call tickets. The results have shown good agreement with those from laboratory speech links set up to simulate the practical conditions and this has encouraged the use of conversational test results for predicting the satisfactoriness of future network plans when they are being examined (*see* Section 8.4).

3.8 METHODS BASED ON INSTRUMENTAL MEASUREMENTS

This section is concerned with methods for determining the transmission performance of a speech link without employing any subjects and depending only upon measurements made with instrumental devices. It is worth referring first to the four categories of assessment test method listed in Section 3.1.2, namely:

1. Examination and description of the speech link itself
2. Measurement of the information capacity of the paths of the speech link
3. Examination of conversations that take place over the speech link
4. Examination of the participants in such conversations.

Methods for conducting assessments according to each of these categories with the aid of subjects have been described and discussed in the preceding sections. In principle, instrumental determinations can also be performed according to the same categories.

Category 1 embraces all those methods by which an estimate of the transmission performance of the speech link is based on a physical description in terms of measured quantities such as loss, level of noise reaching each participant's ear, propagation time, distortions, etc. Methods in this category have been referred to in Sections 3.2.4.1 (calculation of loudness ratings), 3.3.5 (calculation of articulation scores) and 3.4.4 (calculation of listening opinion scores). The category can also be considered to include the use of fundamental information showing relationships between transmission performance scores obtained from past tests using subjects and physical measures of the quantities concerned. This category requires no further comment here.

Category 2 is concerned with measurements on an actual example of the speech link to be assessed. One such measurement would be one in which a sending part of the instrumental equipment emits test signals and a receiving part determines how closely they resemble the original when they reach the other end. A simpler form of device would limit itself to the examination of one principal path of a speech link at a time; it could then only be expected to assess the effects of 'listening' factors. A relevant criterion of transmission efficiency is the degree of correlation between signals at input and output of the transmission system after certain permissible transformations have been made to one or both these signals to maximise this cross-correlation. The arrangement is shown in principle in Figure 3.37. The principle can be reduced to a practical procedure by adopting a suitable simplified model of the speech transmission process in which suitably defined test signals are used to represent elements of speech. In such a case, the transformations referred to include such simple processes as adjustment of relative magnitudes of the signals at input and output of the system under test and matching them in phase.

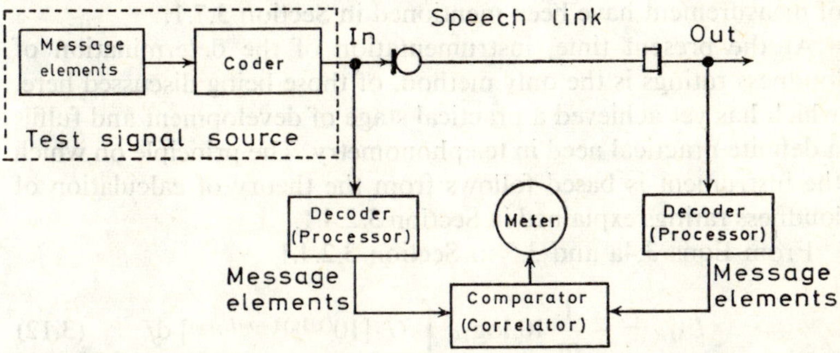

Figure 3.37. Speech communication efficiency meter based on comparison between input and output signals after identical processing of each

Testing devices based on such principles are described in the published literature (Refs* 13, 14, 15, 18). In these, the test signal is either a set of sinusoidal tones at a number of frequencies present simultaneously or a single tone swept in frequency across the band. Another method uses actual speech and performs a cross-correlation between the patterns of zero crossings at the two ends of the path under test. Other possibilities are the use of Gaussian signals of suitably chosen spectra, perhaps representing a selection of speech formants.

Methods based on Category 2 are certainly feasible but it seems likely that considerable difficulty would be experienced in achieving sufficient sensitivity for application to speech paths as good as those found in ordinary telephone connections. Reported applications seem to have been to rather poor quality speech paths.

Assessment methods can be imagined which are based on Category 3, although the prospects seem rather bleak. Certain changes in observable features of the conversational speech signals can be

* These references and others in this section appear in Section 3.8 of the Bibliography.

associated with increases in certain forms of degradation and the prospects of basing an assessment method on these are reviewed elsewhere (Ref. 21). Speech levels emitted from telephone microphones tend to increase as the speech link becomes more degraded and certain patterns in the alternation of talking and listening roles in conversation can be associated with increase of the propagation time.

Examination of the participants (Category 4) by physical measuring instruments is a line of enquiry that does not yet seem to have been followed in the telephone assessment field. Conceivable types of measurement have been mentioned in Section 3.7.1.

At the present time, instrumentation of the determination of loudness ratings is the only method, of those being discussed here, which has yet achieved a practical stage of development and fulfils a definite practical need in telephonometry. The principle on which the instrument is based follows from the theory of calculation of loudness ratings explained in Section 3.2.4.1.

From Eqns 3.4a and 3.5 in Section 3.2.4.1:

$$\overline{L_{ME}} = -\frac{1}{m} 10 \log_{10} \int_0^\infty G \cdot [10^{(1/10)(-mL_{ME})}] \, \mathrm{d}f \qquad (3.12)$$

$\overline{L_{ME}}$ can be expressed in terms of the ratio of sound pressures at the mouth reference point and the ear reference point, that is:

$$L_{ME} = 10 \log_{10} \frac{p_{\text{in}}^2}{p_{\text{out}}^2} \qquad (3.13)$$

Substituting Eqn (3.13) in Eqn (3.12);

$$\overline{L_{ME}} = -\frac{1}{m} 10 \log_{10} \int_0^\infty G \cdot \left[\frac{p_{\text{out}}^2}{p_{\text{in}}^2}\right]^m \mathrm{d}f \qquad (3.14)$$

Suppose that an artificial mouth, closely representative of the acoustical impedance characteristics of real mouths, is fed with a test signal consisting of a sinusoid varying in frequency smoothly and periodically with time at a definite rate so that $\mathrm{d}f/\mathrm{d}t$ can be arranged in a special manner as explained below. Let the r.m.s. sound pressure, p_{in}, from the artificial mouth remain constant as the frequency changes. Applying this as an input to a linear speech path, for which the value of $\overline{L_{ME}}$ is to be determined will produce in a suitable artificial ear, properly representative of the acoustical impedance characteristics of real ears, a sound pressure p_{out} which is varying with frequency in the same manner.

Because the test signal varies smoothly in frequency with time, the variable in Eqn 3.14 can be changed from frequency to time, thus:

$$\overline{L_{ME}} = -\frac{1}{m}\frac{1}{p_{\text{in}}^2}\,10\log_{10}\int_0^T (p_{\text{out}})^{2m}\,F(t)\,\mathrm{d}t \qquad (3.15)$$

where $F(t) = G.(\mathrm{d}f/\mathrm{d}t)$ and T is the period of variation in frequency of the test signal. p_{in} has been removed from the integral because it has been made independent of f and t. By choosing the rate of variation in frequency with time of the test signal $(\mathrm{d}f/\mathrm{d}t)$ to be proportional to $1/G$, the term $F(t)$ becomes constant and can also be removed from the integral, thus:

$$\overline{L_{ME}} = -\frac{1}{m}\frac{G\,(\mathrm{d}f/\mathrm{d}t)}{p_{\text{in}}^2}\,10\log_{10}\int_0^T (p_{\text{out}})^{2m}\,\mathrm{d}t \qquad (3.16)$$

The integral in Eqn 3.16 is simply the mean value, averaged over time, of the $2m$ power of the sound pressure in the artificial ear; instrumentation for this purpose is, in principle, quite simple although certain practical difficulties arise, common to all present uses of existing artificial ears (*see* Section 3.2.4.2).

The constant terms and the logarithm can readily be absorbed in the scaling of the device used to indicate the mean value of $(p_{\text{out}})^{2m}$.

In Figure 3.13, curve v, it is shown that a good approximation to G as a function of frequency is provided by the simple relationship:

$$10\log_{10} G = \text{constant} - 10\log_{10} f$$

or $G = \text{constant}\ (1/f)$

To render $F(t)$ constant, $G.(\mathrm{d}f/\mathrm{d}t)$ must be constant and so

$\mathrm{d}f/\mathrm{d}t = \text{constant}\ (1/f)$ which is satisfied by making $\log f$ proportional to t.

Instrumentation based on this principle has been designed by Braun (Ref. 10) and is commercially available*. Two limitations in its use must be noted, however. First, present available artificial mouths and artificial ears do not perfectly represent their real counterparts under the conditions present in subjective tests or in real use of telephone sets. Secondly, the purpose of this equipment is strictly limited to reproducing loudness ratings of speech paths and it is not intended to provide a complete measure of the communication efficiency of a speech link or complete telephone connection.

* The equipment is often referred to by its name in German, Objektiv Bezugsdämpfungsmesser (OBDM for short). This may be translated as Objective Reference Equivalent Meter, abbreviated to OREM.

4. EFFECTS OF SPECIFIC FACTORS

4.1 INTRODUCTION

4.1.1 General

The previous chapter was devoted to descriptions of various assessment methods which may be employed to assemble the information needed in planning the transmission performance of telephone networks. The present chapter is concerned with presenting the assembled information in the forms needed when studying practical problems. The problems considered arise whenever substantial improvements are made in existing types of transmission equipment or new technologies render entirely new forms practicable. Under the circumstances envisaged, it is usually both desirable and possible to make some use of the existing body of fundamental information but it is also frequently necessary to undertake certain new assessment studies to reveal the effects of factors not previously examined or not studied in sufficient detail. The information given in the present chapter has been chosen to provide that which seems likely to remain of interest and to indicate the forms in which new information ought usefully to be presented.

For reasons explained in Chapter 7, the ultimate purpose in presentation of transmission performance data is to enable estimates to be made of the proportion of telephone calls, over connections within the telephone network, that will be judged satisfactory by the customers.

In Section 3.1.2, the factors liable to cause customers to experience difficulty in conversing over telephone connections were categorised according to the nature of the difficulty; three types of factor were distinguished, namely: 'listening', 'talking' and 'conversational'. These were referred to in Section 3.1.2 as Types A, B and C respectively and Sections 4.2, 4.3 and 4.4 below make use of the same distinctions. Another practical rather than functional distinction is also needed, namely according to whether or not the effects of a given factor are, in principle, controllable by the telephone operating organisation in the network planning process.

Such factors as loss, distortion and circuit noise are directly attributable to the functioning of particular items of telephone plant which are clearly the responsibility of the telephone organisation offering the service; such factors will be termed 'X factors' and it is mainly with these that the following sections of this chapter are concerned.

The telephone operating organisation cannot, of course, control the physical environment in which telephone conversations are conducted (e.g. level and character of room noise and other disturbances) although such factors (called here Y factors) may have profound effects on the satisfactoriness of telephone conversations. Similarly, there can scarcely be any control over the physiological and psychological characteristics (including what might be termed 'telephonic ability') of the telephone users when these are members of the general public; the non-physical characteristics outside the control of the telephone planner are termed here Z factors.

Telephone network planning cannot, however, ignore the effects of Y and Z factors but must treat them as statistical variables, largely independent in their dispersions of those of the X factors. In practice, Y factors are taken into account by assuming them to be of moderate intensity but to allow sufficient margin in planning so that when a customer encounters somewhat adverse Y factors conversation will still be possible, although the difficulty will be somewhat increased. Z factors must, because of the very wide dispersion of their effects (customers have a very wide range of vocal levels and of telephonic ability), be treated with some care; they are usually taken into account by associating all assessment information with suitable values of dispersion as well as presenting 'average' values applicable to a median person.

The various X factors that require study may have different impacts on the economics of the situation. For example, it has usually been possible in the past (and may still sometimes remain so in the future) to save cost by permitting larger losses, higher levels of circuit noise and more attenuation/frequency distortion; when the saving is fairly smoothly related to the consequential increased amounts of these degradations, good engineering practice consists of striking a suitable balance between cost and performance. With certain other factors, however, such as crosstalk, intermittent noise or interruptions, their effects must be reduced to sufficiently small amounts as to be practically unnoticeable; if this were not done, incidence of these troubles would be liable to interpretation by the customers as faulty connections. As the standard of performance expected by telephone users becomes higher, factors may move from the first category to the second.

Although quantitative measures of performance can be ascribed only to complete connections, many of the practical problems arise in the design or specification of particular elements of telephone connections. For example, design of a new telephone set, choice of maximum lengths of subscribers' lines, maximum losses in junction or trunk circuits, design of a new kind of transmission system, all introduce problems that must be studied with proper attention to the transmission environment of the item concerned so that its influence on the characteristics of relevant complete telephone connections can be understood. These cases are best tackled by first choosing a few hypothetical examples of connections which are representative of relevant classes of real connections likely to be affected by the performance of the item under consideration. Such typical connections are termed 'hypothetical reference connections' (HRCs) and they must be so chosen that attention can with confidence be focused on their performance in the knowledge that satisfying the transmission performance of these will ensure that all other classes of connections that may be affected will also be satisfactory. The choice of HRCs is governed very much by the transmission, switching and routing plan in force for the network as well as by the characteristics of the item under study; further discussion of this matter is given in Chapter 7.

Figure 4.1 shows a small selection of HRCs which are frequently of use. Figure 4.1a shows a maximally adverse (from considerations of routing in accordance with the present CCITT recommendations) international connection. Figure 4.1b shows a national connection routed over the maximum number of circuits normally permitted in the telephone network of the United Kingdom. Figure 4.1c shows a local junction connection between customer stations connected to exchanges in the same area; transmission problems may arise here due to insufficient overall loss or excessive sidetone, particularly when the junction loss is very small. For any given study, each transmission element of the HRC will need to have associated with it appropriate particulars such as the loss, the variation of loss, the circuit noise generated by it, any distortions, etc.. (See Section 7.3 for further information on HRCs.)

Although HRCs provide the focus for study and it is the proportion of unsatisfactory calls associated with each that is required,

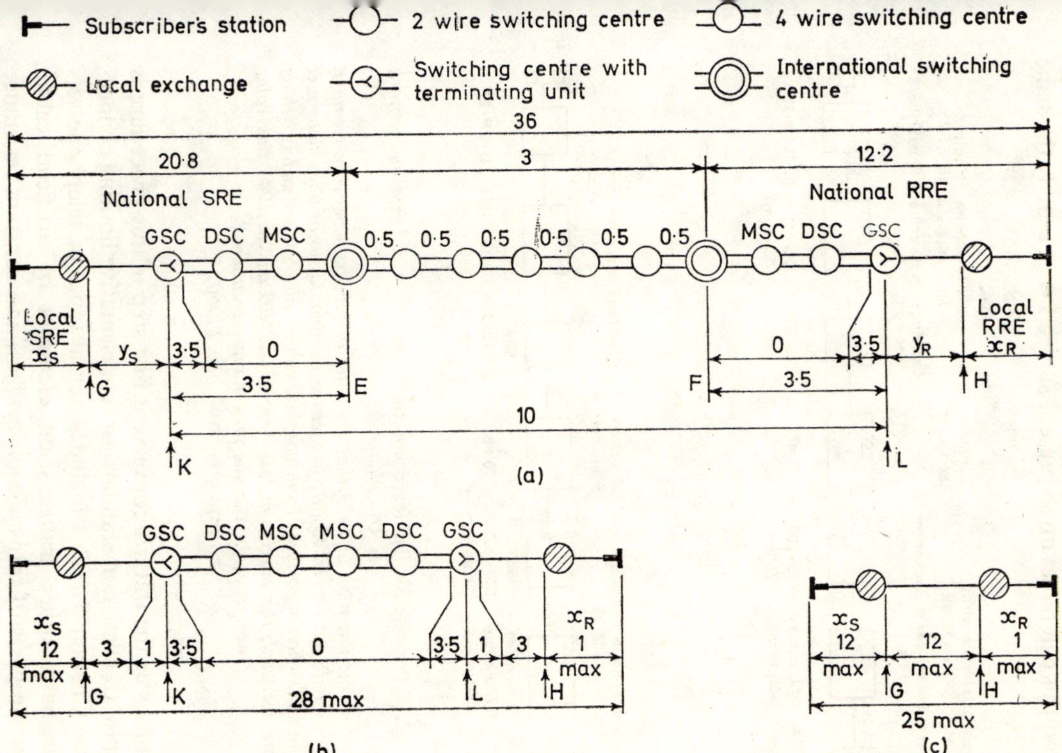

Figure 4.1. Hypothetical reference connections. (a) Long international connection nominal ORE *36 dB. (b) National trunk connection over transit network, nominal* ORE *38 dB max. (c) Local junction connection.*

Notes: 1. Reference equivalents and losses above (a) are maximum values in accordance with CCITT *recommendations. Those below show a possible subdivision; 2. Reference equivalents and losses associated with (b) and (c) are in accordance with British practice* it is not possible to conduct actual assessment tests over exact reproductions of all the relevant HRCs. Analysis of the losses, levels of circuit noise, distortions, etc. of individual elements enables the essential features of an equivalent complete connection to be described and this can be set up to present subjects with exactly the same

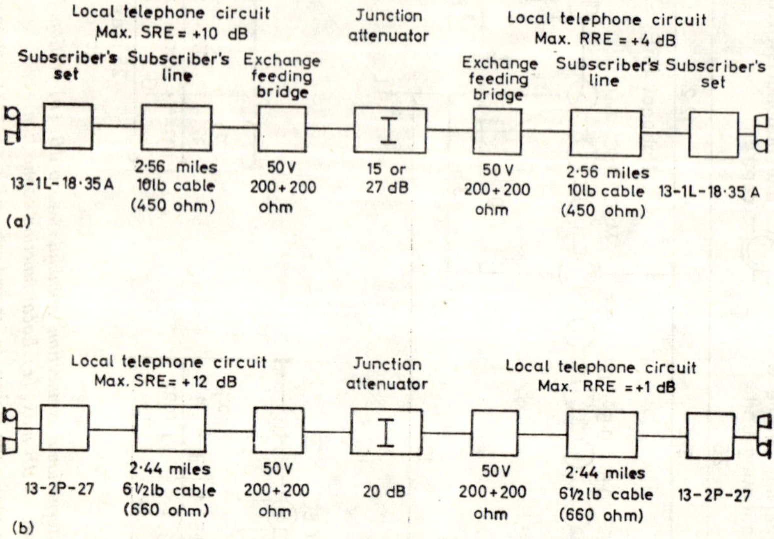

Figure 4.2. Representative limiting connections used for UK telephone network (a) Old plan (1933–1960).

Note: It was the intention in 1933 that the total of all losses between terminal exchanges should not exceed 15 dB; in fact the nominal losses of lines alone met this value and, when departures from nominal values, mismatches and switching losses are included, the maximum loss becomes 27 dB and in 1947 this value was recognised as the proper representation.

(b) New transit network plan (1960)

impressions as the HRC. A convenient first step in this procedure is to adopt a transmission standard or representative limiting connection (*see* Chapter 7). This includes actual types of telephone sets and simulates other transmission elements by artificial cable, appropriate parts of transmission bridges, and constant resistance (usually 600 Ω) attenuators to represent losses of amplified circuits.

Figure 4.2 shows some examples of the representative limiting connections (RLCs) used by the United Kingdom Post Office. These represent the worst (i.e. limiting) telephone connections permitted according to the network transmission plan.* An RLC contains a pair of identical local telephone circuits (LTCs) each representing the worst transmission permitted in the local telephone network. These limiting local telephone circuits (LLTCs) are joined together through an attenuator of 600 Ω10° image impedance to represent the maximum value of the sum of all losses in junctions and trunks in a maximally adversely routed connection. This total must include mismatch losses and losses introduced by switching equipment and exchange wiring. Figures 4.3a, b and c give detailed particulars of limiting local telephone circuits used in the United Kingdom at various periods since 1947. Representative limiting connections, when associated with room noise at a suitable level at each end, provide a transmission performance datum with reference to which the performance of any other connection can be expressed. The RLC currently in force provides a basis for telephone network design and can be used for framing installation instructions as explained in Chapter 7; it can also be used as a convenient and realistic means of reproducing, under laboratory conditions, a range of connections of which the degree of satisfactoriness is well established because of practical experience of very similar assemblies of items in the actual telephone network.

Any complete connection similar to those shown in Figure 4.2, but set up to represent transmission conditions more favourable than limiting, will be termed a *Representative telephone connection* (RTC) when considered as an example in laboratory studies. It is frequently desirable to use, in such RTCs, approximately average subscribers' lines and to vary the loss of the attenuator over a wide range to represent losses occurring in practice. RLCs and RTCs always contain practical telephone sets of types that are well established in the real telephone network and are therefore of known satisfactoriness.

Various terms are used to distinguish speech links, real and hypothetical, used in transmission performance studies and these are briefly summarised below:

Speech link—the most general term explained in Section 3.1.1.

Reference system—a highly stable and readily specifiable speech link used for laboratory studies, see Section 3.1.3.

* Details are given in Chapters 7 and 8 of the evolution and performance of these and Transmission Standards from about 1905. The current RLC is shown in Figure 7.4 which appears in Section 7.2.

Hypothetical reference connection (HRC)—a hypothetical speech link the constitution of which is chosen to represent a definable class of real telephone connections, *see above* in present section.

Representative limiting connection (RLC)—a speech link, capable of being set up in the laboratory, the constitution of which represents the most adverse connections permitted in the telephone network, see above in present section. RLCs have, in the past, also been termed *transmission standards*', *see* Section 3.1.3, but this is now deprecated.

Representative telephone connection (RTC)—similar to a RLC but representing any less unfavourable class of real telephone connections. RTCs have, in the past, also been termed *standard speech links*, *see* Section 3.1.3, but, like *transmission standard*, this term is now deprecated. An RTC is a laboratory realisation of an HRC.

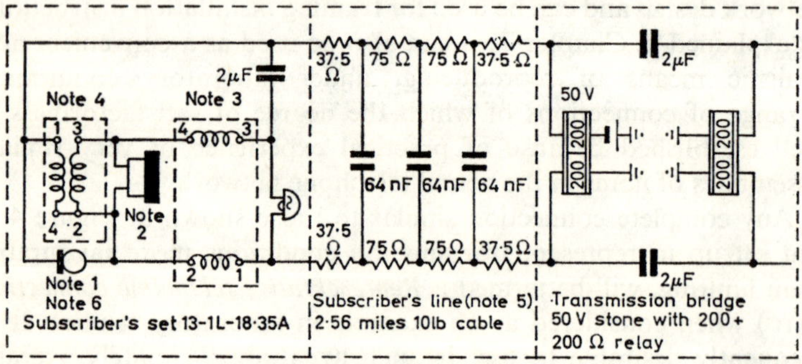

(a)

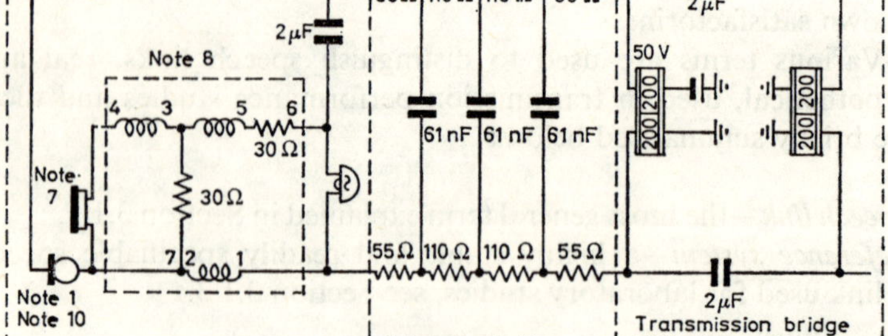

(b)

Intermediate reference system (IRS)—a stable and specifiable speech link used for laboratory studies (see reference system) but having characteristics, such as the form of the transducers and the frequency response of sending and receiving parts, similar to commercial telephone speech links. An intermediate reference system (IRS) thus falls logically between a reference system and a repre-

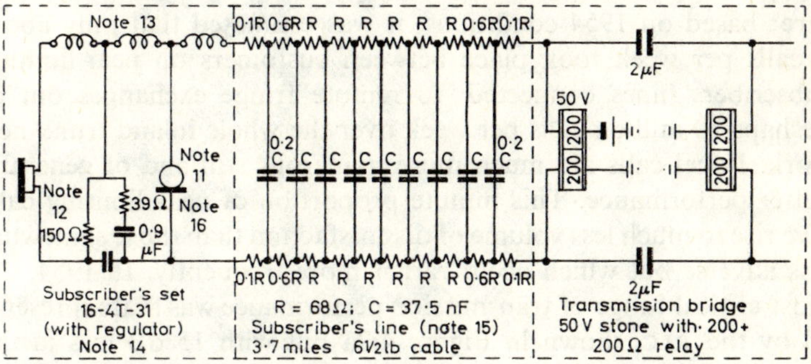

(c)

***Figure 4.3.** Limiting local telephone circuits chosen at various dates to represent the worst transmission performance of a local telephone circuit. (a) chosen in 1947; (b) chosen in 1957; (c) chosen in 1969.*

Notes: **1.** *Transmitter inset no. 13 fitted in telephone No. 164 handle, which is identical in shape to that illustrated in figure 5.2. (a).* **2.** *Receiver inset No. 1L fitted in telephone No. 164 handle.* **3.** *Induction coil No. 18 (non-antisidetone).* **4.** *Transformer No. 35A, which provides a certain amount of sidetone suppression.* **5.** *Subscriber's line loop resistance 450 Ω (2.56 miles), 10 lb cable (176 Ω/mile, 75 nF/mile, image attenuation at 1600 Hz 2.24 dB/mile).* **6.** *Microphone resistance 52 Ω, microphone current 42.5 mA, line current 55 mA.* **7.** *Receiver inset No. 2P fitted in telephone No. 164 handle.* **8.** *Induction coil No. 27 (antisidetone),* **9.** *Subscriber's line loop resistance 660 Ω (2.44 miles), 6½ lb cable (270 Ω/mile. 75 nF/mile, image attenuation at 1600 Hz 2.76 dB/mile).* **10.** *Microphone resistance 52 Ω, microphone and line current 44 mA.* **11.** *Transmitter inset No. 16 fitted in handset No. 3. This handle is similar in shape to those illustrated in Figures 5.2b and d but has a length intermediate between these examples.* **12.** *Receiver inset No. 4T fitted in handset No. 3.* **13.** *Induction coil No. 31.* **14.** *A regulator is fitted but details are not shown here. At the line current appropriate to this LLTC (32 mA) the effect of the regulator is to worsen transmission by no more than 0.5 dB for each direction (sending and receiving).* **15.** *Subscriber's line loop resistance 1000 Ω (3.7 miles), 6½ lb. cable (270 Ω/mile, 75 nF/mile, image attenuation at 1600 Hz, 2.76 dB/mile.* **16.** *Microphone resistance 105 Ω, microphone and line current 32 mA.* **17.** *All individual components are selected to lie near the middle of their respective ranges of manufacturing tolerance. For this reason it is not possible to set up a limiting local telephone circuit until all items have been in full scale production for sufficient time to enable proper selection of items to be made; this is especially true of microphone and earphone*

sentative telephone connection. The use of IRSS is referred to in Section 4.1.4; reference is also made to their use in Section 3.2.2 which includes a description of an IRS.

4.1.2 Composition of real telephone connections

Telephone calls over connections represented by the RLC are very rare; based on 1954 conditions, it was estimated that only about 4 calls per week took place between customers on near limiting subscribers' lines connected to remote fringe exchanges out of perhaps 10 million calls per week over the whole inland trunk network. Local calls are much more numerous still and of generally better performance. This minute proportion of near limiting calls give rise to much less volume of dissatisfaction than those somewhat less adverse but which occur rather more frequently. In 1954, the most critical range of transmission performance was that represented by the RLC shown in Figure 4.2a but with 15 dB loss in the attenuator instead of 27 dB. At the present time, the proportion of near limiting connections is probably about the same as in 1954 but the performance of calls over such connections is now much improved; the plant now conforms with an RLC which contains about 7 dB less overall loss (*see* Figure 4.2b).

Although it is necessary to ensure that reasonable satisfaction is obtainable under near limiting conditions, it is clearly not sufficient to confine attention solely to such situations. Knowledge is needed of the distribution of the composition of actual telephone connections set up in the telephone network and involving the transmission item under investigation. Statistical distributions of transmission losses and other features of particular circuits and transmission elements is discussed in Sections 8.2 and 8.3 and the plans according to which circuits are switched together to form different overall connections are mentioned in Chapter 7. It is necessary here, however, to outline the manner in which various numbers of junction and trunk circuits are assembled to form connections of different degrees of complexity so that the methods used to define the circuit losses and the levels of circuit noise associated with each circuit can be understood.

Figure 4.4 shows various arrangements of circuits connected together via 2 wire or 4 wire switching. The points *G* and *H* (with various suffixes) represent the switching points within the terminal local exchanges of each connection shown. Figure 4.4a shows the

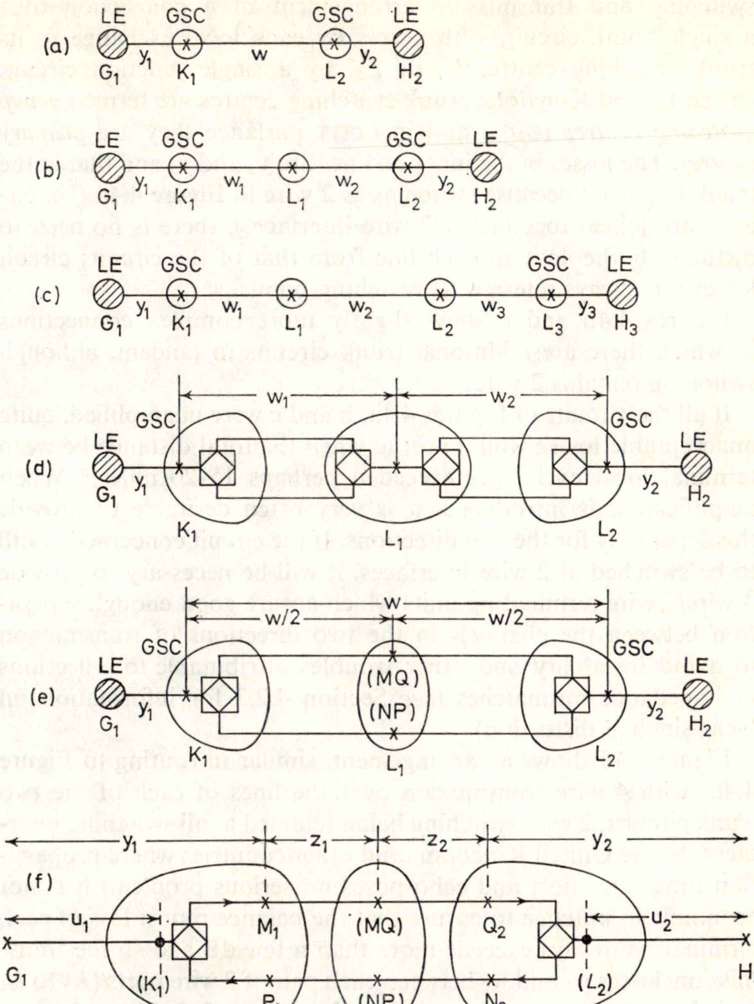

Figure 4.4. Arrangements of circuits with 2 wire and 4 wire transmission and switching. (a) Simple arrangement of two junctions and one trunk circuit; all transmission and switching 2 wire. (b) Arrangement with two trunk circuits in tandem; all transmission and switching 2 wire. (c) As b, but with three trunk circuits in tandem. (d) As b, but transmission of trunk circuits is 4 wire; all switching 2 wire. (e) As b, but transmission between GSCs is 4 wire and switching of trunk circuits at intermediate switching centres is 4 wire. (f) As b, but with both transmission and switching 4 wire at both ends of each trunk circuit. Note: Some advantage is gained by use of dedicated balances at the 2 wire/4 wire terminations, see text

switching and transmission arrangement of a connection over a single trunk circuit, with access by each local exchange to its trunk switching centre, K_1 or L_2, by a single junction circuit. In the United Kingdom, trunk switching centres are termed *group switching centres* (GSCs) and in CCITT parlance they are *primary centres*. The losses in the junction lines are y_1 and y_2 and that in the trunk line is w; because switching is 2 wire in Figure 4.4a (i.e. circuits are joined together at 2 wire interfaces), there is no need to distinguish the loss in each line from that of the circuit; circuit losses are always referred to switching points.

Figures 4.4b and c show slightly more complex connections in which there are additional trunk circuits in tandem, although switching remains 2 wire.

If all the circuits of Figures 4.4a, b and c were unamplified, quite unacceptable losses would accrue when the total distance between terminal local exchanges exceeded perhaps 15–20 miles.* When amplification is introduced, it is very often desirable to provide this separately for the two directions. If the circuit concerned is still to be switched at 2 wire interfaces, it will be necessary to provide 4 wire/2 wire terminating units which ensure good enough separation between the channels in the two directions of transmission to avoid instability and other troubles attributable to reflections at impedance mismatches (*see* Section 4.2.3 for information on 'near singing' distortion).

Figure 4.4d shows an arrangement, similar in routing to Figure 4.4b, with 4 wire transmission over the lines of each of the two trunk circuits, 2 wire switching being retained at all switching interfaces. In the United Kingdom (and other countries where propagation times are short and echo poses no serious problem) it is not economic in practice to ensure that the balance return loss of each termination always exceeds more than a few dB and so the transmission losses w_1 and w_2 between each pair of 2 wire ports (K_1 to L_1 and L_1 to L_2) cannot be made less than about 2–3 dB; with great care in matching all impedances, 0.5 dB can be achieved. Replacement of the 2 wire switching arrangement at L_1 by a 4 wire arrangement, as shown in Figure 4.4e, would reduce the loss between K_1 and L_2 an amount equal to w_1 or w_2.†

* Cable is assumed; overhead lines would extend the distance.

† In practice, the total loss between terminal 2 wire points of a connection containing several 4 wire switched circuits in tandem cannot quite be reduced to that of a single circuit; some additional stability margin must be allowed to cater for fluctuations in gain of amplifying and other active equipment and for

Figure 4.4f shows an arrangement, also similar to Figure 4.4b so far as routing is concerned, in which not only is the interface between the two trunk circuits made with 4 wire switching but also the trunk circuits are switched 4 wire at the group switching centres M_1/P_1 and Q_2/N_2. In this figure $(MQ)/(NP)$ represents the switching interface between the two trunk circuits and the points G_1 and H_2 represent the remote terminals of the junction circuits. In Figure 4.4f, the junctions are each terminated at the respective group switching centres with a 2 wire/4 wire terminating unit to provide the 4 wire interface for switching at M_1/P_1 and N_2/Q_2. The losses of the junction circuits (measured between switching points, at which traffic access is obtainable) are y_1 and y_2 whereas the lines alone provide only u_1 and u_2. z_1 and z_2 are the losses of the 4 wire trunk circuits. All losses are nominally the same in the two directions of transmission.

In the case illustrated in Figure 4.4f, the 2 wire/4 wire terminating units, including their balance impedances, are permanently associated with particular lines and so the balances can be individually matched to that of the line. The requirements for stability are that the return loss measured from point P_1 to point M_1 must be at least w dB even when complete reflection (short circuit or open circuit) takes place at G_1. Taking the requirement to be a minimum return loss between P_1 and M_1 of 7 dB and making reasonable assumptions concerning the impedances and mismatches likely in practice, with $u_1 = 6$ dB, y_1 can be made not less than 6 dB. As compared with the arrangement shown in Figure 4.4e, this represents an improvement in overall loss of $3\frac{1}{2}$ dB for each end so treated.

4.1.3 Definitions of relative transmission levels

Every active transmission element in a telephone connection has a limited dynamic range and, to reduce the effects of circuit noise generated therein, it is desirable to ensure that the transmitted signal should load the equipment as fully as possible without exceeding the load capacity. To ensure that all such items operate under the best loading conditions, it is the practice to define a relative level at the input to each active element (e.g. amplifier, modulator,

tolerances in lining up the circuits. The presence of attenuation/frequency distortion will also reduce the stability margin because the midband gain may be greater than that at the reference frequency at which the circuits are lined up.

etc.). At a point of 0 dB relative level (0 dBr), it is common practice to arrange matters so that the long term mean power per telephone channel is 32 μW; by convention it is assumed that 22 μW can be devoted to telephone speech, the remainder being allowed for associated signalling tones, leakage of carrier tones, echoes, etc. On average, speech during telephone conversations activates each channel only about one quarter of the total time and so the long term mean power of speech while active can be permitted to be as high as 88 μW in each channel. Speech powers, measured while each particular participant in a conversation is active, are distributed approximately lognormally; that is, when speech levels are expressed in decibels (dBm) relative to a datum of, say, 1 mW, the values are distributed according to a Gaussian distribution function. Assuming a standard deviation of speech level of about 5 dB, to account for the dispersion between individual persons using a handset telephone and for some dispersion of the loss between the subscriber's station and the junction side of his local exchange, the mean value of the speech levels needs to measure −13.5 dBm for the long term mean power of speech while active to be 88 μW, and the overall long term mean power, averaged over inactive as well as active periods, to be 22 μW. To avoid overloading an active element transmitting such signals in a single speech channel, the load capacity, as defined by the maximum power level of a sinusoidal test signal, must be at least about +2 dBm at a point of 0 dB relative level. Naturally, if the active element is transmitting a signal composed of a combination of more than one telephony channel, other considerations will apply and the load capacity needed to respect an overall long term mean power of 22 μW per telephone channel must be determined accordingly.

Each trunk circuit shown in Figures 4.4a, b and c is considered to have an individual point of 0 dB relative level (0 dBr) located at its 2 wire input port. Thus in Figure 4.4b, considering transmission from left to right, the point K_1 is at 0 dB relative level so far as the circuit K_1 to L_1 is concerned and, for the same circuit, the point L_1 is at a relative level of $-w_1$ dB. The point L_1 is at 0 dB relative level for the circuit L_1 to L_2, L_2 is at 0 dB relative level for the circuit L_2 to L_3, although L_2 is at $-w_2$ so far as the circuit L_1 to L_2 is concerned, and so on. These conventions are to ensure that none of the circuits, K_1 to L_1, L_1 to L_2, L_2 to L_3, etc. is overloaded when a source of speech signals is connected to its input via the smallest amount of loss. If the connection shown in Figure 4.4c were not a switched one, but all trunk lines were permanently connected together, the source of speech signals at G_1 would always be separated from, say, L_1 by the loss w_1 and so L_1 could be treated as at $-w_1$ so far as the line L_1 to L_2 is concerned and L_2 could be treated as at $-w_1 - w_2$ so far as the line L_2 to L_3 is concerned.

It will be observed from the foregoing that the direct connection, at a switching point, of an output port of relative level -a to an input port of relative level $-$b corresponds to the introduction into the connection of a loss of a$-$b.

The same rules can readily be applied to the case shown in Figure 4.4d because the 4 wire trunk lines are interconnected at 2 wire interfaces.

When 4 wire switching is employed, as in the case shown in Figure 4.4e, it is necessary to consider the allocation of relative levels at the 4 wire switching interface; this problem is illustrated by Figure 4.5.

Figure 4.5a shows a connection through four trunk circuits with 4 wire switching at three interfaces. The minimum value of w is governed by the requirement for stability when the minimum number of trunk circuits are included; as already explained, this value could be as small as 2–3 dB. As the number of trunk circuits included in a connection is increased the minimum permissible value of loss between 2 wire points at opposite extremities becomes greater and, for five trunk circuits, the minimum loss between points K_1 and L_2 must be about 7 dB. These restrictions can be respected in various ways subject to the following two inequalities:

$$w \geqslant 3 \text{ dB (say)} \quad \text{and} \quad w + z_1 + z_2 + z_3 + \ldots \geqslant 7 \text{ dB (say)}$$

One solution, used in the United Kingdom, is to make $w = 7$ dB and all the zs equal to 0 dB. Another solution, used in the United States of America, is to make $w = 4$ dB and to allocate various positive values to the zs depending partly on the lengths of the circuits.

Whichever solution is adopted within a national network, the connection will remain symmetrical between points K_1 and L_2 as shown by the display of relative levels in Figure 4.5a. The distinction between the relative levels for *each circuit* and those for the *complete connection* is also made clear in the figure. Figures 4.5b and c show the effects of reducing the number of trunk circuits in the connection to two in which cases the relative levels for circuits become identical to those for the connection.

Some complications ensue when interconnection is made between two national networks planned on different values of w.

For such connections, virtual switching points are considered having nominal levels of −3.5 dB for the input port of any international circuit and −4 dB for their output ports. Connecting

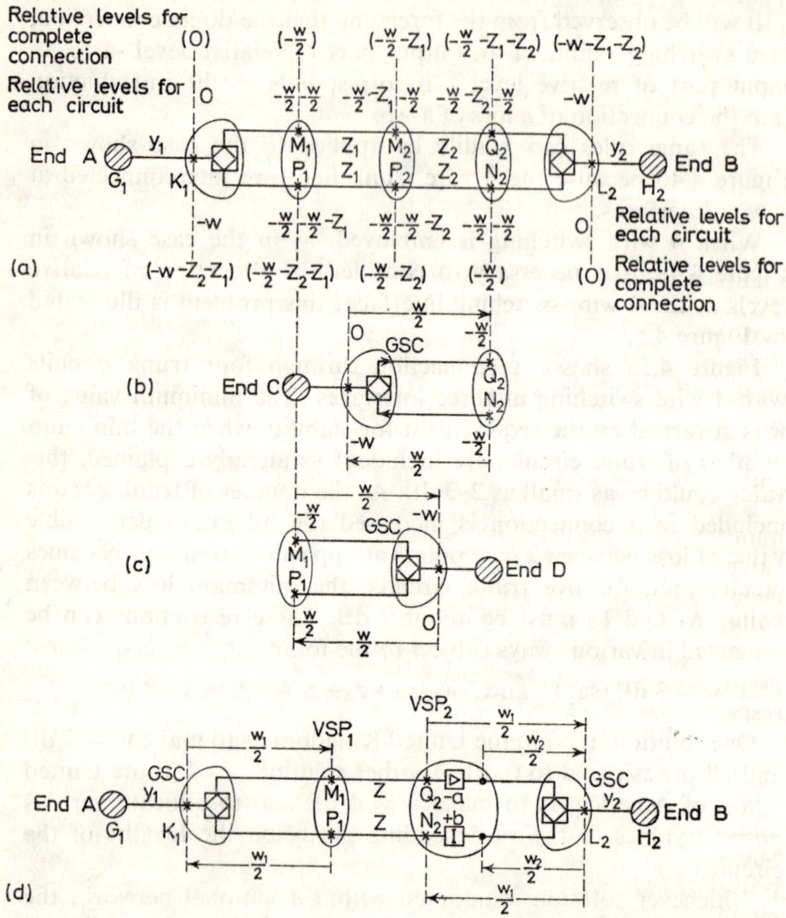

Figure 4.5. Definitions of relative levels in 4 wire circuits. (a) A connection through four trunk links. (b) An alternative circuit switched at Q_2/N_2. (c) An alternative circuit switched at M_1/P_1. (d) Interconnection of two national networks designed for different values of w

the −4 dB output port of the international circuit to a −3.5 dB input port of a national network or of another international circuit has the desired effect of introducing an extra loss of 0.5 dB for each international circuit in the connection. This arrangement is consistent with the use of $w = 7$ dB for national networks and so no problem arises where this is the practice. Figure 4.5d shows the interconnection arrangement when one national network uses $w_1 = 7$ dB but the other uses a value, w_2, that is different. End A can directly respect the levels at its virtual switching point (VSP_1) because these are identical to those used for switching inland circuits. End B, however, needs to insert gain in one direction and loss in the other as shown, so that $a = b = \frac{1}{2}(w_1 - w_2)$. This has no effect upon stability or echo (because these involve the round trip loss, which is unchanged) but it does involve a slight asymmetry, equal to $w_1 - w_2$; this will, however, in practice, have a negligible effect upon the conversational performance of the connection (*see* Section 4.4.1). The asymmetry will also affect the crosstalk performance slightly but this will be negligible in practice.

Other methods to ensure that an adequate margin is provided against instability, etc. are possible, without introducing excessive loss in some connections. For example, the method of 'pad switching' can be used; this and other methods are explained in the CCITT Handbook on National Automatic Networks to which reference should be made for more detailed information.

Allocation of relative levels to respective transmission elements enables the level of circuit noise generated within that element to be defined. The principle is as follows:

Suppose that the relative level at the input to the element is $-a$ dBr and that the transmission loss of the element is z dB; assume that the level of circuit noise generated by the element is quoted as $-b$ dBmOp. This is to be interpreted as indicating that the level of circuit noise, measured with a psophometer (CCITT Recommendation P.53), is $-b$ dBm when referred to a point in the circuit concerned of 0 dB relative level. It should be noted that the psophometer is calibrated in millivolts but these are commonly converted to dBm on the assumption that the measurement is made across a 600 $\Omega/0°$ termination. In the case quoted, the actual level of noise found at the terminated output of the transmission element in question would be $-b-z-a$ dBmp. The sound pressure level attributable to this noise when it reaches the ear of the listener must take account of the actual transmission losses between the output of the element considered and the listener's telephone, together with the receiving sensitivity of that telephone. For this purpose, the relative levels for the complete connection under consideration must be determined, as shown by the bracketed symbols displayed in Figure 4.5a.

4.1.4 Laboratory representations of telephone connections

The information given in Sections 4.1.2 and 4.1.3 will enable the test results obtained on laboratory reference systems and representative telephone connections to be related to those applicable to any given hypothetical reference connection (HRC) being studied, such as one of those illustrated in Figure 4.1. In the laboratory, the total circuit noise can usually be represented by noise from a single source injected at a convenient point in each direction of transmission of the test speech link; the level must, of course, be chosen to correspond, at the listener's ear, to that of the combined noise from a number of separate sources in the HRC. Similarly, all the attenuation/frequency distortion contributed individually by separate circuits can usually be combined and represented by distortion from a single simulating network for each direction of transmission. Care must be taken in laboratory arrangements to ensure that speech and circuit noise are respectively subjected to the proper attenuation/frequency distortions.

A laboratory form of speech link suitable for adapting to represent a very wide range of HRCs is shown in Figure 4.6a. In this form, practical telephone sets and practical transmission items of certain kinds can be included; this simplifies the problem of relating the results from tests in the laboratory to situations in real telephone connections. If unidirectional transmission devices are not already included in the two channels of the 4 wire part of the connection, blocking amplifiers must be inserted to ensure that transmission is confined to the proper directions only. Any circuit noise required must be injected at suitably chosen points so that the desired level reaches each listener and the spectrum of this noise has been appropriately affected by any distorting elements. Adaptations of the basic arrangement shown in Figure 4.6a will be described later when dealing more specifically with certain factors.

The arrangement just described suffers from the disadvantage that precise specification of the characteristics of the telephone sets used is rather difficult and usually it is necessary to employ a large stock of microphones and earphones so that proper averages of their performances are maintained. This is not a very serious problem when the results are to be applied for the same type of telephone set as was actually used in the laboratory tests but, when more fundamental information is to be assembled for general

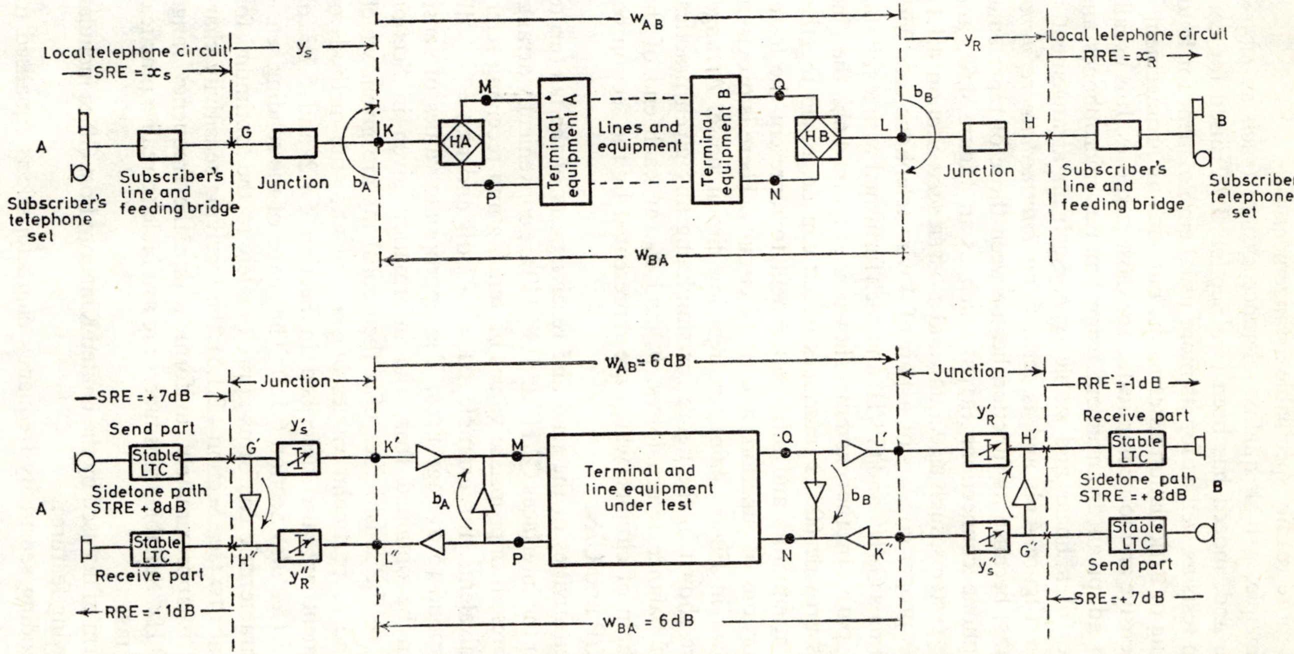

Figure 4.6. Telephone connections arranged for laboratory use. (a) A laboratory arrangement of a telephone connection containing practical telephone sets and other items. Send reference equivalent (SRE) $= x_S$ *: average can be taken as 7 dB for older and current type sets. Receive reference equivalent* (RRE) $= x_R$ *: average can be taken as −1 dB for older type sets and −4 dB for current types. (b) Intermediate reference system* (IRS) *used to represent a telephone connection in laboratory tests*

application to a variety of types of telephone set, it is preferable to use a more readily specifiable arrangement.

In principle, a high quality reference system (such as ARAEN) could be and, indeed has been (*see* Section 3.7.2) used for conversation tests as well as for its more usual employment for loudness rating or articulation assessments but such an arrangement is very expensive and too complicated for most purposes. It is usually better to adopt an arrangement using handset microphones and earphones of a linear and stable type. Such an arrangement is shown in Figure 4.6b and is termed an *intermediate reference system* (IRS) because its function lies between those of representative telephone connections (RTCS), such as in Figure 4.6a, and reference systems, such as ARAEN and NOSFER (*see* Section 4.1.1).

The IRS proper consists of a pair of 4 wire telephone sets with line terminals G′/H″ and G″/H′ and each furnished with a suitable sidetone path. In the example shown in Figure 4.6b, the line terminals provide 600 Ω interfaces which can be joined together through attenuators and other equipment to represent the losses and distortions of junctions and trunk circuits. There is provision (not shown in Figure 4.6b) for injecting circuit noise. Arrangements are shown in the figure for simulating any given practical values of balance return losses, b_A and b_B, at each end of the 4 wire chain of circuits, which are represented by the section between M/P and Q/N.

The sensitivities of the send and receive parts of the IRS (up to and from the interfaces G′/H″ and G″/H′) are conveniently arranged to represent appropriate values of sending and receiving reference equivalents, for example SRE = +7 dB and RRE = −1 dB to correspond to average LTCS. The appropriate values of sensitivity can be obtained from the information given in Section 3.2.2 where the sensitivities of a theoretical IRS are given, which correspond to particular values of SRE and RRE; the methods of measurement are also described in Section 5.1.2 and 5.1.3 of Chapter 5 (*see also* Section 6.1.2). The shape of the sending sensitivity characteristic conforms approximately to the 'optimum' for a practical LTC (*see* Section 4.2.1.5); the receiving sensitivity characteristic is approximately uniform within the frequency range 300–3400 Hz. Suitable transducers are available to realise practical forms of IRS.

The form of IRS used by the United Kingdom Post Office includes the following feature:

The sending sensitivity/frequency characteristic is equalised to provide uniform sensitivity over the range 300–3400 and the equalised microphone and amplifier combination is then associated with a network to provide the desired shape of sending characteristic. The frequency response of the network used is shown in Figure 4.7.

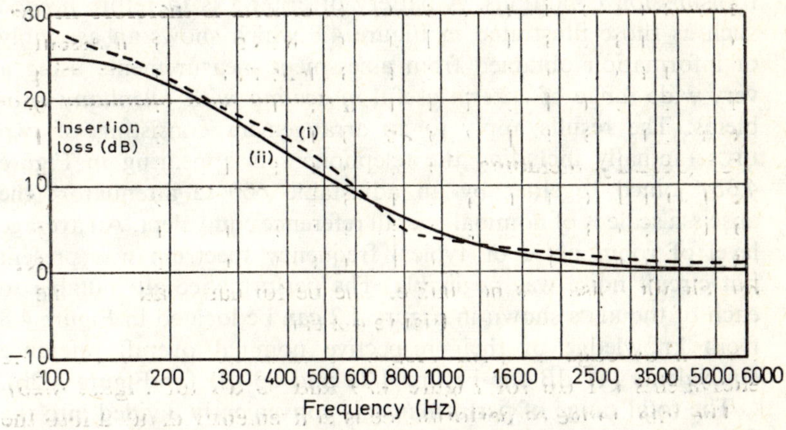

Figure 4.7. Insertion loss/frequency characteristic of equaliser for IRS. *(i) Objective characteristic from Figure 5.5a located to conform as well as possible with curve ii. (ii) Insertion loss of equaliser (2 sections) designed to approximate curve i in shape*

4.1.5 Choice of assessment criteria

The transmission effects of the various factors under consideration can be expressed in any of several ways using various assessment criteria. For some factors, it may be necessary to use special criteria but it is usually convenient to adopt a common set of criteria so that the relative magnitudes of the effects of different factors can be compared with each other.

The purpose for which the assessment is required has, of course, an important bearing on the choice of criteria; it will be assumed here that the results are to provide the quantitative basis for planning a telephone network. A variety of X factors (*see* Section 4.1.1) have to be catered for in detail; Y factors (especially room noise) will, in most cases, be assigned a single representative value; Z factors (speaking, listening and conversational behaviour) will be treated as statistical variables proper account of which can be taken by use of conversational assessment methods or by making suitable allowances.

The telephone connections provided by a public network range widely in performance and assessment criteria best suited for treating problems concerning near limiting transmission conditions are usually unsuitable for dealing with relatively favourable transmission conditions. A battery of criteria is therefore needed such as those illustrated in Figure 4.8 which shows an assembly of information obtained from assessment measurements using a very wide range of criteria useful in dealing with telephony problems. The results apply to an arrangement consisting of two LTCS (actually including the telephone sets appearing in Figure 4.3b) joined together by an adjustable 600 Ω attenuator; the abscissa scale is of nominal overall reference equivalent. An average level of room noise of typical frequency spectrum was present but circuit noise was negligible. The performance attributable to each of the RLCS shown in Figure 4.2 can be located in Figure 4.8 from knowledge of their respective nominal overall reference equivalents (41 dB for Figure 4.2a and 33 dB for Figure 4.2b).

The total range of performance is conveniently divided into the following classes:

(a) Where the speech signals heard over each speech path are nearly indistinguishable from those heard over a noiseless direct airpath. (Telephone connections are not intended to aim at such a high degree of excellence.)

(b) Where the effort needed to understand speech under one way, listening only conditions (i.e. without any return speech path) is no greater than would be needed over a noiseless direct airpath.

(bc) Where some effort may be needed to understand speech under one way conditions but no more effort is needed to converse than would be so over a noiseless direct airpath. (Introduction of the return speech path enables the talker to alleviate effort by the listener by adapting his talking behaviour to assist his partner.)

(c) Where some effort is needed to converse but the amount required is insufficient to cause any extra time to be needed to perform the task that forms the purpose of the conversation over the connection.

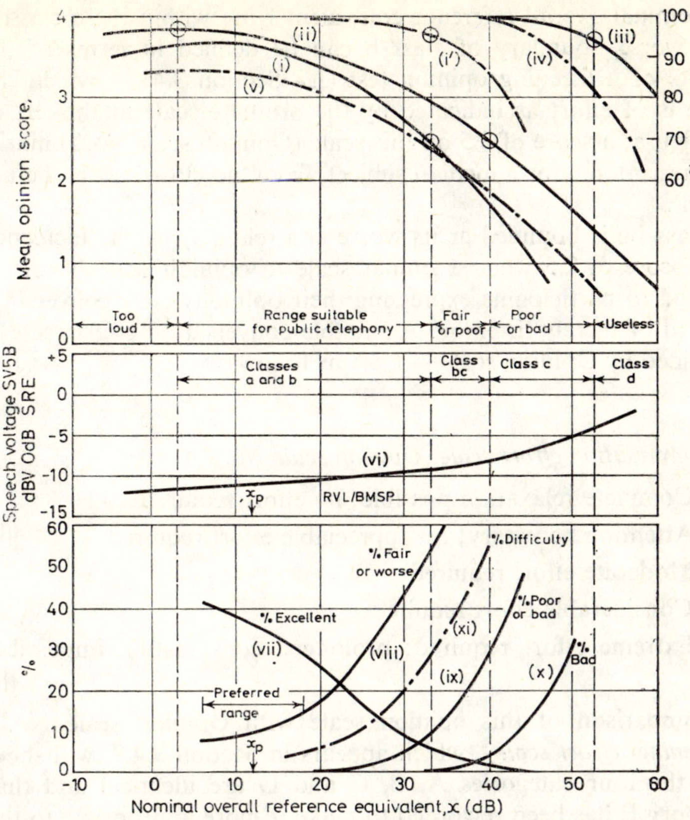

Figure 4.8. Assessments of telephone connections according to various criteria Note: Room noise 50 dBA; circuit noise negligible. (i) Listening opinion score (0S7) (i') Immediate appreciation % (ii) Conversation opinion score (0S8) (iii) Message rate efficiency % (iv) Word articulation (trained crew) % (v) Conversation opinion score (Excellent . . . Bad) (0S9) (vi) Speech voltage

(d) Where the transmission performance is so degraded that the difficulties encountered in conversing can be overcome only by devoting extra time to the task.

The boundaries between these classes (referred to elsewhere as intercategory boundaries) are marked in Figure 4.8 together with an indication of the onset of excessively loud speech being received from the other participant (*see* Section 4.2.1.5). Classes a and b represent the range where the connection provides substantially the conditions of ordinary direct conversation, the preferred range of nominal overall reference equivalent lying within these classes. The worse boundary of class b can be defined in terms of the results of a listening opinion test (*see* Section 3.4) in which the scale is of effort as indicated by the ordinate scale at the left of the figure; a score of 2.5 on this scale (Opinion scale No. 7) marks the transition,* for a median subject, from 'no effort' to 'moderate effort'.

Class bc is bounded at its worse end (class c) by the incidence of a score of 2.5 when a similar scale to opinion scale No. 7 is applied to participants expressing their opinions after conversing, instead of merely listening. A suitable conversation effort scale is provided by Opinion scale No. 8A as follows:

Conversation effort scale, Opinion scale No. 8A

A Complete relaxation possible; no effort required

B Attention necessary; no appreciable effort required

C Moderate effort required

D Considerable effort required

E Extreme effort required; prolonged conversation impossible

Comparison of this opinion scale with Opinion scale No. 7 *(listening effort scale)*, which appears in Section 3.4.2, will show that the four categories A, B, C and D are identical and that category E has been reworded to make it more appropriate to the conversational situation. Early work using a conversation effort scale employed exactly the same wordings as Opinion scale No. 7 and that version was entitled Opinion scale No. 8. Opinion scale No. 8A has also been referred to in publications as Opinion scale No. 11.

The boundary between classes c and d is determined by measuring the time required to perform a defined task over the connection and comparing this with the time required to perform a similar task using a very good speech link (say one in class a or b). The term *message rate efficiency* is given to the ratio of these two times and a percentage of 95 is taken to define the c/d boundary (*see* Section 3.6).

* Alternatively, this transition can be marked by a score of 95% in an 'immediate appreciation' test; this is indicated by curve *i'*.

The points along the scale of nominal ORE at which the above mentioned boundaries are located, as shown in Figure 4.8, applies, of course, only for the conditions stated. Any appreciable circuit noise present, or introduction of appreciable distortion will tend to move the boundaries to the left in the figure, this movement in most cases being greatest for those lying farthest to the right. Corresponding changes will take place if the room noise is different.

Most practical problems concerned with telephony over public networks relate to telephone connections falling within classes b and bc. Although the assessment methods mentioned are very suitable for defining the boundaries, they are not very convenient for expressing results for network planning purposes. To define a boundary precisely, it is desirable for the score to change as much as possible at that boundary for a given small increase in degradation; use of a common criterion for the whole range of practical interest in telephony demands a more gradual change in score as the degradation is increased. For telephone planning purposes, the most useful general assessment methods have proved to be those based on interrogation of participants, obtaining their opinions either according to the categories: excellent–good–fair–poor–bad or in terms of 'difficulty', where the participants are asked to reply yes or no to the question 'Did you, or the person who spoke to you, have any difficulty in talking or hearing over the connection?' (Opinion scale No. 9 and Opinion form No. 12, *see* Section 3.7.2).

Figure 4.8 includes results showing the percentages of the opinions 'excellent', 'fair, poor or bad', 'poor or bad' and 'bad' and the percentage who replied 'yes' to the question on 'difficulty'. It will be noticed that '% poor or bad' and '% difficult' are particularly convenient in indicating the approach to the worse boundary of class b, which is an important feature.

When degrading factors in addition to overall loss are present, the assessment scores shown in Figure 4.8 deteriorate and, to a first approximation, the relationships between the scores according to the various criteria remain constant.* Caution must, however, be exercised when 'unnatural' types of degradation are present. Loss, noise and certain simple varieties of attenuation/frequency distortion can be regarded as 'natural' because they are already familiar to everyone from experience in everyday conversation without a telephone. Severe nonlinear distortion, abnormally long

* But *see* Section 3.7.2.

propagation times and mutilation of signals by voice operated switching devices should be treated as 'unnatural' in this context. Under these circumstances, the various criteria ought to be treated as though they referred to quantities that are, to some extent, independent of each other.

4.2 LISTENING FACTORS

4.2.1 Loss, noise and attenuation/frequency distortion

4.2.1.1 General theoretical framework

Listening factors are those whose direct effect consists of modifying the speech signals after leaving the mouth of one participant and before reaching the ear of the other. In telephone connections, the most important members of this class are loss, circuit and room noise and attenuation/frequency distortion, which are linear effects and are treated in the present section. Nonlinear distortions are treated in Section 4.2.2 and group delay/frequency distortion (another linear effect) is considered in Section 4.2.3.*

Understanding of the influences upon the transmission of speech of the presence of loss, noise and attenuation/frequency distortion is assisted by the diagram shown in Figure 4.9. Curve i shows the frequency spectrum of speech measured at a mouth reference point close to the talker's lips (*see* Section 2.1.3.4). When this speech sound pressure arrives at the listener's ear through a telephone speech path of about optimum mouth to ear loss the frequency spectrum at the entrance to his ear canal will approximate to that shown in curve ii. Curve iv shows the minimum sound pressure levels for speech components to be audible in the absence of any noise and assuming normal hearing for young adults. Curve iii shows the speech sound pressure at the ear when the signals have traversed an approximately limiting commercial telephone speech path.

* Another practical distinction between the distortions treated in Sections 4.2.2 and 4.2.3 is that the former operate in the dimension of amplitude and the latter in that of time. This bears some relationship to the classification used in Section 6.5 to distinguish different types of speech transmission economy system.

Figure 4.9 is useful in illustrating the mouth to ear loss L_{ME}, of a speech path, which is the vertical interval between curve i on the one hand and curve ii or curve iii on the other. Furthermore the speech to threshold ratio, Z, referred to extensively in Sections 3.2, 3.3 and 3.4 can also be observed as the interval between curve ii or iii on the one hand and curve iv on the other.

The ordinate scale of Figure 4.9 uses a conventional sound intensity datum of 1 pW/m² per Hz which is (under certain theoretical conditions, see sections 2.1.3.2 and 2.2.3.1) equivalent to a datum sound pressure level of 20 μN/m² referred to a bandwith of 1 Hz.

In ordinary telephony, circuit noise is usually present and conversations normally take place in the presence of a certain amount of ambient noise. Noise can be taken into consideration by replacing

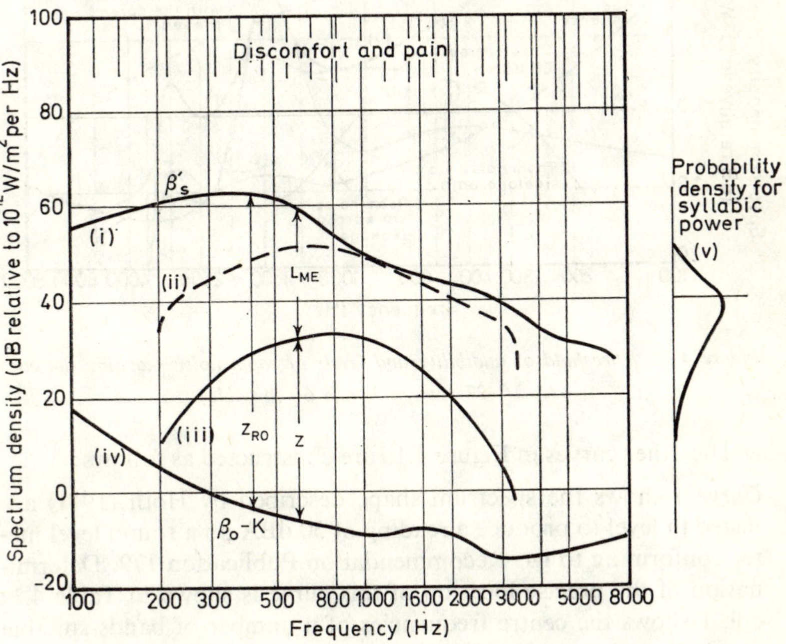

Figure 4.9. Sound pressure at the ear while listening to speech. (i) Speech at 25 mm from lips (Reference Vocal Level). (ii) Telephone speech at approximately optimum level. (iii) Telephone speech over approximately a limiting connection. (iv) Hearing threshold for continuous spectrum sounds (one ear, young persons, sound pressure measured under well fitted earphone)

the 'silent' threshold shown in curve iv of Figure 4.9 by the threshold masked by the total noise reaching the listener's ear; this can be constructed as shown for one example in Figure 4.10.

The threshold of audibility in silence shown in Figure 4.10 is a replica of curve iv in Figure 4.9 and has been derived from curve iii of Figure 2.12 and Table 2.20; the curve shows $\beta_0 - K$ and represents the threshold of audibility for continuous spectrum sounds heard via an earphone applied to one ear.*

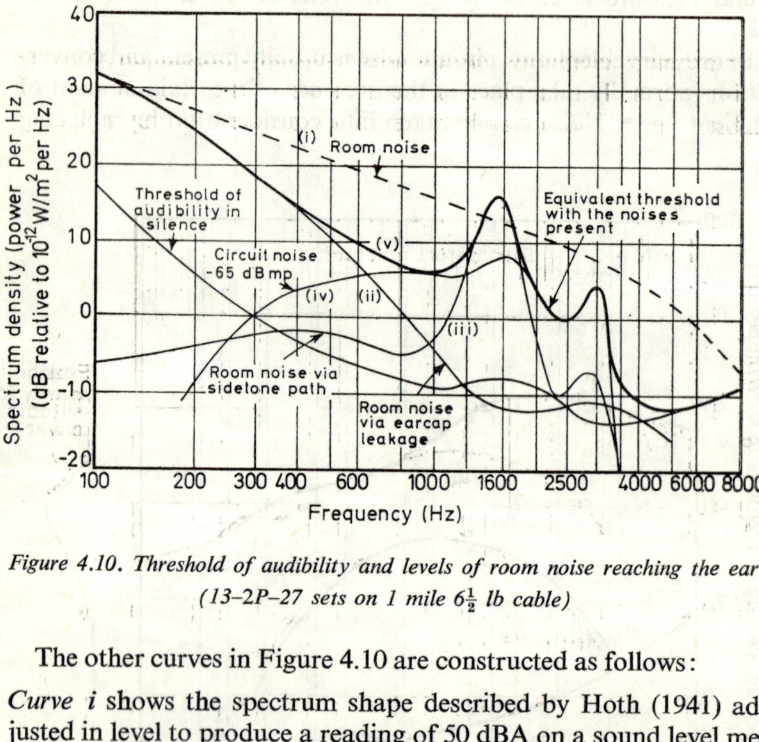

Figure 4.10. Threshold of audibility and levels of room noise reaching the ear. (13–2P–27 sets on 1 mile 6½ lb cable)

The other curves in Figure 4.10 are constructed as follows:

Curve i shows the spectrum shape described by Hoth (1941) adjusted in level to produce a reading of 50 dBA on a sound level meter conforming to IEC Recommendation Publication 179. Determination of the precise location of this curve is shown in Table 4.1: col. 1 shows the centre frequencies of a number of bands suitable for the calculation; col. 2 shows the spectrum shape relative to an arbitrary datum (actually approximately correct); col. 3 shows the relative weights (A weighting) at the frequencies chosen; col. 4

* The numerical values used for curves i and iv in Figure 4.9 appear in Section 4.5.2. *see* Table 4.21.

Table 4.1. CALCULATION OF ROOM NOISE LEVEL (dBA)

Frequency f Hz	*Spectrum* (*Approx.* 50 dBA) β'_N dB	*Weight* IEC *179** dB	*Wtd. level* dB	*10* $\log_{10} \Delta f$	*Level repres. by* Δf	*Wtd. power*	*Adj. to* 50 dBA
(1)	*(2)*	*(3)*	*(4)*	*(5)*	*(6)*	*(7)*	*(8)*
100	+32.0	−19.1	+12.9	+19	+31.9	1550	+32.4
200	+27.2	−10.9	+16.3	+20	+36.3	4260	+27.6
300	+24.3	−7.0	+17.3	+20	+37.3	5370	+24.7
400	+22.3	−4.8	+17.5	+20	+37.5	5620	+22.7
500	+20.7	−3.2	+17.5	+20	+37.5	5620	+21.1
600	+19.5	−2.2	+17.3	+21.8	+39.1	8120	+19.9
800	+17.4	−0.8	+16.6	+23.0	+39.6	9100	+17.8
1000	+15.8	0	+15.8	+23.5	+39.3	8500	+16.2
1200	+14.2	0.6	+14.8	+24.8	+39.6	9100	+14.6
1600	+12.5	1.0	+13.5	+25.7	+39.2	8320	+12.9
2000	+10.9	1.2	+12.1	+26.5	+38.6	7230	+11.3
2500	+9.2	1.3	+10.5	+27.0	+37.5	5620	+ 9.6
3000	+7.8	1.2	+ 9.0	+27.0	+36.0	3980	+ 8.2
3500	+6.3	1.1	+ 7.4	+27.0	+34.4	2750	+ 6.7
4000	+5.0	1.0	+ 6.0	+28.8	+34.8	3010	+ 5.4
5000	+2.2	0.5	+ 2.7	+30	+32.7	1860	+ 2.6
6000	−0.8	0.2	− 0.6	+31.8	+31.2	1320	−0.4
8000	−7.0	−1.1	− 8.1	+33	+24.9	310	−6.6
					Sum	916402	

$10 \log_{10}$ Sum = +49.6 dB

* Using A weighting of IEC Recommendation Publication 179

= col. 2+col. 3; col. 5 gives the allowance, in dB, for the width of the frequency bands chosen; col. 6 = col. 4+col. 5; col.7 is the antilog of 1/10 col. 6. The sum of the values in col. 7 gives the total weighted intensity and converting this to dB gives the value that would be read on the sound level meter when the sound applied to it conformed to col. 2. This result is 0.4 dB lower than the desired value of 50 dBA and so the desired spectrum density, shown in col. 8, is obtained by adding 0.4 dB to col. 2.

Curve ii shows the spectrum density of the room noise after it has leaked into the listener's telephone ear past the earcap. This curve is obtained by subtracting the acoustical attenuation of the earcap seal against the ear from curve i. Suitable values of attenuation for hard telephone earcaps are given in Figure 5.9.

Curve iii shows the spectrum density of the room noise after it has been picked up by the microphone of the listener's telephone set and transmitted through the sidetone path to the earphone on which he is listening. This curve is also obtained by subtraction of the air to air attenuation of the sidetone path to the listener's ear, from the spectrum density of the room noise shown in curve i. Determination of appropriate values for the air to air attenuation of the sidetone path is rather difficult, especially with carbon microphones, but a suitable method is explained in Section 6.1.4.

Curve iv shows the spectrum density of noise reaching the listener's ear from electrical sources in the telephone connection. Such noise

Table 4.2. CALCULATION OF SPECTRUM DENSITY OF CIRCUIT NOISE AT A GIVEN PSOPHOMETRIC LEVEL (dBmp)

Frequency *f* Hz	*Spectrum shape (arb. level)* dBV	*Psoph. weight* dB	*Wtd. level* dBV	*10* $\log_{10} \Delta f$	*Level,* dBV, *represented by freq. interval* Δf	*Wtd.* V^2 *per interval*	*Spectrum at −65* dBmp dBV
(1)	*(2)*	*(3)*	*(4)*	*(5)*	*(6)*	*(7)*	*(8)*
100	−12.5	−41.00	−53.5	+19	−34.5	—	−112.1
200	−7.7	−21.0	−28.7	+20	−8.7	0.1	−107.3
300	−2.5	−10.6	−13.1	+20	+ 6.9	4.9	−102.1
400	−1.1	− 6.3	− 7.4	+20	+12.6	18.2	−100.7
500	−0.5	− 3.6	− 4.1	+20	+15.9	38.9	−100.1
600	−0.2	− 2.0	− 2.2	+21.8	+19.6	91.2	− 99.8
800	0	0	0	+23.0	+23.0	200.0	− 99.6
1000	0	+ 1.0	+ 1.0	+23.5	+24.5	281.8	− 99.6
1250	0	− 0.20	− 0.20	+24.78	+24.58	287.1	− 99.6
1600	0	− 1.68	− 1.68	+25.75	+24.07	255.3	− 99.6
2000	0	− 3.00	− 3.00	+26.54	+23.54	225.9	− 99.6
2500	0	− 4.20	− 4.20	+27.00	+22.80	190.6	− 99.6
3000	0	− 5.60	− 5.60	+27.00	+21.40	138.0	− 99.6
3500	−12.5	− 8.5	−21.0	+27.0	+6.0	4.0	−112.1
4000	< −30	−15.0	< −45	+28.8	< −16	—	< −130
5000	< −30	−36.0	—	+30.0	—	—	—
6000	< −30	< −36.0	—	+31.8	—	—	—
8000	< −30	< −43.0	—	+33.0	—	—	—
					Sum	1736.0	

Total V^2 is equivalent to +32.4 dBVp
or +34.6 dBmp

If we want the spectrum density equivalent to −65 dBmp, we must subtract 99.6 dB from col. 2 and this is shown in col. 8.

is largely attributable to modulation and multiplexing in junction and trunk equipment (line and terminal) and commonly has (at its source) a uniform spectrum density. When such noise reaches the ear it will have been subjected to bandwith limitation and attenuation/ frequency distortion in circuits through which it has been transmitted and in the receive path of the local telephone circuit, including the telephone set earphone. The level of circuit noise attributable to a particular circuit in a connection is expressed in terms of the reading on a circuit noise meter (psophometer) the characteristics of which are recommended by the CCITT (Recommendation P. 53); Table 4.2 shows the calculation of the electrical spectrum density of noise corresponding to a given psophometric level, the principles of these calculations being analogous to those shown in Table 4.1 which was concerned with room noise. The acoustical spectrum density of the circuit noise when it reaches the ear can be

Table 4.3. CALCULATION OF ACOUSTICAL SPECTRUM DENSITY OF CIRCUIT NOISE AT A GIVEN PSOPHOMETRIC LEVEL

Frequency Hz	*Electrical spectrum at −65* dBmp dBV	*Receiving sensitivity*		*Acoustical spectrum −65* dBmp		*Acoustical spectrum −65* dBmp *referred to 0* dB RRE	
		13-2P-27	*16-2P-31*	*13-2P-27*	*16-4T-31*	*13-2P-27*	*16-4T-31*
(1)	*(2)*	*(3)*	*(4)*	*(5)*	*(6)*	*(7)*	*(8)*
100	−112.1	—	—	—	—	—	—
200	−107.3	+ 4.8	+ 1.3	− 8.5	−12.0	− 9.5	−17.8
300	−102.1	+ 8.8	+11.7	+ 0.7	+ 3.6	− 0.3	− 2.2
400	−100.7	+10.2	+15.3	+ 3.5	+ 8.6	+ 2.5	+ 2.8
500	−100.1	+10.8	+17.1	+ 4.7	+11.0	+ 3.7	+ 5.2
600	− 99.8	+11.4	+17.8	+ 5.6	+12.0	+ 4.6	+ 6.2
800	− 99.6	+11.8	+18.5	+ 6.2	+12.9	+ 5.2	+ 7.1
1000	− 99.6	+11.2	+17.7	+ 5.6	+12.1	+ 4.6	+ 6.3
1250	− 99.6	+11.5	+17.5	+ 5.9	+11.9	+ 4.9	+ 6.1
1600	− 99.6	+13.7	+17.7	+ 8.1	+12.1	+ 7.1	+ 6.3
2000	− 99.6	+ 9.2	+12.4	+ 3.6	+ 6.8	+ 2.6	+ 1.0
2500	− 99.6	+ 5.5	+ 5.4	− 0.1	− 0.2	− 1.1	− 6.0
3000	− 99.6	+10.0	+ 9.6	+ 4.4	+ 4.0	+ 3.4	− 1.8
3500	−112.1	− 4.1	+12.0	−22.2	− 6.1	−23.2	−11.9
4000	< −130	—	− 1.6	—	< −35	—	< −41

Notes (1) The receiving sensitivities apply, for both sets, when connected to 1 mile 6½ lb cable and a 50 V, 200+200 Ω feeding bridge.

(2) The values of RRE are: 13–2P–27 set, −1.0 dB
16–4T–31 set, −5.8 dB

calculated by allowing for the receiving sensitivity/frequency characteristic of the local telephone circuit. Table 4.3 shows this calculation for the 13-2P-27 and 16-4T-31 (with regulator) telephone sets on subscribers' lines of 1 mile $6\frac{1}{2}$ lb cable and 50 V, 200+200 Ω feeding bridges. Cols 3 and 4 show the receiving sensitivities of the local telephone circuits in dB relative to 1 N/m^2 per V; cols 5 and 6 show the spectrum at the ear when a psophometric level of −65 dBmp is applied to each local telephone circuit. Col. 5 is given by the sum of the values in cols 2 and 3+94 dB; the last term being needed to locate the spectrum relative to a datum of 20 $\mu N/m^2$. Col. 6 is obtained in the same way from cols 2 and 4. It will be noticed that the spectrum density at the ear is higher for the 16-4T-31 set than for the 13-2P-27 and this is due to its greater sensitivity; its RRE is equal to −5.8 dB as compared with −1.0 for the latter. For certain purposes, it is convenient to standardise the level of circuit noise so that equal numbers represent approximately subjectively equivalent amounts of noise; this can be done by referring the psophometric level to the input to a receive end adjusted to 0 dB RRE. This operation is illustrated in cols 7 and 8 which show the spectrum density for levels of circuit noise respectively of −65+RRE dBmp applied to the two local telephone circuits. These acoustical levels are, of course, the same as would be obtained if noise at a level of −65 dBmp were applied respectively to receiving ends having the same shape of sensitivity/frequency response as those considered but adjusted in sensitivity (equally at each frequency) to 0 dB RRE.

According to CCITT Recommendation P.53, levels of circuit noise on telephone channels are expressed in terms of readings of a psophometer, which is scaled in millivolts. If the measurement is made using the high impedance input condition connected across a 'through' or 'terminated' channel, the results are psophometric p.d.s and they can be converted to milliwatts by taking account of the nominal impedance of the channel on which the noise is being measured. In some cases psophometric e.m.f. is quoted, the values (in millivolts.) then being twice those of the psophometric p.d. In the United States of America, levels of circuit noise are measured with a circuit noise meter scaled in dB relative to a datum of 1 pW (10^{-12} W), the frequency weighting being approximately the same as that of the psophometer. The readings are then in units termed dBrnc, −60 dBmp being equivalent to 30 dBrnc. Reference* 34 gives a full explanation.

* *See* Section 4.2.1 of Bibliography.

Having assembled the spectrum densities of noise reaching the listener's ear from the various sources and through the various paths concerned, it remains to determine the amount by which the total noise at the ear will mask the silent threshold. The principle is explained in Section 2.2.3.1 but the calculation needed to determine the threshold in the presence of the noise consists simply of performing a power summation operation on the noise spectra and treating the silent threshold as though it represented the spectrum of basic, physiological, noise; the result appears in curve v of Figure 4.10.

Returning now to consider Figure 4.9 but imagining curve iv of that figure replaced by curve v of Figure 4.10, it can be seen that increasing the transmission loss of a telephone connection or allowing the total noise reaching the listener's ear to increase has the result that Z will be made smaller. The transmission performance achieved, as judged in a listening test is governed, to a considerable extent, by the values of Z, as a function of frequency; a theoretical model, based on the foregoing framework, is described in Section 3.4.4.

Before leaving the family of curves shown in Figure 4.9, it is important to note that the various curves in the figure apply to long term mean values of the quantities but speech signals are of rather wide dispersion; the inset curve on the right shows the probability density for speech averaged over 125 ms intervals which represent, approximately, the powers of individual syllables. It will be seen that speech at the preferred listening level results in the lowest levels of syllabic power being just clear of the threshold of audibility; as the speech level is reduced by extra loss in the speech path, an appreciable proportion of the syllables will fall below the threshold and will consequently become unintelligible.

Although the deleterious effects of circuit noise become apparent to a listener only after he has heard some speech from his respondent, it is also important to ensure that the level of circuit noise reaching telephone customers during the setting-up stages of a call is not excessive; customers are liable to abandon a call if they consider the noise high enough to interfere with speech at the level they are accustomed to in telephone conversations. When the room noise is 50 dBA, a circuit noise level of about −52 dBmp at the input to a 0 dB RRE receive end will be regarded as disturbing to 50% of subjects when they hear it in the absence of any speech (*see* Section 4.2.1.6).

The following sections present some results of assessment tests in a form useful for telephone network performance studies. The effects of varying the amount of room noise, circuit noise and attenuation/frequency distortion are illustrated by results of tests in which the overall loss was also varied over a very wide range.

4.2.1.2 Listening assessment scores

In order to permit comparison of the effects of factors like loss, noise and distortion, which are liable to cause difficulty in listening (type *A* factors, *see* Section 3.1.2) with those liable to cause difficulty in talking (type *B* factors) or specifically when conversing and the talking and listening roles are alternating rapidly (type *C* factors), it is necessary to express all results in terms of conversation assessment scores. Ideally, all assessment tests ought to be of the conversation type (*see* Sections 3.6 and3.7) but this would be impracticably laborious for the very wide range of test conditions to be covered when studying the effects of various combinations of loss, noise and attenuation/frequency distortion; listening only tests (*see* Section 3.4) must therefore be pressed into service and the results suitably converted to those that would have been obtained from a proper conversation test. Naturally some cross-checking is needed to confirm that the conversion method used is correct for the type of test conditions under consideration.

Figure 4.11 shows, in schematic form, an arrangement of laboratory equipment suitable for studying the effects on transmission performance of overall loss, circuit noise and distortion. Figure 4.11a shows the equipment set up for conversation tests between two locations, Station *A* and Station *B*. The order of transmission items in each path is important and that shown seems most convenient for the present purposes. The amplifiers, *D* and *E*, are inserted to confine the speech to its proper paths and may be of 0 dB gain or adjusted to offset the losses of the 4 wire/2 wire terminating units. Noise levels should be set up to produce defined values when measured at the points *H* (for the direction *A* to *B*) and *G* (for the direction *B* to *A*); knowledge of the receive reference equivalents of the local ends* at *B* and *A* enable the levels of circuit noise to be referred to the input to a receive end of 0 dB RRE. Speech voltages will usually need to be measured during each conversation and the

* The local ends shown in Figure 4.11 may be composed of practical telephone sets, subscribers' lines and feeding bridges (local telephone circuits) or special laboratory equipment to form the equivalent of a 2 wire local telephone circuit. *See also* Section 4.1.4.

points *L* and *M* are suitable; knowledge of the send reference equivalents of the local ends (up to *L* and *M* respectively for *A* and *B*) enables the speech volume results to be referred to the output of a send end of 0 dB SRE. The item *K*, inserted in each direction of trans-

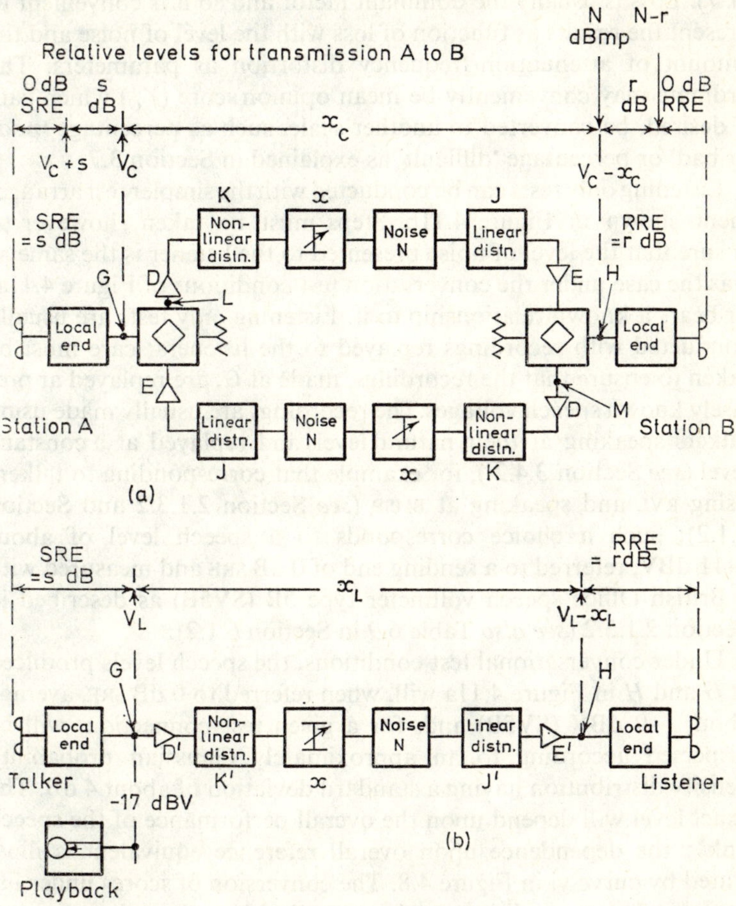

Figure 4.11. Laboratory circuits for assessment tests. (a) Conversation test. (b) Listening test

mission, is responsible for nonlinear distortion* and it may need to be coordinated with the level of speech at its input terminals; this is discussed in Section 4.2.2 below.

* An example of such equipment is the modulated noise reference unit (MNRU) described in Section 4.2.2.1.

Conversation tests are usually conducted in such a way that the subjects are all exposed to the full range of performance provided by the combinations of loss (setting of x), noise level (setting of N) and attenuation/frequency distortion (defined by the characteristics of J). Loss is usually the dominant factor and so it is convenient to present the results as function of loss with the level of noise and the amount of attenuation/frequency distortion as parameters. The ordinate may conveniently be mean opinion score (Y_C) which can, if desired, be converted to another scale, such as percentage 'poor or bad' or percentage 'difficult' as explained in Section 3.7.

Listening only tests can be conducted with the simpler test arrangement shown in Figure 4.11b. Steps must be taken, however to ensure that the level of noise presented to the listener is the same as was the case under the conversation test conditions of Figure 4.11a, or bears a known relationship to it. Listening only tests are usually conducted with recordings replayed to the listeners; care must be taken to ensure that the recordings, made at G, are replayed at precisely known speech voltages. The recordings are usually made using talkers speaking at their natural levels and replayed at a constant level (*see* Section 3.4.2), for example that corresponding to talkers using RVL and speaking at BMSP (*see* Section 2.1.3.2 and Section 5.1.2); such a choice corresponds to a speech level of about −11 dBV, referred to a sending end of 0 dB SRE and measured with a British Office speech voltmeter type 5B (SV5B) as described in Section 2.1.3.2 (*see also* Table 6.9 in Section 6.1.2).

Under conversational test conditions, the speech levels produced at G and H in Figure 4.11a will, when referred to 0 dB SRE, average about −9 dBV (SV5B) and, for a given test connection, will be dispersed according to an approximately Gaussian probability density distribution having a standard deviation of about 4 dB. The exact level will depend upon the overall performance of the speech link*; the dependence upon overall reference equivalent is illustrated by curve vi in Figure 4.8. The conversion of scores under listening only test conditions to those applicable under conversational test conditions must, of course, take into account the differences in speech level when this is referred to a point common to the speech paths under the two conditions; this is explained in Section 4.2.1.3 below.

* The speech level sent to line by participants in a conversation depends, amongst other factors, on the sidetone, the overall loss, the level of circuit and room noise and the distortions in the connection; further information is given in Section 4.2.1.3.

Figures 4.12 and 4.13 show some loss functions obtained under listening test conditions using Opinion scales Nos. 4A (loudness preference) and 7 (listening effort). The methods used are described in

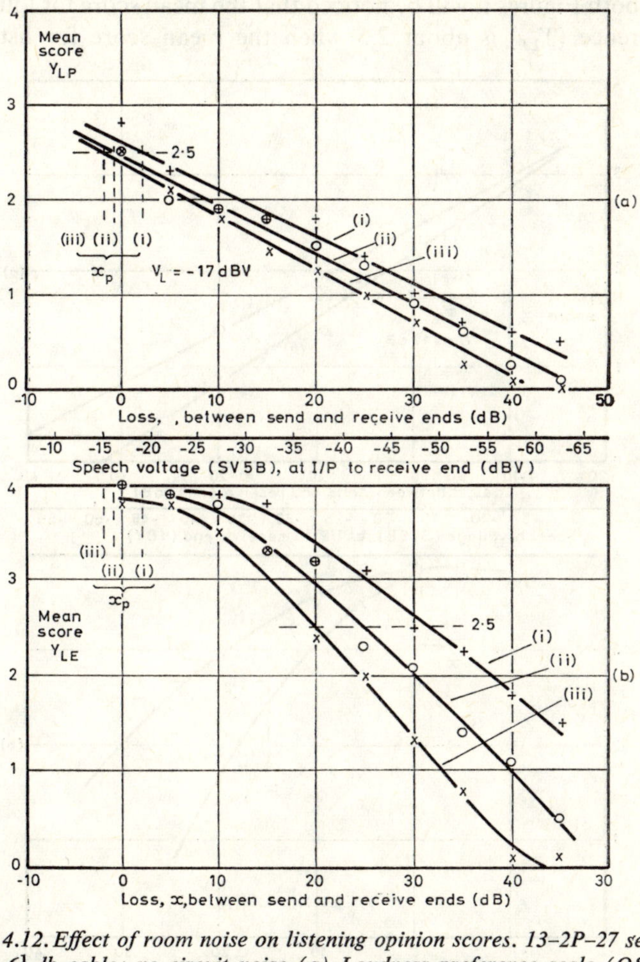

Figure 4.12. Effect of room noise on listening opinion scores. 13–2P–27 sets on 1 mile $6\frac{1}{2}$ lb cable; no circuit noise (a) Loudness preference scale (OS4A). (b) Listening effort scale (OS7). (i(No room noise. (ii) 50 dBA room noise. (iii). 60 dBA room noise

Section 3.4.2. In Figures 4.12 and 4.13, the speech level at point *G* in Figure 4.11b from the recordings was −17 dBV (SV5B); in these tests the SRE of the local ends were 7 dB.

The loss functions shown in Figures 4.12 and 4.13 can often be represented by equations of the following forms:

$$Y_{LP} = A + Bx \tag{4.1}$$

$$Y_{LE} = \alpha + \beta(x - \gamma)^2 \tag{4.2}$$

In both Figures it will be noticed that the mean score for loudness preference (Y_{LP}) is about 2.5 when the mean score for listening

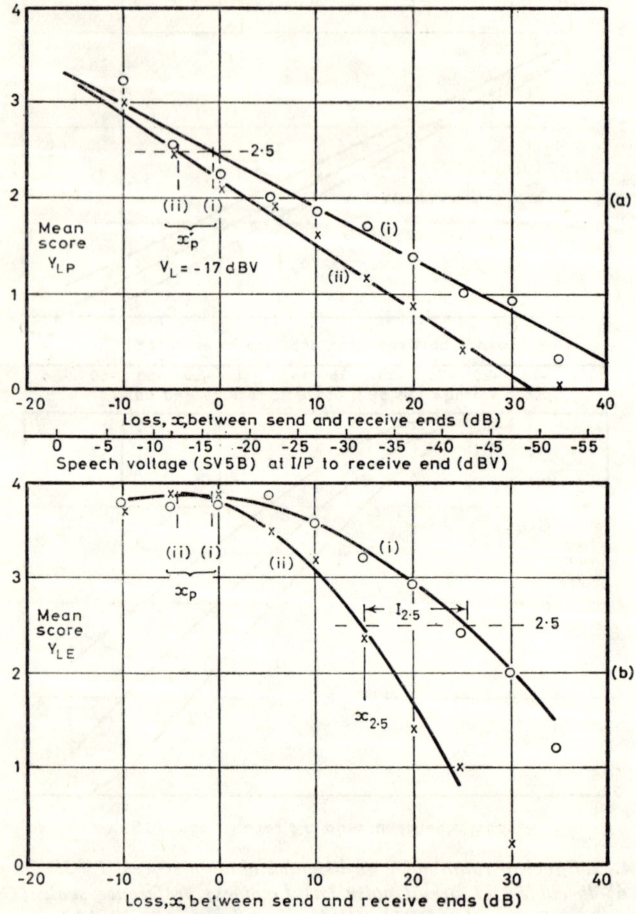

Figure 4.13. Effect of circuit noise on listening opinion scores. (13–2P–27 sets on 1 mile $6\frac{1}{2}$ lb cable; 50 dBA room noise) (a) Loudness preference scale (OS4A). (b) Listening effort scale (OS7). (i) No circuit noise. (ii) Circuit noise −44 dBmp referred to input to receive end of OdB RRE.

Note: The following parabolae have been fitted to the listening effort scale results:

(i) $Y_{LE} = 3.84 - 0.00147(x + 4.5)^2$

(ii) $Y_{LE} = 3.88 - 0.00336(x + 5.3)^2$

effort (Y_{LE}) is at its highest value*; the corresponding speech levels at the input to the receiving end are in the range −13 to −19 dBV (SV5B) which amount to the range −12 to −18 dBV (SV5B) when referred to 0 dB RRE (the RRE of the receiving end actually used was −1 dB).

It will also be noticed that listening effort scores (Y_{LE}) are more sensitive to increases in level of room noise and circuit noise than are loudness preference scores (Y_{LP}); this is particularly noticeable at the larger values of overall loss. The effects of room noise and circuit noise are much greater when the overall loss is substantial than when it is near the preferred value; in fact, at the value of overall loss that produces the preferred listening level, the effect on Y_{LE} of the circuit noise seems to be very slight (the highest level of circuit noise included in these tests was −44 dBmp referred to 0 dB RRE)†. This result enabled the complete set of test scores (Y_{LE}) to be represented by a family of parabolae (*see* Eqn 4.2) having a common maximum and differing only in their curvature (represented by the parameter β in Eqn 4.2). The complete test series included several values of circuit noise up to −44 dBmp referred to 0 dB RRE and analysis yielded a smooth curve of impairment, $I_{2.5}$, at the score of $Y_{LE} = 2.5$ (*see* Figure 4.13b). Figure 4.14 shows $I_{2.5}$ as a function of the level of circuit noise; Figure 4.14a shows an intermediate stage in this analysis in which a mathematically defined curve is fitted to the values of $V_0 - x_{2.5}$, the speech level at the listening local end when $Y_{LE} = 2.5$. The curve of impairment given in Figure 4.14b, together with the information concerning the location of a common maximum of the parabolae (namely estimated values of α and γ in Eqn 4.2), considerably assists conversion of these listening only test results to corresponding conversation test scores as shown in Section 4.2.1.3. (The final estimates yield $x = -1.3$ dB and $Y_{LE} = 3.86$ as the location of this common maximum.)

Introduction of appreciable attenuation/frequency distortion (which may be considered to embrace bandwith limitation) has a different effect on loss functions from that caused by noise. Consideration of Figure 4.9 and the ideas explained in Section 4.2.1.1 will

* This conclusion seems to apply generally including under loudspeaker listening conditions, provided that appreciable noise is present; when room noise and circuit noise are absent, however, other tests show the maximum value of Y_{LE} to correspond to a score of about 2.0 for Y_{LP}.

† This conclusion should be treated with some caution; recent experience suggests that the effects of circuit noise at preferred listening levels becomes appreciable at levels of circuit noise in excess of −55 dBmp when referred to 0 dB RRE (*see* Section 3.4.4).

suggest that the result of cutting off part of the band of frequencies transmitted between mouth and ear is likely to be at least as great under low loss conditions (*see* curve ii in Figure 4.9) as under the rather lossy conditions represented by curve iii in that Figure. In fact, as is shown in Figure 4.15 this is so, the reduction in Y_{LE} being about the same over a very wide range of overall loss when a

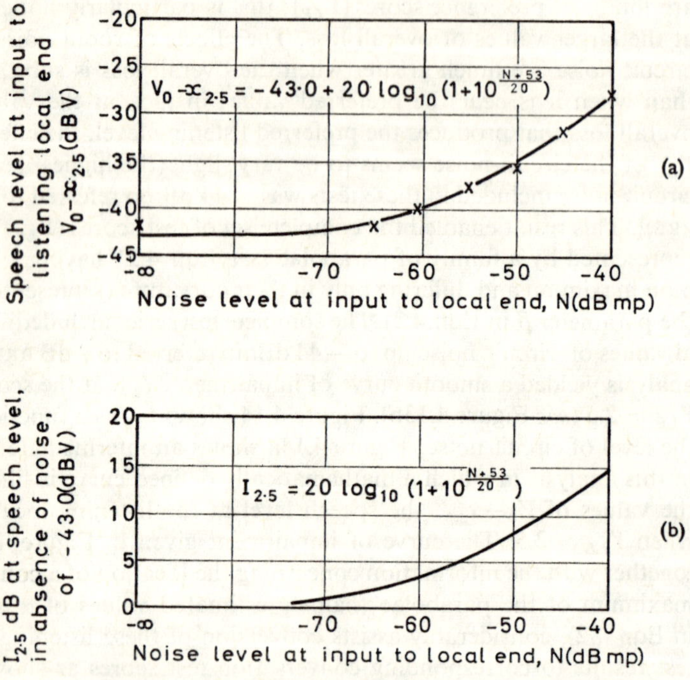

Figure 4.14. Impairment due to circuit noise as a function of white noise level at local end. (a) Speech level for score (OS No.7) of 2.5. (b) Impairment.

Notes: Receive reference equivalent for the local end used was −1.0 dB. Hoth spectrum room noise present at 50 dBA

2020 Hz low pass filter is inserted in a telephone speech path. The effect cannot (as was possible for circuit noise) be simply specified by the impairment at a defined value of Y_{LE}, the effect also on the maximum (both vertically and horizontally) must be considered.

The results for a range of attenuation/frequency distortion, as illustrated in Figure 4.15, however, show considerable changes in α and γ when the amount of distortion is changed, while β varies comparatively little. It has been found, by examination of results from several experiments, that α and γ can be related in a definite manner to $I_{2.5}$ which then becomes a useful measure of the subjective effect of the distortion. These relationships apply for limitation of bandwidth (high pass, low pass or both) and when attenuation/frequency distortion is confined mainly to parts of the frequency range below 800 Hz and above 2000 Hz, i.e. the region of minimum loss lies within this range and is fairly level. The following relationships have been found to apply sufficiently accurately for a substantial body of test results.

$$\Delta\alpha = \alpha_2 - \alpha_1 = 24\beta_1 I_{2.5} \quad (4.3)$$

(Note that β is negative)

$$\Delta\gamma = \gamma_2 - \gamma_1 = -0.7 I_{2.5}\,. \quad (4.4)$$

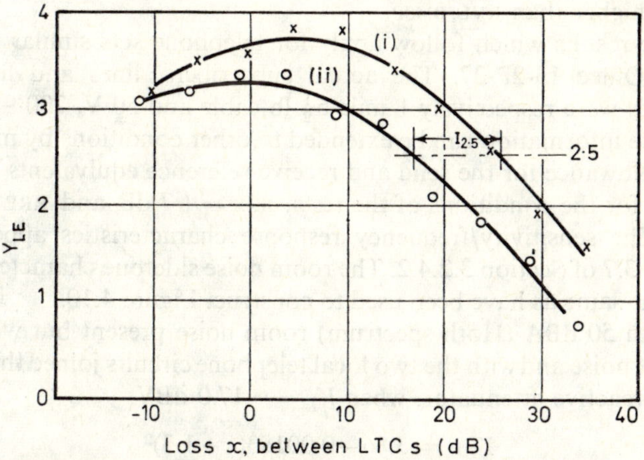

Figure 4.15. Effect of attenuation/frequency distortion on listening opinion score, Y_{LE} (effort scale). 13–2P–27 sets on 1 mile $6\frac{1}{2}$ lb cable; 50 dBA room noise. Notes: 1. Curve i is for the full frequency range and ii applies for LP cut off at 2020 Hz. 2. During these listening tests V_L was held at −17 dBV (SV5B). 3. The fitted curves are defined by (i) $Y_{LE} = 3.72 - 0.00247(x-3.8)^2$, (ii) $Y_{LE} = 3.25 - 0.00186(x+3.0)^2$

α_1, β_1 and γ_1 apply to the speech path free from any distortion; α_2, β_2 and γ_2 apply to the same speech path with the distortion but with the same midband loss (measured, for example, at 1600 Hz). The value of β_2 can be derived from $I_{2.5}$ by solving the appropriate equations formed by substituting the values for the other parameters in Eqn 4.2.

The results from a large body of test results have been summarised in terms of the foregoing expressions and form the present basis* for making network planning studies of the United Kingdom telephone network.

The tests were made with the arrangements shown in Figure 4.11; for listening tests, the speech level (SV5B) at point G (Figure 4.11b) was maintained (by means of recordings) at -17 dBV but during conversation tests the speech level at the corresponding point G (Figure 4.11a) depended upon the individual talker as well as the overall loss, circuit noise and total distortion of the connexion. The average speech voltage at G during conversations over fairly good connections was of the order -17 dBV (SV5B), which is slightly higher than that produced under controlled talking conditions using RVL and BMSP. For connections likely to occur commonly in practice, the speech level can be taken as averaging about 2 dB higher than RVL/BMSP.

The results which follow apply for telephone sets similar to the Post Office 13-2P-27. The actual subscribers' lines and feeding bridges were respectively 1 mile $6\frac{1}{2}$ lb cable and 50 V, 200+200 Ω but the information can be extended to other conditions by making due allowance for the send and receive reference equivalents which were, for the conditions of the tests, SRE $= +7$ dB and RRE $= -1$ dB. The sensitivity/frequency response characteristics appear in Table 3.7 of Section 3.2.4.2. The room noise sidetone characteristics are the same as have been used to construct Figure 4.10.

With 50 dBA (Hoth spectrum) room noise present but with no circuit noise and with the two local telephone circuits joined through a nonreactive attenuator, when $V_L = -17.0$ dBV,

$$Y_{LE} = 3.86 - 0.00183(x+1.3)^2 \qquad (4.5)$$

As the circuit noise is increased, the value of $I_{2\cdot5}$ (*see* Figure 4.13) varies in accordance with the expression:

$$I_{2.5} = 20 \log_{10}\left(1+10^{\frac{N+53}{20}}\right) \qquad (4.6)$$

where N is the psophometrically weighted level of circuit noise entering the receive end at H (Figure 4.11) and the units of N are dBmp. Some values of $I_{2.5}$ calculated from Eqn. 4.6 are shown in Table 4.4.

From the information in Table 4.4, it can be deduced that the

* The tests were conducted during the period 1957–64 and the then current 13-2P-27 telephone sets were used; this set is now obsolete and so the information may need some revision to allow for improved types of set.

Table 4.4. IMPAIRMENT DUE TO CIRCUIT NOISE AT $Y_{LE} = 2.5$

Level of circuit noise		$I_{2.5}$ dB	β ($\alpha = 3.86$ $\gamma = -1.3$ dB)
At H dBmp	*Referred to 0* dB RRE *receive end* dBmp		
(1)	*(2)*	*(3)*	*(4)*
None	None	0	−0.00183
−70	−69 dBmp	1.1	−0.00199
−65	−64	1.9	−0.00211
−60	−59	3.2	−0.00235
−55	−54	5.1	−0.00277
−50	−49	7.6	−0.00351
−45	−44	10.9	−0.00507

Notes 1. For the listening tests, $V_L = -17.0$ dBV (SV5B) and RRE $= -1.0$ dB
2. With circuit noise absent, $x_{2.5} = 26.0$ dB

preferred listening level, $V_L - \gamma$, is −15.7 dBV at the point *H* in Figure 4.11b which is equivalent to −14.7 dBV when referred to 0 dB RRE. The listening level that produces a listening opinion score of 2.5 is −43.0 dBV at point *H* or −42.0 dBV when referred to 0 dB RRE. The last column in the table gives the values of β corresponding to each level of circuit noise so that the parabolae relating Y_{LE} to x_L can be determined.

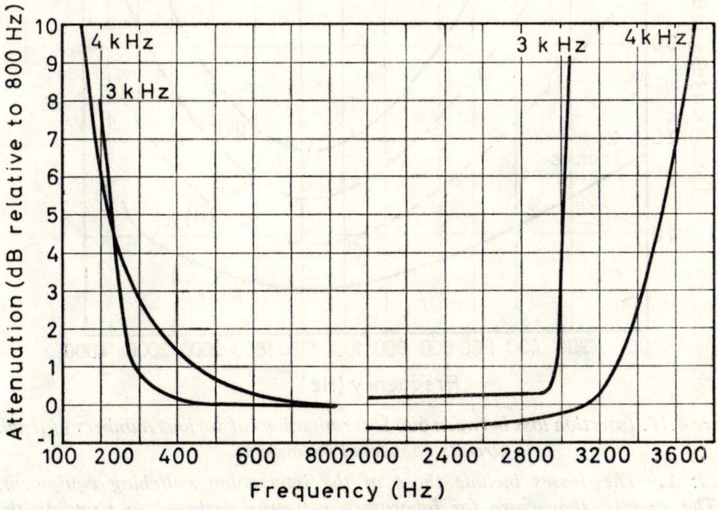

Figure 4.16. Attenuation/frequency characteristics of 3 and 4 kHz spaced telephone channels (single audio–audio links)

Unfortunately treatment of attenuation/frequency distortion is rather more complicated than that of circuit noise. There is no simple method for associating an appropriate value of $I_{2.5}$ with a given attenuation/frequency response although the calculation method indicated in Section 3.4 may provide useful information when direct assessment by a listening opinion test is not possible. Some examples are given from results of actual assessment tests.

1. Elimination of the band of frequencies from 212 to300 Hz: 2 dB
2. Elimination of the band of frequencies from 2850 to 3400 Hz: 1 dB
3. Elimination of the band of frequencies from 2020 to 3400 Hz: 9 dB
4. Insertion of one 3 kHz telephone channel into a connection already containing one 4 kHz channel (*see* Figure 4.16 for attenuation/frequency characteristics of such channels): 2 dB
5. Insertion of two 3 kHz channels: 3 dB

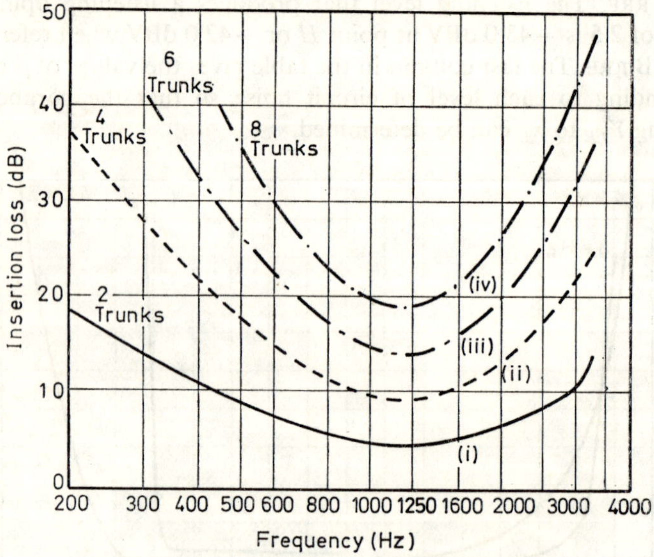

Figure 4.17. Insertion loss between 600 Ω terminations of various numbers of trunk circuits connected in tandem.

Notes: **1.** *The losses include those of the intervening switching equipment.* **2.** *The results shown are for laboratory networks designed to simulate the shapes of the attenuation/frequency distortion characteristics found in service and so the absolute values of loss (at 800, 1000 or 1600 Hz) should be disregarded*

6. insertion of four 3 kHz channels: 4 dB
7. increasing the number of 4 kHz channels until the attenuation/frequency distortion reaches the limits given in CCITT Recomendation G. 132; this would be attained when perhaps 12 channelling equipments of various improved types were connected in tandem, (*see* Table 6.24): 3 dB
8. introducing attenuation/frequency distortion corresponding to curve i in Figure 4.17: 2 dB
9. introducing twice the distortion of 8, see curve ii in Figure 4.17: 5 dB
10. introducing three times the distortion of 8 see curve iii in Figure 4.17: 7 dB
11. introducing four times the distortion of 8, see curve iv in Figure 4.17: 9 dB

The relationships given in Eqns 4.3 and 4.4 have been used to determine the parameters α, β and γ for certain cases of attenuation/frequency distortion the amount of which is expressed in terms of $I_{2.5}$; Table 4.5 shows the results. Naturally, this procedure cannot be carried beyond the extent that results in α approaching the value of 2.5.

Table 4.5. PARAMETERS OF EQN 4.2 RELATING Y_{LE} TO x_L FOR VARIOUS VALUES OF IMPAIRMENT DUE TO ATTENUATION/FREQUENCY DISTORTION

$I_{2.5}$ dB	α ($\Delta\alpha = 24\beta_1 I_{2.5}$)	β	γ ($\Delta\gamma = -0.7I_{2.5}$)
(1)	*(2)*	*(3)*	*(4)*
0	$3.86 = \alpha_1$	$-0.00183 = \beta_1$	$-1.3 = \gamma_1$
2	3.77	−0.00178	−2.7
4	3.68	−0.00173	−4.1
6	3.60	−0.00169	−5.5
8	3.51	−0.00163	−6.9
10	3.42	−0.00156	−8.3

Notes 1. These results apply for $V_L = -17.0$ dBV (SV5B) and RRE $= -1.0$ dB
2. With no distortion, $x_{2.5} = 26.0$ dB

The information provided above (especially that in Tables 4.4 and 4.5) is of considerable assistance in assembling the fundamental data for planning telephone networks; before it can be used for this purpose, however, it must be converted to a form applicable to conversational conditions.

4.2.1.3 Relationship between listening and conversation test results

As has already been explained in Section 4.1.5, the scores from conversation tests are affected differently by transmission degradations from those derived from listening only tests and so the results from the two types of test can be related only empirically. For example, as illustrated in Figure 4.8, even use of a common form of opinion scale (e.g. OS No. 7 and 8) does not yield directly comparable scores; allowance for any difference in the level of speech reaching the ears of the participants does not itself resolve the discrepancies. To derive a satisfactory relationship between results from listening only and conversational tests, it is necessary to perform in a properly coordinated manner pairs of test series of the two kinds using common speech links and the conversation tests must include very careful measurements of speech level at a suitable point in each test arrangement that can be accurately identified under the two test situations.

A large series of conversation tests was conducted in 1957 (at about the same period as that in which the listening tests referred to in Section 4.2.1.2, above, were conducted) and the same basic telephone speech links were used, namely local telephone circuits composed of 13-2P-27 telephone sets on 1 mile $6\frac{1}{2}$ lb cable and 50 V 200+200 Ω feeding bridges. The conversation tests were conducted in accordance with the description given in Sections 3.6 and 3.7 and the Opinion scale used was No. 9 (excellent–good–fair–poor–bad). The corresponding listening tests used Opinion scale No. 7, based on listening effort. Three hypotheses were formulated to provide a framework for determining relationships between listening and conversation test results; these seemed self-evident but their validity was also checked as part of the investigation.

Hypothesis (a) The optimum listening level under conversation conditions is the same as that under listening only conditions.

Hypothesis (b) Given listening levels on connections that differ in kind but are equally acceptable for conversation will yield equal listening opinion scores under listening only conditions.

Hypothesis (c) Subjects use equal talking levels on different connections provided that they are equally acceptable for conversation and are equal in sidetone characteristics.

By invoking these hypotheses and using the results of corresponding conversation and listening only tests, it is possible to obtain a simple formula which can be used to estimate conversation test results from those obtained under listening only conditions.

For the condition of 50 dBA room noise but no circuit noise and with the two local telephone circuits connected together through nonreactive loss, the conversation test results can be represented by the following two expressions:

$$Y_C = 3.35-0.00108\,(x_C+2.5)^2 \quad (4.7)$$

$$V_C = -18.0+0.066\,x_C+0.057\,\exp\,(0.092x_C) \quad (4.8)$$

where Y_C is the conversational mean opinion score (os No. 9) and V_C is the speech level (dBV (SV5B)) at G and H averaged over the talkers at the respective ends A and B*. x_C is the loss, dB, between points G and H shown in Figure 4.11a.

Eqn 4.5 gives the corresponding listening test mean opinion scores, Y_{LE}, using os No. 7 when V_L (*see* Figure 4.11b) is held at -17 dBV (SV5B). The required relationship is derived in

Table 4.6. DERIVATION OF A RELATIONSHIP BETWEEN Y_C AND Y_{LE} FOR EQUAL LEVELS OF SPEECH REACHING THE EAR

Conversation conditions				*Listening only conditions*	
Y_C	x_C dB	V_C dBV	V_C-x_C dBV	x_L *required to produce* V_C-x_C *when* $V_L=-17$	Y_{LE}
(1)	*(2)*	*(3)*	*(4)*	*(5)*	*(6)*
3.35	−2.5	−18.1	−15.6	−1.4	3.86
3.20	+9.3	−17.3	−26.6	+9.6	3.64
3.00	15.5	−16.7	−32.2	15.2	3.36
2.80	20.1	−16.3	−36.4	19.4	3.08
2.60	23.9	−15.9	−39.8	22.8	2.80
2.40	27.2	−15.5	−42.7	25.7	2.53
2.20	30.1	−15.1	−45.2	28.2	2.26
2.00	32.9	−14.7	−47.6	30.6	2.01
1.60	37.8	−13.7	−51.5	34.5	1.52
1.20	42.1	−12.5	−54.6	37.6	1.09

* Even when the measurements are made at a fixed point in a given connection, the speech level depends not only on the individual participants (the dispersion attributable to personal differences is about 2.8–4.1 dB, according to the type of telephone set) but also on the task in which they are engaged. It is also reported that speech levels increase as the distance apart of the participants increases.

Table 46; it should be noted that the results available relate to 13-2P-27 telephone sets (similar information derived from tests using more modern types of telephone set is lacking).

The first column in Table 4.6 above shows selected values of Y_C; the entries in cols 2 and 3 are obtained respectively by solving Eqn 4.7 for x_C and substituting this value of x_C in Eqn 4.8. Col. 4 gives the mean level of speech arriving at the input to the listener's local telephone circuit during conversation. Col. 5 is obtained by substracting the value in col. 4 from -17 dBV (SV5B), which is the speech level used at G (Figure 4.11b) during listening tests. The value of Y_{LE} in col. 6 is obtained by substituting the value in col. 5 into Eqn 4.5. By computing a more extensive set of values and interpolating, Table 4.7 was derived giving Y_C in terms of Y_{LE} for the same listening level.

Table 4.7. CONVERSATION MEAN OPINION SCORE, Y_C, AS A FUNCTION OF LISTENING MEAN OPINION SCORE (OS7), Y_{LE}, WHEN THE SAME LISTENING LEVELS APPLY

Y_{LE}	*0.0*	*0.1*	*0.2*	*0.3*	*0.4*	*0.5*	*0.6*	*0.7*	*0.8*	*0.9*	*1.0*
0						0.50	0.64	0.77	0.89	1.00	1.11
1	1.11	1.21	1.30	1.40	1.49	1.60	1.67	1.75	1.83	1.91	1.99
2	1.99	2.07	2.15	2.23	2.30	2.38	2.45	2.53	2.60	2.67	2.74
3	2.74	2.82	2.89	2.96	3.02	3.10	3.17	3.24	3.31	(3.38)	(3.45)

Note: The last two entries, shown in brackets, are by extrapolation.

Many practical problems in telephony require the prediction of conversational test results from those of listening tests, which are much easier to conduct. In order to use the information given in the table above, it is necessary to know the values of loss, x_C, under conversational conditions that will, on average, produce known values of speech level at the listener. The information given in Eqn 4.8 can, to some extent, be generalised by standardising the speech level to that obtained from the output of a 0 dB SRE sending end and by using nominal overall reference equivalent, X, instead of x_C, the loss between the local telephone circuits. Substituting the appropriate values yields the equation:

$$V_C + \mathrm{SRE} = -11.4 + 0.066X + 0.033 \exp(0.092X) \quad (4.8a)$$

Unfortunately, even for a given type of telephone set and after standardisation to 0 dB SRE, the speech voltage does not depend only on the nominal overall reference equivalent; for example the level of circuit noise reaching the listener and the amount of distortion suffered by the speech signals during transmission over the connection are also important. It may, however, be assumed with reason* that the additional factors mentioned will affect the speech level and conversational opinion score in the same proportion as does the nominal overall reference equivalent. A further degree of standardisation of information concerning speech levels can therefore be achieved by relating V_C+SRE to the conversational mean opinion score Y_C. Table 4.8a shows this relationship derived from extensive data available for 13-2P-27 telephone sets. The generality of the relationship is restricted to (i) telephone sets broadly similar to those mentioned, in particular having similar features such as sidetone, sending sensitivity/frequency response, physical dimensions and other particulars relating to the handset and (ii) subjects whose telephonic experience is fairly similar to those used in the tests, which took place in the United Kingdom in 1957. Subsequent improvements in telephone connections since that date are likely to cause subjects now to apply somewhat more stringent criteria and thereby may reduce the values of Y_C by a small amount.

Table 4.8. V_C+SRE AS A FUNCTION OF Y_C

(a) 13-2P-27 sets (from 1957 data)

Y_C	*0.0*	*0.1*	*0.2*	*0.3*	*0.4*	*0.5*	*0.6*	*0.7*	*0.8*	*0.9*	*1.0*
0											
1	−4.8	−5.1	−5.5	−5.8	−6.1	−6.4	−6.7	−6.9	−7.2	−7.5	−7.7
2	−7.7	−7.9	−8.1	−8.3	−8.5	−8.7	−8.9	−9.1	−9.3	−9.6	−9.8
3	−9.8	−10.1	−10.3	−10.6							

(b) 16-4T-31 sets (with regulators) (1969 data)

Y_C	*0.0*	*0.1*	*0.2*	*0.3*	*0.4*	*0.5*	*0.6*	*0.7*	*0.8*	*0.9*	*1.0*
0											
1	−8.7	−8.8	−9.0	−9.1	−9.2	−9.3	−9.5	−9.6	−9.7	−9.8	−10.0
2	−10.0	−10.1	−10.2	−10.4	−10.5	−10.6	−10.8	−10.9	−11.0	−11.1	−11.3
3	−11.3	−11.4	−11.5	−11.7	−11.8						

Note: The above information should not be used when the conversational opinion score is likely to be affected by excessive loudness of received speech or of sidetone. (Both these factors cause the speech level to *fall* as the mean opinion score decreases).

* This assumption is equivalent to invoking Hypothesis (c), stated above.

For convenience values of V_C+SRE have been tabulated also for the current 16-4T-31 (with regulator) telephone set from tests during 1969 but less extensive than those of 1957; the later results are given in Table 4.8b. Some even more recent tests using linear handset microphones (such as might be used in future 'electronic' telephone sets) have yielded broadly similar results to Table 4.8b so far as the rate of change of speech level with conversational opinion is concerned but more information is needed on the absolute values of V_C+SRE. Clearly the sidetone characteristics of the telephone set have an important effect on the relationships given in Table 4.8 and some information concerning this is included in Section 3.5 and in Section 4.3.1.

Measurement of the speech level from a local telephone circuit under carefully controlled speaking conditions (vocal level and talking distance and position of microphone relative to the mouth fixed) does not give the same value of speech level as that produced under free conversation conditions. For example, using RVL and BMSP, the speech level from a 13-2P-27 set referred to 0 dB SRE is, by calculation −11.8 dBV (SV5B) whereas, under the same controlled conditions, the 16-4T-31 (with regulator) set yields −10.8 dBV (SV5B); it will be noticed from Table 4.8 that the relative magnitudes under conversational conditions are in the reverse order to this. The discrepancy may, in this case, be due to the lower sidetone reference equivalent of the 16-4T-31 telephone set (3.8 dB as compared with 10.6 dB for the 13-2P-27).

The results of measurements of vocal level (using a microphone for measuring the speech sound pressure at a point fixed relative to the talker's lips), show that, under telephonic conversational conditions, subjects use vocal levels somewhat greater, on average, than RVL.*

This applies even when the overall loss is at about the preferred level. Furthermore, the increase in speech level, measured electrically on the circuit used for conversation, as the communication efficiency becomes worse, is caused mainly by movement of the microphone of the handset towards a better location in front of the mouth and only slightly is the increase attributable to any further raising of the vocal level.

An example of the use of the foregoing information will now be given.

* Perhaps RVL+5 dB when an average level of typical room noise is present. See also the free speaking vocal levels quoted in Section 2.1.

From Table 4.4 it is found that the values of Y_{LE} for a telephone speech path composed of 13-2P-27 sets on 1 mile $6\frac{1}{2}$ lb cable and 50 V, 200+200 Ω feeding bridges with a level of circuit noise of −60 dBmp at the input to the listening local telephone circuit are given by the expression:

$$Y_{LE} = 3.86 - 0.00235(x_L + 1.3)^2 \quad (4.9)$$

when $V_L = -17$ dBV (SV5B).

Table 4.9 shows, in col. 2, a few values of Y_{LE} calculated from Eqn 4.9.

Table 4.9. ESTIMATION OF Y_C AS A FUNCTION OF x_C FROM Y_{LE} AS A FUNCTION OF x_L

x_L $V_L = -17$ dBV SRE = +7 dB	Y_{LE}	Y_C *Estimated from* Y_{LE}	V_C+SRE dBV	$V_C - V_L$ dB	$x_C = x_L + (V_C - V_L)$ dB	*Measured* Y_C *(after smoothing)*
(1)	*(2)*	*(3)*	*(4)*	*(5)*	*(6)*	*(7)*
0	3.86	3.35	−10.9	−0.9	−0.9	3.35
10	3.56	3.14	−10.1	−0.1	9.9	3.15
20	2.79	2.59	−8.9	+1.1	21.1	2.61
30	1.56	1.64	−6.8	+3.2	33.2	1.67

Col. 2 is the listening mean opinion score for the value of loss between LTCS given in col. 1; col. 3 is obtained from col. 2 using the conversions given in Table 4.7; col. 4 is obtained from col. 3 and Table 4.8; col. 5 is obtained by subtracting V_L+SRE from the speech level given in col. 4; col. 6 is self-evident. The result of these computations is Y_C, given in col. 3 as a function of x_C, appearing in col. 6. If particular values of x_C are required, it will be necessary to do some interpolation. As a check, col. 7 shows some measured results which apply for the values of x_C in col. 6; these are from tests mentioned in Section 3.7.

In the above example, actual measured values of Y_{LE} were available; where this is not so but reliable information on sensitivity/frequency characteristics is available, it may be possible to make use of the method described in Section 3.4.4 for estimating Y_{LE} by calculation; an example is given in Section 4.2.2.1.

4.2.1.4 Conversational assessment scores

The information used at present (June, 1970) by the United Kingdom Post Office for network planning studies was derived from one large series of conversation tests, conducted during 1957 in the manner described in Sections 3.6 and 3.7; a wide range of overall reference equivalents was covered with two values of injected circuit noise, namely: none, and a level equivalent to −59 dBmp referred to the input to a 0 dB RRE receive end. These results were supplemented by several series of listening tests (*see* Section 4.1.2) each devoted to extending the range of circuit noise types and levels and to assessing the effects of certain types of attenuation/frequency distortion, including the effects of various low pass and high pass filters. The effects of nonlinear distortions have also been studied in a similar manner which is discussed below (Section 4.2.2). The results for a wide range of combinations of overall reference equivalent and levels of circuit noise are summarised in Table 4.10 which shows values of conversational mean opinion score on the scale 'excellent–good–fair–poor–bad'. The results for combinations of overall reference equivalent and 'impairment' due to attenuation/frequency distortion are summarised in Table 4.11 on the same basis. Combinations of all three factors mentioned can be treated from these tables in a manner described in Section 4.2.4.

The information given in Tables 4.10 and 4.11 was assembled by the method described in Section 4.2.1.3 above; slightly different speech voltage data were, however, used* and this accounts for slight disagreements between the results in these tables and certain values given in the section above.

For many purposes (*see* Chapters 7 and 8), it is desired to use conversational opinion data in the form of percentage 'unsatisfactory' instead of as mean opinion scores. Tables 4.12 and 4.13 have therefore been prepared showing percentage 'poor' or 'bad' opinions as functions of nominal overall reference equivalent, level of circuit noise and amount of impairment. The relationship between percentage 'poor' or 'bad' and mean opinion score was taken from Table 3.17 in Section 3.7.

* By assuming a linear relationship between V_C and x_C, the relationship between Y_C and x_C will also become parabolic and so the conversion from Y_{LE} to Y_C becomes a purely algebraic one of changing the values of the three parameters α, β and γ. The relationship $V_C = -18.0+0.066x_C$ was used for the connections free of circuit noise and without impairment, instead of Eqn 4.8; i.e. the last (nonlinear) term of that equation was ignored.

Table 4.10. CONVERSATIONAL MEAN OPINION SCORES ('EXCELLENT–GOOD–FAIR–POOR–BAD') AS A FUNCTION OF NOMINAL OVERALL REFERENCE EQUIVALENT AND LEVEL OF CIRCUIT NOISE (50 dBA ROOM NOISE PRESENT)

Nom ORE *x* dB	*Level* (dBmp) *of circuit noise, N*												
	None	−70	−68	−66	−64	−62	−60	−58	−56	−54	−52	−50	−48
16	3.18	3.17	3.17	3.16	3.16	3.15	3.14	3.13	3.12	3.10	3.08	3.06	3.02
17	3.15	3.14	3.13	3.13	3.12	3.12	3.11	3.10	3.08	3.06	3.04	3.01	2.97
18	3.12	3.10	3.10	3.10	3.09	3.08	3.07	3.06	3.04	3.02	3.00	2.96	2.91
19	3.09	3.07	3.07	3.06	3.05	3.04	3.03	3.02	3.00	2.97	2.94	2.90	2.85
20	3.06	3.03	3.03	3.02	3.01	3.00	2.99	2.97	2.95	2.92	2.89	2.84	2.78
21	3.02	2.99	2.99	2.98	2.97	2.96	2.94	2.92	2.90	2.87	2.83	2.78	2.72
22	2.98	2.95	2.95	2.94	2.93	2.91	2.89	2.87	2.85	2.81	2.77	2.71	2.63
23	2.94	2.91	2.90	2.89	2.88	2.86	2.84	2.82	2.79	2.75	2.70	2.64	2.56
24	2.90	2.86	2.85	2.84	2.83	2.81	2.79	2.77	2.73	2.69	2.64	2.56	2.47
25	2.85	2.81	2.80	2.80	2.78	2.76	2.74	2.71	2.67	2.62	2.56	2.49	2.38
26	2.80	2.76	2.75	2.74	2.72	2.70	2.68	2.65	2.61	2.55	2.49	2.40	2.29
27	2.75	2.70	2.70	2.69	2.67	2.64	2.62	2.58	2.54	2.48	2.41	2.32	2.20
28	2.70	2.65	2.64	2.63	2.61	2.58	2.55	2.52	2.47	2.40	2.33	2.23	2.10
29	2.65	2.60	2.58	2.57	2.54	2.52	2.49	2.45	2.39	2.32	2.24	2.13	1.99
30	2.59	2.54	2.52	2.51	2.48	2.45	2.42	2.37	2.32	2.24	2.16	2.04	1.88
31	2.53	2.48	2.46	2.44	2.41	2.38	2.34	2.30	2.24	2.16	2.06	1.94	1.77
32	2.47	2.41	2.39	2.38	2.34	2.31	2.27	2.22	2.16	2.07	1.97	1.83	1.65
33	2.41	2.34	2.32	2.31	2.27	2.24	2.19	2.14	2.07	1.98	1.87	1.72	1.53
34	2.35	2.27	2.25	2.23	2.20	2.16	2.11	2.06	1.99	1.88	1.77	1.60	1.41
35	2.28	2.20	2.18	2.16	2.12	2.08	2.03	1.97	1.89	1.78	1.66	1.49	1.28
36	2.21	2.12	2.10	2.08	2.04	2.00	1.95	1.88	1.80	1.68	1.55	1.37	1.14
37	2.14	2.05	2.03	2.01	1.96	1.91	1.86	1.79	1.70	1.58	1.44	1.25	1.00
38	2.06	1.97	1.95	1.92	1.87	1.83	1.77	1.70	1.60	1.47	1.33	1.12	0.86
39	1.99	1.89	1.86	1.84	1.79	1.74	1.67	1.60	1.50	1.36	1.21	0.99	
40	1.91	1.81	1.78	1.75	1.70	1.64	1.58	1.50	1.39	1.25	1.09		

Note: The level, *N*, of circuit noise is referred to the input to a receive end of 0 dB receive reference equivalent.

In a corresponding manner Tables 4.14 and 4.15 show percentage 'difficult' opinions as functions of the three independent variables, the conversion from mean opinion score having been performed with the aid of Table 3.18 in Section 3.7.

Practical transmission performance problems usually involve the presence of a combination of loss, circuit noise, impairment and, possibly, other forms of degradations. Fundamental data cannot be assembled to cover all such possible combinations and so some

Table 4.11. CONVERSATIONAL MEAN OPINION SCORES ('EXCELLENT–GOOD–FAIR POOR–BAD') AS A FUNCTION OF NOMINAL OVERALL REFERENCE EQUIVALENT AND 'IMPAIRMENT' DUE TO BANDWIDTH LIMITATION OR ATTENUATION/FREQUENCY DISTORTION (50 dBA ROOM NOISE PRESENT BUT NO CIRCUIT NOISE)

Nom ORE *x* dB	*Impairment, I,* dB										
	0	*1*	*2*	*3*	*4*	*5*	*6*	*7*	*8*	*9*	*10*
16	3.18	3.14	3.09	3.04	3.00	2.96	2.91	2.86	2.82	2.78	2.73
17	3.15	3.10	3.07	3.01	2.97	2.94	2.89	2.81	2.77	2.75	2.71
18	3.12	3.07	3.03	2.98	2.94	2.90	2.85	2.78	2.74	2.71	2.67
19	3.09	3.04	2.99	2.94	2.90	2.87	2.80	2.75	2.70	2.67	2.62
20	3.06	3.01	2.96	2.91	2.86	2.81	2.76	2.71	2.66	2.62	2.57
21	3.02	2.96	2.92	2.87	2.82	2.76	2.71	2.67	2.62	2.57	2.52
22	2.98	2.92	2.88	2.82	2.77	2.71	2.67	2.62	2.57	2.52	2.47
23	2.94	2.88	2.83	2.78	2.73	2.67	2.62	2.57	2.52	2.47	2.42
24	2.90	2.84	2.79	2.74	2.68	2.62	2.57	2.52	2.47	2.42	2.37
25	2.85	2.80	2.74	2.68	2.63	2.58	2.52	2.47	2.42	2.37	2.32
26	2.80	2.74	2.69	2.63	2.58	2.52	2.47	2.41	2.37	2.32	2.26
27	2.75	2.69	2.64	2.58	2.52	2.47	2.42	2.36	2.31	2.26	2.21
28	2.70	2.64	2.58	2.52	2.47	2.42	2.36	2.30	2.25	2.20	2.15
29	2.65	2.58	2.53	2.46	2.41	2.36	2.30	2.25	2.19	2.14	2.09
30	2.59	2.53	2.47	2.41	2.35	2.30	2.24	2.18	2.13	2.08	2.02
31	2.53	2.47	2.41	2.34	2.29	2.24	2.18	2.12	2.07	2.01	1.96
32	2.47	2.41	2.34	2.28	2.22	2.16	2.11	2.06	2.00	1.94	1.89
33	2.41	2.34	2.28	2.21	2.15	2.09	2.04	1.99	1.93	1.88	1.82
34	2.35	2.27	2.21	2.14	2.09	2.02	1.97	1.92	1.86	1.81	1.75
35	2.28	2.21	2.14	2.08	2.01	1.96	1.90	1.84	1.79	1.74	1.68
36	2.21	2.14	2.07	2.00	1.94	1.88	1.83	1.78	1.72	1.66	1.61
37	2.14	2.06	2.00	1.93	1.87	1.81	1.75	1.71	1.64	1.58	1.53
38	2.06	1.99	1.92	1.85	1.79	1.74	1.68	1.63	1.56	1.51	1.45
39	1.99	1.91	1.84	1.77	1.71	1.66	1.60	1.55	1.48	1.43	1.37
40	1.91	1.84	1.76	1.70	1.63	1.57	1.52	1.46	1.40	1.34	1.29

Note: 'Impairment' is measured in a listening opinion test (effort scale) at a mean score of 2.5; the corresponding score in a conversation test would be obtained, in the absence of circuit noise and distortion, at a nominal overall reference equivalent of about 32 dB.

simple rules must be used to estimate the effects of combinations of degradations from information available on, say, combinations of pairs of factors. A method used for this purpose, using data from the foregoing tables to estimate the effects of combinations of the three factors covered therein is described in Section 8.4 when dealing with some practical applications to telephone network design problems. The method requires certain assumptions concerning the independence of the effects of the various factors; although these assumptions are believed quite justifiable for the particular factors concerned in the example, further information is needed on the validity of applying it to combinations of 'listening', 'talking' and 'conversation' degradations.

Table 4.12. PERCENTAGE 'POOR' OR 'BAD' OPINIONS AS A FUNCTION OF NOMINAL OVERALL REFERENCE EQUIVALENT AND LEVEL OF CIRCUIT NOISE (50 dBA ROOM NOISE PRESENT)

Nom ORE *x* dB	*Level* (dBmp) *of circuit noise, N*												
	None	−70	−68	−66	−64	−62	−60	−58	−56	−54	−52	−50	−48
16	0.22	0.24	0.24	0.26	0.26	0.28	0.30	0.32	0.35	0.37	0.46	0.52	0.67
18	0.35	0.40	0.40	0.40	0.43	0.46	0.49	0.52	0.59	0.67	0.75	0.94	1.24
20	0.52	0.63	0.63	0.67	0.71	0.75	0.79	0.89	0.99	1.17	1.41	1.85	2.4
22	0.84	0.99	0.99	1.04	1.10	1.24	1.41	1.58	1.76	2.11	2.6	3.3	4.5
24	1.32	1.67	1.76	1.85	1.94	2.11	2.3	2.6	3.0	3.6	4.4	5.9	8.0
26	2.2	2.7	2.8	2.9	3.2	3.4	3.7	4.2	4.8	6.1	7.5	9.8	13.5
28	3.4	4.2	4.4	4.5	4.8	5.4	6.1	6.8	8.0	9.8	12.1	15.7	21.0
30	5.2	6.3	6.8	7.0	7.7	8.6	9.4	10.8	12.4	15.3	18.5	23.9	32.0
32	8.0	9.6	10.1	10.5	11.8	12.8	14.2	16.1	18.5	22.4	27.3	34.6	44.4
34	11.4	14.2	15.0	15.7	16.8	18.5	20.6	22.9	26.8	32.0	37.8	46.6	57.5
36	16.4	20.2	21.0	22.0	23.9	25.8	28.4	32.0	36.2	42.8	50.0	59.5	70.6
38	22.9	27.3	28.4	29.9	32.5	34.6	37.8	41.7	47.2	45.3	61.6	71.5	
40	30.4	35.7	37.3	39.0	41.7	45.0	48.3	52.7	58.5	65.6	72.7		

Note: The level, *N*, of circuit noise is referred to the input to a receive end of 0 dB receive reference equivalent.

Table 4.13. PERCENTAGE 'POOR' OR 'BAD' OPINIONS AS A FUNCTION OF NOMINAL OVERALL REFERENCE EQUIVALENT AND 'IMPAIRMENT' DUE TO BANDWIDTH LIMITATION OR ATTENUATION/FREQUENCY DISTORTION (50 dBA ROOM NOISE PRESENT BUT NO CIRCUIT NOISE)

Nom ORE *x* dB	*Impairment, I,* dB										
	0	*1*	*2*	*3*	*4*	*5*	*6*	*7*	*8*	*9*	*10*
16	0.22	0.30	0.43	0.59	0.75	0.94	1.24	1.67	2.0	2.4	3.0
20	0.52	0.71	0.94	1.24	1.67	2.11	2.7	3.3	4.0	4.7	5.7
24	1.32	1.85	2.3	2.9	3.7	4.7	5.7	6.8	8.0	9.4	10.8
28	3.4	4.4	5.4	6.8	8.0	9.4	11.1	13.1	15.0	16.8	18.9
32	8.0	9.6	11.8	13.9	16.1	18.5	20.6	22.9	25.8	28.9	31.4
36	16.4	19.3	22.4	25.8	28.9	32.0	34.6	37.3	40.6	43.9	46.6
40	30.4	34.1	38.4	41.7	45.6	48.9	51.6	54.8	58.0	61.1	63.6

Note: 'Impairment' is measured in a listening opinion test (effort scale) at a mean score of 2.5; the corresponding score in a conversation test would be obtained, in the absence of circuit noise and distortion, at a nominal overall reference equivalent of about 32 dB.

Table 4.14. PERCENTAGE 'DIFFICULT' OPINIONS AS A FUNCTION OF NOMINAL OVERALL REFERENCE EQUIVALENT AND LEVEL OF CIRCUIT NOISE (50 dBA ROOM NOISE PRESENT)

Nom ORE *x* dB	*Level* (dBmp) *of circuit noise, N*												
	None	*−70*	*−68*	*−66*	*−64*	*−62*	*−60*	*−58*	*−56*	*−54*	*−52*	*−50*	*−48*
16	4.0	4.2	4.2	4.4	4.4	4.6	4.8	5.0	5.1	5.5	5.9	6.2	7.0
18	5.1	5.5	5.5	5.5	5.7	5.9	6.0	6.2	6.6	7.2	7.4	8.4	9.6
20	6.2	6.8	6.8	7.0	7.2	7.4	7.6	8.1	8.6	9.4	10.2	11.5	13.5
22	7.9	8.6	8.6	8.8	9.1	9.6	10.2	10.7	11.2	12.4	13.9	16.0	19.0
24	10.0	11.0	11.2	11.5	11.8	12.4	13.2	13.9	15.3	16.8	18.6	21.8	25.6
26	12.8	14.2	14.6	14.9	15.7	16.4	17.2	18.2	19.6	22.1	24.7	29.0	34.6
28	16.4	18.2	18.6	19.0	19.6	20.9	22.1	23.4	25.6	29.0	32.4	37.8	44.4
30	20.5	22.5	23.4	23.8	25.1	26.5	28.0	30.5	33.0	37.3	41.2	47.8	56.5
32	25.6	28.5	29.5	30.0	31.9	33.6	35.7	38.2	41.2	46.2	51.5	59.3	68.5
34	31.4	35.7	36.8	37.8	39.2	41.2	43.9	46.7	50.5	56.5	62.4	71.0	79.5
36	38.6	43.4	44.4	45.6	47.8	50.0	52.7	56.5	60.8	67.0	73.3	81.1	89.0
38	46.7	51.7	52.7	54.4	57.1	59.2	62.4	66.0	71.0	76.9	82.7	89.6	
40	54.0	60.3	61.9	63.4	66.0	69.0	71.9	75.6	80.3	85.5	90.4		

Note: The level, N, of circuit noise is referred to the input to a receive end of 0 dB receive reference equivalent.

Table 4.15. PERCENTAGE 'DIFFICULT' OPINIONS AS A FUNCTION OF NOMINAL OVERALL REFERENCE EQUIVALENT AND 'IMPAIRMENT' DUE TO BANDWIDTH LIMITATION OR ATTENUATION/FREQUENCY DISTORTION (50 dBA ROOM NOISE PRESENT BUT NO CIRCUIT NOISE)

Nom ORE *x* dB	*Impairment, I,* dB										
	0	*1*	*2*	*3*	*4*	*5*	*6*	*7*	*8*	*9*	*10*
16	4.0	4.8	5.7	6.6	7.4	8.4	9.6	11.0	12.1	13.5	15.3
20	6.2	7.2	8.4	9.6	11.0	12.4	14.2	16.0	17.9	19.3	21.4
24	10.0	11.5	13.2	14.9	17.2	19.3	21.4	23.4	25.6	28.0	30.5
28	16.4	18.6	20.9	23.4	25.6	28.0	30.9	34.0	36.8	39.2	41.8
32	25.6	28.5	31.9	35.1	38.2	41.2	43.9	46.8	50.0	53.0	55.7
36	38.6	42.3	46.2	50.0	53.0	56.5	59.0	61.7	64.8	68.0	70.5
40	54.8	58.5	62.7	65.8	69.5	72.3	74.5	77.3	79.9	82.3	84.1

Note: 'Impairment' is measured in a listening opinion test (effort scale) at a mean opinion score of 2.5; the corresponding score in a conversation would be obtained, in the absence of circuit noise and distortion, at a nominal overall reference equivalent of about 32 dB.

4.2.1.5 Definition of an optimum telephone connection

Because of economic considerations and because the conditions under which most telephone calls are made would not allow full advantage to be taken, the speech links provided by telephone connections cannot be perfect. In practice, therefore, it is better to study optimisation of those characteristics the values of which can be chosen in design and to carry this out against a background of features which cannot be controlled by the telephone operating organisation. Clearly the characteristics of speech and hearing of ordinary citizens have to be accepted but other practical features of the situation are the presence of a certain amount of room noise, the existence of bandwidth limitation, nominally to the frequency range 300–3400 Hz, and possibly a small amount of circuit noise which is unavoidable. Distortions, both linear and nonlinear should clearly be reduced as much as is economically possible (rather than be 'optimised'); the same reasoning applies to propagation time, echo and perhaps sidetone*. The most conspicuous features that lend themselves without serious complications to optimisation are the overall loss and the attenuation/frequency response from mouth to ear, and these will be treated here. Complicated optimisation problems arise when the relative amounts of two undesirable factors have to be decided, such as transmission loss and echo; these require consideration of aggregates of calls over connections possible within complete networks and not just those over a single connection (*see* Chapter 8).

Figures 4.12 and 4.13 show results of loudness preference tests made under listening only conditions, i.e. using a fixed speech level from the sending end. If speech levels under conversation conditions are assumed to be distributed according to a Gaussian distribution function and listeners are assumed to behave in the same way under conversation as under listening only conditions when a given level of speech sound pressure enters their ears, the values of loss at which scores of 2.5 and 1.5 are obtained with the scale used (Opinion scale No. 4, *see* Section 3.4.2), will approximate closely to the median values for the dichotomies 'too loud'/'not too loud' and 'too quiet'/'not too quiet'. Making allowances for the differences in speech level at the sending end of the connection under conversational and listening test conditions, and for the sending and receiving reference equivalents of the local telephone circuits used, yields the median values of nominal overall reference equivalent shown in Figure 4.18. The results given apply for room noise, Hoth spectrum, at a level of 50 dBA but with no circuit noise on the telephone connection; the effects of changing these

* *See* Sections 4.3 and 4.4.

conditions can be seen in Figures 4.12 and 4.13. The curves drawn in Figure 4.18 through the median values show how persons might be expected to vote after conversing with participants chosen at random (so that the speech levels sent from the other end are properly distributed). The slopes of these graphs (defined by standard deviations) have not been measured directly but are estimated

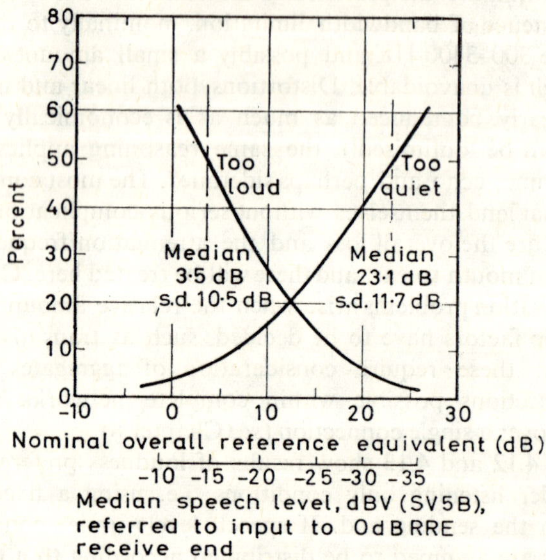

Figure 4.18. Distribution of loudness preferences with room noise 50 dBA. No circuit noise.

Notes: **1.** *Median 'too loud' and 'too quiet' from listening tests (Scores 2.5 and 1.5 in tests similar to those illustrated by Figure 4.13a).* **2.** *Standard deviations are estimated from conversation tests using 5 category scales so as to include effect of speech level distribution.* **3.** *Relationship between abscissa scales is from information used in Table 4.6*

from other conversational test results in which a five category scale (not based on loudness) was used; they are, nevertheless believed reliable within a tolerance of about $\pm 1\frac{1}{2}$ dB on the values of standard deviation.

Clearly, subjects vary widely in their opinions of what is too loud and too quiet; at best, only about 60% can be said to be 'satisfied', namely when the nominal overall reference equivalent is about 12 dB. It must also be noted that, under conversational conditions in the presence of 50 dB room noise* (as well as under listening only test conditions), subjects continue to give higher scores as the nominal overall reference equivalent is reduced below 12 dB down to at least as far as 3½ dB. The 'optimum' nominal overall reference equivalent may therefore be somewhat lower than the value of 12 dB which approximately minimises 'too loud'+'too quiet'. Further study is needed to determine the effects, if any, of subjects' conditioning and expectations and the influence of improvements in quality (nonlinear distortion and sensitivity/frequency response) of the telephone sets. Taking all the evidence that is available on this matter, it must be concluded that nominal overall reference equivalents from about 6 to 18 dB comprise a range of 'indifference' provided that circuit noise, sidetone, echo and other degrading features are insignificant. Naturally, further information is needed when non-handset types of telephone are being considered (*see* Sections 5.2 and 5.3).

The problem of choosing the best sensitivity/frequency response for a telephone connexion is susceptible to analytical treatment and this will now be explained.

During the early period of telephony, the frequency responses of items forming parts of connections were dictated by the limitations of the contemporary technology. It was soon realised that the range 600–1600 Hz was about the most important 1000 Hz and particular efforts were made in this range to ensure ample sensitivity for the transducers and to control losses as much as possible. Section 8.1 shows the sensitivities of limiting local telephone circuits at various epochs of the United Kingdom telephone network. As it became technically and economically feasible, frequency responses have improved and circuit losses been reduced to make better use of the range of frequencies that modern multiplex transmission systems are designed to transmit, namely from about 200 Hz to almost 4000 Hz. The feasibility of treating certain speech transmission problems analytically (*see* Sections 3.2.4, 3.3.5 and 3.4.4) has made it possible to determine the best frequency response for a speech path under certain given restrictions. A study was conducted in 1947 to determine the best frequency response for telephone microphones so that the maximum value of articulation score would be attained under certain rather adverse circumstances.

* When room noise is absent the maximum listening opinion score on the effort scale (os No. 7) tends to agree with the middle of the 'preferred' listening level range and not with the upper extreme, as is the case when room noise is present.

The results of this are described in Section 5.1.2 and have been used to set design objectives for new telephone microphones developed during the period 1947 to about 1965, when it became clear that the problem ought to be re-examined in the light of more modern developments.

It is now possible to analyse the problem rather more generally and it can be shown that the best frequency response for adverse circumstances, which will allow only a rather poor 'best' performance is not necessarily the best for more favourable conditions.

In Section 3.2.4, it is quoted that the loudness of received speech is a monotonically increasing function of the value given by an expression of the form:

$$\text{Loudness} = \text{constant} \times \int_0^\infty Q(Z)\, B'\, df \tag{4.10}$$

In Section 3.3.5, the articulation score is quoted as a monotonically increasing function of the value given by an expression of the form:

$$\text{Band articulation} = \int_0^\infty P(Z)\, B'\, df \tag{4.11}$$

In Section 3.4.4, it is stated that the listening opinion score can be related monotonically to the value given by an expression of form:

$$\text{Listening opinion index} = A \cdot D \int_0^\infty P(Z)\, B'\, df \tag{4.12}$$

The multiplier A is little short of unity except when the listening level is uncomfortable, either because it is too faint or too loud, and the multiplier D falls below unity only when the circuit noise is exceptionally loud. Ignoring A and D and replacing $P(Z)$ by the ratio $Q(Z)/Q_{max}$, where Q_{max} is a constant for a particular criterion, the three expressions can be considered as special cases of the general expression:

$$\text{Performance index} = \int_0^\infty Q(Z)\, B'\, df \tag{4.13}$$

in which the term $Q(Z)$ is a 'growth' function, peculiar to the criterion, and B' is a frequency weighting factor which differs somewhat according to the criterion (*see* Sections 3.2, 3.3 and 3.4). The form of the 'growth' function is indicated in Figure 3.19, which appears in Section 3.4.4, and convenient mathematical approximations are discussed below.

Considering a narrow frequency band, Δf Hz wide, the contribution to the performance index is given by:

$$Q(Z)\,B'\,\Delta f \tag{4.14}$$

The speech power in the same frequency band is:

$$W'\,\Delta f \tag{4.15}$$

Where W' is the spectrum density (say μW/Hz) of the speech signal.

If it is desired to economise as much as possible in the speech power transmitted at a defined interface in a telephone speech path, for example at the input to trunk circuits at GSCs to avoid overloading long distance transmission equipment, it is best to distribute the speech power over the frequency bands that make up the complete range of frequencies transmitted in such a way that any given small increase in speech power causes the same increase in performance index whichever the frequency band to which it is allocated*. Analytically, the condition is satisfied when:

$$B'\,\frac{\mathrm{d}Q(Z)}{\mathrm{d}W'} = \text{constant} \tag{4.16}$$

It is convenient to treat the 'growth' function $Q(Z)$ as a function of W'/W_0' instead of Z, where:

$$Z = 10\log_{10}(W'/W_0')$$

and W_0' is the spectrum density of the noise at the interface under consideration. If the noise is not sufficiently great to mask the threshold of audibility completely, W_0' must be taken as the equivalent threshold of audibility for continuous spectrum sounds (*see* Section 2.2.3.1).

A general expression can be used for the growth function, as follows:

$$Q(W'/W_0') = \text{constant} + \text{constant}\,(W'/W_0')^{1-k} \tag{4.17}$$

from which:

$$\frac{\mathrm{d}Q(W'/W_0')}{\mathrm{d}(W'/W_0')} = \text{constant}\,(W'/W_0')^{-k} \tag{4.18}$$

* This theorem is due to Ackroyd who also suggested the method of analysis.

and

$$\frac{dQ(W'/W_0')}{dW'} = \text{constant } W'^{-k} \cdot W_0'^{k-1} \tag{4.19}$$

Substituting Eqn 4.19 in Eqn 4.16 and rearranging, the most economic spectral distribution of speech power at the interface where the limitation in power transmission applies is:

$$W' = B'^{1/k} W_0'^{(k-1)/k} \text{ constant} \tag{4.20}$$

Converting these values of power, in say μW, to power levels in dB relative to 1 μW:

$$10 \log_{10} W' = (1/k)\, 10 \log_{10} B' + \frac{k-1}{k}\, 10 \log_{10} W_0' + \text{constant} \tag{4.21}$$

This expression shows the best spectral distribution of speech power as a function of the frequency weighting factor B' and the spectral distribution of noise representing the masked hearing threshold when this has been referred to the point of potential overload by taking account of the losses from that point up to the ear of the listener (including the electroacoustical sensitivity of the earphone).

The expression given in Eqn 4.21 is particularly useful in defining the best frequency response for telephone microphones. The spectral distribution of speech as it leaves the mouth can be denoted by W_S' and so, to secure the desired distribution, W' at the potential overload interface, the sending sensitivity of the speech path, in dB relative to 1 V per N/m², as a function of frequency, from the talker's mouth up to the point in the speech path at which W' applies, must be equal to:

$$S_{opt} = 10 \log_{10} W' - 10 \log_{10} W_S' \tag{4.22}$$

In Chapters 2 and 3 the symbol β_S' has been used for $10 \log_{10} W_S'$.

The value of the parameter k used in the above expressions depends on the criterion used and the part of the range of sensation level, Z, concerned. Choice of k therefore depends on the order of overall performance of the speech link under study. It is convenient, here, to refer to Figure 3.19 where $P(Z)$ is shown as a function of Z; $P(Z)$ can be considered as the ratio $Q(Z)/Q_{max}$ where Q_{max} is the upper asymptote of $Q(Z)$. Five parts of the range of Z can usefully be distinguished for the present purpose, as follows:

1. When $Z <$ about -10 dB ($W'/W'_0 <$ about 0.1). The speech is then very close to the threshold of audibility and the contribution to the subjective sensation is directly proportional to the power within each frequency band of the signal. This implies that $k = 0$.

2. In the low signal level region when speech is only just audible and the articulation is somewhat restricted, it has been shown by Ackroyd that, for articulation, $k = 0.75$. The same index is, as can be seen by reference to Section 3.2, applicable for the criterion 'loudness' (including 'reference equivalent') when the speech is at a comfortable listening level, say $Z > +20$ dB).

3. When Z lies between about -10 and -4 dB, $Q(Z)$ is approximately proportional to log (W'/W'_0) from which it will be seen that $k = 1$. The same value of k would apply according to the 'information' criterion suggested by Lalou.

4. The upper region in Figure 3.19 from $Z = -4$ dB upwards is represented by the expression $Q(Z) = Q_{max}\{1-(W'/W'_0)^{-0.3}\}$ from which it is seen that $k = 1.3$. This is a very wide range, of considerable importance in practical telephony problems, where the listening effort opinion scale is relevant.

5. When Z exceeds about $+40$ dB, $Q(Z)$ has practically reached its asymptote (within about 2%) and the value can be assumed practically independent of any further change in W'/W'_0. The value of k can then be taken as infinite.

For those problems where the values of Z lie within a single range, as defined above, for all the frequency bands in the range of frequencies transmitted, the following expressions apply for the optimum shape of the spectral density of the signal at the point of potential overload. These are obtained by substituting the appropriate value of k in Eqn 4.21 and ignoring constants.

1. Threshold of audibility; $k = 0$

$$10 \log_{10} W' = (\text{a large number})\,(10 \log_{10} B' - 10 \log_{10} W'_0)$$

2. Articulation and loudness ratings; $k = 0.75$

$$10 \log_{10} W' = (1.33)\, 10 \log_{10} B' - (0.33)\, 10 \log_{10} W'_0$$

3. Information criterion; $k = 1$

$$10 \log_{10} W' = 10 \log_{10} B'$$

4. Listening opinion scores (effort scale); $k = 1.3$

$$10 \log_{10} W' = (0.77)\ 10 \log_{10} B' + (0.23)\ 10 \log_{10} W_0'$$

5. An exceptionally high standard where $Q(Z) \doteqdot Q_{max}$ within the range of frequencies transmitted; k = infinity

$$10 \log_{10} W' = 10 \log_{10} W_0'$$

In practice 2 and 4 are of particular interest and examples follow of their application.

Table 4.16. DETERMINATION OF OPTIMUM SENDING FREQUENCY RESPONSE TO MAXIMISE LOUDNESS OF RECEIVED SPEECH; LISTENING IN SILENCE

Frequency Hz	*10 log₁₀ B′*	*1.33 10 log₁₀ B′*	*10 log₁₀ W′₀*	*0.33 10 log₁₀ W′₀*	*Optimum 10 log₁₀ W′ (3)–(5)*	β_S'	*Optimum* S_{MG} *(6)–(7)*
(1)	*(2)*	*(3)*	*(4)*	*(5)*	*(6)*	*(7)*	*(8)*
100	−1.4	−1.9	+39.0	+12.9	−14.8	+ 5.5	−20.3
200	+0.8	+1.1	+22.2	+ 7.3	− 6.2	+10.6	−16.8
300	+1.0	+1.3	+11.8	+ 3.9	− 2.6	+12.0	−14.6
400	+1.0	+1.3	+ 7.4	+2.4	− 1.1	+12.1	−13.2
500	+1.0	+1.3	+ 4.7	+ 1.6	− 0.3	+11.0	−11.3
600	+0.8	+1.1	+ 3.4	+ 1.1	0	+ 8.7	− 8.7
800	+0.4	+0.5	+ 1.1	+ 0.4	+ 0.1	+ 3.4	− 3.3
1000	0	0	0	0	0	0	0
1250	−0.5	−0.7	+ 0.5	+ 0.2	− 0.9	− 2.3	+ 1.4
1600	−1.2	−1.6	+ 1.0	+ 0.3	− 1.9	− 5.2	+ 3.3
2000	−1.9	−2.5	0	0	− 2.5	− 7.6	+ 5.1
2500	−2.8	−3.7	− 2.5	− 0.8	− 2.5	− 9.6	+ 7.1
3000	−3.5	−4.7	− 5.0	− 1.6	− 3.1	−11.5	+ 8.4
3500	−4.3	−5.7	− 3.7	+ 1.2	− 6.9	−13.9	+ 7.0
4000	−5.1	−6.8					

Notes: 1. Each column has been normalised to 0 dB at 1000 Hz.
2. The values of 10 $\log_{10} B'$ apply for audibility, *see* Section 3.2
3. The values of 10 $\log_{10} W_0'$ are derived from $\beta_0 - K$ with allowance for loss from the point at which limitation in mean power is assumed to the listener's ear. The receive sensitivity of this section of the speech path has been based on the transmission characteristics of an audio trunk channel.
4. The values of β_S' apply for average male and female subjects speaking freely.

Table 4.16 shows the determination of optimum sending frequency response to maximise loudness (or articulation under rather poor transmission conditions). Listening in silence has been assumed but some remarks on the influence of room and circuit noise are given later. In the table, cols 2–6 show the derivation of the levels of the best mean speech spectrum, 10 $\log_{10} W'$, at the point G at which limitation in mean power is to be respected; the point G might, in practice, be located at the output of the LTC or at the input to the network of trunk circuits, viz. the Group Switching Centre (GSC)*. The values of 10 $\log_{10} W'_0$ which apply for silent listening conditions are given by the threshold of audibility for continuous spectrum sounds ($\beta_0 - K$) referred back from the entrance to the ear canal (where the signal is acoustical) to the point G (where the signal is electrical).

$$10 \log_{10} W'_0 = \beta_0 - K - S_{GE}$$

where S_{GE} is the receiving sensitivity (dB relative to 1 N/m^2 per V) from the interface G to the listener's ear. The optimum shape for S_{GE} is, naturally, open to debate and the present optimising procedure is for the sending sensitivity only, subject to a given shape for S_{GE}. In practice it is difficult to justify any but a uniform sensitivity for receiving and this will be assumed here; departure from uniformity seems likely to have undesirable effects on cross-talk and other interference.

The sending sensitivity S_{MG} (from the mouth of the talker to the point G) is obtained by subtracting the speech spectrum levels at the mouth (β'_S) from the optimum signal spectrum at G; the results of this operation are shown in col. 8.

Table 4.17 shows the determination of optimum sending frequency response to maximise listening opinion score (effort scale). Listening in silence has again been assumed and this table follows the same arrangement as Table 4.16; only the coefficients used in cols (3) and (5) differ. In this case the equation for optimising 10 $\log_{10} W'$ cannot be applied in the frequency range where the receive sensitivity has fallen severely because this causes the value of Z to fall outside the range for which $k = 1.3$ applies.

Comparison of the conclusions from Tables 4.16 and 4.17 show the following:

For conditions where the best attainable performance is poor, or to minimise the overall reference equivalent, it is more profitable to provide *less* sensitivity where the threshold rises and to devote the power saved to frequency regions where it gives more return. For higher grade performances, however, the coefficient of 10 $\log_{10} W'_0$ becomes positive and so extra sensitivity can be devoted with

* Reference should be made to Figure 4.6a (Section 4.1); the direction A to B is being considered.

Table 4.17. DETERMINATION OF OPTIMUM SENDING FREQUENCY RESPONSE TO MAXIMISE LISTENING OPINION SCORE (EFFORT SCALE); LISTENING IN SILENCE

Frequency Hz	$10 \log_{10} B'$	0.77 $10 \log_{10} B'$	$10 \log_{10} W'_0$	0.23 $10 \log_{10} W'_0$	*Optimum* $10 \log_{10} W'_0$ $(3)+(5)$	β'_S	*Optimum* S_{MG}	*Practical shape*
(1)	*(2)*	*(3)*	*(4)*	*(5)*	*(6)*	*(7)*	*(8)*	*(9)*
100	−1.8	−1.4	+39.0	+9.0	+7.6	+ 5.5	+2.1	—
200	−0.2	−0.2	+22.2	+5.1	+4.9	+10.6	−5.7	−15.1
300	+0.7	+0.5	+11.8	+2.7	+3.2	+12.0	−8.8	−11.8
400	+1.1	+0.8	+7.4	+1.7	+2.5	+12.1	−9.6	−10.9
500	+1.1	+0.8	+4.7	+1.1	+1.9	+11.0	−9.1	−9.7
600	+1.0	+0.8	+3.4	+0.8	+1.6	+8.7	−7.1	−7.4
800	+0.5	+0.4	+1.1	+0.3	+0.7	+3.4	−2.7	−2.7
1000	0	0	0	0	0	0	0	0
1250	−0.7	−0.5	+0.5	+0.1	−0.4	−2.3	+1.9	+1.9
1600	−1.7	−1.3	+1.0	+0.2	−1.1	−5.2	+4.1	+4.2
2000	−3.3	−2.5	0	0	−2.5	−7.6	+5.1	+5.1
2500	−5.4	−4.2	−2.5	−0.6	−4.8	−9.6	+4.8	+5.0
3000	−7.3	−5.6	−5.0	−1.2	−6.8	−11.5	+5.7	+4.8
3500	−8.9	−6.9	+3.7	+0.9	−6.0	−13.9	+7.9	−6.4
4000	−10.0	−7.7						−25

Notes: 1. *See* Table 4.16
2. The values of $10 \log_{10} B'$ apply for listening effort, *see* Section 3.4
3. and 4. *See* Table 4.16
5. Col. 9 gives a more practical shape than col. 8, likely to be achievable economically.

profit to overcoming, partially, the raised threshold. (Except, of course, where the threshold is raised severely by cutoff.)

Interesting consequences arise when the other three criteria are employed. For example, 1, which would be relevant if it were desired to make a very attenuated speech signal as audible as possible. This leads to the conclusion that the response giving the most audible crosstalk (but perhaps not the most intelligible) would be one in which all the speech power is concentrated in a very narrow band centred at about 1000 Hz. It is also interesting to note that early telephone microphones and earphones did produce this kind of response and so were particularly appropriate to the circumstances of their time, when overall losses were very severe indeed (*see* Section 8.1).

For criterion 3, the coefficient of $10 \log_{10} W'_0$ is zero and so, for this case, the threshold has no influence and the optimum signal spectrum at the point of potential overload is the same shape as

$10 \log_{10} B'$. On the contrary, for criterion 5, $10 \log_{10} B'$ has no influence and the optimum signal spectrum should match the shape of the threshold $10 \log_{10} W'_0$, i.e. the signal to threshold ratio should be independent of frequency over the range of frequencies effectively transmitted.

In spite of these differences, the results shown in Tables 4.16 and 4.17 together with the results, not shown here, of applying the other three criteria reveal broadly similar shapes of optimum sending sensitivity. Naturally the sensitivity of a telephone set alone must take into account any frequency dependent losses between its terminals and the point *G*.

In all practical cases, room noise and circuit noise are likely to be present and these will have the effect of raising the threshold of audibility, $10 \log_{10} W'_0$. The effects of room noise sidetone, earcap leakage and practical levels of circuit noise of approximately uniform spectrum on the line combine to shift the threshold by an amount of the order 15–20 dB; furthermore the amount of shift is unlikely to vary greatly in different parts of the range of frequencies transmitted (say 200–3500 Hz). In such circumstances, the effect on the optimum shape of sending frequency response of the presence of room and circuit noises will be negligible. In any case, it will be observed by study of Tables 4.16 and 4.17 that departures from uniformity in the threshold shift will produce effects on the optimum shape of $10 \log_{10} W'$ having opposite senses in these two cases.

The conclusion could, therefore, reasonably be drawn that the best sending sensitivity response from the talker's mouth up to the point *G* represented either by the output from the local exchange or the GSC according to the transmission plan, ought to approximate in shape to that shown in col. 8 of Table 4.17. It may well be necessary, however, to depart somewhat from this precise shape for practical reasons, such as limitations imposed in designs of transducers, and so slightly modified values have been entered in col. 9 of the table to ease such problems. This shape is being used for further study. Because differences in subscribers' lines will affect the response, it is perhaps reasonable to regard the chosen shape of sensitivity/frequency characteristics as applying for the most commonly occurring subscribers' lines; this will, of course, result in the response with very short lines having some 'top lift' and that with limiting lines being subjected to appreciable higher frequency attenuation distortion.

A practical sensitivity/frequency response for receiving, assumed nominally uniform, ought to include some attenuation/frequency limitation, for example as shown in Figure 3.10c which appears in Section 3.2. Note that Figure 3.10b shows a sending sensitivity/frequency characteristic different from that arrived at above; the response shown in Figure 3.10b was derived from an earlier study intended to optimise performance when losses and noise restrict that attainable very severely (*see* Section 5.1.2).

4.2.1.6 Disturbing effect of noise in the absence of speech

While a customer is waiting for a connection to be established he may be exposed to some circuit noise; if the noise is excessive he may abandon the call although, if he had persisted, the noise would not have been high enough in level to degrade the conversation seriously. It is therefore desirable to have some information on the levels of circuit noise likely to be regarded as disturbing under such circumstances.

The following results were obtained from a listening experiment in which naive subjects were exposed to a succession of 5 s bursts of circuit noise and required to vote on a 7 grade scale of audibility and disturbance. Several types of circuit noise were included and some differences in their 'potency' were observed.

For each type of circuit noise, reproduced from a recording, the successive bursts were reproduced at different levels chosen at random (without replacement) from a set of ten at 6 dB intervals. Each 5 s listening period was indicated by a signal lamp. Subjects listened under silent ambient conditions and in the presence of 50 dB room noise (40 dB weighting on ASA sound level meter); in silence the level of circuit noise ranged within each sequence of ten bursts of the same kind of noise from -85 dBmp to -31 dBmp, referred to the input to a 0 dB RRE receive end; with the room noise the range was -79 to -25 dBmp. Listening was on a 13-2P-27 telephone set with 1 mile $6\frac{1}{2}$ lb subscriber's line and 50 V, $200+200\ \Omega$ feeding bridge; the microphone of this set was 'live' so that the proper room noise sidetone conditions would be present. The RRE of this local telephone circuit is -1 dB but the results have been standardised by referring all levels of circuit noise to 0 dB RRE.

Eight different kinds of noise were used but 'white' noise appeared three times as frequently in the experimental design as each of the others; thus there were ten noise conditions in the experiment and ten subjects were used for listening in accordance with a Latin

Square design. Each noise was reproduced at the required level measured with psophometric frequency weighting but with an indicating instrument of 1.7 s integration period, instead of the normal 100 ms integration period of the psophometer. Steps were also taken to ensure that the level of each kind of noise varied little between such integration periods.

The scale on which the listeners voted used the following categories:

Opinion scale No. 5.

A Inaudible: noise completely undetectable
B Just audible: noise can just be detected by listening carefully
C Slight: noise detectable, but not disturbing
D Moderate: noise slightly disturbing
E Rather loud: noise causes appreciable disturbance
F Loud: noise very disturbing, but call would be continued
G Intolerable: noise so loud that the call would be abandoned, or operator asked to change the line.

The results for the 'white' noise conditions are shown in Figures 4.19 and 4.20. Figure 4.19 shows the mean scores for ten subjects each giving three independent judgements; the votes have been scored 0 for *A*, 1 for *B*, etc. to 6 for *G* and mean scores calculated for each noise setting. These results show how, on average the sensation of disturbance increases in intensity as the level of the noise increases. Taking votes for *A*, *B* and *C* as representing absence of disturbance and *D*, *E*, *F* and *G* as the presence of disturbance, the threshold of disturbance can be located by the level of noise corresponding to a mean score of 2.5, namely −60.1 dBmp and −52.1 dBmp for silent and 50 dB room noise listening respectively. Corresponding levels have been determined for the other kinds of circuit noise and their differences from the level for 'white' noise gives their 'potencies'; when the threshold level is less than that for 'white' noise the potency is positive. The more commonly occurring kinds of telephone circuit noise were found to possess approximately 0 dB potency; these were: rectifier noise (having components at 300, 600, 900 etc. Hz), intermodulation noise from an FDM system with somewhat raised loading and with pilots removed, noise from a contemporary carrier system (1956) with normal loading and with its pilots. The other types of noise (perhaps better described as 'interference') and their potencies were as follows:

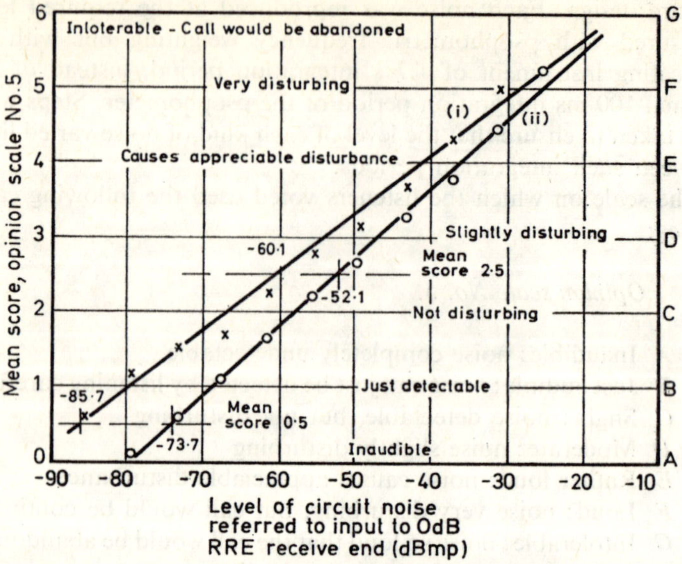

Figure 4.19. Opinion of subjects hearing circuit noise in the absence of any speech via the telephone connection.

Notes: **1.** *Curve i relates to listening in the absence of any room noise; curve ii applies in the presence of Hoth Spectrum room noise at 50 dBA level.* **2.** *See text for full description of categories of opinion scale*

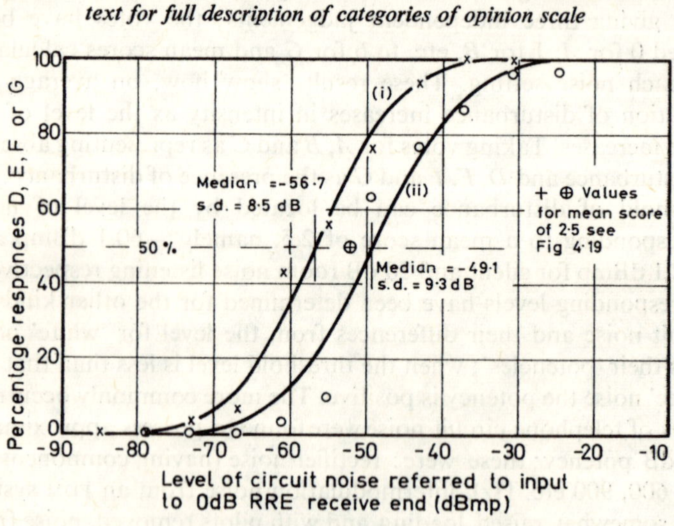

Figure 4.20. Distribution of subjects' opinions at the threshold of disturbance due to circuit noise in the absence of any speech signals.

Note: Curve i relates to listening in the absence of any room noise; Curve ii applies in the presence of 'Hoth Spectrum' room noise at 50 dBA level

Speech interference from a broadcast channel; 6–7 dB more potent than 'white' noise.

Music interference from a broadcast channel: 9 dB more potent when heard in silence and 6 dB in 50 dB room noise.

In order to separate the effect of the intelligibility of the speech from its intensity, it was also used 'reversed', i.e. the tape replayed in the reverse direction. The reversed speech was only 2–3 dB more potent than the 'white' noise so that the intelligibility was of some importance.

The threshold of audibility (mean score 0.5) of all examples was the same except the music interference, the threshold of audibility of which was 3–4 dB lower than that of the others. (The particular example of light music used was of particularly penetrating quality.)

In studying the merits of telephone network plans, it is commonly necessary to determine the proportion of customers who will suffer disturbance and for this purpose it is necessary to express the results in the form shown in Figure 4.20. The proportion of subjects who voted *D*, *E*, *F* or *G* is plotted against the level of the circuit noise and Gaussian probability integral functions have been fitted to the data by probit analysis. The fitted functions enable the median tolerance to circuit noise before it is declared disturbing to be estimated, together with a measure of the rate of increase in proportion which can be expressed as the equivalent standard deviation of the Gaussian distribution. Only the results for 'white' noise have been included in the figure but the results for the three replications have been combined. It will be noticed that the medians differ very slightly from the noise levels corresponding to a mean score of 2.5 shown in Figure 4.19.

4.2.2 Nonlinear distortion

4.2.2.1 Instantaneous nonlinearity

Nonlinear distortions likely to be present in telephone connections can be considered, so far as telephony is concerned (this may not apply for data), as being classified into two types as follows:

1. Those that are instantaneous and involve no storage. In such cases, the relevant transmission element causing the nonlinear distortion can be represented by a single transfer characteristic which specifies the relationship between simultaneous

instantaneous values of output and input amplitudes. Such a transfer characteristic may be continuous or can consist of discontinuous, step-like, sections such as is the case with amplitude quantisation.

2. Those that do involve storage and can be represented by a device of which the gain changes, relatively slowly compared with the Nyquist rate, in accordance with the recent past history of the input amplitudes. Speech has quasistationary properties (*see* Section 2.1.3) and so it is possible to introduce, with certain advantages under appropriate conditions, nonlinearity of this type without producing any appreciable nonlinear distortion of the first type defined above. Compandors of the syllabic type and constant volume devices exploit these harmless features of nonlinearity making use of suitable storage and averaging arrangements.

The present Section is devoted to consideration of the first type of nonlinear distortions and the second type are treated in Section 4.2.2.2.

Until the advent of quantising in transmission systems such as those using pulse code modulation (PCM), nonlinearity of the instantaneous type was encountered in telephony only as a defect in amplifiers, modulators, etc., and as an essential feature of the operation of carbon microphones. Shortcomings in design are liable to cause such distortion and these must be controlled by suitably framed clauses in specifications of equipment and commonly simple measurement of harmonic or intermodulation distortion products is sufficient. Such measurements do not, and are not required to, bear any direct fundamental relationship* to the subjective performance of telephone speech paths containing the relevant elements. Such information as is available on acceptable amounts of harmonic and intermodulation distortions when testing carbon microphones is included in Section 5.1.2.

More topical are considerations of the main effects on transmission of the instantaneous nonlinearity introduced by PCM systems. This matter now requires the same disciplined study as has been accorded to the other listening degradations already treated in Section 4.2.1. Because it is a listening degradation, it is convenient first to consider the effects of nonlinearity in PCM systems (mainly quantising distortion) on subjects in listening only

* Establishment of such a relationship for high fidelity systems has been suggested by Wigan (see Bibliography).

tests; the treatment here, therefore, parallels as far as is possible that given above to circuit noise and impairment due to attenuation/frequency distortion and bandwidth limitation.

The distortion products caused by an instantaneously nonlinear element are somewhat like circuit noise, because they cause some masking of audibility over the whole frequency range, but, unlike circuit noise, their magnitudes are, in certain important respects, correlated with those of the signal. For example, when the signal ceases, the distortion products disappear; when the signal amplitude increases the distortion products, also, in general increase. Because of these properties, nonlinear distortion products of a given long term mean power are less audible (and therefore cause less impairment) than a constantly applied circuit noise of the same long term mean power, the same level of speech being present in both cases. An interesting demonstration of this is provided by the results, shown in Table 4.18, of the following experiment. A listening test

Table 4.18. EFFECT ON DETECTABILITY OF DISTORTION PRODUCTS OF INTRODUCING DELAY IN THE DISTORTION PRODUCTS

Delay in distortion products relative to speech ms	*Median threshold of detectability, speech to distortion ratio* dB
0	22.2
30	26.7
60	30.7
∞	38.6

arrangement was set up using the modulated noise reference unit (MNRU) which is a device (see below) constructed to simulate nonlinear distortion from a practical PCM system. The simulated distortion products from the MNRU can be taken from an output separate from that which transmits the undistorted speech; normally the simulated distortion products are combined by direct addition to the undistorted speech, but, for this experiment, an all-pass delay device was inserted in the path of the simulated distortion products before combination with the speech. The amount of delay was adjustable. The experiment consisted of determining the threshold of detectability of the distortion for values of delay from zero upwards. With zero delay, the correlation between distortion products and the speech signal is maximum and, as the delay increases, the correlation is reduced until, with very long delay, the correlation becomes zero; as the delay is increased, the long term mean power of the distortion products is, of course, unaffected.

The results given in Table 4.18 show that, when the correlation is maximum the speech masks the detectability of the simulated distortion products by at least 16 dB more than when the simulated distortion products are uncorrelated. It is also interesting to note that the threshold of detectability for uncorrelated simulated distortion products is about the same as that found when 'white' noise is used (both speech and noise being restricted to the telephone band 300–3400 Hz).

The nature of the functional relationship existing between the short term mean power of the distortion products and that of the speech is, of course, important, as also is the frequency spectrum of the distortion products. An important example, of considerable relevance to quantising in practical PCM transmission systems, is the case of distortion products that are uniform in frequency spectrum (within the range of the signal being transmitted) and directly proportional (in short term mean power) to the signal. A device constructed to simulate this form of distortion consists of a modulator fed with speech and random noise to produce noise instantaneously proportional in amplitude to that of the speech (but randomly related in sign). Some results from tests using this device (termed the modulated noise reference unit, MNRU) are given below. The MNRU can be controlled to adjust the ratio of the power of the signal to that of the modulated noise and, when this ratio is expressed in dB, it is termed the Q value of the setting of the MNRU. The equipment used a specially designed modulator so that the ratio, at a given setting, remains constant over a very wide range of instantaneous amplitudes.

The results of listening comparison tests, using sentence material and never repeating a sentence to any given listener (*see* Section 3.4.2), yields a relationship between signal to injected circuit noise ratio, G_n, and Q as shown in Figure 4.21. These results were obtained with the listening level maintained at −28 dBV (SV5B) when referred to the input to a receive end of 0 dB receiving reference equivalent; as can be seen by comparison with the information given in Section 4.2.1.2, this level is about 10 dB lower than the listening level that yields the highest listening score (Effort scale OS No. 7). However, the threshold of detectability of the distortion simulated by the MNRU corresponds, over a wide range of listening levels, to a setting of $Q = 20.3$ for 50% of the experienced listeners used (see Figure 4.22), the standard deviation of their individual thresholds being 3.4 dB; it may therefore be assumed that the relationship shown in Figure 4.21 can be used over the range of listening levels from −18 dBV to −48 dBV referred to the input

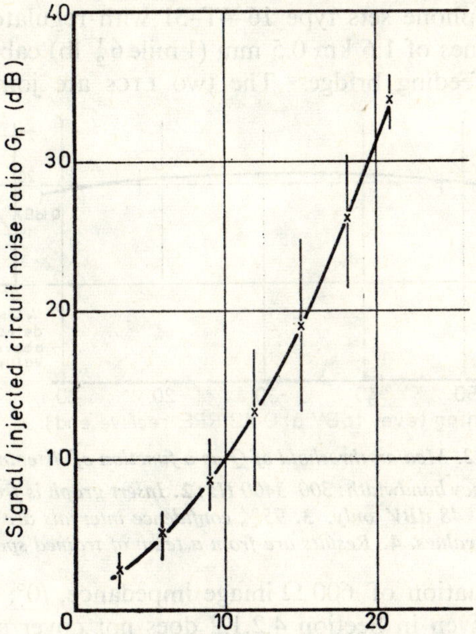

Figure 4.21. Equivalent speech/injected circuit noise ratio, G_n, as a function of Q, the setting of the modulated noise reference unit (MNRU).

Notes: **1.** *Frequency bandwidth: 300–3400 Hz;* **2.** *Listening level: −28 dBV (SV5B) at 0 dB* RRE *receive end;* **3.** *Equation of curve:* $G_n = 1.07 + 0.285Q + 0.0602Q^2$; **4.** *Experimentally determined points are marked together with their 95% confidence intervals*

to a 0 dB RRE receive end, i.e. from about the optimum listening level down to 30 dB lower than this. Note that this statement does not imply that the *performance* of a speech path under listening conditions is independent of listening level, far from it, but only that the *rate of exchange* between signal to injected circuit noise ratio and Q is independent of listening level.

The foregoing relationship can be used to estimate the listening opinion score (Effort scale OS No. 7) for any given combination of Q and listening level by use of information, using this scale, obtained with injected circuit noise combined with variation in listening level. As a further step, the listening opinion scores can be converted to mean opinion scores as obtained from conversation tests by application of the method described in Section 4.2.1.3. The complete procedure is illustrated in Table 4.19.

The primary data contained in cols 1, 2 and 3 of Table 4.19 apply for telephone sets type 16–4T–31 with regulators, used on subscribers' lines of 1.6 km 0.5 mm (1 mile $6\frac{1}{2}$ lb) cable and 50 V, 200+200 Ω feeding bridges. The two LTCs are joined by non-

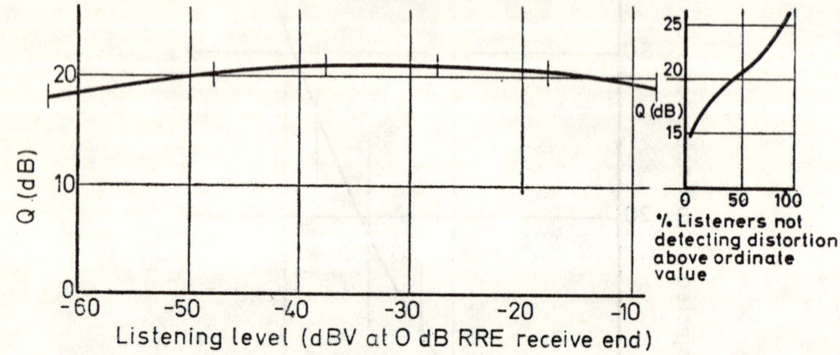

Figure 4.22. Median threshold of Q as a function of listening level.
Notes: **1.** *Frequency bandwidth: 300–3400 Hz;* **2.** *Insert graph is from results over range −18 to −48 dBV only.* **3.** *95% confidence intervals are marked at the experimental values.* **4.** *Results are from a team of trained speech testers*

reactive attenuation of 600 Ω image impedance, $\underline{/0°}$. Because the information given in Section 4.2.1.2 does not cover a sufficiently wide range of levels of injected circuit noise for the present example, use has been made of the calculation method mentioned briefly in Section 3.4.4 to estimate the required listening opinion scores; Table 3.7 in Section 3.2.4.2 gives the sending and receiving sensitivities of these LTCs and Section 3.2.4.3 contains calculations of the sending and receiving reference equivalents* quoted in the Notes appended to Table 4.19. Col. 4 in the table shows the level of speech reaching the listener, standardised by referring this level to the input to a receive end of 0 dB RRE. The level of circuit noise, referred to the same point and expressed in dBV, without psophometric weighting but measured in a 300–3400 Hz frequency band, appears in col. 5. The ratio of speech power to that of injected circuit noise, both referred to the listener's ear, is denoted by G_n

* The value of RRE used in Table 4.19 differs very slightly from that quoted in Section 3.2.4.3 because the sensitivity at 4000 Hz was ignored.

and is given in col. 6. The values of Q in col. 7 are obtained from G_n by use of the relationship given in Figure 4.21. At this stage in the

Table 4.19. DETERMINATION OF CONVERSATIONAL MEAN OPINION SCORES (Y_C) AS A FUNCTION OF Q AND NOMINAL ORE, GIVEN LISTENING MEAN OPINION SCORES (Y_{LE}) AS FUNCTIONS OF OVERALL TRANSMISSION LOSS AND LEVEL OF CIRCUIT NOISE

x_L dB	N, dBmp, *at rec.* LTC	Y_{LE}	V_R, dBV *at 0* dB RRE	N, dBV, *at 0* dB RRE	G_n dB	Q dB	Y_C	V_C+ SRE dBV	*Nom* ORE dB
(1)	(2)	(3)	(4)	(5)	(6)	(7)	(8)	(9)	(10)
20	−40	0.91	−32.5	−34.3	1.8	3.0	1.01	−8.7	23.8
25	−45	0.95	−37.5	−39.3	1.8	3.0	1.06	−8.8	28.7
15	−40	1.50	−27.5	−34.3	6.8	7.6	1.60	−9.5	18.0
20	−45	1.67	−32.5	−39.3	6.8	7.6	1.73	−9.6	22.9
25	−50	1.59	−37.5	−44.3	6.8	7.6	1.66	−9.6	27.9
10	−40	1.90	−22.5	−34.3	11.8	11.3	1.91	−9.9	12.6
15	−45	2.10	−27.5	−39.3	11.8	11.3	2.07	−10.1	17.4
20	−50	2.16	−32.5	−44.3	11.8	11.3	2.12	−10.2	22.3
25	−55	2.14	−37.5	−49.3	11.8	11.3	2.05	−10.1	27.4
10	−45	2.43	−22.5	−39.3	16.8	15.2	2.32	−10.4	12.1
15	−50	2.60	−27.5	−44.3	16.8	15.2	2.45	−10.6	16.9
20	−55	2.68	−32.5	−49.3	16.8	15.2	2.51	−10.7	21.8
25	−60	2.43	−37.5	−54.3	16.8	15.2	2.32	−10.4	27.1
10	−50	2.88	−22.5	−44.3	21.8	16.8	2.66	−10.8	11.7
15	−55	3.09	−27.5	−49.3	21.8	16.8	2.81	−11.0	16.5
20	−60	2.90	−32.5	−54.3	21.8	16.8	2.67	−10.8	21.7
25	−65	2.65	−37.5	−59.3	21.8	16.8	2.49	−10.6	26.9
10	−55	3.34	−22.5	−59.3	26.8	18.8	2.98	−11.2	11.3
15	−60	3.28	−27.5	−54.3	26.8	18.8	2.95	−11.1	16.4
20	−65	3.16	−32.5	−49.3	26.8	18.8	2.86	−11.1	21.4
10	−60	3.48	−22.5	−54.3	31.8	20.4	3.08	−11.4	11.1
15	−65	3.43	−27.5	−59.3	31.8	20.4	3.04	−11.3	16.2
10	−80	3.72	−22.5	−74.3	51.8	≮22	3.25	−11.6	10.9
15	−80	3.58	−27.5	−74.3	46.8	≮22	3.16	−11.4	16.1
20	−80	3.28	−32.5	−74.3	41.8	≮22	2.95	−11.1	21.4
25	−80	2.86	−37.5	−74.3	36.8	≮22	2.64	−10.8	26.7

Notes: Cols 1, 2 and 3 are the primary data, x_L being the loss between LTCs having SRE = = +7.1 dB and RRE = −5.4 dB

The values of Y_{LE} apply for speech at RVL and the handset at BMSP; under these conditions the speech voltage at the output of the sending LTC is −17.9 dBV. This is denoted by V_L.

Col. 4 = $V_L - x_L -$ RRE

Col. 5 = col. 2 −2.2+2.5−RRE (−2.2 is to convert from dBm to dBV and +2.5 is to convert from psophometric weighting to 'flat' measurement in the band 300–3400 HZ)

Col 6 = col. 4 − col. 5

Cols. 7 is obtained from col. 6 and the information in Fig 4.21.

Cols. 8, 9 and 10 are obtained from cols. 3 and 4 as explained in Section 4.2.1.3 using Tables 4.7 and 4.8b contained therein.

The conclusions from the determination are given by cols 7, 8 and 10.

computations, col. 3 gives listening opinion score (Opinion scale No. 7) as a function, jointly, of listening level (col. 4) and Q (col. 7).

The conversion from listening opinion scores (col. 3) to conversation opinion scores (col. 8) and determination of the values of nominal ORE (col. 10) that correspond, under conversation conditions, to received speech levels as given in col. 4 is explained in Section 4.2.1.3.

The information given in col. 8 of Table 4.19 has been used, with some interpolation, to plot graphs of Y_C as functions of Q for

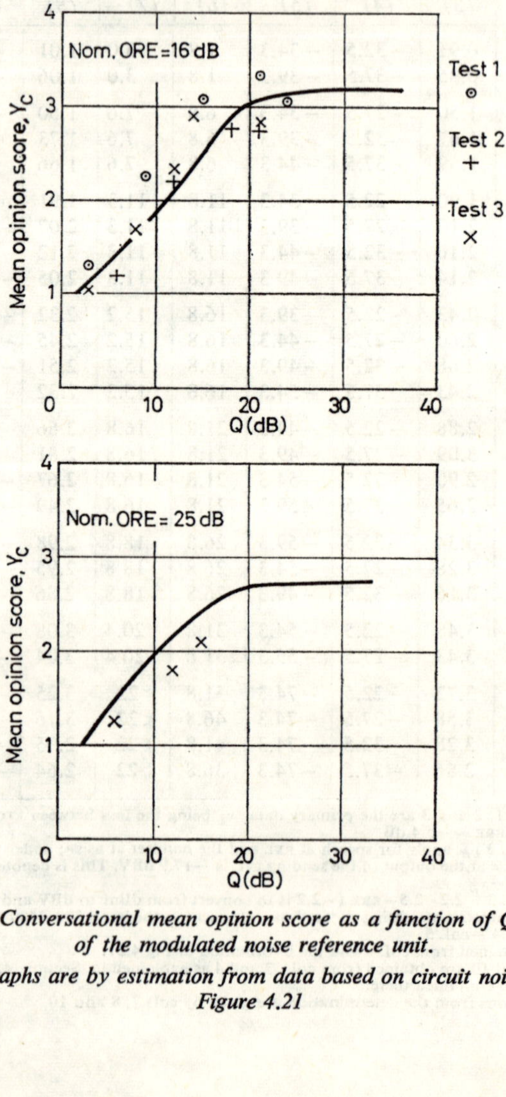

Figure 4.23. Conversational mean opinion score as a function of Q, the setting of the modulated noise reference unit.

Note: The graphs are by estimation from data based on circuit noise and use of Figure 4.21

values of nominal ORE of 16 and 25 dB. These are shown in Figure 4.23 which also contains some points obtained from three conversation experiments; the agreement between estimation and experiment is considered satisfactory, especially in view of the many differences in telephone equipment which had to be allowed for in the estimations. The experimental results apply for a laboratory speech link arranged as in Figure 4.6 which appears in Section 4.1.

The foregoing justifies treatment of problems in which nonlinear distortion is an important transmission feature by use of equivalences of quantising distortion to suitable levels of injected circuit noise as derived from the relationship between G_n and Q given in Figure 4.21. Use can then be made of fundamental information relating to telephone connections concerning opinion scores as a function jointly of nominal overall reference equivalent and level of circuit noise present; Tables 4.10, 4.12 and 4.14 contain examples of such information. An example of the application of the principles described here is given in Section 6.2.

4.2.2.2 Nonlinearity involving storage

A transmission element having some storage arrangements can 'remember' features of the past variations in input amplitude with time. To a certain degree, carbon telephone microphones possess this property and no doubt this has contributed to the difficulties that have been experienced in relating instrumentally determined measures of nonlinear distortion in such devices with their acceptability as parts of telephone sets. A deliberate introduction of nonlinearity with storage can serve to improve the transmission performance of a speech link suffering from noise arising in a main transmission element between terminal equipments, e.g. in a cable or radio relay system. Examples of such applications are given in Section 6.5, under the heading of 'amplitude compression or limiting'; advantage is taken of the characteristics of speech and hearing perception to deceive the ear–brain mechanism into believing that the connection is quiet because noise is being reduced or suppressed when speech is absent or low level segments are being transmitted. When loud speech segments are being transmitted the accompanying noise is much greater but this is not noticed because it is masked by the speech.

The performance of such devices when the storage time constants or integration periods are comparable with the shorter speech segments (say 10–20 ms) can be defined in terms of the respective levels of circuit noise during 'speech off' and 'speech on' conditions. For example, the 'companding advantage' of a syllabic compandor can be expressed as a simple function of 'speech off' noise level and 'speech on' signal to noise ratio.

Let N, dBmOp, be the level of circuit noise on a transmission system without the compandor; and let S, dBmO, be the level of mean power of the speech at the input to the compressor part of the compandor. Compressor and expander are arranged so that each has a loss of 0 dB when the input level of a test tone is U, dBmO; U is chosen to ensure that the transmission system is properly loaded by the compressed speech signals and the value of U is termed the 'unaffected level'.

While speech is present at level S, it will be raised in level by the compressor by an amount $\frac{1}{2}(U-S)$ dB, assuming that the degree of compression and expansion obeys a 2:1 law when expressed in dB. The speech to noise ratio will therefore be improved to this extent by the compressor and the improvement in 'speech on' performance will be unaffected by transmission through the expander. During 'speech off' periods, only the expander has any influence and its effect is to reduce the level of circuit noise reaching the listener by $U-N$, so improving the ratio of speech while present to circuit noise during the silent intervals in speech by this amount.

Listening comparison tests in which subjects were asked to choose between speech paths with and without compandors when different levels of circuit noise were present have shown that the effective improvement by introducing a 2:1 compandor is equal to $\frac{1}{3}$ (improvement in speech on signal to noise ratio) $+\frac{2}{3}$ (improvement in speech off circuit noise level); the effective improvement being the difference between the actual level of circuit noise with the compandor present and the subjectively equivalent level of circuit noise in the absence of the compandor.

Therefore, taking $N=-30$ dBmOp (corresponding to a very noisy circuit), $S=-12$ dBmO (approximately a median talker using a long distance connection with average local lines) and $U=-6$ dBmO (which results in practically no change in speech power loading when the compressor is introduced):

$$\text{Improvement} = \frac{U-S}{6}+\frac{2(U-N)}{3} = \frac{5U}{6}-\frac{S}{6}-\frac{2N}{3}\ \text{dB}$$

$=-5+2+20 = 17$ dB which represents a very substantial improvement.

Naturally, certain problems may arise if the circuits being compandored are numerous and when there are appreciable frequency bandwidth limitations in them; use of compandors is therefore justified only when the improvement needed is substantial and will offset the disadvantages. Note also that the transmission of non-speech signals may be worsened because the device is designed to exploit the statistical and other peculiarities of speech and hearing.

4.2.3 Group delay/frequency distortion

When extensive changes in transmission loss as a function of frequency are present in a transmission element, it is likely that this will be accompanied by changes in phase at frequencies adjacent to the loss discontinuity. Such group delay/frequency distortion is found in loaded cable lines and in the channel filters of multiplex transmission systems. Group delay/frequency distortion has not been subjected to such systematic study as attenuation/frequency distortion and no comparable body of information exists; the CCITT has recommended (G.133) that the distortion (relative to the delay at 1000 Hz) should not in a complete 4 wire chain of circuits, exceed 60 ms at 300 Hz and 30 ms at 3400 Hz but typical values for 12 pairs of FDM channelling equipments in tandem may amount to the following:

300 Hz	50 ms
400	35
2000	14
3000	22
3400	41

Figure 4.24 shows the delay distortion/frequency characteristics of typical United Kingdom Post Office pairs of channel filters, the attenuation/frequency distortion of which is shown in Figure 4.16.

Group delay/frequency distortion, when severe, may cause transmitted speech to appear to be accompanied by an echo; when the distortion is mainly at the upper part of the transmitted frequency range, the effect can be described a 'ringing'; distortion at the lower frequency part of the range causes the speech to be 'blurred'. Somewhat similar subjective effects are produced by 'near singing' distortion present when the stability margin of a 4 wire transmission circuit is inadequate. Near singing distortion manifests itself by 'rolls' in the attenuation/frequency characteristic and by 'listener echo', i.e. repeated reproduction of certain parts of the fre-

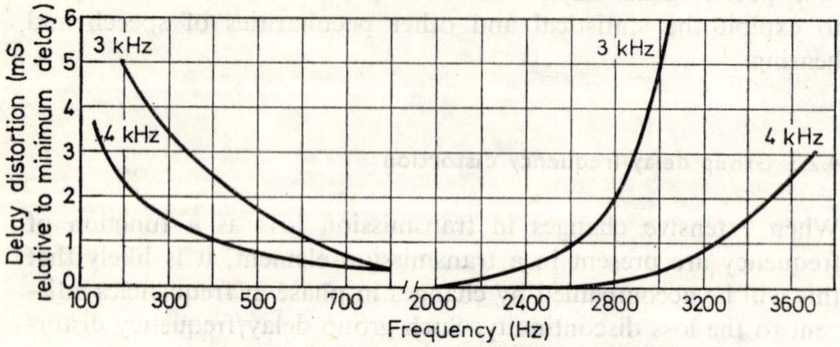

Figure 4.24. Delay distortion/frequency characteristics of 3 and 4 kHz spaced telephone channels (single audio–audio links)

quency band, decaying relatively slowly. If, at a given frequency, the voltage received after passing completely round the go and return loop is $1/r$ of that at the input to the loop, the 'stability', M, at that frequency, assuming the most unfavourable change in phase, is $10 \log_{10} r$ dB. The frequencies at which maxima and minima occur in the attenuation/frequency characteristic of the circuit will depend upon the delay round the loop but the spread, in dB, of these rolls is given by:

$$D = 20 \log_{10} \frac{r+1}{r-1}$$

and the minimum value of the ratio of signal power transmitted to the power of the listener echo, expressed in dB, is given by:

$$E = 20 \log_{10}(r-1)$$

A few values calculated from these expressions are shown in Table 4.20.

It has been suggested that satisfactory transmission performance of telephonic speech requires a minimum value of 3 dB for M.

Table 4.20. AMPLITUDE OF ATTENUATION/FREQUENCY 'ROLLS' AND RATIO OF SIGNAL TO LISTENER ECHO AS FUNCTIONS OF STABILITY

Stability		*Amplitude,* dB, *of 'rolls'*	*Ratio,* dB, *of signal to listener echo*
M, dB	*r*	*D*	*E*
0	1	∞	−∞
1.5	1.4	15.6	−8
3	2	9.6	0
5	3.2	5.6	6.8
6	4	4.4	9.6
9	8	2.2	16.9

4.2.4 Combination of the effects of different listening factors

Any general statement concerning the manner in which the effects of different degradations interact with each other is bound to be very complicated if the various factors have comparable effects; in most practical problems in telephony, however, only transmission loss is likely to cause a substantial proportion of telephone customers to have difficulty and so a great deal of simplification of the treatment is possible.

Only the effects of listening degradations are considered here, although this is only part of the more general problem which must embrace also the talking and conversational degradations. Specifically the factors considered here are transmission loss (l), circuit noise (n), attenuation/frequency distortion (d) and quantising distortion (q). Under present network conditions, transmission loss is almost always the dominant degrading factor.

No very systematic study of this matter has been made in the context of telephony performance but some information on similar matters has been published relating to the visual appreciation of still, coloured pictures projected on a television screen and suffering from various degradations in combinations. The same theoretical models that were developed for this purpose may have some relevance in telephony but this has not been examined in detail; for the present, it seems sufficient to adopt the following simple statistical model to cater for practical combinations in telephony of the four listening factors mentioned above.

It is assumed that, for a given telephone connection containing a particular combination of the four factors, dissatisfaction with one factor is independent of dissatisfaction with any of the others. This leads to a model in which four basic probabilities are postulated which are constant for a given telephone connection; these are as follows:

p_l = probability of dissatisfaction due to transmission loss
p_n = probability of dissatisfaction due to circuit noise
p_d = probability of dissatisfaction due to attenuation/frequency distortion
p_q = probability of dissatisfaction due to quantising distortion

The events associated with each probability, i.e. dissatisfaction, are assumed to occur independently; this does not, however, imply that the probabilities which govern their frequencies of occurrence do not change together if, for example, the overall transmission loss is changed. Indeed changing the overall loss affects not only p_l but also p_n and p_d and to some extent p_q.

Let P_{ln} be the probability of dissatisfaction due to transmission loss or circuit noise when both are present.

P_{ld} be the probability of dissatisfaction due to transmission loss or attenuation/frequency distortion when both are present.

and P_{lq} be the probability of dissatisfaction due to transmission loss or quantizing distortion when both are present.

In accordance with the assumptions concerning independence of events, it follows that:

$$P_{ln} = p_l + p_n - p_l p_n$$

$$P_{ld} = p_l + p_d - p_l p_d$$

and

$$P_{lq} = p_l + p_q - p_l p_q$$

In practical problems (*see*, for example, Section 6.2), P_{ln}, P_{ld} and P_{ld} can be determined as well as p_l; P_{lndq}, however, which is used to denote the total probability of dissatisfaction when all four factors are present, cannot usually be determined directly from information available and so must be estimated from the model. According to the assumptions:

$$P_{lndq} = p_l + p_n + p_d + p_q - (p_l p_n + p_l p_d + p_l p_q + p_n p_d + p_n p_q + p_d p_q) + p_l p_n p_d + p_l p_n p_q + p_l p_d p_q + p_n p_d p_q - p_l p_n p_d p_q$$

The single probabilities are given by the expressions:

$$p_n = \frac{P_{ln} - p_l}{1 - p_l} \quad p_d = \frac{P_{ld} - p_l}{1 - p_l} \quad p_q = \frac{P_{lq} - p_l}{1 - p_l}$$

Examples of the application of these formulae appear in Sections 6.2 and 8.4.

4.3 TALKING FACTORS

Chief of the features of telephone connections that may cause difficulty to the participants in their roles as talkers are sidetone and talker echo. Fundamental information is therefore needed on the minimum losses tolerable for sidetone and echo paths; this should include some suitable method of averaging the losses when they are dependent on frequency. In the case of echo, the round trip delay time is also an important parameter. Some rather detailed results of tests are included in Section 3.5, which treats the problems of method for assembling suitable information; the present section includes only that which is generally useful in transmission planning of telephone networks.

4.3.1 Speech sidetone

The transmission loss of a sidetone path, from the mouth to the ear of the person who is talking, is usually expressed for telephone engineering purposes in terms of sidetone reference equivalent, that is as the loss introduced into the NOSFER which results, under certain specified test conditions, defined in CCITT Recommendation P. 73, in the same loudness of speech sounds perceived by a listener as that produced by the sidetone path. If actual measurements of sidetone reference equivalent (STRE) are not possible or results are not available, the STRE can be calculated given appropriate sensitivities as functions of frequency for microphone and earphone, together with knowledge of the characteristics of the electrical portion of the telephone set and the line plant to which it is connected. Such calculations are covered by the information given in Section 3.2.4 (which explains the calculation of loudness ratings from the mouth to ear losses) Section 5.1.5.2 (which explains the method for defining the electrical characteristics of the sidetone path of a telephone set) and Section 6.1.4 (which explains the method for determining the mouth to ear loss of the sidetone path allowing for the characteristics of the line plant). It must be emphasised that the sidetone loss (STRE) of a telephone set may be critically dependent on the line characteristics and so no direct subjective determination of STRE according to Recommendation P. 73 under a single arbitrary line condition can give a proper indication of the sidetone performance of a telephone set when used in a switched telephone network; the line conditions differ from one station to another and even from one call to another for a given telephone station.

Expressing speech sidetone path losses as single numbers according to loudness rating determinations, tacitly implies the assumption that all speech sidetone paths of given loudness rating (e.g. STRE) are equally acceptable, whatever their frequency responses

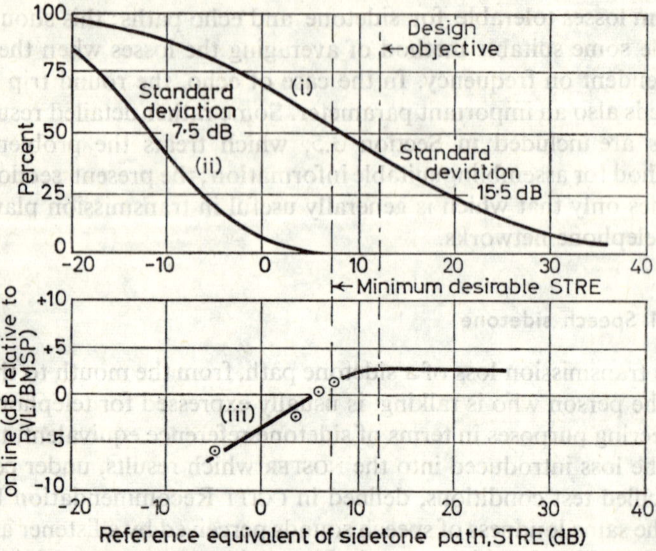

Figure 4.25. Effect of variation of sidetone reference equivalent (STRE) *on detectability, objectionableness and speech voltage. (i) Percentage of talkers able to detect sidetone after reciting a 15 second passage in a silent environment. (ii) Percentage of talkers objecting to sidetone under the same conditions as i. (iii) Speech voltage from a sending end relative to that produced when* RVL *and* BMSP *are used. Observations made during conversations over connections of about 20 dB nominal* ORE *with 50 dBA room noise present but with no circuit noise*

and other characteristics; this assumption has only limited empirical validation although it seems, prima facie, to be quite reasonable. For the present no practical alternative has been proposed to use of the loudness criterion for weighting the relative importance of transmission losses at the various frequencies. For this reason fundamental information on the effects of varying the sidetone losses of telephone sets is conveniently presented as functions of STRE.

Figure 4.25 shows the percentage of customers likely to be able to detect the presence of sidetone and of those likely to find it objectionable; also shown is the relationship between speech output from a handset telephone and STRE. The speech voltage information has been derived from tests on a modern type of telephone set (with its sidetone path arranged for controlled reduction and increasing of its normal sidetone loss) and show the output relative to that from the same telephone when spoken into at RVL and BMSP. The speech output from a telephone set of which suitable sending sensitivity values are available can be calculated as shown in Table 6.9. Such estimations require to be checked wherever possible by means of suitable conversation tests because the result can be appreciably affected by the design of the telephone handset.

The information shown in Figure 4.25 can be used to set design objectives for telephone sets; for example, in the United Kingdom STRE of 12 dB is considered a reasonable objective, although difficult to attain under all line conditions, and 7 dB is a lower bound below which the STRE should not be allowed to fall except under fairly rarely occurring circumstances. Information from tests by the American Telephone and Telegraph Company (Bell Telephone Laboratories) show somewhat more severe effects on 'objectionableness' than those shown in curve ii of Figure 4.25 (median threshold further to the right, standard deviation about 6 dB) but a less steep relationship between speech voltage and STRE than that shown in curve iii of Figure 4.25. It must be remembered that the results can be affected considerably by the methods used to conduct the tests, see Section 3.5.

Some remarks on the subject are contained in Section *E* of CCITT Recommendation P. 21 (G. 121). In view of the increased importance of sidetone in telephone sets having greater sensitivities, the subject certainly requires more study. For the present, however, the information given in curve iii of Figure 4.25 may be used to determine sending allowances for sidetone paths of appreciably lower loss than the design objective mentioned above.

4.3.2 Talker echo

In certain classes of telephone connection, namely when the mean one way propagation time of the trunk circuits exceeds about 12 ms, echo presents an important network design problem and information is needed showing what percentage of customers is likely to experience echo that is objectionable to them. Figure 4.26 shows a pair of sigmoid curves ('detectable' and 'objectionable') that apply for a round trip delay of 50 ms. The shapes of these curves and their relative displacements do not change appreciably for round trip delays in the range 15 to at least 60 ms but their joint location along the scale of nominal overall reference equivalent of the echo path

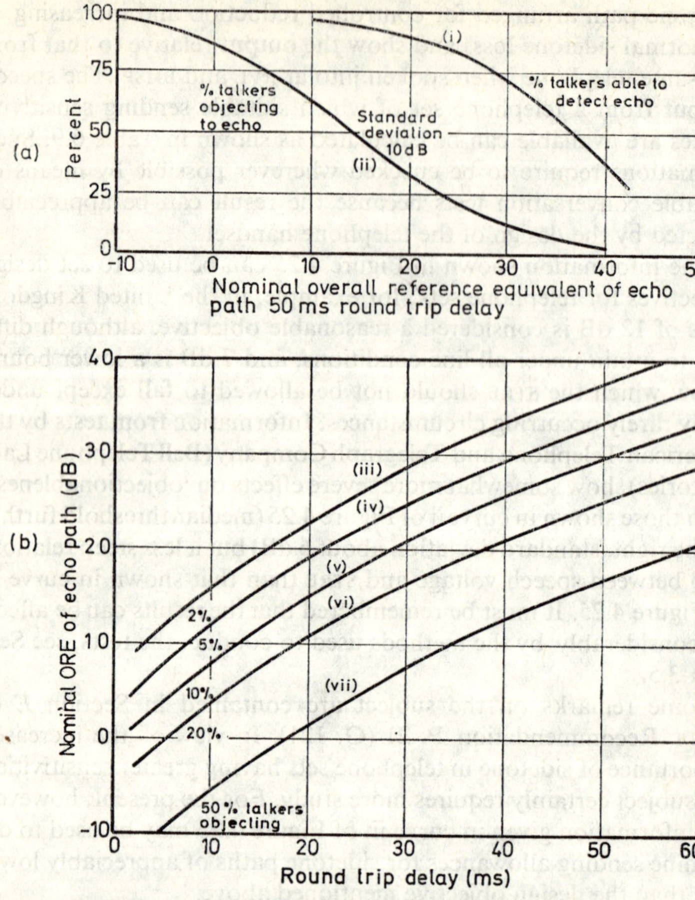

Figure 4.26. Effect of talker echo on detectability and objectionableness (a) Percentage detecting and objecting. (b) Relationship between nominal ORE *of echo path and delay*

(SRE + RRE + losses of intervening lines) is a function of the round trip delay. Information is given in Section 3.5 that shows how the median threshold of objectionableness varies with roundtrip delay and this, together with the information contained in curves i and ii of Figure 4.26, enables the fundamental information to be presented in the convenient form shown in curves iii, iv, v, vi and vii.*

* It will be found that the values from curve ii are not exactly consistent with those from curves iii, iv, v, vi and vii; this is because different subjects were used and because the latter curves include information from several other experiments.

4.4 CONVERSATIONAL FACTORS

4.4.1 Asymmetry

One participant in a conversation may be able to hear the other very well but be unable to make his speech understood due, for example, to a considerably greater loss in the direction from him to the respondent than in the other direction. Clearly this participant will rate the connection less favourably than if the other participant can hear him as well as he can hear the distant participant. It might also be supposed that the person who has to listen to speech over the worse direction path would rate this less highly because the other person is not experiencing the same difficulty as he is; however, this is not so as the results in Figure 4.27 show.

The upper part of this figure reproduces some results of laboratory conversation tests conducted by the L. M. Ericsson Co., Stockholm, Sweden in a manner similar to that described in Section 3.6.2. and using the Opinion scale No. 9, as described in Section 3.7.2, to determine conversational mean opinion scores. The transmission loss of the connection was made different in the two directions and the results can be compared with those for symmetrical connections of various losses. Curve i shows the conversational mean opinion score, Y_C, for symmetrical conditions as a function of the nominal overall reference equivalent. Curves ii and iii show the results when the loss from *A* to *B* was kept constant at 42 dB nominal overall reference equivalent and that from *B* to *A* varied over the range 0 to 42 dB; curve ii shows the results for the participants at *A* and curve iii shows those for *B*. *A*'s opinion (curve ii) is slightly reduced by the larger loss in the direction *A* to *B* but *B*'s opinion (curve iii) was hardly affected by the smaller loss in the direction *B* to *A*. The speech levels measured at the outputs of the respective telephone sets are shown in curves iv, v and vi; it will be noticed that the speech level of *B* is raised in the asymmetrical cases as well as that of *A*, although *A* can hear *B* quite well without *B* talking any louder. This effect in which the subjects try to share their difficulties has also been observed in other tests (Ref 1)*.

The lower part of Figure 4.27 shows similar results from tests conducted in the same way but in the laboratories of the Swedish Board of Telecommunication. Curve vii shows the results for symmetrical connections; the assymmetry in these tests was, however, arranged so that the mean loss of the two directions remained constant at about

* See Section 4.4.1 of the Bibliography.

22 dB. Amounts of asymmetry, that is differences between the losses in the two directions, up to ±44 dB were introduced. The results for the asymmetrical cases are plotted against the loss (nominal overall reference equivalent) in the direction *B* to *A*; a value of 22 dB on the abscissa scale therefore corresponds to zero asymmetry, 12 dB represents 20 dB asymmetry, etc, as indicated in the figure. It will be seen that for asymmetry within the range ±20 dB, *A*'s opinion is affected only by the loss from *B* to *A* and *B*'s opinion only

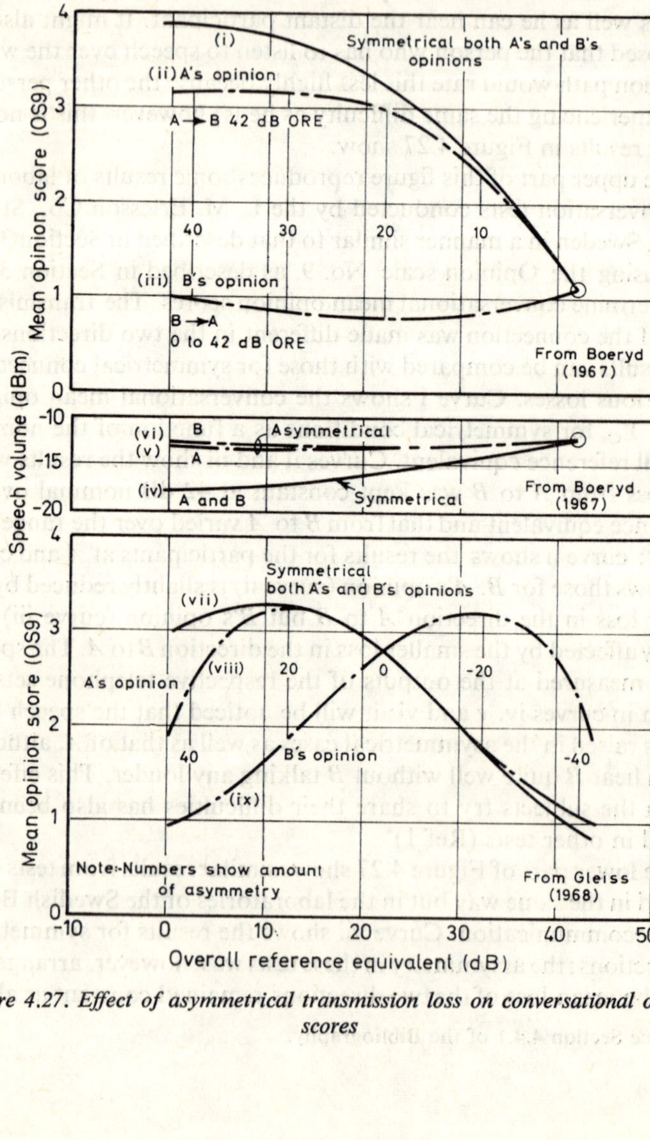

Figure 4.27. Effect of asymmetrical transmission loss on conversational opinion scores

by that from *A* to *B* as shown in curves viii and ix. Even when the asymmetry is very large, the opinion of the less fortunate participant (who suffers by receiving the more attenuated speech) is not worse than it would have been had the connection been symmetrical and of loss equal to the worse direction of the asymmetrical case.

Asymmetry can arise from other factors than line transmission loss. For example, the telephone sets in different countries have somewhat different relationships between their sending and receiving sensitivities (or reference equivalents) and so asymmetry of as much as 8–10 dB can exist when particular types are interconnected over an international circuit. Naturally, within each country, the connections will remain symmetrical notwithstanding differences in the quantity SRE–RRE. Certain other circumstances can increase asymmetry of loss by a few decibels and additional asymmetry can be caused by circuit noise or room noise being different for the two participants. Even when the connection is physically symmetrical and the room noise is the same at both ends, talkers may emit their speech at levels differing by as much as 15 dB; participants will experience the effects of this in ordinary, non-telephone conversation and accept the situation and this probably accounts for their willingness to accept quite large physical asymmetries in a telephone connection.

4.4.2 Propagation time

In the very early days of telephony, connections were necessarily short and 'end to end' propagation times were quite negligible. When longer distance telephony first became possible, heavily loaded cables were employed and the loading reduced the speed of transmission; at this period, before amplification was feasible, circuit losses were still quite large and echoes were hardly troublesome as compared with the direct effect of the propagation time which caused any expected reply to reach the enquirer later than he expected and so could confuse the participants. At this period a limit of 250 ms was set for the mean one way propagation time (MOPT). When amplification of loaded cables became possible, circuit losses were reduced and echo became an important design consideration, requiring the development and introduction of voice switched echo suppressors. The introduction of high velocity transmission systems lessened the importance of studying the effects of long propagation times because only a few tens of ms could then be introduced into even very long distance connections; the short delayed echoes then present could not be ignored but were easily controlled by quite simple forms of voice switched echo suppressor.

In the later 1950s, the prospect of satellite communication systems appeared and interest in studying these matters was revived. By 1965, when the first commercial satellite system for public telephony was inaugurated, studies had shown that the propagation time of a single hop synchronous orbit satellite system would be acceptable, provided that suitable types of echo suppressor were used. In the course of this early work, it was found that the direct effect of increasing the propagation time (in the absence of any echo, a condition achievable in a complete 4 wire laboratory speech link

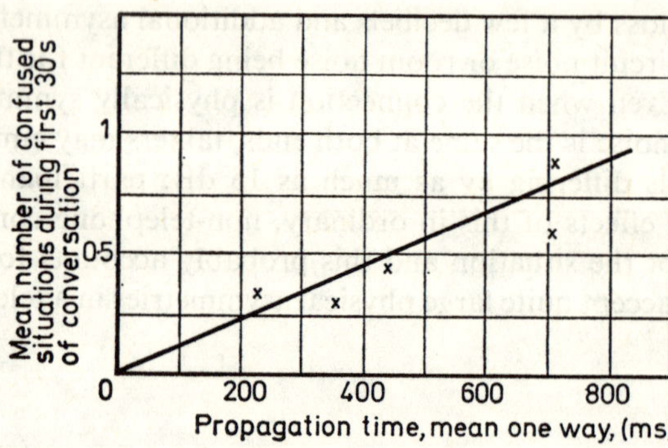

Figure 4.28. Incidence of confused situations in conversation over connections having long propagation times.
Note: The connections used were laboratory speech links free from echo and without echo suppressors

but not yet practicable in a real network) could be observed in certain features of the structural pattern of conversations. Figure 4.28 shows the frequency with which confused situations occur which can readily be identified as explained in Section 2.3.2. The incidence shown in the figure applies for the first 30 s of the conversation; after the first minute, the incidence is much reduced. Later studies have shown, however, that troublesome situations occurring at the beginning of a conversation have much less influence on the opinions of the participants than is the case when the same type of trouble occurs after the first minute.

After the first synchronous orbit satellite communication system (Early Bird) had been put into public service, a very extensive series of experiments was conducted. During the first six months of service, a sample of the customers who had made calls between Europe (including the United Kingdom) and the United States of America were interviewed and asked various questions concerning their opinions of the call they had just completed, including whether they or their respective partners had experienced any difficulty in conversing. The results are shown as Curve i in Figure 4.29; it will be seen that laboratory tests (conducted later) support these 'field' test results. The Early Bird results and the laboratory tests referred

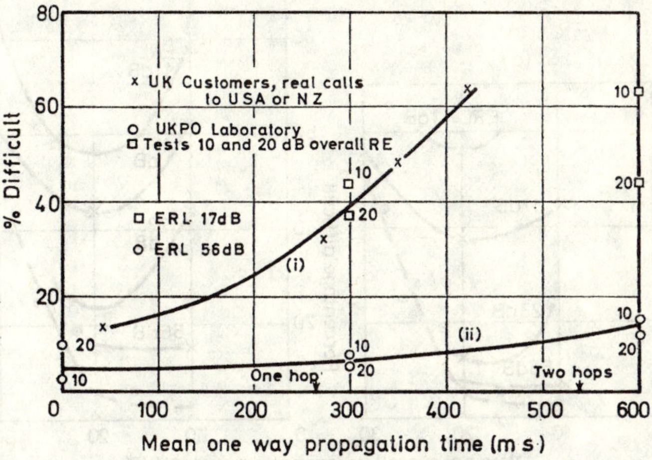

Figure 4.29. Effect of long propagation time on percentage of users who experience difficulty.

Notes: **1.** ***Propagation times are marked for synchronous orbit satellite circuits with negligible terrestrial extensions.*** **2.** ***Curve i applies for average echo return losses of about 17 dB, curve ii applies for echo return losses greater than about 30 dB.*** **3.** ***Echo suppressors of the voice switching type were present in all cases***

to apply for speech links having not only the value of propagation time indicated in the abscissa of the figure but also practical imperfections in echo return loss which necessitated use of a new type of voice switched echo suppressor with certain limitations in its performance.

The later laboratory tests have shown that increasing the echo return losses of speech links to very high values considerably improves the performance, as judged by the subjects' opinions concerning any difficulty they experienced*. This is illustrated by curve ii in

* This does not, of course, imply necessarily that their behaviour was not adversely affected (*see* Figure 4.28).

Figure 4.29. It is interesting to note that the increase in echo return loss eliminates the bursts of echo released by the echo suppressors just after double talks but does not eliminate the 'chopping' which also occurs occasionally; it may be concluded that subjects' opinions are little affected by such chopping but are seriously affected by the echo bursts. (Some early experimental work on this subject also yielded results that support the proposition that a moderate

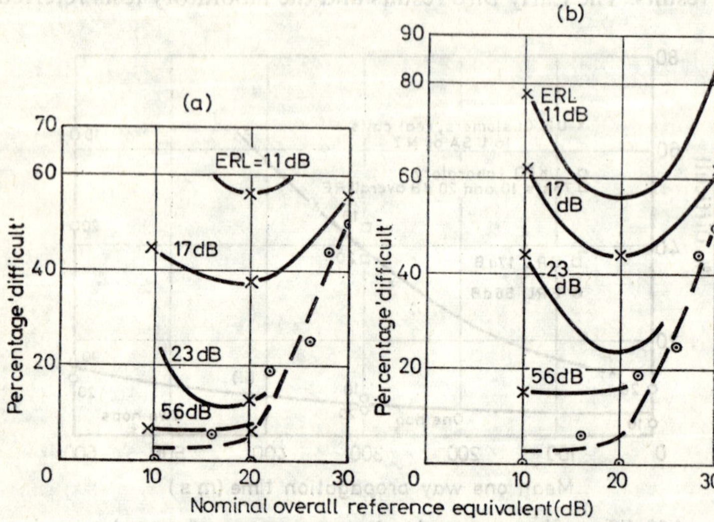

Figure 4.30. Effect of echo return loss (ERL) on percentage of users who experience difficulty in conversation. (a) MOPT = 300 ms, (b) MOPT = 600 ms
Notes: **1.** *Encircled points and broken line apply for negligible propagation time.* **2.** *Shapes of curves for 11 and 23 dB ERL are assumed from the information available for 17 dB.* **3.** *The results for 300 and 600 ms MOPT apply for circuits furnished with echo suppressors of the present voice switching type*

amount of chopping is quite acceptable under certain circumstances, *see also* Section 4.4.3).

Under the conditions when rather long propagation times are present, it may become economically feasible to apply rather sophisticated means for improving echo return loss and it then becomes necessary to determine how large this must be. Figure 4.30 shows the results when an existing type of voice switched echo suppressor is also present; when the overall transmission loss is fairly high, the echo return loss does not need to be particularly great (although greater than is at present possible by conventional means) but low overall losses require very high echo return loss values. (Nominal overall reference equivalent values in the range 6–18 dB are likely to be the aim for modern telephone networks).

At the present time the maximum recommended value of mean one way propagation time (MOPT) for international connections is 400 ms (CCITT Recommendation G. 114 (P.14)).

4.4.3 Voice operated devices

Under certain conditions there is a need for, or some advantage can be gained by, discrimination between periods when speech is 'present' or the talker is 'active' and those when speech is absent or the talker inactive. Measurement of speech voltage commonly requires this facility (*see* Section 2.1.3.2) and examples of the use of such discrimination in service include suppression of echoes, provision of extra sensitivity in loudspeaking telephones, economy of channel capacity by 'time sharing', stabilisation of connections subject to wide fluctuation in transmission loss and economising in carrier power for certain kinds of multiplex system. The specific problems associated with each of these applications are discussed elsewhere: echo suppression and antisinging devices are mentioned in Section 6.3; loudspeaking telephones are treated in Section 5.2; TASI and other time sharing systems are included in Section 6.5. A central problem in designing all such devices is that of detecting reliably and rapidly the presence of speech.

The detection of speech, as an electrical waveform, in the presence of interfering noise must be treated as a statistical problem the best solution of which depends on the purpose for which the detection is required. Three different features can be distinguished in the various applications mentioned above and these may require differing solutions of the problem. The three types of application are as follows:

1. Measurement of an activity factor in system loading problems; no great urgency exists in reaching each decision.
2. Detection of the onset of a speech utterance to control the action of connecting the talker to a free channel; some urgency exists in reaching a decision but delay of 10–20 ms is quite acceptable.
3. Detection of the onset of a speech utterance to control the action of disconnecting a previously connected silent channel; if the silence on this channel is not to be disturbed by even a very short 'spike' of speech the action taken as a result of the detection must be very rapid indeed, possibly in less than 1 ms.

For 1, the statistical problems do not need to be solved in 'real' time, especially if recordings are available.

2 is exemplified by the problems arising in designing a time sharing economy system (like TASI) where it is important that the talker shall be connected fairly quickly but disconnection, when he ceases, of the channel he has been using is relatively non-urgent. In this application it is better to err on the side to taking action (connecting) when speech is not really present than of not taking action when speech really is present.

An important example of 3 is the problem of suppressing echo. When only one participant becomes active, action to suppress must be taken very rapidly but removal of the suppression may be delayed without disadvantage. If speech is present in both directions simultaneously, it is better to err by not suppressing, because suppression then has the disadvantage of 'chopping' wanted speech; when only one talker is active, suppression of his speech by falsely interpreting noise on the other channel as speech must also be avoided. Detection thresholds for echo suppression therefore tend to be placed somewhat higher in level than those of a TASI speech detector.

Voice switching in a loudspeaking telephone or in an antisinging device is particularly difficult to design because the action taken on detection is two-fold, namely to connect (or facilitate) the channel in one direction and to disconnect that in the other without at any time allowing both to be connected.

The actual onset of a speech utterance is very difficult to define rigorously because speech sounds differ considerably in their relative levels and they do not suddenly reach their full intensities. In the range of levels from 23 to 13 dB below the level of the long term mean power of the speech while active (23–13 dB below the SV5B reading), the rate of rise of intensity level is, on average, about 0.3 dB/ms. Setting the threshold of an envelope type detector at a level 15 dB below the level of the long term mean power of the speech while active is usually sufficiently low to ensure negligible loss of initial parts of speech sounds and, for most sounds, a further delay in connection of up to 30 ms can be tolerated with such detection; exceptions to this are *h* and *s* which will not tolerate delay of even as little as 6 ms after activation of such a detector.

With envelope detection, the smoothing time constant for the control signal may be about 12–20 ms and this will considerably smooth the fluctuations in any constant level Gaussian noise so that the threshold setting of the speech detector (defined in terms of the r.m.s. level of a sinusoidal test tone) need not exceed the r.m.s. level of the noise by more than a small margin. Speech sounds, when considered over short periods of a few tens of ms, have probability density distributions of amplitude differing little from that of Gaussian noise and so, at the moment of detection (i.e. passage of the detector control signal through the threshold of the detector) the problem is one of trying to distinguish two Gaussian signals, of which one is rising in level at perhaps 0.3 dB/ms and requires to be detected as rapidly as possible. If no smoothing is applied to the control signal and detection is performed, not on the envelope but on the instantaneous waveform, the threshold must be made higher to avoid false operation by peaks in the noise (by perhaps 10 dB) and the speech, approximated as above by a Gaussian signal, will also have peaks higher by the same amount; the result will be practically no change in the operate-time or reliability of the detector. It is true that the minimum level of long term mean power of speech that will just operate the detector (after a long enough wait) will be much less with the instantaneous waveform type of detection but this is not relevant to any realistic detection problem requiring action in real time. Use of the envelope type of detector has the advantage that, under practical conditions where short duration spikes of noise may be present, false operation may be less likely to occur.

When speech has been detected and the appropriate action taken, it is necessary to include as part of each continuous utterance any short duration silences which are structural to the speech sounds being emitted (see Section 2.1.3.2). This can be achieved by providing a sufficient 'hangover' in the speech detector so that it does not become released during such intervals; a hangover of at least 100 ms is necessary. To alleviate the consequences of false operation by a short duration impulse of interference, it is the practice in TASI (and other similar) speech detectors to arrange that the full extent of hangover is provided only if the detector has remained activated (i.e. input level above threshold) continuously for at least a certain minimum period (50 ms in the case of TASI).

4.4.4 Handsfree telephony

At the present time, the telephone network is designed to have characteristics suitable for 'bread and butter' subscriber's station equipment, namely handset telephones with characteristics lying within fairly narrow limits. Other types of subscriber's station equipment are available but they are designed to fit in with the network characteristics which have been dictated by those of the overwhelming majority of ordinary handset telephones. The provision of handsfree telephony facilities is one example of such 'minority' use of the telephone network.

When the telephone is used for certain tasks, it is clearly advantageous for one participant to have both his hands free from holding the telephone handset; one example is the operator of a switchboard or data sending set, another is a person needing to consult files of such bulk or dispersion that he has to walk about in his telephone location. The first example would be suited in most cases by use of a head mounted microphone and earphone unit (or 'headset telephone', *see* Section 5.3). In the second example, it might be sufficient to provide only a loudspeaker so that the person concerned could hear while away from his desk telephone but would be willing to resume use of the handset for talking to his correspondent; in many situations, however, both-way conversation needs to be maintained while moving about.

While handsfree telephone facilities remain in the minority, connections over public circuits are very likely to be between a handset telephone at one end and a handsfree telephone at the other; rarely, except on PBX and private tie-line calls, will two handsfree telephones be connected together.

The characteristics necessary, and even the physical form of the equipment, to provide the best performance will differ considerably from one condition of use to another; it is therefore not surprising to find the performances of currently available types of handsfree telephone fall short of what is desirable when they are used under at least some of the conditions under which the customer expects to be able to use his equipment.

At present, handsfree telephone facilities are commonly thought of in terms of the so-called 'loudspeaking telephone' which seeks to enable the customer to hear and to speak to his correspondent with both his hands free and to permit him at least some measure of movement. To fit in with the characteristics of current telephone network plant, the characteristics and performance of loudspeaking telephone sets must conform with certain requirements (not yet all properly defined, much less universally agreed upon) and these are indicated in Section 5.2.

Some information concerning the fundamental transmission performance requirements is available from conversational and other laboratory experiments and this is presented below; first, however, it is necessary to adopt some arbitrary definitions of sending and receiving sensitivity for the cases of a microphone located some distance from the talker's lips and for listening to speech received from a loudspeaker a little way from the listener. The relative location of mouth, microphone, ears and loudspeaker used here as defined in Figure 4.31.

A loudspeaking telephone installation cannot be treated entirely separately from the room in which it is used; the rooms used for these experiments were treated to reproduce typical reverberation times likely to be found in a well carpeted and curtained office, although the dimensions of the test rooms were small by office standards, even for single person occupation. The rather modest distances from mouth to microphone and from loudspeaker to ears which have been adopted here do, however, reduce the influence of the room characteristics on the transmission performance from the very severe effects introduced by attempts to employ such equipment in a relatively untreated room and adopting much greater talking and listening distances.

The sending parts of the experimental equipment used consisted of directional (cardioid) high quality microphones having free field sensitivity/frequency responses that were practically level. The sensitivity of the microphone and associated amplifier, up to the 'line' terminals, was defined relative to that of the ARAEN sending end by comparing the speech voltage produced at the line terminals with that from the ARAEN when both are fed simultaneously with speech from an artificial mouth while maintaining the prescribed talking distances. The relative levels of speech voltage at the line terminals then yield a sending rating relative to ARAEN which may be treated as roughly approximating to a loudness rating. The reference equivalent of the ARAEN sending end is -2.5 dB and so the results can be converted approximately to units of reference equivalent by subtracting 2.5 dB.

The receiving parts of the experimental equipment consisted of a 10 in dia. domestic loudspeaker unit of fairly high quality mounted on a 2 ft square (600×600 mm) baffle board. Under free field conditions the loudspeaker and associated amplifier was almost uniform in frequency response over the speech frequency range, at least from 200–5000 Hz. The receiving sensitivity, from the 'line' terminals, was defined by measuring the sound pressure level at the point to be occupied by the listener's ear while speech was injected at the line terminals. The sound pressure level was measured by use of the sending end of ARAEN, which has a defined value of sensitivity practically independently of the frequency. This value of sound pressure level was compared numerically with that produced theoretically in a real ear applied to the receiving end of ARAEN when the same injected level of speech is applied

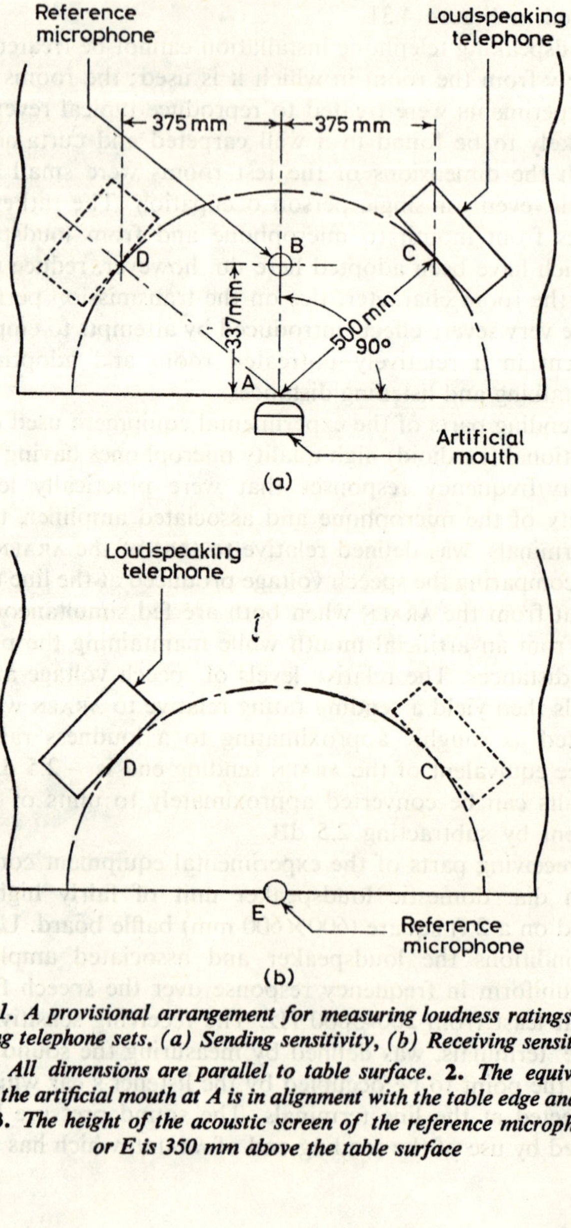

Figure 4.31. *A provisional arrangement for measuring loudness ratings of loud-speaking telephone sets. (a) Sending sensitivity, (b) Receiving sensitivity.*

Notes: **1.** *All dimensions are parallel to table surface.* **2.** *The equivalent lip position of the artificial mouth at A is in alignment with the table edge and 350 mm above it.* **3.** *The height of the acoustic screen of the reference microphone at B or E is 350 mm above the table surface*

to the line terminals. An allowance was made, based on a theoretically determined value, for the obstruction effect present when a real subject is listening to the loudspeaker. The final result has been treated as a loudness rating relative to ARAEN although this must be qualified because only single ear listening is implied by the method measurement. The reference equivalent of the receiving end of ARAEN is +2.8 dB and so the results can be converted notionally to reference equivalents by adding 2.8 dB.

The definitions of sending and receiving sensitivities given above require further study before general adoption can be proposed but they do serve to permit comparison of test results.

The experiments were of two kinds, namely:

1. Listening tests to determine preferred listening levels and listening opinion scores appropriate to controlled talking conditions;
2. Conversation tests to determine talking behaviour and conversational opinion scores.

Telephone connections using handset telephones were also included in the same experiments for control purposes.

The sending and receiving ratings of each end of the equipment could be set up so that their total value (relative to ARAEN send and receive ends, respectively) amounted to a minimum of 31 dB without instability via the complete loop including both loudspeaker to microphone local paths. For these experiments the sending and receiving ratings were set to +20.5 dB relative to ARAEN (+18 dB 'SRE') and +15 dB relative to ARAEN (+18 dB 'RRE'); this provided a reasonable margin against instability and the separation of ratings between sending and receiving was arbitrary but it did permit either end of the loudspeaking telephone equipment to be replaced by a typical handset local telephone circuit without excessively noticeable asymmetry. In the conversation tests only loudspeaking telephone to loudspeaking telephone and handset to handset combinations were included and so only the total overall rating was important. The listening tests were conducted using recorded speech and the results were expressed in terms of speech voltages referred to 0 dB receive ratings and so these were also independent of the actual values of sending and receiving ratings used.

During the conversation tests the speech voltages produced at the output of the sending parts of the loudspeaking telephone sets were measured and the results referred to 0 dB sending ratings*. The values of V_C+'SRE' for the loudspeaking telephone sets were as follows when associated with the conversational mean opinion scores (Y_C) (using Opinion scale No. 9 *see* Section 3.7.2).

Without room noise,	$Y_C = 3.58$, V_C+'SRE' =	−4.6 dBV (SV5B)
	3.38	−5.0
	3.04	−3.4
	2.62	−3.4
With 50 dBA room noise,	$Y_C = 2.84$, V_C+'SRE' =	−1.2 dBV (SV5B)
	2.37	−1.0
	1.79	−0.8

Presented in this manner the results with 50 dBA room noise can be compared with those in Table 4.8 in Section 4.2.1.3 and it will be seen that the speech voltages are some 9–10 dB higher at a given value of Y_C. During the conversation tests, vocal levels were measured by means of an auxiliary microphone and from the information gained it can be stated that the participants were emitting speech some 5.5 dB higher in vocal level; the remainder of the 9–10 dB increase in speech voltage sent to line is accounted for by the use, by the subjects, of somewhat closer mouth to microphone distances than those used to determine the sending ratings of the loudspeaking telephone sets (as shown in Figure 4.31). The conversational opinion scores, Y_C, are shown in Figure 4.32 plotted as functions of nominal overall reference equivalent (for the handset connections) or 'notional' nominal overall reference equivalent (for the loudspeaking telephone connections). In this figure, the points are from the conversation tests and the graphs are derived from the results of the listening opinion tests according to the method described in Section 4.2.1.3. The transmission loss between the handset local telephone circuits that corresponds to the values of nominal ORE shown are indicated in a subsidiary scale. Additional subsidiary scales are also included indicating the values of transmission loss between loudspeaking telephone sets which correspond to the values of nominal 'ORE'

* Corresponding measurements were made on the handset connections and the speech voltages obtained were as expected from other experiments. The conversation opinion scores, however, were much lower than expected (see comment later).

shown; one of these applies to the experimental non voice-switched arrangement used in these tests and the other is an estimate for a practical voice switched loudspeaking telephone set.

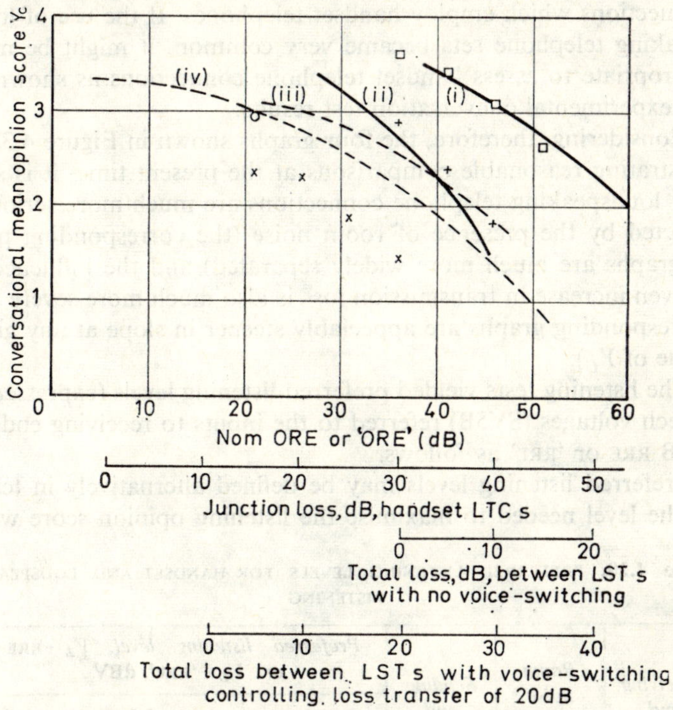

***Figure 4.32.** Conversational mean opinion scores for connections containing loudspeaking telephones and handset local telephone circuits.*

Legend

□ LSTS *RN0*	*Curve (i)*	*Derived from listening tests*
+ LSTS *RN50*	*Curve (ii)*	
⊙ *Handset* LTCS *RN0*	*Curve (iii)*	
× *Handset* LTCS *RN50*	*Curve (iv) From Table 4. 10 but in good agreement with derivation from listening tests*	

The conversation test scores are reasonably well predicted from the listening opinion test results in the case of the loudspeaking telephone connections (curves i and ii) but those for the handset connections are considerably lower than those predicted; this discrepancy is believed to be due to the contrast, within the conversation tests, between the 'quality' of the handset connections and that of the superior quality sound emitted from the loudspeakers. It is therefore, perhaps, reasonable to ignore these handset test results when making appraisements comparing the performances of loudspeaking telephone sets used as 'minority' instruments with those of the overwhelming majority of telephone connections which employ handset telephones. If the use of loudspeaking telephone sets became very common, it might be more appropriate to assess handset telephone connections as shown by the experimental conversation test results.

Considering, therefore, the four graphs shown in Figure 4.32 as illustrating reasonable comparisons at the present time, it is seen that loudspeaking telephone connections are much more seriously affected by the presence of room noise (the corresponding pairs of graphs are much more widely separated) and the influence of a given increase in transmission loss is also much more severe (the corresponding graphs are appecially steeper in slope at any given value of Y_C).

The listening tests yielded preferred listening levels (expressed as speech voltages (SV5B) referred to the inputs to receiving ends of 0 dB RRE or 'RRE' as follows.

Preferred listening levels may be defined alternatively in terms of the level needed to maximise the listening opinion score when

Table 4.20. PREFERRED LISTENING LEVELS FOR HANDSET AND LOUSPEAKER LISTENING

Receiving end	*Room noise level*	*Sending end*	*Preferred listening level, V_L − RRE or V_L − 'RRE', dBV*		
			max score OS No. 7	*score 2.5 OS No. 4A*	*score 2.0 O S No. 4A*
Handset LTC	None	Handset LTC	−28.6	—	−28.7
		LST	−28.9	—	−29.4
	50 dBA	Handset LTC	−17.6	−18.4	—
		LST	−23.8	−18.8	—
LST	None	Handset LTC	−28.3	—	−31.2
		LST	−33.8	—	−33.4
	50 dBA	Handset LTC	−21.0	−22.7	—
		LST	−24.9	−24.0	—

Opinion scale No. 7 (listening effort) is used or in terms of a particular score when the loudness preference opinion scale (Opinion scale No. 4A) is used. When room noise is present a score of 2.5 on OS4A usually yields results in good agreement with the 'minimum listening effort' criterion (*see* Section 4.2.1.2); in the absence of any room noise, however better agreement was obtained in these experiments by use of a score of 2.0. The results for the handset LTC (13-2P-27 set on 1 mile, $6\frac{1}{2}$ lb cable) agree reasonably well with those quoted in Section 4.2.1.2 which were derived from different test series. From the above table it may be concluded that the best listening level with a loudspeaking telephone set is some 2–4 dB lower than with a handset telephone.

Taking the sending and receiving results together, it follows that 'optimum' LST-LST connections will have ratings about 13 dB greater than those of handset telephone connections. It must be noted that this statement is based on the measurement of loudness ratings for loudspeaking telephone sets as defined above and conversion of the values relative to ARAEN into 'notional' reference equivalents by adding the reference equivalents of ARAEN.

Taking a compromise value of optimum listening level (V_L – 'RRE') as −27 dBV and assuming that a LST of 0 dB 'SRE' yields a speech voltage of −3 dBV, it follows that the 'best' value of nominal 'ORE' for a loudspeaking telephone connection will be about 24 dB. This is some 8–10 dB less than can be achieved in a LST having no voice switching; by introducing voice controlled logic to transfer loss from the send direction of transmission to the receive, as the participant changes his role in conversation from listening to talking, it becomes possible to extend the maximum line loss over which satisfactory operation is possible. Naturally, if both participants talk simultaneously, at least one will not be heard; furthermore, to avoid switching taking place unnecessarily during structural silences, hysteresis or operational delay (or both) must be used in the speech detector or its logic. Some chopping is therefore inevitable; this will usually be acceptable, however, if no more than about 20 dB loss is transferred and provided that the conversation is not too vigorous (*see* Section 2.3). 20 dB of loss transferred by the voice switching will enable the loss of the line associated with each loudspeaking telephone to be increased by 10 dB and the sum of 'SRE' + 'RRE' to be reduced by 20 dB. In this way the optimum nominal 'ORE' can be secured with some 10–12 dB transmission loss in the lines connecting the two LSTs together; reasonably acceptable results will be obtained in silence with some 25 dB loss.

The foregoing information can be used to make some proposals for desirable values of 'SRE' and 'RRE' for loudspeaking telephone sets. Taking 34 dB as the minimum value of 'SRE'+'RRE' to avoid instability in the absence of any voice transferred loss and assuming that 20 dB loss is to be transferred, the value of 'SRE'+'RRE' for a voice switched LST can be set at 14 dB. To comply with the equivalent of a difference of 9 dB between SRE and RRE for handset LTCs, the difference 'SRE'−'RRE' for an LST needs to be 16 dB; this follows from a 10 dB benefit in sending and a 3 dB benefit in receiving attributable, as explained above, to higher vocal level, relatively closer mouth to microphone distance (than that used to determine 'SRE') and lower acceptable listening level when using a LST as compared with a handset connection. Suitable values of 'SRE' and 'RRE' then are found to be +15 dB and −1 dB respectively. Section 5.2 shows these values in comparison with values quoted for various real LSTs and according to specifications produced by various organisations.

4.5 CROSSTALK

4.5.1. General

Figure 4.33 shows two telephone connections coupled, undesirably, by a crosstalk path. The participants in a conversation between *A* and *B* may be disturbed by speech overheard from a conversation between *C* and *D*, the crosstalk path being via $P-Q$. Not only may this crosstalk be troublesome to the users of the $A-B$ connection but also the privacy of those using the $C-D$ connection may be infringed. The crosstalk path shown in the Figure (corresponding to near end crosstalk) is only one example of the variety of routes by which the two connections may be undesirably coupled (*see* Figure 3.2 in Section 3.1.1, Section 6.1.5 and Section 6.3).

Very frequently, provision of adequate crosstalk separation between telephone channels imposes some economic penalty and so it is important to ensure that no unnecessary expenditure is incurred in providing more than sufficient attenuation in crosstalk paths. The question: 'How much is sufficient?' is difficult to answer. If the cost of doing so is small, speech via crosstalk paths should be made inaudible except in a very small proportion of calls. (The problem must be treated statistically.) More commonly, cost is important and so the requirements for privacy are given most consideration and 'intelligibility' is used as the relevant criterion rather than 'audibility'. Under some conditions, other criteria may be more appropriate; for example several degrees of audibility may be considered as follows:

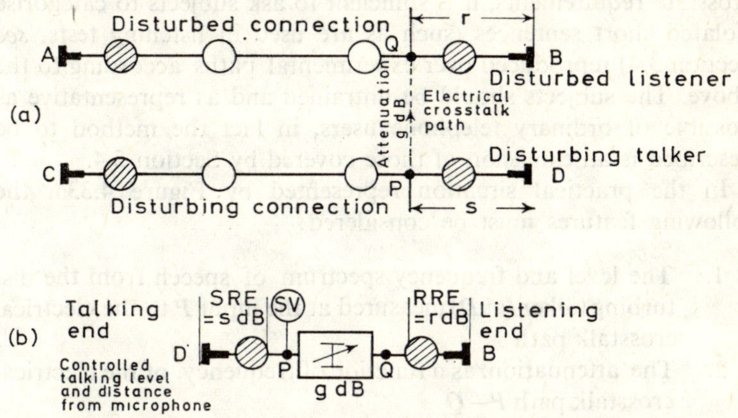

Figure 4.33. Composition of crosstalk paths and arrangement for laboratory tests. (a) Definition of crosstalk path. Notes: Crosstalk path shown from a talker at D to a listener at B represents near end crosstalk. The same electrical path PQ could cause crosstalk from C to B or from D to A; these cases would represent far end crosstalk. (b) A laboratory arrangement for obtaining fundamental data

1. Speech crosstalk audible but neither objectionable nor intelligible
2. Speech crosstalk objectionable although not appreciably intelligible
3. Speech crosstalk intelligible
4. Speech crosstalk disturbing (because it interferes with wanted speech)

Degree 2 may be relevant when the crosstalk heard is scarcely intelligible (perhaps because it is distorted) but has a speech-like rhythm and this may make the hearer fear that he is being overheard. Degree 4 might apply if the hearer is not interested in what he overhears, does not fear being overheard and is concerned only with its disturbing effect on his own conversation.

In the information that follows, it is assumed that a three category assessment will suffice, namely:

A Not audible (or not detectable)
B Audible but not intelligible
C Intelligible

In assembling fundamental information to guide designers on crosstalk requirements, it is sufficient to ask subjects to categorise isolated short sentences (such as are used in listening tests, *see* Section 3.4) reproduced over experimental paths according to the above. The subjects should be untrained and as representative as possible of ordinary telephone users, in fact the method to be described is an extension of those covered by Section 3.4.

In the practical situation represented by Figure 4.33a, the following features must be considered:

1. The level and frequency spectrum of speech from the disturbing talker (at *D*) measured at the input *P* to the electrical crosstalk path
2. The attenuation, as a function of frequency, of the electrical crosstalk path $P-Q$
3. The sensitivity, as a function of frequency, of the receiving equipment from *Q* to the ear of the listening participant at B
4. The listening acuity of the participant at B, including the masking effects of any noise from the circuit and any ambient noise at his location. The participant's motivation is also important, i.e. his desire or willingness and ability to concentrate in listening to low intensity sounds.

In the test situation, the arrangement shown in Figure 4.33b is suitable. Provided that the sending and receiving ends of this experimental path (*D* to *P* and *Q* to *B*) are reasonably representative of ordinary telephone equipment, the overall loss, from *D* to B, may be expressed in terms of nominal overall reference equivalent, $s+g+r$. Of course, if the overall mouth to ear loss (L_{ME}) is known, the overall loudness rating can be calculated as explained in Section 3.2.4.3. (As shown below (Section 4.5.2) knowledge of L_{ME} enables the threshold of audibility for speech to be calculated.)

The experiment consists of varying the nominal overall reference equivalent over a wide range (by varying *g*) and determining the percentage of subjects who find speech introduced at *D* audible (or detectable) when they listen at *B*; curve i in Figure 4.34a shows some typical results. In the same experiment, the percentage voting 'intelligible' can also be determined as shown by curve ii in this figure. If the experiment is conducted in such a manner that the speech voltage at *P* in Figure 4.33b, is held constant, the sigmoids shown in Figure 4.34a will represent the distribution in listening acuity of the subjects.

The talkers, at *D*, are either required to emit speech at a definite level, so that the speech voltage at *P* remains fixed or the talkers'

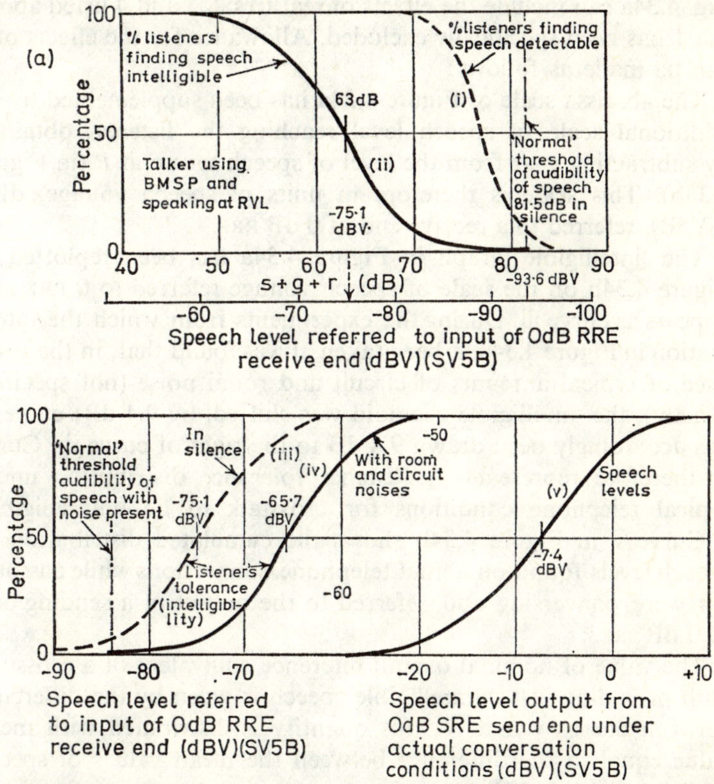

***Figure 4.34.** Results of laboratory tests and their presentation for practical applications. (a) Typical form of results from a laboratory test using constant speech level from talker. (b) Cumulative distributions of listener's intelligibility tolerance and of speech levels on disturbing channel*

speech is recorded and then replayed at a constant level. For the example illustrated in Figure 4.34a, the speech voltage, V_P, at *P* (Figure 4.33b) was such that V_P+s was -12.1 dBV when referred to 0 dB SRE.

A suitable range of levels of circuit noise (injected into the speech path so that the level at *Q* is known for each setting) must be used and the room noise at the listening station must be set at suitably representative values. The results shown in Figure 4.34a apply without any circuit noise or room noise; when appreciable levels of noise are present the curves are moved towards the left, i.e. the nominal overall reference equivalent is reduced.

The information obtained as described above and illustrated in Fig. 4.34a can include the effects of features 2, 3 and 4 listed above but 1 has been specifically excluded. Allowance for the effects of 1 can be made as follows:

The abscissa scale of Figure 4.34a has been supplemented by an additional scale of speech level reaching the listener, obtained by subtracting $g+r$ from the level of speech set up at P (in Figure 4.33b). This scale is therefore in units of speech voltage, dBV (SV5B), referred to a receive end of 0 dB RRE.

The 'intelligible' graph in Figure 4.34a has been replotted in Figure 4.34b on the scale of speech voltage referred to 0 RRE and appears as curve iii. During the experiments from which the information in Figure 4.34 has been taken, it was found that, in the presence of typical amounts of circuit and room noise (not specified exactly), the 'intelligible' sigmoid was shifted, by 9.4 dB; curve iv has accordingly been drawn 9.4 dB to the right of curve iii. Curve iv therefore represents the listener tolerance distribution under typical telephone conditions for crosstalk to be unintelligible.

Curve v in Figure 4.34b shows the cumulated distribution of speech levels found on actual telephone connections while customers were conversing and referred to the output of a sending end of 0 dB SRE.*

The value of nominal overall reference equivalent of a crosstalk path providing just unintelligible speech is given by the difference between curves v and iv; this quantity is distributed with mean value equal to the difference between the mean values of speech level (curve v) and listener tolerance (curve iv) and variance equal to the sum of the variances of these two distributions. The resulting distribution is shown in Figure 4.35a. It can usually be assumed that each of the sigmoids and distributions referred to above is Gaussian; where this may not be so, numerical convolution will be necessary; note, however, that convolution of two non-Gaussian distributions will result in a distribution more nearly Gaussian than its components.

* The values quoted here were measured at the input to international circuits and allowance made for the estimated SRE in actual service between the talkers and the point of measurement. The levels are rather high compared with more recent information given in Section 6.3.

The graph in Figure 4.35a gives the percentage of combinations of talker (on the disturbing channel) and listener (on the disturbed channel) that will result in crosstalk being intelligible to listeners under typical practical conditions for any given value of nominal

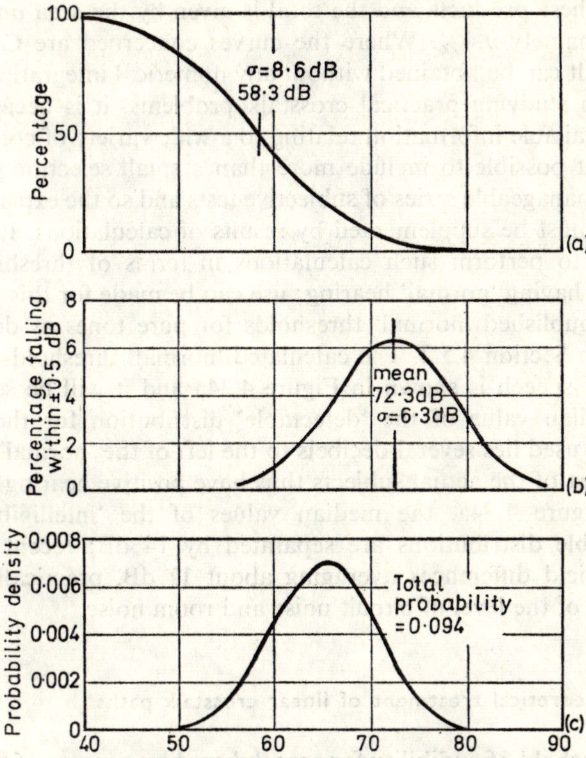

***Figure 4.35.** Combination of the effects of distributions of talker levels and nominal overall reference equivalents of crosstalk paths. (a) Percentage of combinations of talker and listener resulting in crosstalk intelligible to the listener. (b) Distribution of nom.* ORE *for an example of actual crosstalk paths. (c) Probability that a given nom.* ORE *will occur and give rise to intelligible crosstalk*

overall reference equivalent of the crosstalk path. Thus, a crosstalk path of 72 dB nominal ORE corresponds to a probability of intelligible overhearing of 5%.

In practice, crosstalk paths have distributed nominal ORE, partly because crosstalk attenuation differs from one cable to another (or one modulating equipment to another) and partly because sending and receiving reference equivalents, *s* and *r* in Figure 4.33, are distributed. The graph in Figure 4.35b shows an example of such a distribution of crosstalk nominal ORE. The total probability of intelligible overhearing is obtained by integrating the products of the ordinates in the graphs of Figure 4.35a and b; Figure 4.35c shows these products and the total is given by the area under this curve, namely 9.4%. Where the curves concerned are Gaussian, the result can be obtained without any numerical integration.

When studying practical crosstalk problems, it is necessary to have available information relating to a wide variety of conditions. It is not possible to include more than a small selection of these in any manageable series of subjective tests and so the experimental results must be supplemented by results of calculations. It is convenient to perform such calculations in terms of thresholds for persons having 'normal' hearing; use can be made for this purpose of the published 'normal' thresholds for pure tones as described below in Section 4.5.2. The calculated 'normal' threshold of audibility of speech is shown in Figure 4.34a and it will be seen that the median value of the 'detectable' distribution for the actual subjects used lies several decibels to the left of the 'normal' threshold; most of the actual subjects thus have positive hearing losses.*

In Figure 4.34a, the median values of the 'intelligible' and 'detectable distributions are separated by 14 dB; recent experiments yield differences averaging about 11 dB, practically independent of the level of circuit noise and room noise.

4.5.2 Theoretical treatment of linear crosstalk paths

The threshold of audibility for speech heard by persons of 'normal' hearing can be calculated by use of the theory explained in Section 3.2.4.1, as follows:

The loudness, λ, is given by the expression:

$$\lambda = 64 \int_0^\infty Q(Z)B' \, df$$

where Z is the sensation level, B' is a loudness frequency weighting factor and $Q(Z)$ is given by the expression:

$$Q(Z) = \text{constant } (10^{Z/10})^m$$

*** To be considered clinically deaf, the subjects hearing loss must exceed at least 30 db.**

At the very low levels involved with crosstalk, the constant m should be taken as 1 and so, using the logarithmic form,

$$10 \log_{10} Q(Z) = \text{constant} + Z$$

By substituting the appropriate values of Z as a function of frequency in these expressions and performing the integration, it is possible to determine the loudness λ of a given speech signal reaching the ear of a listener. When λ exceeds a critical value (close to 1) the speech will be audible; determination of λ for a given situation therefore enables the amount of loss needed to reduce speech to inaudibility to be calculated. It is not necessary to calculate λ but only $\bar{Z}$ as the following example shows.

The calculation of $\bar{Z}$ and λ are shown in Table 4.21. Col. 2 gives the speech spectrum corresponding to reference vocal level (RVL) and col. 3 gives $\beta_0 - K$, the threshold for continuous spectrum sounds applicable for persons of 'normal' hearing; Z_{RO} in col. 4, which is the difference col. 2 minus col. 3, is the sensation level for zero mouth to ear transmission loss. The mouth to ear loss, L_{ME}, for the combination of sending and receiving ends indicated is shown in col. 5; this has been obtained from the sensitivity data for these LTCS given in Table 6.11 which appears in Section 6.1.2. The sending and receiving reference equivalents, SRE and RRE, quoted in the table have been calculated from the sending and receiving sensitivities in the manner described in Section 3.2.4.3.

The values of sensation level, Z, obtained from a talker (at RVL and using BMSP) when the sending and receiving ends are connected directly together are given in col. 6 of Table 4.21; these are obtained by subtracting L_{ME} from Z_{RO}. Col. 7 contains the frequency weighting factors appropriate to audibility (*see* Figure 3.11 in Section 3.2.4.1 for B'; Table 2.20 in Section 2.2.3.1 gives $10 \log_{10} B' \Delta f$). Col. 8 is the sum of the two preceding columns and col. 9 is the antilog of 1/10 the entry in col. 8. The sum of the values in col. 9 is the weighted mean of $Q(Z)$ which can (because the parameter m has been taken as 1) be equated to $10^{\bar{Z}/10}$. The value of $\bar{Z}$ follows, together, if desired, with λ.

The next stage is to determine the value of $\bar{Z}$ (or λ) that applies at the threshold of audibility. Table 2.22 which appears in Section 2.2.3.1 shows the determination of $\bar{Z}$ and λ for continuous spectrum sound consisting of band limited Gaussian noise. Similar determinations can also be made for speech sounds using published information and these are shown in the Table 4.22 together with results from a recent series of experiments specially designed for the purpose.

Table 4.21. CALCULATION OF LOSS TO REDUCE SPEECH TO THRESHOLD OF AUDIBILITY IN THE ABSENCE OF ANY NOISE

Frequency f Hz	β'_S dB	$\beta_0 - K$ dB	Z_{RO} dB	L_{ME} (BMSP) dB	$Z = 10 \log_{10} Q(Z)$	$10 \log_{10} B' \Delta f$	$10 \log_{10} Q(Z) B' \Delta f$	$Q(Z) B' \Delta f$
(1)	(2)	(3)	(4)	(5)	(6)	(7)	(8)	(9)
100	+55.5	+17.5	+38.0			−18.5		
200	+60.6	+5	+55.6	+15.5	+40.1	−15.3	+24.8	300
300	+62.0	0	+62.0	+3.6	+58.4	−15.1	+43.3	21 600
400	+62.1	−3	+65.1	−0.4	+65.5	−15.1	+50.4	109 600
500	+61.0	−5	+66.0	−2.6	+68.6	−15.1	+53.5	224 000
600	+58.7	−6	+64.7	−4.1	+68.8	−13.5	+55.3	339 000
800	+53.4	−8	+61.4	−7.4	+68.8	−12.7	+56.1	407 000
1000	+50.0	−9	+59.0	−10.3	+69.3	−12.6	+56.7	467 000
1250	+47.7	−8.5	+56.2	−13.1	+69.3	−11.8	+57.5	562 000
1600	+44.8	−8	+52.8	−11.4	+64.2	−11.6	+52.6	182 000
2000	+42.4	−9	+51.4	−4.6	+56.0	−11.5	+44.5	28 200
2500	+40.4	−11.5	+51.9	+4.1	+47.8	−11.9	+35.9	3 900
3000	+38.5	−14	+52.5	−2.4	+54.9	−12.6	+42.3	17 000
3500	+36.1	−13.5	+49.6	−0.3	+49.9	−13.4	+36.5	4 500
4000	+33.9	−13	+46.9	+20.5	+26.4	−12.4	+14.0	
5000	+30.7		+43.2			−12.5		
6000	+29.1		+40.6			−12.0		
8000	+28.4		+37.4			−12.8		

$$\overline{Q(Z)} = \text{Sum} = 2\,366\,100$$

$$= 10^{\bar{Z}/10}$$

$$\bar{Z} = 63.7 \text{ dB}$$

$$\lambda = 64\,\overline{Q(Z)} = 1.52 \times 10^8$$

Sending end: 16–4T–31 (with regulator) on 1 mile 6½ lb cable SRE = +7.1 dB
Receiving end: 16–4T–31 (with regulator) on 1 mile 6½ lb cable RRE = −5.8 dB
Junctions and Trunks: none

Notes β'_S applies for the speech spectrum obtained from male and female talkers at their natural levels; the absolute level has, however been adjusted to correspond to RVL.

$\beta_0 - K$ applies for 'normal' hearing.

From Table 4.21, it has been shown that, with the sending and receiving ends connected directly together the value of $\bar{Z}$ is +63.7 dB; to reduce the speech to inaudibility, $\bar{Z}$ must be reduced to −20 dB; therefore the additional loss between sending and receiving ends necessary to reduce the speech to inaudibility is +63.7−−(−20) = 83.7 dB. With this loss inserted the nominal overall

Table 4.22. SUMMARY OF INFORMATION ON Z AND λ APPLICABLE WHEN THE SOUND IS AT THE THRESHOLD OF AUDIBILITY

Source of information	*Type of sound*	*Conditions*	$\bar{Z}$, dB	λ
Hawkins and Stevens (1950)	Noise	In silence on earphone	−14.8	2.1
Fletcher and Galt (1950)	Speech	In silence both ears distant source	−13.8	2.8
Hawkins and Stevens (1950)	Speech	In noise on earphone	−14.7	2.2
United Kingdom Post Office (1970)	Speech	Without and with noise	−18 to −22 mean −20.6	1.0 to 0.4
Value used for present calculations			−20	0.6

reference equivalent is $+7.1+83.7+(-5.8) = 85.0$ dB. It must be noted that this result applies for a talker speaking at RVL and adopting BMSP and for listeners having 'normal' hearing. Furthermore this estimate has been made in the absence of any circuit noise or room noise.

When noise is present the values of Z at each frequency are reduced by the amount by which the presence of noise has shifted the threshold of audibility. As explained in Section 2.2.3.1, the effect on the hearing threshold of the presence of a continuous spectrum noise can be expressed in terms of the resulting masking audiogram, denoted by M as a function of frequency. The expression for M may be written:

$$M = 10\log_{10}\{10^{(\beta_0-K)/10}+10^{N'/10}\}-(\beta_0-K)$$

The first part of this expression is the absolute value of the threshold when it is masked by noise of spectrum density N' and is obtained by performing a 'power addition' operation on the quantities β_0-K and N' which are in dB. The shift in threshold, M, is then given by the difference between the masked threshold and β_0-K which is the unmasked threshold. Table 4.23 illustrates the procedure with an example. Col. 2 shows the unmasked threshold, col. 3 shows the spectrum density of the noise* and col. 4 gives the result of converting cols 2 and 3 to power ratios, adding and restoring to dB. Col. 5 is the difference between col. 4 and col. 2. This table also includes the results of calculating *M* for several levels of circuit noise. The weighted means, $\overline{M}$, are not obtained directly from the information in this table and will be referred to later.

Table 4.23. DETERMINATION OF SHIFT IN THRESHOLD, *M*, WHEN CIRCUIT NOISE IS PRESENT (ROOM NOISE ABSENT)

Frequency Hz	$\beta_0 - K$ dB	Circuit noise −65 dBmp referred to 0 dB RRE: Spectrum density N' at ear dB	Masked threshold dB	Threshold shift	−85 dBmp *M* dB	−80 dBmp *M* dB	−75 dBmp *M* dB	−55 dBmp *M* dB
(1)	(2)	(3)	(4)	(5)	(6)	(7)	(8)	(9)
100	+17.5	—	—	0	0	0	0	0
200	+5	−17.8	+5.0	0	0	0	0	0.2
300	0	−2.2	+ 2.1	2.1	0	0.1	0.3	8.5
400	−3	+2.8	+ 3.8	6.8	0.1	0.5	1.4	15.9
500	−5	+5.2	+ 5.6	10.6	0.4	1.3	3.1	20.3
600	−6	+6.2	+ 6.5	12.5	0.7	1.8	4.2	22.2
800	−8	+7.1	+ 7.2	15.2	1.2	3.1	6.3	25.1
1000	−9	+6.3	+6.4	15.4	1.3	3.2	6.4	25.3
1250	−8.5	+6.1	+ 6.3	14.8	1.1	2.8	5.9	24.6
1600	−8	+6.3	+ 6.5	14.5	1.1	2.7	5.7	24.3
2000	−9	+1.0	+ 1.4	10.4	0.4	1.2	3.0	20.1
2500	−11.5	−6.0	− 5.0	6.5	0.2	0.5	1.3	15.6
3000	−14	−1.8	− 1.6	12.4	0.7	1.9	4.3	22.2
3500	−13.5	−11.9	− 9.6	3.9	0.1	0.2	0.6	11.9
4000	−13	<−41	−13	0	0	0	0	0
Weighted mean, $\overline{M}$, dB				12.5	0.8	2.5	5.0	21.6

Receiving end: 16−4T−31 (with regulator) on 1 mile 6 1/2 lb cable; RRE = 5·8 dB.

Values of $\overline{Z}$ for cases where noise is present could be calculated by following the procedure in Table 4.21 but with the entries in col. 3 substituted by the masked threshold such as exemplified by col. 4 in Table 4.23. When the information required relates to a

* The spectrum density of noise reaching the ear from an electrical source in the circuit is determined as explained in Table 4.3 which appears in Section 4.2.1.1, and from which the values used here have been taken.

series of conditions (including silence), it is simpler to proceed as follows.

In Table 4.21, any change in value in col. 3 will result in exactly the same change in cols 4, 6 and 8. It therefore possible to introduce the shifts in threshold due to masking by noise by subtracting them from the values in col. 8; this is illustrated in Table 4.24, col. 2 of which reproduces col. 8 of Table 4.21. The threshold shift as determined in Table 4.23 appears in col. 3 of Table 4.24 and the remaining columns show the appropriate consequential changes. Z is then calculated as explained in connection with Table 4.21.

Table 4.24. CALCULATION OF LOSS TO REDUCE SPEECH TO THRESHOLD OF AUDIBILITY WHEN CIRCUIT NOISE IS PRESENT (ROOM NOISE ABSENT)

Frequency Hz	*10 log*^{10} *Q(Z) B′ Δf in silence*	*Threshold shift due to circuit noise*, *M*, dB	*10 log*$_{10}$ *Q(Z) B′ Δf with circuit noise*	*Q(Z) B′ Δf with circuit noise*
(1)	(2)	(3)	(4)	(5)
100		0		—
200	+24.8	0	+24.8	300
300	+43.3	2.1	+41.2	13.200
400	+50.4	6.8	+43.6	22 900
500	+53.5	10.6	+42.9	19 500
600	+55.3	12.5	+42.8	19 100
800	+56.1	15.2	+40.9	12 300
1000	+56.7	15.4	+41.3	13 500
1250	+57.5	14.8	+42.7	18 600
1600	+52.6	14.5	+38.1	6 500
2000	+44.5	10.4	+34.1	2 600
2500	+35.9	6.5	+29.4	900
3000	+42.3	12.4	+29.9	1 000
3500	+36.5	3.9	+32.6	1 800
4000	+14.0	0	+14.0	—

Sum 133 200
$= 10^{\overline{Z}/10}$
$\overline{Z} = 51.2$ dB

Sending end 16–4T–31 (with regulator) set on 1 mile 6 1/2lb cable SRE=+7.1
Receiving end 16–4T–31 (with regulator) set on 1 mile 6 1/2lb cable RRE =–5.8dB
Junctions and trunks: none. Circuit noise –65 dBmp referred to 0 dB RRE
Mean shift in threshold due to circuit noise, $\overline{M} = 63.7-51.2 = 12.5$ dB
Therefore nominal ORE to reduce speech to threshold = 85.0–12.5 = 72.5 dB
Note This result applies assuming RVL/BMSP at the sending end and 'normal' hearing at the receiving end. (See Notes under Table 4.21)

The mean shift in threshold of speech due to the presence of noise is given by the change in value of $\overline{Z}$ from that appropriate to silent listening; this is shown at the foot of Table 4.24 as $\overline{M}$. The value of nominal overall reference equivalent needed to reduce speech (at RVL and the talker using BMSP) to inaudibility when noise is present is given by subtracting $\overline{M}$ from the value needed under silent listening conditions, as shown at the foot of Table 4.24. These values are summarised in Table 4.25 for a range of levels of circuit noise; attention is drawn to the qualifications applicable to these results which are stated in the notes appended to Table 4.25. It must also be noted that the thresholds of intelligibility will be about 11 dB lower in nominal overall reference equivalent than those for audibility (detectability).

Table 4.25. CALCULATED VALUES OF NOMINAL ORE NECESSARY TO REDUCE SPEECH TO THRESHOLD OF AUDIBILITY. (FOR 16-4T-31 (WITH REGULATOR) SETS ON 1 MILE $6\frac{1}{2}$ lb CABLE, ROOM NOISE ABSENT)

Level of circuit noise, dBmp, *referred to 0* dB RRE	*Mean shift of threshold due to circuit noise,* $\overline{M}$, dB	*Nominal* ORE dB
(1)	(2)	(3)
None	0	87.0
−85	0.8	86.2
−80	2.5	84.5
−75	5.0	82.0
−65	12.5	74.5
−55	21.6	65.4

Notes 1. A median level of conversational speech voltage 2 dB higher than that corresponding to RVL and BMSP has been assumed. A dispersion of about 4.8 dB standard deviation must also be allowed attributable to the variations between talkers.

2. Subjects with 'normal' acuity of hearing have been assumed; many members of the telephone using public have appreciable hearing loss (without being 'deaf'). In any case a dispersion of hearing acuity must also be taken into account; when pure tone hearing thresholds are being measured, a dispersion of about 6 dB standard deviation is found for a population of subjects having 'normal hearing', *see* Table 2.17 in Section 2.2.3.1.

3. The presence of quite low levels of room noise will shift the threshold (i.e. increase $\overline{M}$ appreciably, especially when the level of circuit noise is low. With 'average' room noise of 50 dBA (Hoth spectrum), the 'no circuit noise' threshold will be raised about 14 dB (depending on the sidetone characteristics of the telephone set); by covering the microphone to exclude room noise from the sidetone path, this increase of the threshold level can be reduced to only about 3 dB (at this level of room noise).

4. The circuit noise has been assumed to have a uniform spectrum but passed through a typical chain of carrier system channel filters.

4.5.3 Nonlinear crosstalk paths

Distortion products, introduced by a nonlinear transmission path will usually reduce the intelligibility of speech; under certain conditions, however, quantisation can introduce considerable amplification and this may enhance the intelligibility of speech signals heard at very low listening levels. This situation can arise in a PCM transmission system when a decision amplitude very nearly coincides with zero; a very small input signal sufficient to pass through the decision amplitude will, under ideal conditions, result in an output of which the amplitude alternates between $\pm\frac{1}{2}h$, where h is the interval between adjacent quantised output amplitudes. If the r.m.s. value of the input signal is considerably less than $\frac{1}{2}h$, considerable gain will have been produced. It is also true that the signal will have been dichotomised and the output will be accompanied by distortion products; the effect of the gain may, however, predominate and render the speech more intelligible than would have been the case if the system had been completely linear.

Practical PCM transmission systems will not be ideal and so coincidence of a decision amplitude with zero will not persist all the time, i.e. some random drift will make it almost equally probable for the zero to be located anywhere between a pair of adjacent decision amplitudes. The trouble will, therefore, be confined to such periods of time that the zero is sufficiently close to a decision amplitude. The intelligibility will also in practice be reduced by the presence of sources of noise, those before the input to the quantiser having a very important effect.

It is not possible here to given more than a brief explanation of the nature of this problem; a full quantitative treatment has not yet been completed.

As well as the exact relative location of zero and the decision amplitudes, the following features of practical situations are important:

1. Location of T_{min} on the scale of transmission level referred to the input to a 0 dB RRE receive end; T_{min} is the level (dBV) of a sinusoidal test voltage whose peak to peak amplitude is equal to the interval between adjacent decision amplitudes near zero. In an ideal system this interval is the same as h, which was defined above. Location of T_{min} on this scale is a function of the characteristics of the quantiser and the loss between its input and the listener.
2. The level of noise at the listener from sources after the quantiser.
3. The level of noise entering the quantiser input.
4. The level of the crosstalk speech entering the quantiser input.

Having located T_{min} on the scale mentioned above, it is convenient to consider the levels of Items 3 and 4 relative to T_{min}. The explanation will now make use of Figure 4.36, the full curve in which applies for a linear crosstalk path and reproduces curve iii from Figure 4.34b; silent listening conditions are first assumed. The corresponding characteristic for a crosstalk path containing, near the listener, a PCM system is shown by the broken curve. It will also be assumed that a decision amplitude is very near zero.

When the crosstalk speech enters the quantiser at any level greater than about $T_{min}+10$ dB, the result will be practically the same as for a linear crosstalk path. When the speech level is comparable (say within ± 6 dB) with T_{min}, the quantising distortion

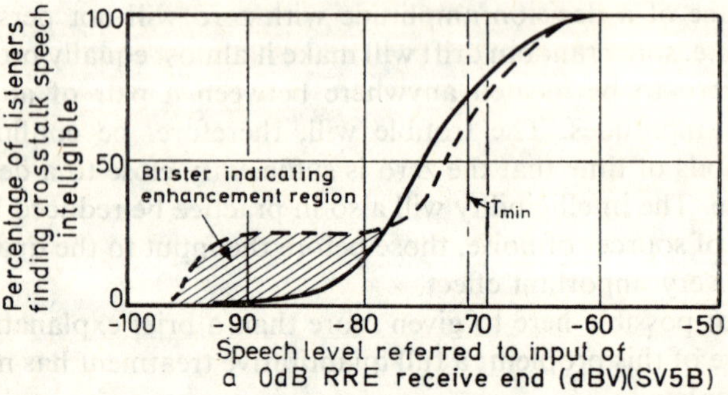

Figure 4.36. *Diagram showing possible effect of nonlinear distortion on the intelligibility of crosstalk.*

Note: T_{min} *is the level of the smallest sinusoidal input signal that will produce an output from the* PCM *system even when displaced midway between two decision amplitudes near zero*

products will reduce the intelligibility as indicated in the figure. When the speech level has been reduced to below $T_{min}-10$ dB, the intelligibility becomes independent of the speech level: this will continue until either noise or distortion produced by the inaccuracies of the system or the residual noise at the input become larger than the input speech signal, when intelligible output from the quantiser will cease. It will therefore be observed that the presence of quantising has raised a 'blister' on the sigmoid which applied for linear transmission conditions and this region represents one in which enhancement of intelligibility could take place.

When noise is present at the output of the quantiser only, the characteristics in Figure 4.36 will be moved to the right and the height of the blister will be reduced. For this effect to become appreciable the level of the noise must exceed about -80 dBV on the scale used in the figure. The characteristic for the nonlinear crosstalk path is, however, more sensitive to noise entering the input of the quantiser and the presence of such noise at a level of −80 dBV would so reduce the width of the blister as practically to eliminate any enhancement.

If it is desired to guard against trouble from crosstalk enhancement even when noise levels are extremely low, this can be done by reducing T_{min}. T_{min} can be reduced by providing finer quantising in the PCM system or by increasing the loss between the input to the PCM system and the listener; either of these actions will move T_{min} to the left in the figure and thereby reduce the height of the blister.

4.6 MISCELLANEOUS

As explained in Section 4.1.1, the foregoing Sections 4.2, 4.3, 4.4 and 4.5 are concerned with information required to estimate the effects on customer satisfaction of changing the intensity of those transmission factors susceptible to control in planning telephone networks; these have been termed *X* factors. Physical factors outside the control of the telephone network planner, like room noise, have been allowed for generally by assuming 'average' types and intensities likely to apply normally in telephone use; these are called *Y* factors. Under special circumstances, e.g. in noisy call offices, allowances can be made for abnormal *Y* factors. The non-physical factors which cannot be controlled by the telephone network planner (*Z* factors) have been taken into account by ensuring that the conversational performance data are collected from subjects (or customers) exposed to suitably realistic environments. The effects on customer satisfaction of changes in the circumstances under which speech communication takes place can be substantial, as the following information indicates.

These examples illustrate certain changes in behaviour and response by subjects resulting from controlled changes of the environment within experimental situations; it is reasonable to suppose that prior exposure of subjects to different kinds of speech communication experience will also have some effect upon their behaviour and responses within the experiments. Extra-experimental 'conditioning' is extremely difficult to investigate and little reliable and well founded information is available at present.

Figure 4.37 shows the opinions expressed by untrained subjects after conducting two conversations successively over the same physical speech link (but the participants were given no indication that the speech link used for the second conversation was the same as they had used for the first. The first conversation was of three minutes duration on any topic they cared to choose; in fact, they spent a great deal of the time on personal introductions and 'social chat', the subjects in most cases not having been previously acquainted with each other. The second conversation was devoted to the performance of a definite task set by the experimenter. (The task used was that described in Section 3.6 using material such as that illustrated in Figure 3.30.) In the experiment, the two tasks were separated by individual questioning of the two participants by the

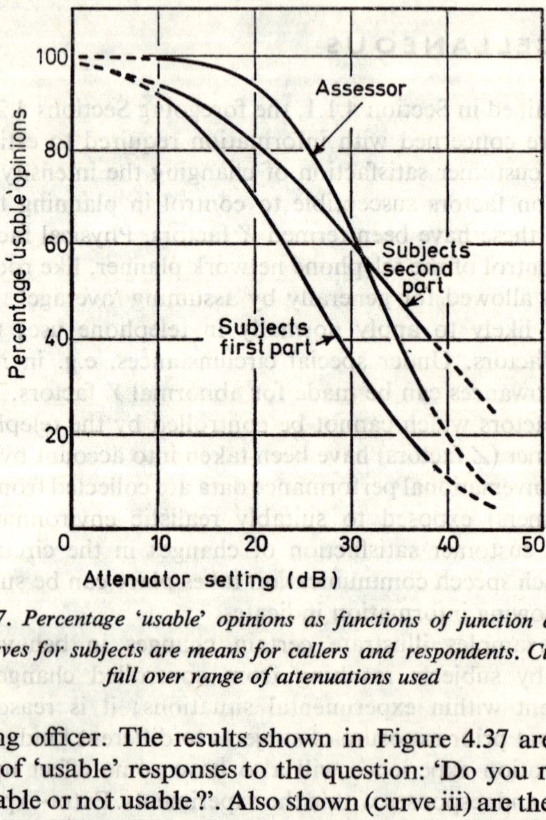

Figure 4.37. Percentage 'usable' opinions as functions of junction attenuation. Notes: Curves for subjects are means for callers and respondents. Curves shown full over range of attenuations used

controlling officer. The results shown in Figure 4.37 are the percentages of 'usable' responses to the question: 'Do you regard the line as usable or not usable?'. Also shown (curve iii) are the opinions of the controlling officer after he had monitored the first (freely chosen topic) conversation. It must be noted that the opinions after the second conversation might have been influenced by the fact that it was after the first one but other evidence suggests that mere order of presentation of test conditions is not very important within conversation experiments. It is therefore believed that the different forms of conversation were mainly responsible for the very different opinion results.

It is also interesting to note that the form and degree of difficulty of the task performed by the subjects has quite a marked effect upon the speech levels they produce during the conversations. The results shown in Table 4.26 are also taken from the same series of experiments as mentioned above but during the early stages when the conversational tasks were being developed. The results quoted apply at an attenuation setting (*see* Figure 4.37) of 30 dB and room noise at a level of about 60 dBA was present at both ends of the connection. The three tasks required conversational cooperation between the participants who were told to complete it as quickly as they could.

Table 4.26. EFFECT ON SPEECH VOLTAGE PRODUCED BY SUBJECTS PERFORMING CONVERSATIONAL TASKS OF DIFFERING DIFFICULTY

Topic for conversation	*Relative difficulty of task*		*Typical results at about 40* dB *nominal* ORE *(60* dBA *room noise)*		
	Mean error rate	*Proportion without error*	*Duration* s	*Percentage unfavourable opinions*	*Speech voltage* dBV†
Freely chosen	—	—	180*	64	0
Task 1	0.39	0.75	230	40 (Tasks 1–3)	+0.6
Task 2	1.05	0.45	240		+1.5
Task 3	1.18	0.30	300		+2.6

* Terminated by the operator † Relative to that for 'Freely-chosen'

Task 1 required identification of five pictorial designs out of sixteen. Titles were provided to both participants to avoid any need for spelling. The designs were similar to those in Figure 3.30 which appears in Section 3.6.

Task 2 required identification of five areas on rather simplified map-like diagrams. Titles usually required spelling. (*See* Figure 3.29 in Section 3.6.)

Task 3 was like Task 2 but the diagrams were more map-like and were rather complex. Most titles required spelling. (*See* Figure 3.28 in Section 3.6.)

Each task was performed by a different group of subjects but they were drawn from the same population. Each conversation task was preceded by a conversation between the same pair of subjects on a topic chosen by themselves the results shown in Table 4.26 for 'freely chosen topic' therefore relate to the whole of the three groups of subjects who performed respectively the three tasks.

The differing difficulties of the three tasks are shown by the values of mean error rate which is the mean number of incorrect identifications out of five; the proportion of conversations free from such

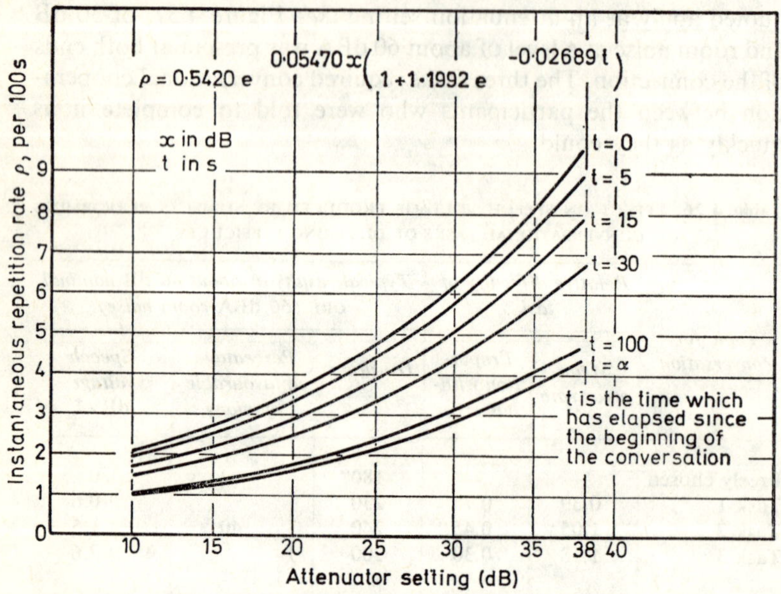

Figure 4.38. Analysis of requests for repetitions. Free conversations; instantaneous repetition rate as a function of time and attenuator setting

incorrect identifications is also shown. Careful scrutiny of the results revealed no significant correlation between the speech voltages and the opinions of the connections expressed by the participants during the conversation tasks. This suggests that the subjects were able to distinguish transmission difficulty from difficulty attributable to the topic; more recent experience has, however, shown a different result.

The results of some rather detailed studies of the incidence of requests for repetition of material not understood are given in Figure 4.38; the conversations concerned took place over laboratory speech links. The time of occurrence of each request was noted and the results classified in 10 s intervals from the beginning of each conversation. This figure shows the fitted surface representing the instantaneous repetition rate as a function of overall loss and time elapsed. It will be seen that the rate of occurrence is greatest at the beginning of the conversation and falls off exponentially as the participants become more accustomed to each other's speech and the topic being discussed. This information shows that observations of behaviour used to assess the transmission performance of any telephone connection must take into account the duration of the

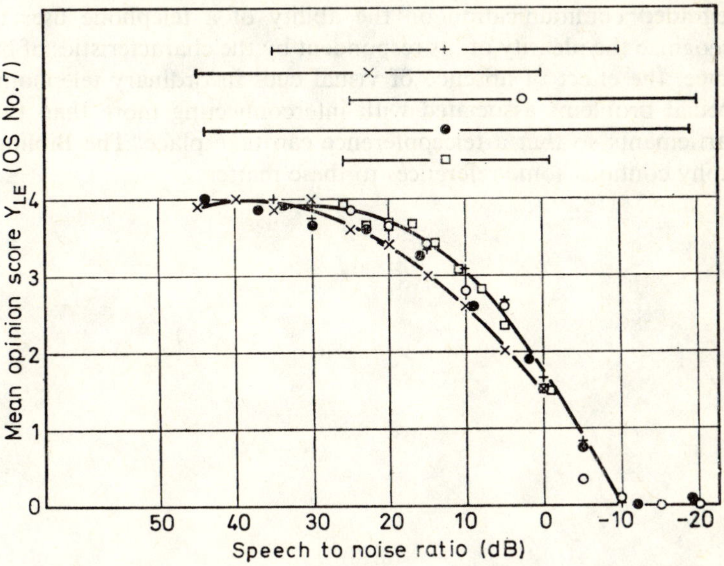

Figure 4.39. Effect of changing the range of test conditions within a listening opinions test

conversation and the time (measured from the beginning of the conversation) at which the observations are made.

Although the sequence in which the individual conversations take place in conversation tests is not usually of much importance, the case is different under listening only test conditions (conducted as described in Section 3.4.2) where the different test treatments are presented in rapid succession to each subject, who hears only five sentences before the treatment is changed. Figure 4.39 shows some results of listening tests conducted with different ranges of test condition within each session. The results apply for the ARAEN degraded by injection of circuit noise (having the same mean spectrum as speech); for each session a different range of settings was used as indicated at the top of the figure. Experimental points are marked for each range and, for the extreme cases, parabolae have been fitted. These results show that there is some tendency for the subjects to change their opinions to suit the range of test conditions presented to them, particularly when very good conditions are included.

In the Bibliography some references are given to published work on the effects of training, verbal context and language.

Other miscellaneous factors of interest include the influence of degraded communication on the ability of a telephone user to recognize the identity of his respondent by the characteristics of his voice, the effect of absence of visual cues in ordinary telephony, special problems associated with interconnecting more than two participants so that a teleconference can take place. The Bibliography contains some references to these matters.

5. TELEPHONE SETS

5.1. HANDSET TELEPHONES

The vast majority of subscribers' station equipment includes handset telephones, of which the three basic transmission elements are: microphone, earphone and electrical circuit. The microphone converts acoustical sound waves into an electrical analogue and the earphone performs the inverse function. The function of the electrical circuit is to couple the microphone to the line terminals, the line terminals to the earphone and to restrict transmission through the path from microphone to earphone, i.e. the sidetone path. The information given here relates to telephone sets required for use on 2 wire subscribers' lines and the examples given apply to present type sets having passive electrical circuits, i.e. induction coils of the 'hybrid' type. Introduction of active devices into the electrical circuit may confer a greater degree of flexibility in design but it does not affect the transmission principles described unless non-linear elements are used.

The characteristics of a telephone set to be used in a public telephone network can be discussed under the following headings and these are used in the sections which follow:

1. Handset dimensions
2. Microphone characteristics
3. Earphone characteristics
4. Sending and receiving losses of the electrical circuit
5. Sidetone suppression provided by the electrical circuit
6. Miscellaneous requirements
7. Overall performance when used for telephone conversation under a range of associated conditions

5.1.1 Handset dimensions

The handle on which microphone and earphone are mounted must be such that the mouthpiece will be suitably located in front of the user's mouth while the earphone is resting comfortably against his ear with a reasonably good acoustical coupling to the ear canal. The length of the handle and the relative positions of microphone and earphone must be chosen carefully so that persons having an appreciable range of head dimensions and shapes will be properly accommodated. Reducing the length of the handset will tend to locate the microphone closer to the mouth, and so improve 'sending', but too short a handset will be uncomfortable for some persons and will cause them to displace the earphone from its proper position on the ear and so degrade 'receiving'. A convenient hand grip must be provided so that the user can hold the handset firmly without his fingers becoming pressed against the cheek. Guidance on these matters is provided by published information on the dimensions of persons' heads.

Since about 1928, when handset telephones of the modern variety were being designed to replace the 'candlestick' type, several sets of measurements of human heads have been made by various telephone organisations and some of these are reported in publications by the CCITT (formerly CCIF). A great deal of the more general biometrical information available is of little use for telephone handset design because it does not provide the most relevant dimensions. Two points on the head must be considered:

1. The centre of the opening of the ear canal, i.e. the 'ear reference point' defined in Section 2.2.1
2. The centre of the lips, i.e. the plane used to define the 'mouth reference point' as described in Section 2.1.1

1 is usually defined as the centre of the circle of contact between a conventional circular earcap, say 55 mm diameter, and a hypothetical plane against which it could be placed; for the purposes of measurement, the earcap is located comfortably against the ear. 2 is defined by the centre of one face of a circular wire ring (a 'lip ring'), 25 mm inside diameter and 1.6 mm thickness, placed lightly against the lips. By using a dummy handset provided with these locating items, the distance, δ, between the defined points can be measured together with the angle, α, between the line joining these points and the plane of the earcap. Figure 5.1 gives

a summary of available information. The most extensive set of results are of measurements made in the United States of America and published by CCIF in 1931 (Ref.* 4); 3889 persons were measured, the sample including women as well as men; the point C in Figure 5.1 shows the mean position ($\delta = 137.1$ mm and $\alpha = 17.7°$) and the enclosed area covers about 80% of the conversations, the boundary having been drawn so that a line tangential to it excludes only about 5%. The points M_1 and F_1 relate to measurements made respectively on males and females in the United King-

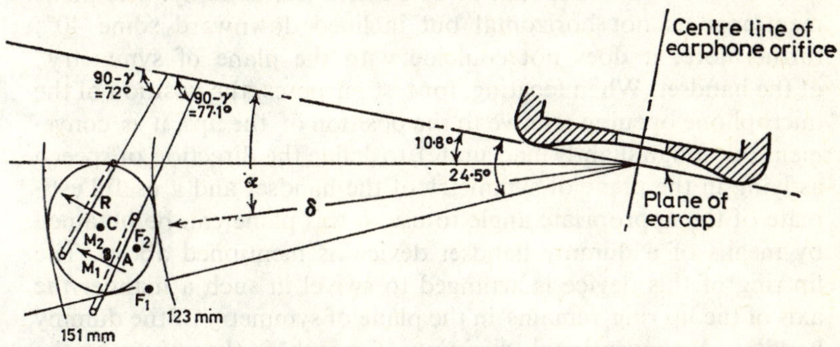

Figure 5.1. Location of lip position relative to opening of ear canal.
Notes: **1.** *Points R and A are located as follows: A $\delta = 136$ mm, $\alpha = 22°$, $\gamma = 12.9°$; R $\delta = 140$ mm, $\alpha = 15.5°$, $\gamma = 18°$.* **2.** *Area shown includes about 80% of the sample of 3889 lip positions;* **3.** *A straight line tangential to the boundary of the area shown excludes 5% of lip positions;* **4.** *Solid line shows plane of lips; broken line shows lip ring 1.6 mm thick;* **5.** *Points A, R, M_1, M_2, F_1 and F_2 are defined in the text*

dom during the period 1932–35; M_2 and F_2 relate to similar measurements made in the United Kingdom in 1954. The sizes of the samples used in these later measurements (40–243) are small and so only the mean values are shown.

It will be seen from Figure 5.1 that female heads are not only smaller than male ones but also yield less reproducible results; this is possibly due to a desire to avoid disturbing the hair or, perhaps, to the wearing of earrings. It will also be seen that α tends to become larger as the dimension δ becomes smaller.

A second angle is required to define the direction in which speech is emitted from the mouth into the mouthpiece of the microphone.

* This and other references in this section appear in Section 5.1 of the Bibliography.

This is very difficult to define precisely but, because sound from the mouth is radiated fairly uniformly over a wide solid angle, precise definition of direction is less important than definition of the relative location of lips and mouthpiece opening of the microphone. A first approximation can be defined as follows.

The plane passing through the two ear openings and through the centre of the lips provides a reference to define the direction in which speech is emitted by a talker. The direction of emission does not, however, lie in this plane but, when the head is erect, it seems reasonable to expect speech to be emitted horizontally. The plane mentioned is not horizontal but inclined downward some 40°; furthermore, it does not coincide with the plane of symmetry* of the handset. When locating, for test purposes, the position of the microphone opening relative to the position of the lips, it is convenient (although slightly inaccurate), to define the direction of speech as lying in the plane of symmetry of the handset and a useful estimate of the appropriate angle to use in this plane can be obtained by means of a dummy handset device as mentioned above. The lip ring of this device is arranged to swivel in such a manner the axis of the lip ring remains in the plane of symmetry of the dummy handset. A conventional 'direction of speech' is then given by the angle, γ, between the axis of the lip ring and the plane of the earcap when both are placed against the respective human organs and the lip-ring is swivelled to fit against the lips†. This angle can be obtained while δ and α are being measured. A more accurate determination of the 'direction of speech' relative to the handset when held by subjects while emitting speech horizontally is described later (Section 5.1.2); the method just described applies, however, for two lip ring positions defined by the CCITT, the more accurate determination being used to define British Modal Speaking Position (BMSP) (*see* Section 5.1.2).

Although, as mentioned above, δ and α prove to be somewhat related in value, γ is uncorrelated with either δ or α and so it may be expressed independently of these quantities. Values of from 12°

* Ideally, perhaps, handsets ought to be designed specifically for right ears or left ears; in practice they are intended to be used on either ear and so they have a plane of symmetry.

† Reference is made in CCITT Recommendations P.45 and P.72 to an angle β which is defined as lying in the plane passing through both ear openings and the centre of the lips. This plane is inclined at an angle θ to the plane in which γ is defined; θ is referred to in Section 5.1.2. In practice, the values attributed to β in the CCITT Recommendations are treated as though they referred to γ; in fact β (as defined by CCITT) differs from γ (as defined here) by only a few degrees.

to 18° have been quoted as mean values of γ from the various available sets of results.

The two lip ring positions shown in Figure 5.1 and defined by the CCITT are as follows: Position R is that used for determining reference equivalent (*see* Section 3.2.2) and Position A is that specified for use when measuring articulation ratings by the AEN method (*see* Section 3.3.3).

The BMSP, used by the United Kingdom Post Office for testing telephone handsets under formalised conditions, is rather similar to Position A and is considered to be more representative of actual, freely held, usage of handset telephones than Position R. Position A is used by the Bell Telephone System. The respective values of δ, a and γ are as follows:

Position R, $\delta = 140$ mm; $\alpha = 15.5°$; $\gamma = 18.0°$
Position A, $\delta = 136$ mm; $\alpha = 22.0°$; $\gamma = 12.9°$

Position R, which has been recommended by CCIF/CCITT since 1934 for measuring reference equivalents of handset telephones (*see* Recommendation P.72), was originally chosen by arranging the results of the 3889 American observations referred to above as a joint distribution of δ and a with cells 5 mm×2° from an arbitrary origin and finding the cell which happened to contain the largest number of observations; this gave a rather crude estimate of the modal value. More careful analysis of the same data yields the *mean* values shown in Figure 5.1.

Position A was derived from a very well diversified series of measurements carefully coordinated by the CCIF during the period 1931–35. The United Kingdom results quoted above were included together with corresponding information supplied by the German, Swedish and Czechoslovak Administrations and others; unweighted means were taken of the results from the various sources. In 1949, when the CCIF was in process of standardising a procedure for measuring articulation ratings (the AEN method), it was decided (Ref. 12) to adopt this position (*see* Recommendation P.45) but, for the sake of continuity, to retain use of Position R for measuring reference equivalent.

Figure 5.2a, b and c show profiles of three differently shaped handsets which yield particularly good results when held freely by either men or women. It will be noted that the opening of the mouthpiece lies, in each case, just outside the 5% boundary of the pistribution of lip positions described above. Tests have shown that altering the form of the mouthpiece has little effect on the perform-

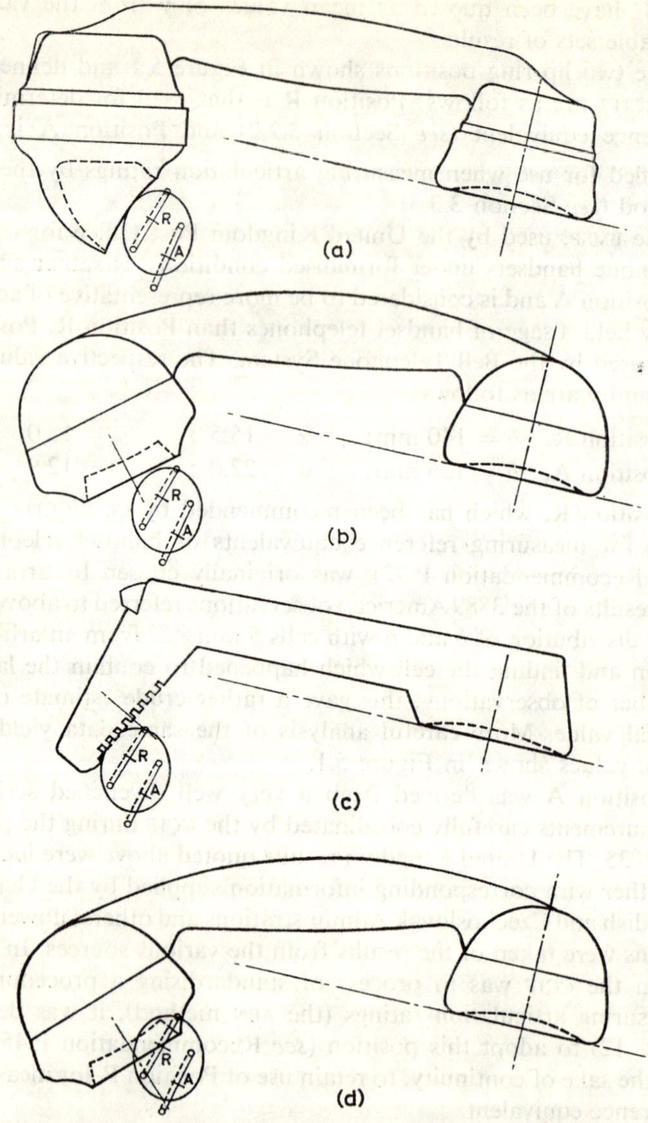

Figure 5.2. Profiles of telephone handsets. (a) Outline of handset having horn type mouthpiece. (b) Handset having flat mouthpiece and handle modified to give best results. Modifications to handle consisted of varying its length and the angle of the earpiece relative to the handle. (c) A modern handset of very light weight. (d) A shortened handset, otherwise similar to b but yielding unsatisfactory results. (Note: Locations of points R and A and associated lip planes are defined in Figure 5.1

ance (sending or receiving, see above) provided that the mouthpiece opening is located as stated above, relative to the distribution of lip positions.

Figure 5.2d shows the profile of an unsuitable handset; this is similar to that shown in Figure 5.2b but the length and angle of the handle has been modified in an attempt to improve sending efficiency by changing the dimensions to bring the mouthpiece opening closer to the point A. Under freely held conditions with subjects conversing there was no improvement in sending performance and, in fact, reception was degraded by about 2 dB. When the handset cannot be fitted comfortably to the head, not only is the coupling to the ear impaired but also the poor fit of the earcap allows room noise more easily to enter the ear. (The conversation tests referred to took place in room noise of 60 dBA level.)

5.1.2 Microphone characteristics

At present, most telephone sets in service on public networks are furnished with microphones of the carbon type. A carbon microphone acts as a modulator of the direct current fed through it so that the electrical speech power produced does not have to be supplied solely from the acoustical speech power applied to its input; this gives carbon microphones a great advantage in sensitivity over sound powered types. Figure 5.3 shows the construction of a modern type of carbon telephone microphone. In addition to its absolute sensitivity, the following other features are important and must be examined so that suitable performance will be obtained:

1. Freedom from excessive changes in sensitivity with variation in attitude of the carbon granule chamber
2. Sensitivity/frequency characteristic
3. Effect on sensitivity of varying the d.c. feeding current
4. Nonlinear effects, including both amplitude distortion and generation of intermodulation products
5. Noise due to burning of the contacts between carbon granules
6. Packing together, or welding, of the carbon granules leading to temporary loss of sensitivity

Carbon microphones constructed in accordance with Figure 5.3 are almost entirely free from attitude effects but some other designs are such that their performances are sensitive to the attitude in which the granule chamber is placed. In these cases it is necessary to know certain angles, relative to the vertical, at which the handset is inclined when held in the usual telephoning position. To obtain

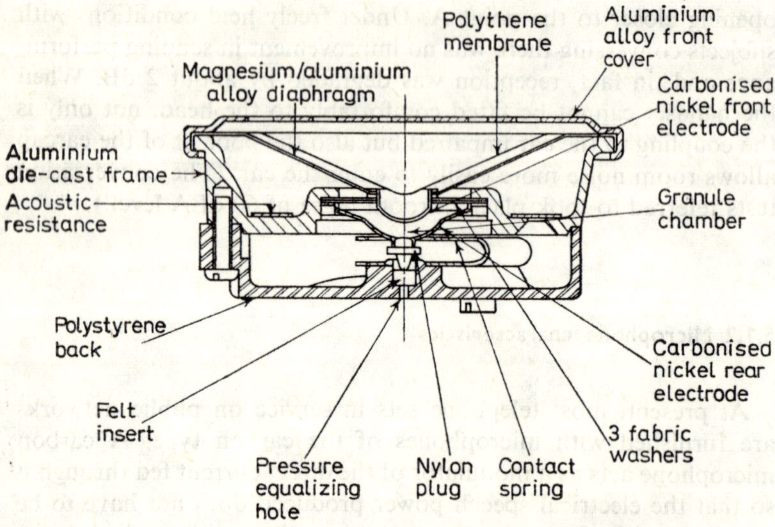

Figure 5.3. Transmitter inset No. 16

this information, the measurements of head dimensions quoted in Section 5.1.1, made by the United Kingdom Post Office, included determinations of two angles which will be referred to respectively as 'earphone rotational angle' and 'handset rotational angle'; these will be defined later.

The dummy handset used for measuring the dimensions of persons' heads (referred to in Section 5.1.1) was fitted also with two spirit levels each mounted on the periphery of a circular ring which could be rotated. One ring was rotatable on an axis through the centre of the earcap with its plane parallel to that of the face of the earcap; the other was in a plane perpendicular to that of the first and with its axis lying in the plane of symmetry of the handset, that is it was rotatable about an axis that might be called the major axis of the handle. Both rings were calibrated to read zero when the plane of symmetry of the handset was horizontal—the angular displacement registered by the first was termed the 'earphone rotational angle' and that by the second the 'handset rotational angle'.

The mean values and standard deviations obtained were as follows:

Sample	Earphone rotational angle, ϕ	Handset rotational angle, θ
100 males	mean 36°; s.d. 7.2°	mean 18°; s.d. 6.7°
40 females	mean 39°; s.d. 7.5°	mean 20°; s.d. 8.0°

Naturally, the 'earphone rotational angle' was such that the handle pointed downwards away from the earphone; the 'handset rotational angle' was positive in the sense that the upper part of the earphone was farthest from the medial plane of the person's head.

When artificial mouths are used for testing telephone handsets, or loudness rating or articulation methods are used (*see* Sections 3.2 and 3.3), it is necessary to standardise a conventional talking position which will not only locate the microphone at a proper distance from the talker's lips (or equivalent lip position in the case of an artificial mouth) but also ensure that the microphone capsule assumes the correct orientation with respect to the horizontal plane; this is particularly important when carbon microphones are tested. An appropriate *distance* is assured by adopting the lip ring location defined by A in Figure 5.1 but, as explained in Section 5.1.1, the *direction* of speech resulting from this definition will be inclined downwards almost 40° below the horizontal when a handset is held at the angles indicated above. The following position has therefore been adopted by the United Kingdom Post Office and is referred to as British Modal Speaking Position (BMSP):

The location of the centre of the face of the lip ring, which defines the equivalent position of the lips, shall be at the point A shown in Figure 5.1, namely $\alpha = 22°$, $\gamma = 12.9°$ and $\delta = 136$ mm as specified in CCITT Recommendation P.45.

The handset shall be mounted so that the following angles apply:

Earphone rotational angle; $\phi = 37°$

Handset rotational angle; $\theta = 19°$

When the handset has been located in the foregoing manner, the lip ring is rotated about a horizontal diameter until the lip ring is vertical and therefore normal to the direction of speech from the mouth. The talker is assumed to emit speech horizontally; the artificial mouth is arranged so that the axis is horizontal.

When carbon microphones are tested, it is usually necessary to measure their sensitivity/frequency characteristics at several values of input sound pressure. It is convenient to use the r.m.s. sound pressure corresponding to Reference Vocal Level (RVL) as one value; this may be set up by ensuring that the level of sound pressure relative to 20 $\mu N/m^2$ is +89.3 dB (−4.7 dB relative to 1 N/m^2) at the mouth reference point defined in Section 2.1.1. This mouth reference point is located on the axis of the mouth (or artificial mouth) and 25 mm from the mouth side of the lip ring (which defines the equivalent lip position of the artificial mouth); the sound pressure is measured in the absence of any obstruction, i.e. before the handset is placed in position. Other values of sound pressures for testing carbon microphones should extend at least over the range from 10 dB below the value indicated to 10 dB above that value; the range must be great enough to ensure that the maximum values of sensitivity are determined as well as those appropriate to the sound pressures of the low level components of speech.

Carbon microphones require suitable 'conditioning' before each series of observations or run of frequencies. This can usually consist of rotating the handset so that the microphone capsule revolves about its axis at least 180°.

The output from the microphone may be measured as the voltage across a suitable load impedance or as the open circuit e.m.f.; carbon microphones require to be supplied with feeding current via an arrangement which preserves the desired alternating current conditions. Using an arrangement which allows open-circuit e.m.f. to be measured, the microphone sensitivity is given by:

$$S_M = 20 \log_{10} \frac{\text{open circuit output e.m.f.}}{\text{free field sound pressure at mouth reference point}}$$

the units being dB relative to 1 V per N/m^2

By expressing the sensitivity in this manner, the effects of differences in the lengths of handset will be appropriately reflected in the results. Under practical conditions of service, considerable departures from the so-called 'modal' position will, naturally, occur and these will be accompanied by variations in vocal level from RVL; the effects of such factors on the speech level appearing on the line are discussed in Section 4.2.1.3 and 6.3.

The 'best' sensitivity/frequency characteristic for telephone connections is discussed in Section 4.2.1.5; to a large extent the desired shape is governed by the sensitivity (S_M) of the microphone as a function of frequency. The 'best' shape also depends on the level of performance at which 'optimisation' is desired; the sensitivity/frequency characteristic set by the United Kingdom Post Office as a design objective in 1947 (and included in Figure 5.5) was determined for the following conditions:

1. The frequency response was optimized to produce the best articulation score for a given speech power produced at the line terminals of the telephone set.
2. The telephone earphone was assumed to be uniform in sensitivity over the frequency range 300–3400 Hz.
3. The optimisation was carried out for rather poor line transmission conditions, namely about 30 dB nominal overall reference equivalent and with attenuation/frequency distortion appropriate to a junction connection over 17 miles of unloaded junction cable.
4. Various levels of room noise (Hoth spectrum) were considered at the listening end, namely: none; 60 dBA (fairly loud); 70 dBA (loud). Circuit noise was ignored.
5. When room noise was present, two different sidetone conditions were considered; it must be remembered that the microphone response itself affects sidetone.

The graphs shown in Figure 5.4 give the results of this study. Each graph shows the calculated optimum sensitivity/frequency response for the microphone (a small allowance, as a function of frequency, has been included for transmission loss from microphone to line); the five graphs have been located so that each would produce about the same speech level at the line as the (then currently used) telephone set 13-2P-27.

Sensitivity/frequency characteristics of telephone microphones can be determined by use of real talkers and the method yields a 'real voice' characteristic. Each subject who talks speaks the same conventional phrase (see Section 2.1.3.2) of about 15 s duration alternately into the test microphone and into a measuring microphone. The output from the test microphone is measured in V and that from the measuring microphone gives the sound pressure (under free field conditions) at the mouth reference point (*see* Section 2.1.1). The voltage measuring equipment contains an arrangement for measuring in frequency bands of one half or one quarter octave. Figure 5.5a shows the results of measurements by the real voice method compared with those made by using an artificial mouth and a sine wave generator; also included is the design objective mentioned above. Figure 5.5a applies for a rather old design of carbon microphone; linear microphones, such as

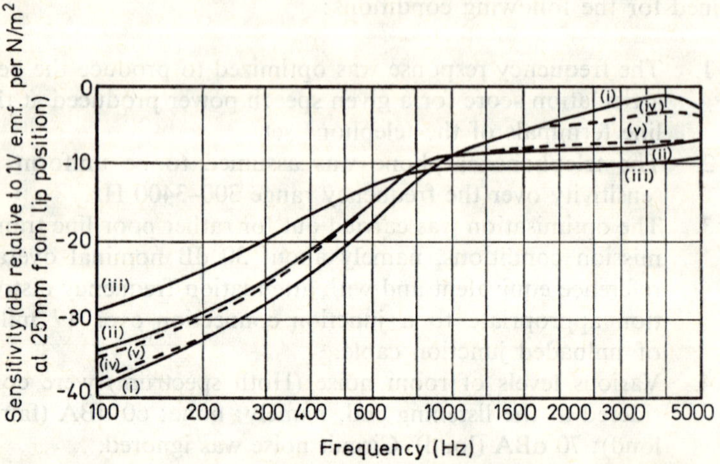

Figure 5.4. Optimum sensitivity frequency characteristics for a telephone microphone. (i) No room noise. (ii) 60 dB room noise, sidetone path typical of 1947 practice. (iii) 70 dB room noise, sidetone path as ii. (iv) 60 dB room noise, sidetone path 10 dB greater loss than ii. (v) 70 dB room noise, sidetone as iv

those likely to become widely used in the future, yield much better agreement between results of real voice and artificial mouth (pure tone) measurements than that shown here. In the case of carbon microphones, the real voice sensitivity characteristics tend to agree fairly well with those obtained with an artificial mouth using a fairly high sound pressure or with such a level of sound pressure as produces, at each frequency, the greatest value of sensitivity. With speech there seems to be a tendency for the higher values of sensitivity to be maintained during momentary low levels of speech pressure as well as during the higher speech pressures occurring at syllabic rate (*see* Section 2.1.3.3). In fact it has been found that the sensitivity/frequency characteristic given by taking the 'upper envelope' of a family of sensitivity curves, such as those illustrated in Figures 5.5b and 5.5c, yields a good estimate of the sensitivity to speech.

Figure 5.5b shows a family of pure tone sensitivity/frequency characteristics for the older type Post Office telephone microphone and Figure 5.5c shows a similar family for the current type which was developed with the above design objective in mind. These characteristics were measured using the Post Office artificial mouth and with the handsets set up in the BMSP as described above.

The sending sensitivies of local telephone circuits given in Section 3.2.4.2 and the sensitivities of telephone microphones denoted by S_M in Section 6.1.2 have, unless otherwise stated, been determined in accordance with the foregoing principles.

Carbon microphones depend for their action upon modulation of direct current fed to them and so their electroacoustical efficiencies vary with the magnitude of the feeding current. In service, this feeding current depends upon the battery voltage, the resistance of the feeding arrangements, the resistance of the telephone set (of which the resistance of the microphone itself forms an important part), and, in the case of telephone sets fed over the subscriber's line from the exchange, the resistance of the subscriber's line and any other apparatus such as PBX equipment. Figure 5.6 (curves ii and iv) shows examples of the change in efficiency as the feeding current through the microphone is altered; the results shown were obtained from loudness rating tests or measurements of speech voltage when spoken into at BMSP and RVL, the talkers using the conventional phrase 'CAN CON BY DACE ALSO'.

The d.c. resistances of carbon microphones also change with the magnitude of the feeding current and curves i and iii of Figure 5.6 show examples of the relationship. In local battery telephone sets, the resistance of the microphone almost completely governs the value of feeding current that flows; in common battery telephone sets, where the feeding current is conducted over the subscriber's line, the microphone resistance is only a small fraction of the total resistance in the path of the feeding current. Carbon microphones suitable for local battery operation must therefore be so designed that their resistance does not fall too much when they heat up because this will tend to produce an unstable situation and, with an unsuitable microphone, may lead to its almost complete destruction.

The resistance, as well as the sensitivity, of carbon microphones depends upon the sound pressure applied and, to some extent, upon its previous treatment. When measuring the talking resistance, a carbon microphone must therefore be held in an appropriate position and spoken into at an appropriate vocal level. The motion of picking up the handset will usually impart a suitable conditioning. The resistance will usually be found to be higher before it has been spoken into than while speech is present or after it has ceased completely; the highest value of resistance, at a given value of feeding current, is important in a common battery telephone set and is termed the 'signalling' resistance (because it affects the signalling relay in the exchange when a call is being originated).

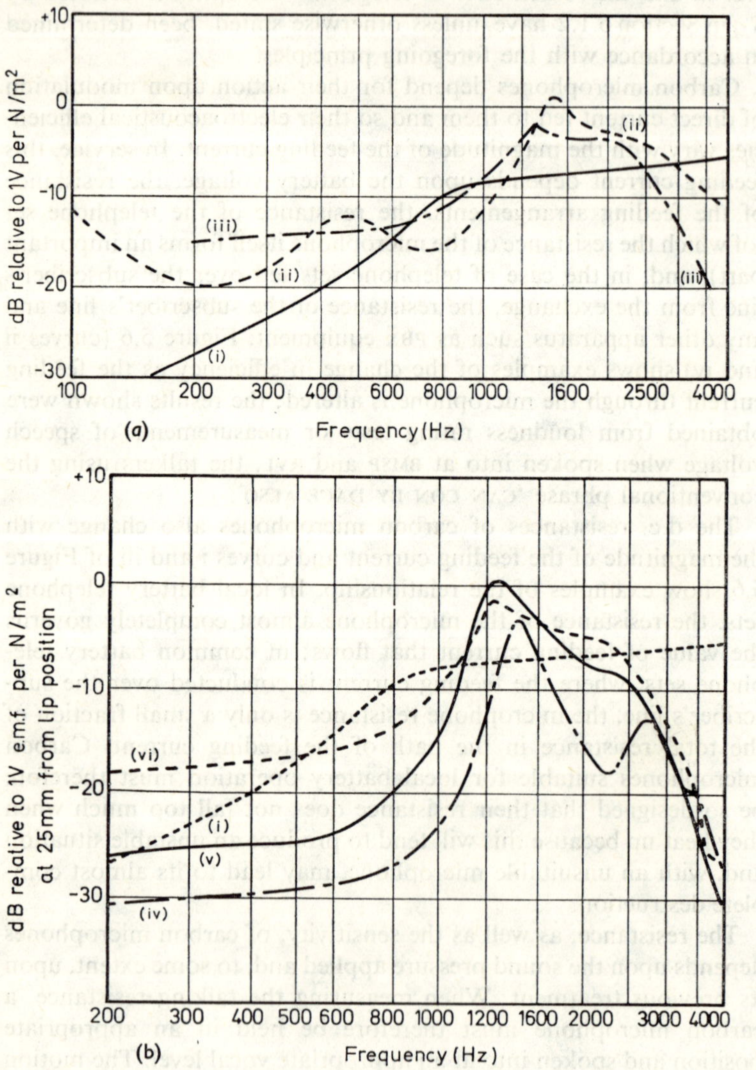

Telephone sets using carbon microphones are usually designed to avoid any d.c. shunt path that could divert feeding current from the microhpone. It can be seen in Figure 5.6 that high microphone resistance is associated with low sensitivity and so any shunting conductance will tend to divert a larger proportion of the current when this is least desired.

The combined effects of changes in feeding current and microphone resistance upon the sensitivity, as the resistance of the subscriber's line is changed, can be represented approximately as a straight line; a typical value of this slope is about 0.26 dB per 100 Ω. Older type microphones may yield rather greater slopes, for example, up to perhaps 0.8 dB per 100 Ω; Section 6.1.3 contains more detailed information (*see* Figures 6.4a and b).

The carbon microphone of the telephone set is probably the

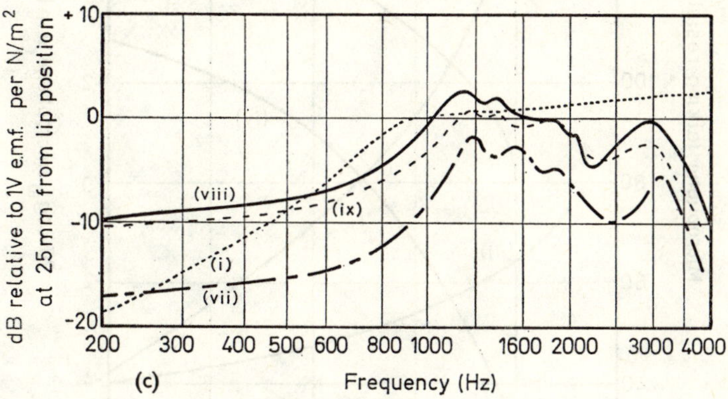

Figure 5.5. Sensitivity/frequency characteristics of telephone microphones. (a) Design objective and real voice characteristic. (i) Design objective. (ii) Transmitter inset No. 13, real voice. (iii) Transmitter inset No. 13, artificial mouth, pure tones, 5 N/m².

Note: Absolute values of sensitivity are approximate only. The following conditions apply: Open circuit e.m.f.; 50 mA feed current; sound pressure referred to point 25 mm in front of lips. (b) Pure tone characteristics for Post Office transmitter inset No. 13. (i) Design objective located to produce same r.m.s. speech e.m.f. as v; (iv) 0.3 N/m²; (v) 1.0 N/m²; (vi) 3.0 N/m².

Note: Open circuit e.m.f.; 50 mA feed current; sound pressure referred to point 25 mm in front of equivalent lip position of artificial mouth. (c) Pure tone characteristics for Post Office Transmitter inset No. 16. (i) Design objective located to produce same r.m.s. speech e.m.f. as (viii). (vii) 0.17 N/m², (viii) 0.57 N/m², (ix) 1.7 N/m².

Note: Open circuit e.m.f.; 47 mA feed current; sound pressure referred to point 25 mm in front of equivalent lip position of artificial mouth

most serious cause of nonlinear distortion in telephone connections; modern carbon microphones are much improved in this respect but appreciable nonlinearity still remains. This feature is part of the price that must be paid for the excellent sensitivity

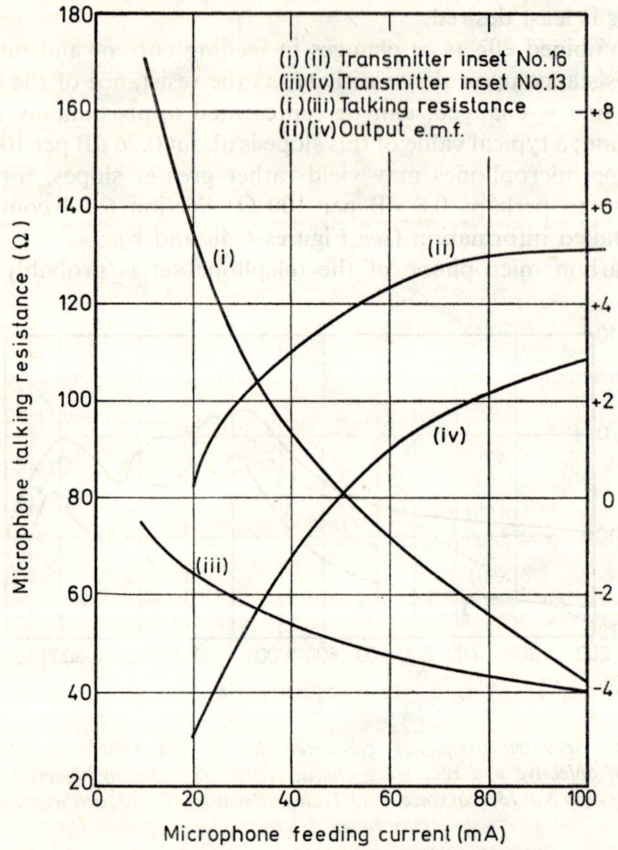

Figure 5.6 Efficiency and resistance of carbon microphones as functions of feeding current

achieved. The nonlinearity reveals itself as amplitude distortion which is illustrated by the changes in sensitivity shown in Figures 5.5b and 5.5c as the input sound pressure is changed; it will be seen that the later type (Transmitter inset No. 16) exhibits less of this type of distortion.

When a participant in a telephone conversation stops talking and listens, the microphone will convey some of the room noise to his ear via the sidetone path. It might be supposed that the 'expansion' effect apparent in the pure tone sensitivity characteristics (Figures 5.5b and 5.5c) at lower levels of sound pressure might confer some advantage by discriminating against any room noise at a modest level which was much lower in level than the speech. (For example room noise at 50–60 dBA as compared with speech sound pressure levels at the microphone of 90 dB or more). This does not, however, appear to be so; the microphone operates at almost normal voice sensitivity when picking up room noise during the intervals between speech.* This phenomenon is illustrated in Section 6.1.4 by some results obtained from measurements on Transmitters inset No. 16; these results confirmed earlier experience with the Transmitter inset No. 13.

The presence in a speech path of a carbon microphone affects the probability distribution of amplitudes of speech waveforms on telephone lines and this must be taken into account when designing transmission equipment (*see* Section 6.3). It is also worth noting that the presence of nonlinear distortion from carbon microphones increases the probability of error in recognising certain phonemes, particularly fricatives. The following results are taken from a special articulation test in which three different telephone microphones were compared. The linear microphone was associated with attenuation/frequency distortion to simulate that of the Transmitter inset No. 13 carbon microphone; the experimental carbon microphone had less nonlinear distortion and was smoother in sensitivity/frequency response. The results shown in Table 5.1 are the percentages of certain phonemes that were recognised correctly as final consonants.

When complex signals are applied to a carbon microphone, intermodulation products are produced; a test using pure tones

Table 5.1

Phoneme	*Linear microphone*	*Transmitter inset No. 13*	*Experimental carbon mic.*
f	59%	41%	77%
v	92	69	91
θ (pith)	54	45	64
s	70	45	76

* No doubt better discrimination against room noise could be obtained by use of a pressure gradient (or 'noise cancelling') microphone.

applied simultaneously at different frequencies is as follows. Two tones are applied, at frequencies of 1600 and 300 Hz with sound pressures respectively (at the mouth reference point) of 0.1 and 3.0 N/m^2; the output at 1300 Hz should be at least 15 dB below that at 1600 Hz. Other methods have been proposed but none has been demonstrated to produce results particularly well correlated with the subjective impressions given by listeners when speech, transmitted by the microphone, is presented to them*.

Because of their mode of action, carbon microphones produce, in their outputs, some noise, even when not being stimulated by any sound input. When the maximum value of feeding current is applied to a carbon microphone from a high impedance source, the noise e.m.f. (unweighted) should not exceed about 100 μV for a carbon microphone having a resistance similar to that shown in curve iii of Figure 5.6.

Static pressure on a carbon microphone affects its performance because this presses the granules together. Under certain conditions, particularly after standing motionless in a quiet environment, the granules can become 'packed' and the sensitivity will then deteriorate seriously. Normally, movement of the handset when it is picked up from the rest will suffice to free the granules. Another cause of sticking together of the granules is 'welding' by passage of a high current (even of very short duration) such as a switching or dialling surge; for this reason, dial and handset cradle switch contacts must be suitably arranged to avoid this trouble.

Clearly replacement of the carbon microphone would be advantageous in telephone sets. Types now under study include pressure modulated solid state devices, electret capacitor microphones and others chosen to give a good prospect of high sensitivity and reliability. It will be necessary to provide some amplification with many such types, especially those that are sound powered. Introduction of active devices in telephone sets opens the way also to control the various impedances associated with the set without the complicated interactions which, at present, limit flexibility.

5.1.3 Earphone characteristics

The earphones of telephone sets are, at present, mostly of the moving iron type; Figure 5.7 shows the construction of a typical modern design. Telephone earphones already possess adequate sensitivity,

* A method developed for measuring quantising distortion might also be considered for this purpose (*see* Section 6.2)

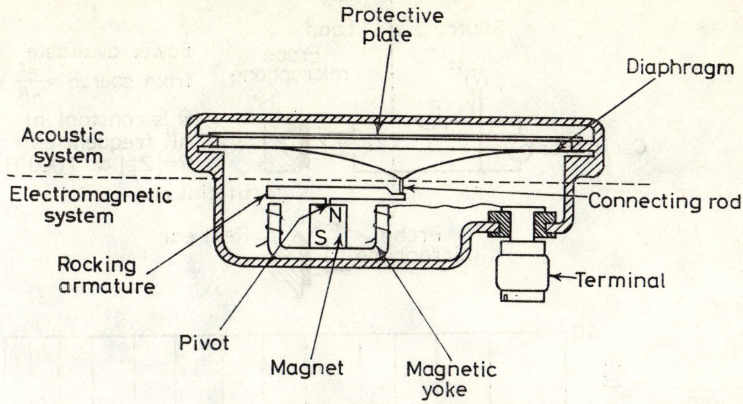

Figure 5.7. Receiver inset No. 4T

thanks largely to developments in magnetic materials, and are practically free from nonlinearity but the rather large positive reactance components of their electrical impedances are a slight disadvantage.

For the purpose of measuring its sensitivity, the acoustical output (sound pressure) from a telephone earphone is determined at the ear reference point by coupling the earphone to a suitable sound pressure measuring device designed to represent the acoustical load of human ears. Such a testing device is termed an artificial ear (*see* Section 2.2.1). The electrical input may be applied as a p.d. across the terminals of the earphone, in which case the sensitivity is given by:

$$S_E = 20 \log_{10} \frac{\text{output sound pressure in N/m}^2}{\text{voltage across input terminals}}$$

the units being dB relative to 1 N/m^2 per V. This method yields the 'constant voltage' sensitivity/frequency characteristic.

Alternatively, the sensitivity can be expressed in terms of the electrical power available from a resistive source of defined resistance; a source of constant resistance equal to the modulus of the impedance of the earphone at 1000 Hz is convenient when it is desired to compare the sensitivities of earphones having different impedances.* This method is illustrated in Figure 5.8a and is

* Other forms of presentation can be used with this object in view. For example, the electrical input to the earphone can be expressed at each frequency of measurement as the real power (watts) or the apparent power (volt-amperes) actually absorbed by the earphone. The so-called 'constant available power' method has the merit of simplicity but the results cannot directly be combined with other transmission losses; only the constant voltage method will be considered here, because it is most suitable for the purpose.

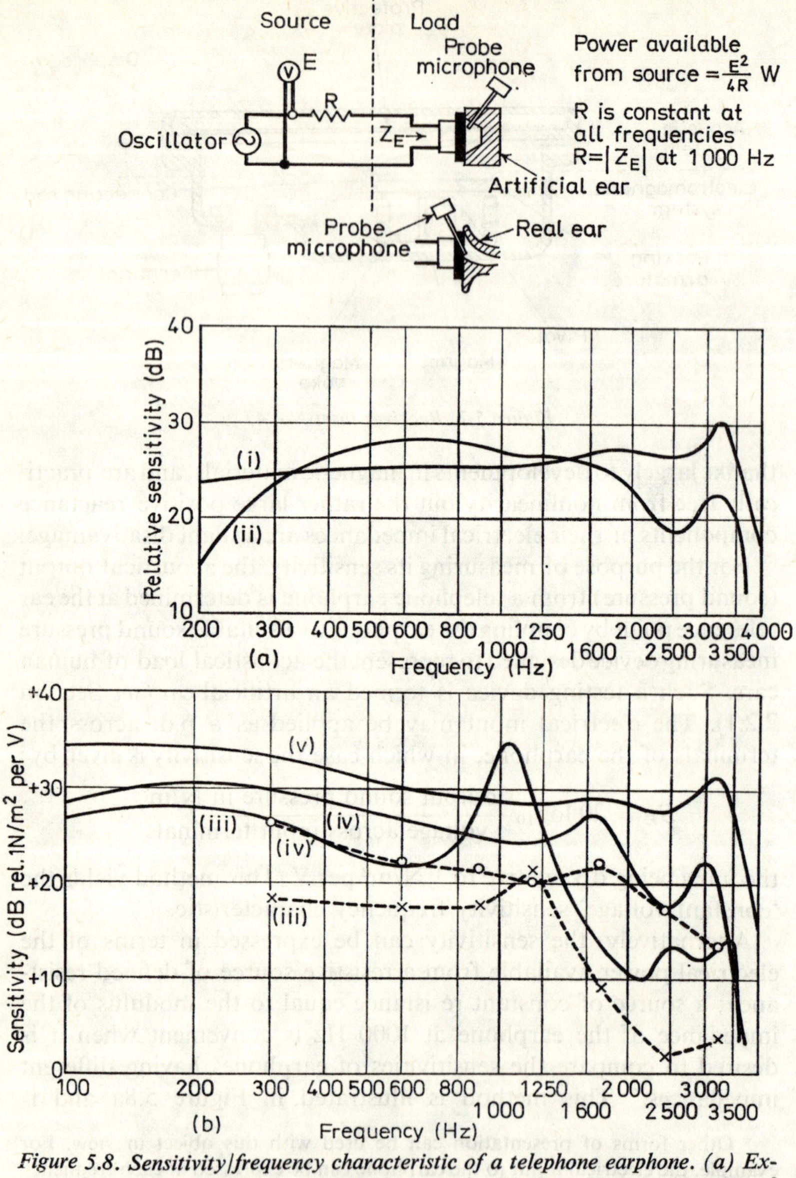

Figure 5.8. Sensitivity/frequency characteristic of a telephone earphone. (a) Expressed in terms of 'constant available power'. (i) Sound pressure in 3 cm³ artificial ear. (ii) Sound pressure in a sample of real ears.

Note: Sensitivity scale is relative to sound pressure of 1N/m² for power of 1 mW available from source. This is commonly denoted by 'dB relative to 1 N/m² per (mW)$^{1/2}$'. (b) Measured on the BS 2042 artificial ear and expressed in terms

termed the 'constant available power' method; this figure also shows (curve i) the constant available power sensitivity/frequency characteristic of a telephone earphone of the type shown in Figure 5.7; the output sound pressure was measured with the BS 2042 artificial ear (*see* Section 2.2.1). Curve ii of Figure 5.8a shows the results of measuring the sound pressure (with a probe tube microphone) in a sample of real ears with the handset held by the subject against his ear as he would normally use it during a telephone conversation. When the earphone is very carefully fitted to real ears and all leaks eliminated, the acoustical loading is fairly well represented by that of an artificial ear but, as can be seen in Figure 5.8a, there can be considerable differences between the real ear sound pressure and that measured in an artificial ear.

The constant voltage sensitivity/frequency characteristics (values of S_E) of three designs used successively are shown in Figure 5.8b; these are all of about the same electrical impedance (see Table 5.2) and so they illustrate fairly the successive improvements in sensitivity and attenuation/frequency distortion.

Figure 5.8b also includes some sensitivity information obtained by the 'audibility threshold' method; this method consists of relating the level of a pure tone signal applied to the earphone under test, to render the tone just audible to a listener, to the level of the minimum audible sound pressure at the same frequency and for the same subject, as measured with a suitable calibrated audiometer. The broken graphs. iii′ and iv′ in Figure 5.8b show results of such determinations for the same items as those to which iii and iv apply; it will be seen that these subjectively determined sensitivities are, on average, about 6 dB lower than those determined using the BS2042 artificial ear, the difference being almost independent of frequency over the range 300–3400 Hz.

The receiving sensitivities of local telephone circuits given in Section 3.2.4.2 and the sensitivities of telephone earphones denoted by S_E in Section 6.1.2 have been determined by applying corrections as explained above to the values given by measurements with an artificial ear.

---

of applied p.d. with some results by the subjective threshold method for comparison.

(iii) Post Office Receiver inset No. 1L } *(iii)′ and (iv)′ show results by subjective*
(iv) Post Office Receiver inset No. 2P } *threshold method*

(v) Earphone of same design as Post Office Receiver inset No. 4T but of different nominal impedance

Table 5.2. IMPEDANCES, Z_E, OF SOME TELEPHONE EARPHONES

Frequency Hz	*Receiver inset No. 1L (iii)* Ω	*Receiver inset No. 2P (iv)* Ω	*Rocking Armature Receiver (v)* Ω	*Receiver inset No. 4T* Ω
200	135 +48°	120 +53°	100 +53°	43 +50°
300	171 +52	159 +58	138 +56	57 +55
400	202 +52	191 +60	172 +56	70 +57
500	239 +53	228 +61	214 +57	85 +58
600	284 +52	283 +62	255 +56	98 +58
800	337 +50	351 +64	302 +56	125 +55
1000	399 +52	420 +63	354 +54	152 +48
1250	452 +55	497 +61	400 +47	174 +39
1600	524 +57	538 +60	453 +43	150 +42
2000	604 +58	640 +63	499 +42	192 +47
2500	707 +59	770 +62	506 +35	243 +36
3000	815 +58	878 +60	415 +50	159 +45
3500	913 +59	980 +59	530 +58	222 +56
4000	1010 +59	1050 +59	620 +58	252 +56

Note The curves, iii, iv and v in Figure 5.8b to which these earphones relate are indicated in the headings; the Receiver inset No. 4T, also included, forms part of the current United Kingdom Post Office telephone set. Except for its impedance, the Receiver inset No 4T is identical in design to the Rocking Armature Receiver; its sensitivity is given in col. 7 of Table 6.10 in Section 6.1.2.4.

The earcap of a telephone earphone is not only required to provide a comfortable surface for the listener to locate the handset against his ear and to provide as good a seal as possible between earphone and ear but also to exclude as much room noise as possible by providing acoustical loss in the path between the room and the ear canal of the listener. Some measured values of loss provided by earcaps are shown in Figure 5.9; these apply for fairly careful fitting of the earphone against the ear and, under practical telephone conversation conditions, the earcap leakage may be worse than the values shown in the figure, especially if the sidetone or received speech is loud or the shape of the handset inappropriate.

Telephone earphones are designed primarily to produce the desired output sound pressures at the different frequencies when coupled closely to an acoustical impedance representative of the human ear under rather ideal conditions. Better freedom from the effects of changes in coupling between earphone and ear can be achieved by making the acoustical output impedance of the earphone as low as possible; this will also reduce the leakage of room noise into the ear. This has been an important consideration in designing the United Kingdom Post Office Receiver inset No. 4T; by appropriate mechanical and acoustical design, the acoustical impedance is much lower than was the case with the earlier designs of earphone mentioned above (Receivers inset Nos 1L and 2P).

Unlike the microphone, of which its sensitivity/frequency characteristic is susceptible to optimisation (*see* Section 5.1.2), earphones have not been submitted to a well defined optimization procedure. Any reasonably uniform characteristic will be satis-

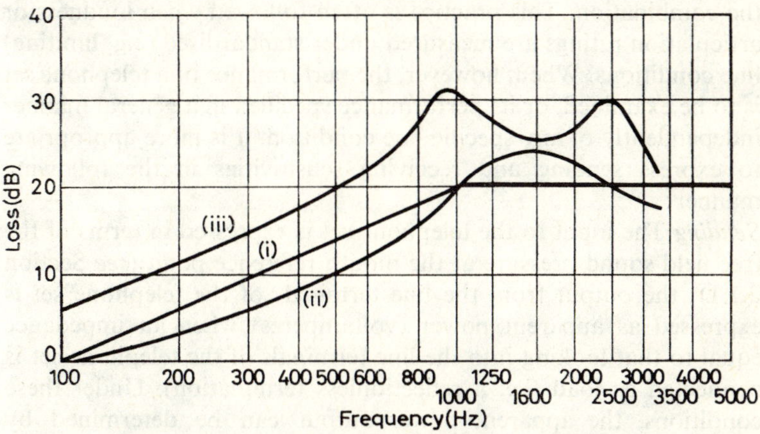

Figure 5.9. Acoustical loss of earcap leakage path to room noise. (i) Loss as used for 1947 determination of best frequency characteristic for telephone microphones. (ii) From results of audibility threshold tests with a hard circular earcap of telephone earphone. (iii) From results of audibility threshold tests with sponge-rubber earpad of ARAEN *earphone*

factory; appreciable departure from an approximately level characteristic might accentuate crosstalk between subscribers' lines in the local network without giving any noticeable advantage in transmission.

Although telephone earphones do not give rise to any appreciable nonlinear distortion, there is some practical advantage in ensuring that the maximum output is less than would cause any discomfort to the user if any severe clicks or impulses were applied.* It may even be desirable to fit some separate nonlinear electrical device (limiter) in the circuit to secure this property.

* Pain is caused when the total r.m.s. sound pressure level exceeds about 120 dB above 20μ N/m^2 (i.e. an r.m.s. sound pressure in excess of 20 N/m^2).

5.1.4 Sending and receiving losses of the electrical circuit

It is possible to express the sending and receiving sensitivities of telephone sets in different ways according to the purpose for which the results are required. In some cases, one particular combination of subscriber's line, local exchange circuit (feeding bridge) and junction may be of particular interest and then it would be convenient to terminate the line terminals of the telephone set with this combination of items or with an impedance equal to that of the combination. This practice is often followed when loudness or articulation ratings are measured under standardised (e.g. limiting) line conditions. When, however, the performance of a telephone set is to be examined, or its performance specified in a general manner independently of any specific line condition, it is more appropriate to express sending and receiving sensitivities in the following manner:

Sending The input to the telephone set is expressed in terms of the free field sound pressure at the mouth reference point (*see* Section 2.1.1); the output from the line terminals of the telephone set is expressed as apparent power (voltamperes) when an impedance equal to that looking into the line terminals of the telephone set is connected as load (i.e. a reflectionless termination). Under these conditions, the apparent power output can be determined by measuring the open circuit e.m.f. (E_C) at the line terminals of the telephone set and the impedance ($Z_C = R_C + jX_C$) looking into these line terminals. The apparent power output is then given by the expression:

$$W_{C/\text{out}} = \frac{E_C^2}{4|Z_C|} \text{ VA}, \quad \text{where} \quad |Z_C| = (R_C^2 + X_C^2)^{1/2} \quad (5.1)$$

In practice, it is usually convenient to determine the sending sensitivity of the complete set by combining the sensitivity of the microphone, determined as described in Section 5.1.2, with a suitably determined loss attributable to the sending direction of the electrical circuit of the telephone set. The arrangement for measuring the transmission loss of the electrical circuit in the sending direction is shown in Figure 5.10a.

The apparent power output from the source, representing the microphone, into a matched load is:

$$W_{M/\text{out}} = \frac{E_M^2}{4|Z_M|} \text{ VA}, \quad \text{where} \quad |Z_M| = (R_M^2 + X_M^2)^{1/2} \quad (5.2)$$

The apparent power output from the line terminals of the telephone is given by Eqn 5.1 and so the sending transmission loss of the electrical circuit of the telephone set is:

$$L_S = 10 \log_{10} \frac{E_M^2}{E_C^2} \frac{|Z_C|}{|Z_M|} \text{dB} \qquad (5.3)$$

$$= 20 \log_{10} E_M - 20 \log_{10} E_C + 10 \log_{10} \frac{|Z_C|}{|Z_M|} \text{ dB} \qquad (5.3a)$$

Receiving. The electrical input to the line terminals of the telephone set is taken as the apparent power (voltamperes) from a source matched to the impedance of the telephone set, namely of impedance equal to Z_C. The input signal can be applied as a p.d. (V_C) across the line terminals of the telephone set; the apparent power (voltamperes) input to the telephone set is then:

$$W_{C/\text{in}} = \frac{V_C^2}{|Z_C|} \text{ VA} \qquad (5.4)$$

The output from the telephone set for the receiving direction is measured as the sound pressure in the ear, or in a suitable artificial ear (*see* Section 2.2.1).

To obtain the receiving sensitivity of the complete set by combining the sensitivity of the earphone, determined as described in Section 5.1.3, with a suitable value of loss attributable to the electrical circuit in its receiving direction, the electrical transmission loss needs to be measured as shown in Figure 5.10b.

The apparent power input to the line terminals is given by Eqn 5.4.

The apparent power input to the earphone is:

$$W_{E/\text{in}} = \frac{V_E^2}{|Z_E|} \text{ VA}, \quad \text{where} \quad |Z_E| = (R_E^2 + X_E^2)^{1/2} \qquad (5.5)$$

The receiving transmission loss of the electrical circuit of the telephone set is therefore:

$$L_R = 10 \log_{10} \frac{V_C^2}{V_E^2} \frac{|Z_E|}{|Z_C|} \text{dB} \qquad (5.6)$$

$$= 20 \log_{10} V_C - 20 \log_{10} V_E + 10 \log_{10} \frac{|Z_E|}{|Z_C|} \text{ dB} \qquad (5.6a)$$

Determination of L_S requires knowledge of appropriate values to use for the source impedance, Z_M, and the earphone impedance Z_E; for carbon microphones, Z_M is resistive and depends upon the value of feeding current in the manner shown in curves i and iii of Figure 5.6. Z_E is shown in Table 5.2 for certain types of earphone used in United Kingdom Post Office telephone sets. Z_C is also required and this must usually be measured in a suitable manner to ensure that the appropriate value of feeding current is flowing; Table 5.3 shows values of Z_C for telephone sets in use in the United Kingdom.

Correspondingly, determination of L_R requires knowledge of the appropriate values of Z_M, Z_E and Z_C.

Table 5.3. VALUES OF IMPEDANCE, Z_C, LOOKING INTO THE LINE TERMINALS OF SOME UNITED KINGDOM POST OFFICE TELEPHONE SETS

Telephone Set	*13-2P-27*		*16-4T-31 without regulator*		*16-4T-31 with regulator*					
Feed current	*44* mA		*32* mA		*32* mA		*64* mA		*101* mA	
Z_M	$52 \angle 0$		$105 \angle 0$		$105 \angle 0$		$67 \angle 0$		$41 \angle 0$	
Frequency Hz	Z_C Ω	$\angle\theta$ °	Z_C Ω	$\angle\theta$ °	Z_C Ω	$\angle\theta$ °	Z_C Ω	$\angle\theta$ °	Z_C Ω	$\angle\theta$ °
200	253	+42	697	+22	698	+21	223	+14	145	+15
300	306	+43	605	+10	615	+10	233	+16	155	+18
400	353	+44	576	+10	590	+10	243	+17	164	+19
500	400	+46	546	+10	564	+ 9	253	+18	172	+20
600	465	+46	552	+12	568	+10	266	+18	182	+21
800	530	+47	555	+13	573	+12	279	+19	191	+22
1000	610	+48	580	+13	596	+13	296	+18	202	+22
1250	712	+46	597	+14	622	+12	305	+18	209	+22
1600	765	+45	587	+15	602	+12	304	+20	211	+26
2000	875	+44	630	+18	638	+15	325	+22	227	+29
2500	987	+42	670	+15	675	+12	335	+23	240	+32
3000	1067	+42	630	+18	630	+14	340	+28	243	+38
3500	1156	+42	688	+24	692	+16	370	+30	268	+40
4000	1217	+42	728	+20	718	+16	383	+32	283	+48

Note Measured with earphones (Receiver inset No. 2P or 4T) having impedances, Z_E, as shown in Table 5.2.

Figure 5.10 shows suitable arrangements for making these measurements, including means for applying the feeding current without disturbing the a.c. conditions in the circuits. Table 5.4 shows some examples of the results of measurements of L_S and L_R; the first two cases, 13-2P-27 and 16-4T-31, are of telephone sets without any regulation and these yield results that depend only slightly upon the value of feeding current, the values shown applying for 'limiting' line conditions. The results shown for the regulated telephone set are, naturally, dependent upon the value of feeding current. The Y values used in designing the unregulated telephone sets are shown and these are explained later (*see* Section

Table 5.4. VALUES OF TRANSMISSION LOSSES, L_S AND L_R, OF ELECTRICAL PARTS OF SOME UNITED KINGDOM POST OFFICE TELEPHONE SETS

Telephone Set	*13-2P-27*		*16-4T-31 without regulator*		*16-4T-31 with regulator*					
Feed current	*44* mA		*32* mA		*32* mA		*64* mA		*101* mA	
Z_M	*52*	/0	*105*	/0	*105*	/0	*67*	/0	*41*	/0
Frequency Hz	L_S dB	L_R dB	L_S dB	L_R dB	L_S dB	L_R dB	L_S dB	L_R dB	L_S dB	L_R dB
200	8.5	7.1	3.9	9.1	4.3	9.1	6.2	9.5	6.1	8.5
300	9.6	5.9	2.3	8.2	2.7	8.3	6.8	8.3	6.7	7.6
400	9.8	4.6	2.0	7.7	2.4	7.8	7.2	7.7	7.2	7.2
500	10.0	3.4	1.8	7.2	2.1	7.3	7.5	7.3	7.7	6.9
600	9.4	3.5	1.8	6.8	2.1	6.9	7.6	6.9	8.3	7.0
800	8.9	3.6	1.7	6.4	2.1	6.5	7.8	7.0	8.9	7.1
1000	8.2	3.4	1.7	6.4	2.1	6.3	8.0	7.4	9.3	7.4
1250	7.5	3.4	1.7	6.6	2.1	6.7	7.9	8.0	9.7	7.9
1600	7.1	3.6	1.7	6.6	2.1	6.6	7.8	7.8	9.9	7.7
2000	6.7	3.8	1.8	6.1	2.1	6.1	8.0	7.9	10.4	8.2
2500	6.3	4.1	1.7	7.0	2.0	7.0	8.0	9.1	10.8	9.4
3000	6.1	4.4	1.9	6.5	2.2	6.4	8.1	8.5	11.1	8.9
3500	6.0	4.6	1.8	6.3	2.0	6.2	8.2	8.7	11.6	9.6
4000	5.9	4.9	1.8	6.6	2.0	6.6	8.4	9.4	11.8	10.3
Y-value	0.6		3.3							
Theoretical loss	4.3	2.0	1.2	6.3						

Note Measured with earphones (Receiver inset No. 2P or 4T) having impedances, Z_E, as shown in Table 5.2 and Z_C as in Table 5.3.

5.1.5.1); the theoretical sending and receiving losses which are related to the Y ratios are shown and it will be observed that the measured values of L_S and L_R exceed the respective theoretical values by a considerable amount in the case of the older, 13-2P-27, set but only slightly for the modern, 16-4T-31, set which uses a much more efficient induction coil than that of older types.

The method of combining values of L_S and L_R with losses attributable to the feeding bridges used in practice and to any given subscriber's line is explained with examples in Section 6.1.2. The manner in which all these losses can be combined with the micro-

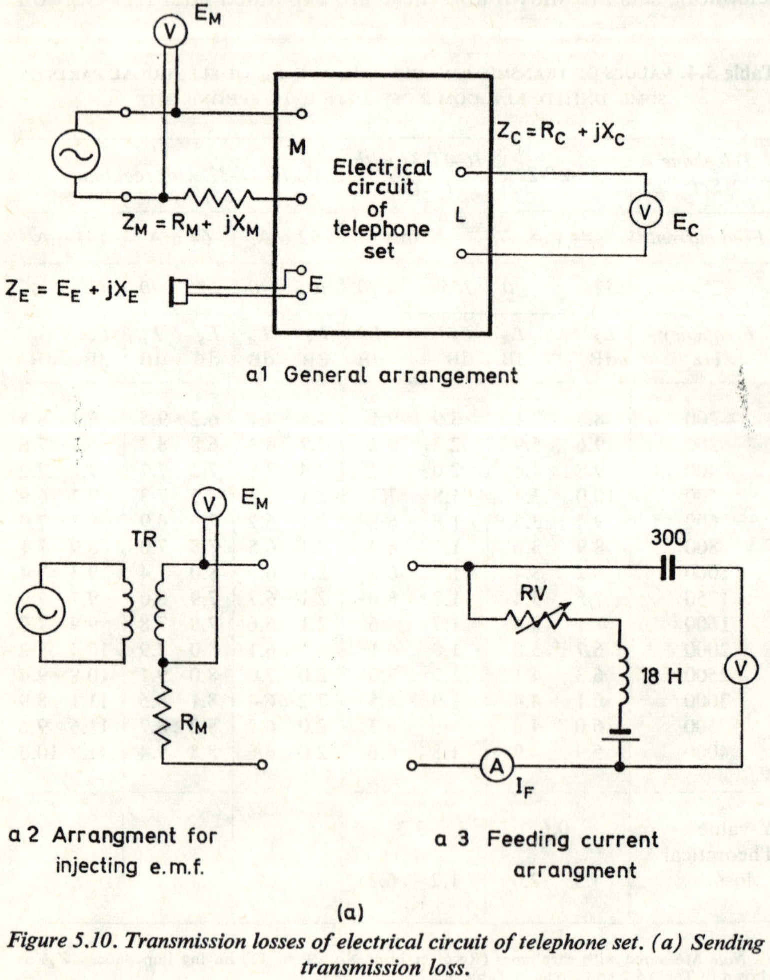

Figure 5.10. Transmission losses of electrical circuit of telephone set. (a) Sending transmission loss.

phone and earphone sensitivities S_M and S_E is also explained in that section.

It should be noted that the definitions of L_S and L_R are practical and yield values of loss that depend upon the impedances Z_M and Z_E of the microphone and earphone that are associated with the

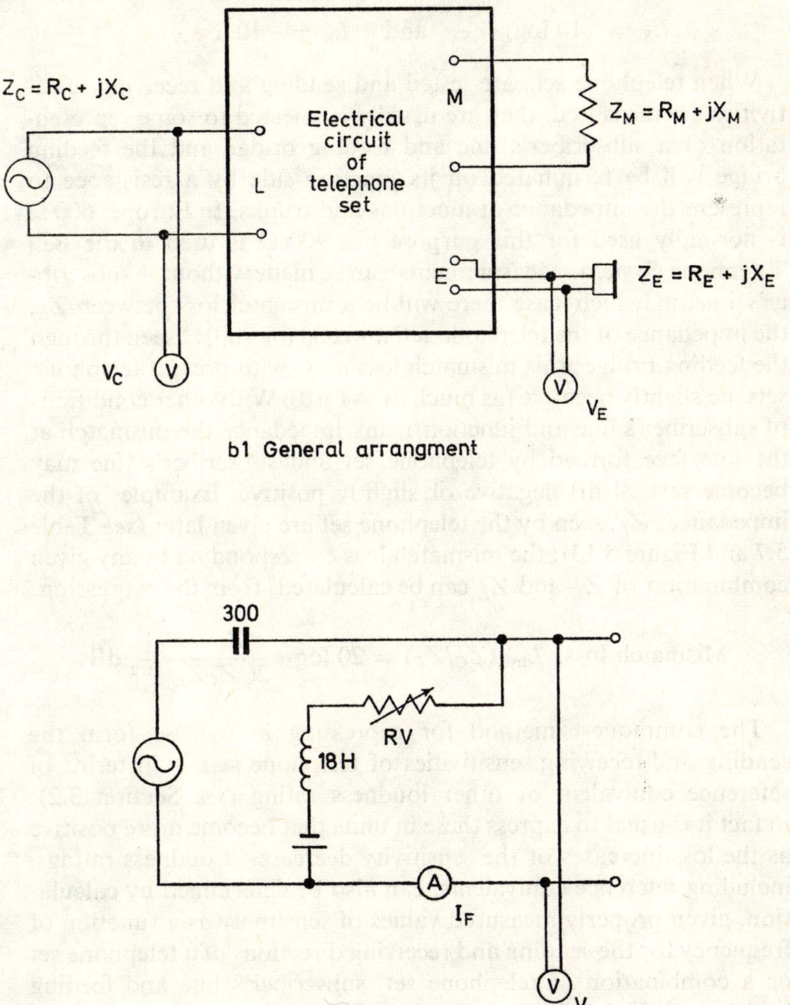

b2 Feeding current arrangment

(b)

Figure 5.10. (b) Receiving transmission loss

particular electrical circuit; the ideal conditions, assuming perfect matching at all ports of the electrical circuit, are considered below (*see* Section 5.1.5.1) and theoretical sending and receiving efficiencies, η_S and η_R, are derived; these are related, under the ideal conditions and assuming lossless components, to L_S and L_R as follows:

$$L_S = -10 \log_{10} \eta_S \quad \text{and} \quad L_R = -10 \log_{10} \eta_R$$

When telephone sets are tested and sending and receiving sensitivities are measured, they are usually connected to some representation of a subscriber's line and feeding bridge and the feeding bridge will be terminated on its junction side by a resistance to represent the impedance of junctions and trunks. In Europe, 600 Ω is normally used for this purpose but 900 Ω is used in the Bell Telephone System. Measurements can be made without a subscriber's line, in which case there will be a mismatch loss between Z_C, the impedance of the telephone set, and 600 (or 900) Ω seen through the feeding bridge; this mismatch loss may, with present telephone sets, be slightly negative (as much as −1 dB). With other conditions of subscriber's line and junction/trunk impedance the mismatch at the interface formed by telephone set and subscriber's line may become several dB negative or slightly positive. Examples of the impedance, Z_L, seen by the telephone set are given later (*see* Table 5.7 and Figure 5.13); the mismatch loss corresponding to any given combination of Z_C and Z_L can be calculated* from the expression:

$$\text{Mismatch loss, } L_{\text{mis}}(Z_C/Z_L) = 20 \log_{10} \frac{|Z_C + Z_L|}{2(|Z_C Z_L|)^{1/2}} \text{ dB}$$

The commonest method for expressing in concise form the sending and receiving sensitivities of telephone sets is in terms of reference equivalent or other loudness rating (*see* Section 3.2); in fact it is usual to express these in units that become more positive as the loss increases or the sensitivity decreases. Loudness ratings, including reference equivalents, can also be determined by calculation, given properly measured values of sensitivity as a function of frequency for the sending and receiving directions of a telephone set or a combination of telephone set, subscriber's line and feeding bridge (called a local telephone circuit) (*see* Sections 3.2.4 and 3.8.). For zero subscriber's line and with a 50 V, Stone transmission

* Appendix 1 gives a chart for the convenient approximate determination of mismatch loss.

bridge, having 2 μF capacitors and 200+200 Ω relay, the sending and receiving reference equivalents of telephone sets suitable for use in public telephone networks are of the following orders of magnitude:

Sending; 0 to +7 dB

Receiving; −6 to −1 dB

The telephone sets referred to in Table 5.4 have the values of reference equivalent under zero subscriber's line conditions shown in Table 5.5.

Table 5.5

Telephone Set	*13-2P-27*	*16-4T-31 without regulator*	*16-4T-31 with regulator*
Sending	+4 dB	−1 dB	+5 dB
Receiving	−3 dB	−9 dB	−6 dB

Values of reference equivalent when subscribers' line are present are given in Section 6.1.3.

5.1.5 Sidetone suppression provided by the electrical circuit

5.1.5.1 Theoretical design of induction coils

'Induction coil' is the name given to the major component of the electrical circuit of a subscriber's set and the basic problems in designing such a circuit are as follows:

1. The microphone must be efficiently coupled to the line terminals.
2. The line terminals must be efficiently coupled to the earphone.
3. The coupling between microphone and earphone associated in the same telephone set must introduce as much loss as possible.*

In the very early telephones, the microphone and earphone were simply connected in series as shown in Figure 5.11a. To allow better

* In some speech communication systems, a problem may exist of ensuring that enough of the talker's speech is fed back to him to give him the impression that the connection is 'live'; this will not become a problem in public telephone networks until completely 4 wire connections are possible.

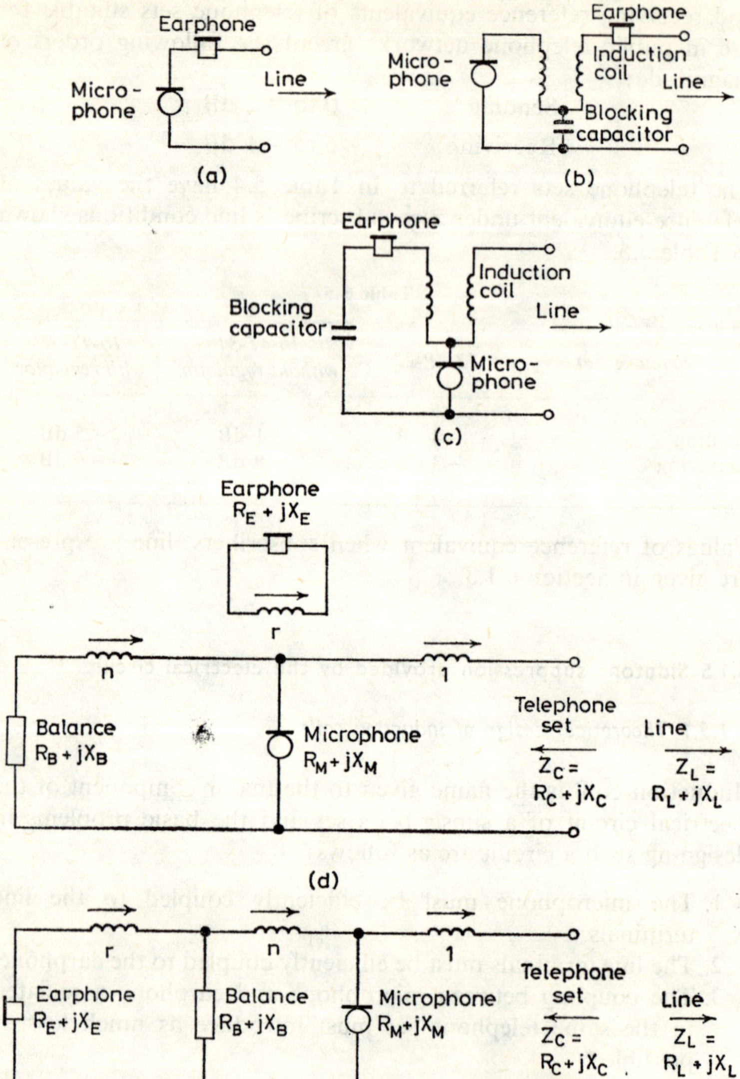

Figure 5.11. Electrical circuits of telephone sets. (a) Microphone and earphone simply in series. (b) Microphone coupled through transformer. (c) Earphone coupled through transformer. (d) Simple hybrid. (e) Modified hybrid

impedance matching, a transformer was introduced as shown in Figure 5.11b or, to avoid passing the feeding current through the earphone, as in Figure 5.11c. None of these arrangements provided any special attenuation in the sidetone path, namely that concerned in problem 3 above. Introducing an additional winding (to form a 'hybrid transformer') and a 'balance network' forms an 'antisidetone' induction coil as shown in Figure 5.11d; a more sophisticated arrangement, possessing further advantages, is illustrated in Figure 5.11e.

The antisidetone induction coil can be considered as a network having four ports; these are connected respectively to microphone, earphone, line and balance. The various impedances are indicated in Figure 5.11 d and e which also shows the turns ratios of the various windings which are placed on a common core of the induction coil.

To simplify the treatment, the case will be considered where the impedances terminating the four ports are all fixed resistances equal to R_M, R_E, R_L and R_B, respectively for microphone, earphone, line and balance. With the simple hybrid arrangement of Figure 5.11d, the conditions for maximum transfer of power when sending and when receiving and when perfect balancing of sidetone is secured are as follows:

$$R_M = \frac{n}{n+1} R_L$$

$$R_E = \frac{r^2}{n+1} R_L$$

$$R_B = nR_L$$

n and r are the ratios of the numbers of turns on the windings as shown in the figure, the line winding being taken as unity. R_M, R_E and R_B have been expressed in terms of R_L.

Theoretical sending and receiving efficiencies η_S and η_R are defined as follows:

$$\eta_S = \frac{\text{power absorbed in the line}}{\text{maximum power available from the microphone}}$$

and

$$\eta_R = \frac{\text{power absorbed in the earphone}}{\text{maximum power available from line}}$$

The overall efficiency when a sending telephone set is directly connected to a receiving one of the same type is given by the product

$\eta_S \eta_R$ and, for this to be a maximum, $\eta_S = \eta_R = 0.5$, corresponding to 3 dB loss of power for sending and for receiving (6 dB loss overall). For the simple hybrid of Figure 5.11d, this condition is satisfied by making $n = 1$; in this design, there is no restriction on r and it can be chosen to suit any particular impedance of earphone.

The ratio η_S/η_R is termed the Y ratio of the induction coil. Although maximum overall efficiency is achieved by making $Y = 1$,

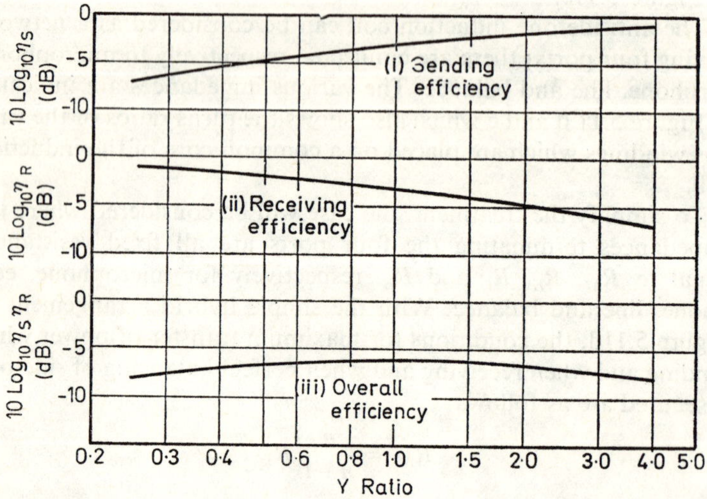

Figure 5.12. Theoretical efficiencies of induction coils as a function of Y ratio

it is often desirable to depart from this condition to take advantage, for example, of excess sensitivity of an earphone by changing the Y ratio to favour sending. Figure 5.12 shows the effects on η_S and η_R and the theoretical overall efficiency $\eta_S\eta_R$, when these are all expressed in logarithmic units (dB). The expressions for η_S and η_R in terms of Y are as follows:

$$\eta_S = \frac{Y}{Y+1} \quad \text{and} \quad \eta_R = \frac{1}{Y+1}$$

In practice, the simple hybrid of Figure 5.11d is inconvenient for use in telephone sets with feeding current supplied from the local exchange. R_M is likely to be less than 100 Ω while R_L will exceed 300 Ω at a design frequency of 1600 Hz; this requires n to be made less than 0.5. For this design, $Y = n$ and so the telephone set could be designed only for a value of Y much less than the theoretical optimum of $Y = 1$. With modern earphones of high sensitivity, it is, moreover, desirable to use a value of Y much greater than 1, so that surplus sensitivity can be transferred from receiving to sending where it is more valuable. A high value of Y thus enables the speech levels at the junction to be maintained with longer subscribers' lines.

The simple hybrid arrangement is quite suitable for a local battery telephone set because microphone and earphone can, in that case, be interchanged; the earphone can, of course, be wound to have any desired value of R_E.

A modified type of hybrid such as shown in Figure 5.11e is usually employed although there are many possible variations of this. For the arrangement shown in this figure, the conditions for maximum transfer of power when sending and when receiving, together with perfect balancing of sidetone are as follows:

$$R_M = \frac{n(n+r)}{(n+1)(n+r+1)} R_L$$

$$R_E = \frac{r(n+r)}{n+1} R_L$$

$$R_B = \frac{nr}{n+r+1} R_L$$

The Y ratio, which was equal to n in the simple hybrid, now becomes:

$$Y = \frac{n(n+r+1)}{r}$$

The relationships between η_S, η_R and Y remain unchanged as in Figure 5.12, the best theoretical overall efficiency being still obtained when $Y = 1$.

In the foregoing treatment, all the impedances have been assumed non-reactive. If the impedances are represented as shown in Figure 5.11d, the expressions given above remain valid for R_M, R_E and R_B but must, for maximum transfer of real power, be supplemented by corresponding expressions for X_M, X_E and X_B. For the simple hybrid, these would be:

$$X_M = -\frac{n}{n+1} X_L$$

$$X_E = -\frac{r^2}{n+1} X_L$$

$$X_B = nX_L$$

If reflectionless matching is required (instead of conjugate matching which maximises the transfer of real power), the signs of the expressions for X_M and X_E must be made positive, instead of negative.

For the modified hybrid shown in Figure 5.11e, the expressions become:

$$X_M = -\frac{n(n+r)}{(n+1)(n+r+1)} X_L$$

$$X_E = -\frac{r(n+r)}{n+1} X_L$$

$$X_B = \frac{nr}{n+r+1} X_L$$

When any induction coil is arranged in the manner described above, the power from the microphone divides between the line and the balance in the ratio $Y:1$, no power entering the earphone (perfect sidetone suppression). Correspondingly, in the receiving direction, the power from the line divides between the microphone (in which it is, of course, wasted) and the earphone in the ratio $Y:1$, no power entering the balance. It can also be shown that the impedance seen by the microphone is equal to:

$$k_1(R_L+jX_L)+k_2(R_B+jX_B)$$

and the impedance looking into the line terminals of the induction coil from the line is equal to:

$$k_3(R_M+jX_M)+k_4(R_E+jX_E)$$

where k_1, k_2, k_3 and k_4 are coefficients dependent only on n and r. By making appropriate substitutions, it follows that the two series components of the impedance seen by the microphone are respectively equal to:

$$\eta_S Z_M \quad \text{and} \quad \eta_R Z_M$$

The first is identifiable with the line impedance when matched to Z_C and the other with the balance impedance Z_B, each being seen through the circuit of the induction coil.

Similarly, the two series components of the impedance seen by the line are equal to:

$$\eta_S Z_C \quad \text{and} \quad \eta_R Z_C$$

The first is identifiable with the impedance of the microphone (Z_M) and the other with the impedance of the earphone (Z_E), each being seen through the circuit of the induction coil.

Thus, it can be seen that, as η_S is made larger (Y ratio increased), the impedance looking into the line terminals of the telephone set (Z_C) becomes controlled more by the impedance, Z_M, of the microphone than by that, Z_E, of the earphone. A large Y ratio therefore helps to reduce the effect on Z_C of the high positive reactance of moving iron earphones (Z_M is, of course, practically non-reactive when carbon microphones are used).

The design problem has been illustrated here in terms of a 'transformer' type of induction coil forming the electrical circuit but many other types of network, including those containing active devices can be used. For example, forms of Wheatstone bridge or a.c. bridge arrangements containing mutual and/or self inductance as well as resistance and capacitance could, in principle be used. At present (1970) most telephone sets use some variant of the hybrid transformer arrangement shown in Figure 5.11e. In the future, some greater flexibility may be imparted by inclusion of active devices arranged to permit greater freedom in providing the most desirable impedance conditions. Provided no nonlinearity is introduced, the external performance of the electrical circuit will be subject to the same relationships as explained here.

5.1.5.2 Sidetone and line impedance

The foregoing information is concerned with the design of the electrical circuit of the telephone set to provide suitable sending and receiving sensitivities and to reduce sidetone to vanishing point when the line terminals of the set are closed with a certain value of line impedance, denoted by Z_{SO}, which is a characteristic of design. So far as the transfer of power from telephone set to line is concerned, it is usually sufficient to carry out the design mainly in terms of a single frequency, say 1600 Hz, and to arrange matters as far as possible so that the sensitivity does not suffer excessively when a departure is made from this design frequency. The sidetone performance, however, requires special consideration because accurate matching of Z_{SO} to the impedance Z_L of the line, although very important, is very difficult to achieve and a suitable compromise must be sought bearing in mind the wide range of Z_L to which the telephone set will be exposed.

Of the three impedances to be considered, Z_C, the impedance looking into the set from the line, and Z_{SO} are determined by the design of the telephone set and are therefore fixed for a given telephone station*. Z_L, however, may vary widely according to the connection. Values of Z_C for typical telephone sets are shown in Table 5.3; the associated values of Z_{SO} are shown in Table 5.6; also included are the values chosen as design objectives during the development of the 16-4T-31 telephone set. Table 5.7 shows some examples of Z_L as functions of frequency and Figure 5.13 shows how Z_L may change, at four particular frequencies, as the length of the subscriber's line varies and the type of junction changes.

Table 5.6. VALUES OF IMPEDANCE, Z_{SO}, THAT, WHEN CONNECTED TO THE LINE TERMINALS OF SOME UNITED KINGDOM POST OFFICE TELEPHONE SETS, CAUSES SIDETONE TO VANISH

Telephone set	*13-2P-27*		*16-4T-31 without regulator*		*16-4T-31 with regulator*						*Chosen in 1954 as objective for design purposes*	
Feed current	*44* mA		*32* mA		*32* mA		*64* mA		*101* mA			
Frequency Hz	Z_{SO} Ω	$/\theta$ °	Z_{SO} Ω	$/\theta$ °	Z_{SO} Ω	$/\theta$ °	Z_{SO} Ω	$/\theta$ °	Z_{SO} Ω	$/\theta$ °	Z_{SO} Ω	$/\theta$ °
200	1135	−90	737	−74	734	−80	1588	−90	2204	−90		
300	891	−82	796	−60	815	−65	1203	−90	1603	−90		
400	759	−71	821	−52	792	−55	1070	−90	1239	−90		
500	671	−62	800	−49	789	−51	1001	−81	1044	−90	1100	−35
600	619	−56	798	−47	785	−48	925	−71	908	−84		
800	561	−46	757	−44	753	−45	827	−63	721	−71		
1000	527	−40	711	−44	713	−44	772	−55	619	−62	803	−35
1250	506	−35	667	−44	663	−43	727	−49	538	−52		
1600	493	−30	609	−44	606	−43	701	−44	470	−41		
2000	490	−28	559	−44	557	−43	661	−40	424	−29	595	−38
2500	488	−27	513	−43	505	−41	626	−36	394	−17		
3000	491	−27	478	−43	475	−41	601	−32	376	−7	511	−40
3500	495	−28	456	−42	444	−40	583	−29	400	−1		
4000	501	−29	438	−41	424	−38	567	−26	400	−0		

Note The measurements were made using Z_M and Z_E as for Table 5.3; in fact, the values used are unimportant. Except when the regulator is present, the values of Z_{SO} are little affected by the magnitude of the feeding current.

* They may, of course, be dependent on the feeding current.

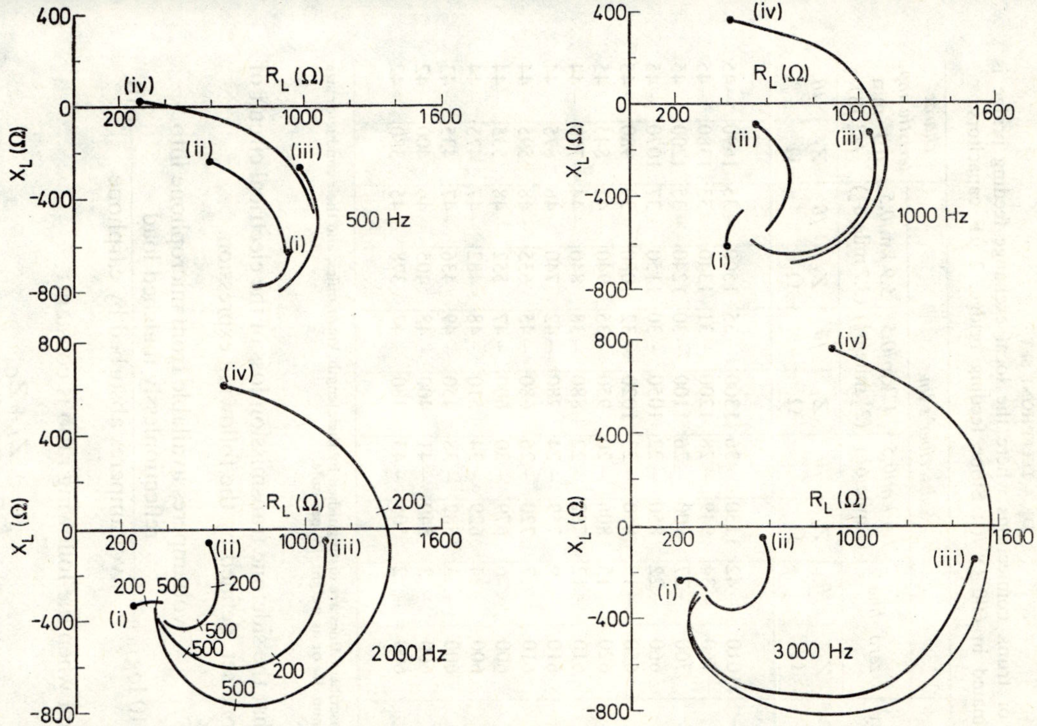

Figure 5.13. Range of line impedance Z_L that may terminate the line terminals of a subscriber's set. (i) Average unloaded junction. (ii) 600 Ω junction. (iii) 1100 Ω junction. (iv) Local call to zero subscriber's line and 13-2P-27 telephone set.

Note: Each graph shows effect of varying resistance of 6½ lb cable subscriber's line from zero to 1000 Ω (numbered end represents zero) some intermediate values of resistance are marked for 2000 Hz. A Stone transmission bridge with 2 μF capacitors was used; with a transformer transmission bridge the zero end of each graph is displaced upwards by a small amount

The sidetone performance of a telephone set is determined by a relationship between these three impedances. It has been shown by

Table 5.7a SOME EXAMPLES OF THE IMPEDANCE, Z_L, SEEN BY THE LINE TERMINALS OF A TELEPHONE SET

Junction or trunk connections where the local exchange feeding bridge is terminated in 600 Ω $\underline{/0}$; Stone feeding bridge, 2 μF capacitors.

Frequency Hz	*Subscriber's line*								*charac-teristic imp.* *0.5* mm (*6½* lb)	
	zero line		*1.6 km/0.5* (*1* mile/*6½*)		*4* km/*0.5* (*2½* mile/*6½*)		*5.9* km/*0.5* (*3.7* mile/*6½*)			
	Z_L Ω	$\underline{/\theta}$ °	Z_L Ω	$\underline{/\theta}$ °	Z_L Ω	$\underline{/\theta}$ °	Z_L Ω	$\underline{/\theta}$ °	Z_L Ω	$\underline{/\theta}$ °
200	1010	−42	1130	−36	1300	−35	1500	−35	1690	−45
300	760	−34	949	−28	1200	−31	1340	−33	1380	−45
400	700	−27	900	−26	1100	−30	1240	−35	1200	−45
500	660	−22	850	−22	1050	−30	1150	−37	1070	−45
600	650	−19	850	−23	1020	−32	1080	−39	980	−45
800	620	−15	806	−20	950	−35	940	−42	845	−45
1000	610	−12	782	−22	880	−38	840	−44	760	−44
1250	610	− 9	759	−24	780	−42	740	−46	675	−44
1600	610	− 7	720	−26	680	−45	635	−48	595	−44
2000	600	− 6	679	−30	600	−47	552	−48	535	−44
2500	600	− 4	629	−34	510	−48	482	−47	475	−44
3000	600	− 3	582	−38	450	−49	436	−47	435	−43
3500	600	− 3	540	−41	400	−48	405	−46	400	−42
4000	600	− 2	505	−43	380	−47	378	−45	380	−42

Note Subscribers' lines are distinguished by their length, km or miles, and their conductor size copper) in mm dia or as weight, lb, per mile.

Wigan that the sidetone transmission loss of the electrical circuit of a telephone set, defined by the following expression,

$$L_{ST} = 10 \log_{10} \frac{\text{voltamperes available from microphone into reflectionlessly matched load}}{\text{voltamperes absorbed by earphone}} \text{dB},$$

is constant when the following ratio is constant:

$$k = \frac{|Z_L + Z_C|}{|Z_L - Z_{SO}|}$$

It has also been shown by Wigan that the locus, on the impedance diagram, of Z_L for constant k is a circle, which can be constructed as shown in Figure 5.14.

Table 5.7b Local connections, through a Stone transmission bridge (2 μF capacitors) feeding current symmetrically to identical subscribers' lines; the subscriber's line remote from the measuring end is terminated with a 16-4T-31 (with regulator) telephone set

Frequency Hz	*Subscribers' lines*					
	zero lines		*1.6* km/*0.5* (*1* mile/*6½*)		*5.9* km/*0.5* (*3.7* mile/*6½*)	
	Z_L Ω	$/\theta$ °	Z_L Ω	$/\theta$ °	Z_L Ω	$/\theta$ °
200	850	−77	1023	−43	2005	−29
300	530	−73	833	−34	1625	−42
400	380	−68	770	−28	1345	−45
500	300	−60	750	−23	1175	−46
600	255	−52	750	−21	1055	−48
800	210	−38	750	−20	873	−49
1000	197	−26	750	−22	757	−48
1250	191	−16	740	−25	667	−47
1600	185	−7	700	−30	587	−46
2000	185	+5	660	−35	527	−46
2500	197	+13	610	−40	474	−45
3000	190	+20	550	−43	434	−45
3500	214	+28	500	−47	404	−45
4000	224	+31	454	−50	378	−45

The sidetone transmission loss, L_{ST}, is given by the following:

$$L_{ST} = L_S + L_R + 20 \log_{10} \frac{|Z_C + Z_{SO}|}{2|Z_C|} + 20 \log_{10} k \text{ dB}$$

The first three terms in this expression for L_{ST} depend only on the design of the electrical circuit of the telephone set and, in practice their sum is usually almost constant and falls within the range 4–6 dB.

Although this form is useful to promote understanding, it is usually more convenient, when treating particular cases, to rearrange the last two terms into elements more familiar to the transmission engineer and this is explained in Section 6.1.4. It can also be shown that the reference equivalent of the sidetone path is related to the sum of the sending and receiving reference equivalents of the telephone set alone (without its line) mainly by the return loss between Z_{SO} and Z_L; *see* Section 6.1.4 **for** elaboration of this.

The values of L_{ST} for a few combinations of telephone set and local line arrangements are shown in Table 5.8. The large improve-

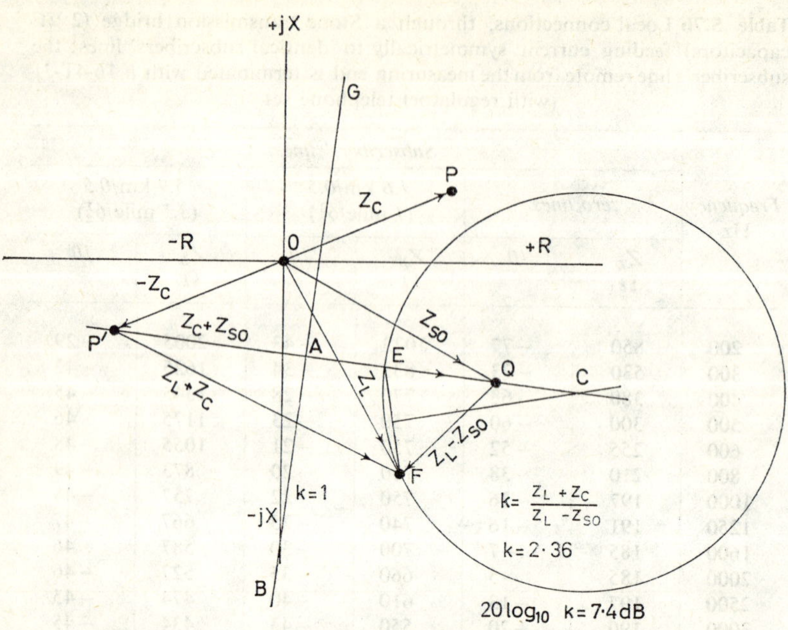

Figure 5.14. Construction of impedance locus for constant sidetone. Note: The construction is as follows: (a) Mark the points P and Q so that OP and OQ represent the vectors Z_C and Z_{SO} respectively. Mark the point P′ so that OP′ represents the vector $-Z_C$. Join P′ to Q and extend beyond Q. (b) Choose any point F (well away from P′Q) so that P′F/QF is equal to the desired value of k. This can be done by striking an arc of one unit radius (on some suitable scale) centred at Q and another arc of radius k units centred at P′ so that the arcs intersect, the point of intersection yields a suitable example of F. (c) Bisect /P′FQ letting the bisector cut P′Q at E. Construct the perpendicular bisector of EF and let this cut the extension of P′Q at C. (d) C is then the centre of the required circle, its radius being CE. (e) The loci for different values of k are a family of circles all centred on the extension of P′Q and of increasing radius as k approaches 1, when k = 1 the locus is the straight line BAG which is the perpendicular bisector of P′Q

ment in suppression of sidetone by the electrical circuit of the 16-4T-31 set as compared with the 13-2P-27 can be observed, except in the case of local connections with zero subscribers' lines, where the impedance Z_L is particularly unfavourable (see Table 5.7b). On the shorter subscribers' lines the regulated 16-4T-31 set gains considerable advantage from its regulation; nevertheless, the impedance match between Z_{SO} and the majority of the line impedances, Z_L, met in practice is much superior to that of the 13-2P-27 set as can be illustrated as follows.

Table 5.8a VALUES OF TRANSMISSION LOSS, L_{ST}, OF SIDETONE THROUGH THE ELECTRICAL PARTS OF SOME TELEPHONE SETS ASSOCIATED WITH CERTAIN LINE CONDITIONS

Junction or trunk connections where the local exchange feeding bridge is terminated in 600 Ω/0°; Stone feeding bridge, 2 μF capacitors.

Telephone set	*13-2P-27*	*16-4T-31 without regulator*	*16-4T-31 with regulator*		
Subs' line	3.9 km/0.5 (2.44 mile /6½)	5.9 km/0.5 (3.7 mile/6½)	*zero*	1.6 km/0.5 (*1* mile/6½)	5.9 km/0.5 (3.7 mile/ 6½)
Feed current	44 mA	32 mA	*101* mA	64 mA	32 mA
Z_M	52 /0° Ω	105 /0° Ω	41 /0° Ω	67 /°0 Ω	105 /0° Ω
Frequency Hz	L_{ST} dB	L_{ST} dB	L_{ST} dB	L_{ST} dB	L_{ST} dB
200	20.1	15.4	25.9	22.5	15.7
300	18.9	18.3	23.6	21.3	18.6
400	18.4	20.1	23.0	21.4	20.4
500	17.9	21.9	22.3	21.4	22.1
600	17.9	24.0	22.8	22.2	24.3
800	17.9	26.1	23.3	23.7	26.5
1000	18.0	28.6	24.7	25.7	29.0
1250	18.8	32.8	26.5	28.1	33.0
1600	19.7	36.0	28.1	31.4	35.1
2000	20.7	34.2	29.9	35.8	33.4
2500	22.1	32.6	32.5	47.1	32.4
3000	23.0	30.4	32.5	39.3	30.9
3500	23.4	28.5	32.7	33.3	29.2
4000	23.4	28.0	33.1	30.7	28.9

Note. Measured with earphones (Receiver inset No. 2P or 4T) having impedances, Z_E, as shown in Table 5.2.

The return loss between Z_{SO} and Z_L is the main controlling factor in determining the amount of sidetone suppression provided by the electrical circuit of the telephone set and, as explained in Section 6.1.4, this accounts for most of the difference between L_{ST} and the sum L_S+L_R. It is therefore instructive to assemble this difference for some combinations of telephone set and line condition as shown in Table 5.9. For the older 13-2P-27 set this quantity (practically the same as the return loss between Z_{SO} and Z_L) hardly exceeds 10 dB and, in the lower frequency range, is only

Table 5.8b Local connections, through a Stone transmission bridge (2 μF capacitors) feeding current symmetrically to identical subscribers' lines; the subscriber's line remote from the measuring end is terminated with a telephone set of the same type as that at the measuring end.

Telephone set	*16-4T-31 with regulator*		
Subs' line	*zero*	*1.6* km/*0.5* (*1* mile/*6½*)	*5.9* km/*0.5* (*3.7* mile/*6½*)
Frequency Hz	L_{ST}	L_{ST}	L_{ST}
200	26.9	21.9	13.9
300	21.7	20.5	16.6
400	18.7	20.3	19.5
500	17.3	20.6	22.1
600	16.7	22.1	24.6
800	17.0	22.8	30.0
1000	18.0	24.3	34.8
1250	19.5	26.7	38.3
1600	21.0	30.1	37.0
2000	22.3	33.1	35.0
2500	25.3	35.8	34.3
3000	26.0	31.9	32.3
3500	28.0	28.8	30.3
4000	30.1	26.8	29.4

Note Feeding currents as for corresponding combinations in Part a of the table.

about 5 dB. With the 16-4T-31 set on the longer line conditions the return loss is some 10 dB better than for the 13-2P-27 set; unfortunately the impedances, Z_L under short line conditions are so unfavourable that the return loss achieved suffers very badly, especially for local calls. This information shows the importance of assigning proper relative importance to the various line conditions met by a telephone set; in the case of this example considerable weight has been given to the long line conditions, where customers will experience greater losses and may experience difficulty on this account, as compared with short line and local call conditions where negligible difficulty is to be expected.

The construction of constant k circles on the impedance plane is useful to illustrate the design problem described above; Figure 5.15 shows some examples where 20 $\log_{10} k$ has been taken as 10 dB. 20 $\log_{10} k$ does not, in practical cases differ very much from $L_{ST}-(L_S+L_R)$ or the return loss between Z_{SO} and Z_L. Comparison of these circles in Figure 5.15 with the range of Z_L shown in Figure

Table 5.9. EFFICIENCY OF SIDETONE BALANCE EXPRESSED AS $L_{ST}-(L_S+L_R)$

	With 600 Ω junction impedance					*Symmetrical local call conditions*		
Telephone set	*13-2P-27*	*16-4T-31 (no reg.)*	*16-4T-31 with regulator*					
Subs' line	3.9 km/ 0.5 mm	5.9 km/ 0.5 mm	*zero*	1.6 km/ 0.5 mm	5.9 km/ 0.5 m	*zero*	1.6 km/ 0.5 mm	5.9 km/ 0.5 mm
Feed current	44 mA	32 mA	101 mA	64 mA	32 mA	101 mA	64 mA	32 mA
Z_M	52 /0 Ω	105 /0 Ω	41 /0 Ω	67 /0 Ω	105 /0 Ω	41 /0 Ω	67 /0 Ω	105 /0 Ω
Frequency Hz	$L_{ST}-(L_S+L_R)$	$L_{ST}-(L_S+L_R)$	$L_{ST}-(L_S+L_R)$	$L_{ST}-(L_S+L_R)$	$L_{ST}-(L_S+L_R)$	$L_{ST}-(L_S+L_R)$	$L_{ST}-(L_S+L_R)$	$L_{ST}-(L_S+L_R)$
200	4.5	2.4	11.3	6.8	2.3	12.3	6.2	0.5
300	3.4	7.8	9.3	6.2	7.6	7.4	5.4	5.6
400	4.0	10.4	8.6	6.5	10.2	4.3	5.4	9.3
500	4.5	12.9	7.7	6.6	12.7	2.7	5.8	12.7
600	5.0	15.4	7.5	7.7	15.3	1.4	7.6	15.6
800	5.4	18.0	7.3	8.9	17.9	1.0	8.0	21.4
1000	6.4	20.5	8.0	10.3	20.6	1.3	8.9	26.4
1250	7.9	24.5	8.9	12.2	24.2	1.9	10.8	29.5
1600	9.0	27.7	10.5	15.8	26.4	3.4	14.5	28.3
2000	10.2	26.3	11.3	19.9	25.2	3.7	17.2	26.8
2500	11.6	23.9	12.3	30.0	23.4	5.1	18.7	25.3
3000	12.5	22.0	12.5	22.7	22.3	6.0	15.3	23.7
3500	12.8	20.4	11.5	16.4	21.0	6.8	11.9	22.1
4000	12.6	19.6	11.0	12.9	20.3	8.0	9.0	20.8

5.13 illustrates the above discussion by showing that they include most of the range, with the notable exceptions of short subscribers' lines especially those connected under local call conditions.

Circles much smaller in radius, corresponding to greater suppression of sidetone, would include much less than the range of line impedances likely to be encountered in practice.

The three sets of circles shown in Figure 5.15 apply for three different values of Z_C (the impedance looking into the telephone set from the line), as follows:

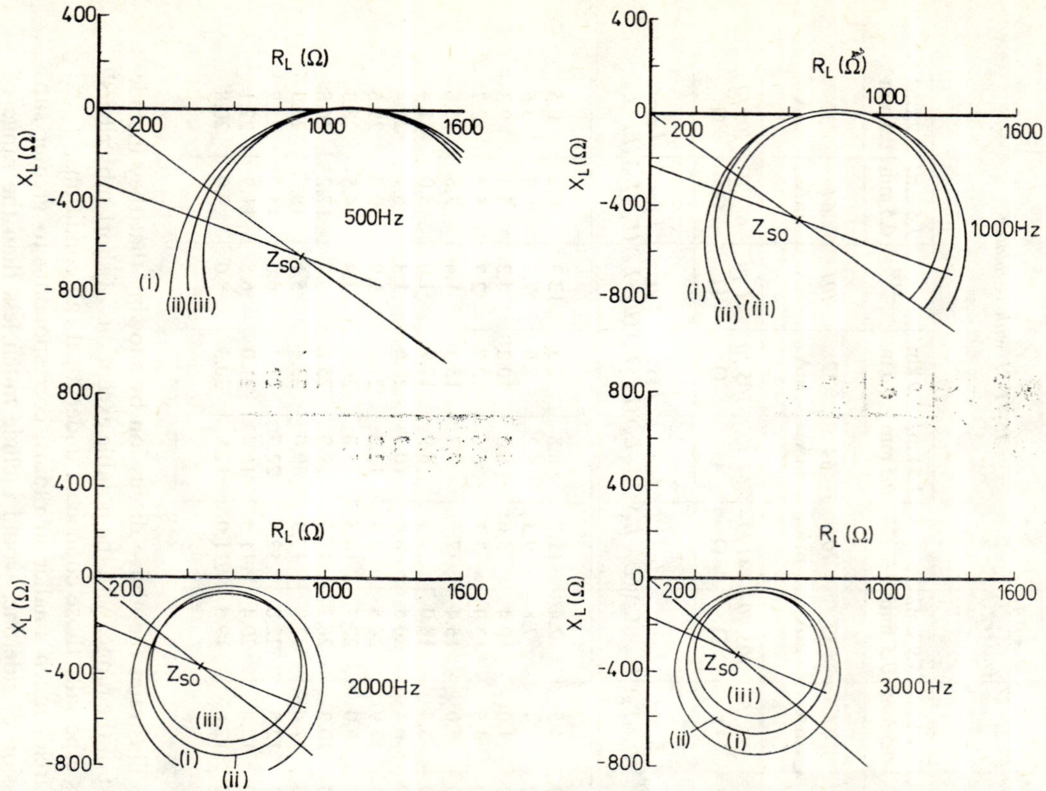

Figure 5.15. Locus of Z_L for $20 \log_{10}k = 10$ dB. (i) Z_C reflectionless match to Z_{SO}, (ii) Z_{S0} resistive and equal to R_{SO}, (iii) Z_C conjugate match to Z_{SO}

(i) Reflectionless matching of Z_C to Z_{SO}, i.e. $R_C = R_{SO}$ and $X_C = X_{SO}$

(ii) Z_C purely resistive and equal to R_{SO}

(iii) Conjugate matching of Z_C to Z_{SO} i.e. $R_C = R_{SO}$ and $X_C = -X_{SO}$

It will be noticed that reflectionless matching yields the largest radius circles for a given value of k.

The best choice for Z_C is difficult to make on account of the many factors influenced by it. Although reflectionless matching is desirable from considerations of sidetone, it must be noted that choice of Z_C for the telephone sets has a considerable influence on the values of Z_L encountered by other telephone sets under short line, local call conditions. Choice of values for Z_C is also affected by considerations of circuit stability and echo when designs for the switched trunk network are being formulated, an important associated impedance in such studies being that of the balance used in the 4 wire/2 wire terminations of 4 wire circuits; usually this is $600\,\Omega\underline{/0^\circ}$. On balance there is much argument in favour of $Z_C = 600\,\Omega\underline{/0^\circ}$, in which case Z_{SO} might need to be located as a compromise between $600\,\Omega\underline{/0^\circ}$ and the characteristic impedance of the most commonly used subscriber's cable, with some consideration given to the continuing existence of loaded cable in junctions (about $1100\,\Omega\underline{/0^\circ}$). Naturally, if expense were no consideration, impedance matching could be performed to maintain all interfaces as 600/600 Ω (say).

5.1.6 Miscellaneous requirements

In addition to the strictly transmission requirements described above, telephone sets for use in public networks must provide facilities for signalling, dialling and ringing. (*See* Refs 20, 21 and 22.) This involves provision of cradle switch contacts which close a d.c. circuit when the handset is lifted off its rest, dial contacts which intermittently interrupt this path an appropriate number of times per second when the dial is operated and connection of a bell or other calling sounder. The bell must be inserted with a capacitor in series with it to prevent closure with a d.c. path across the line terminals when the handset is on its rest; the dial must have off-normal contacts to short circuit the earphone to prevent the user hearing loud pulses while dialling. The off-normal contacts of the dial must also be arranged to short circuit the microphone, not only to reduce the resistance in the dialling path but also to prevent damage to it by the impulses which can impair its performance by 'welding' together the contacts between the carbon granules. All these facilities must be associated with the transmission components in an economical manner and in such a way that any components used in common for different purposes do not result in any adverse interaction between functions.

5.1.7 Overall performance

Although the acceptibility of a given telephone set from the point of view of the customer depends to a considerable extent upon its intrinsic good quality (e.g. freedom from distortions and appropriate shape of handset), many other factors govern customer satisfaction and these depend upon relationships between the characteristics of the telephone set and those of the network of lines, etc. to which it is connected. A telephone set which provides satisfactory service in one telephone network may not be suitable for use in another. Telephone networks differ in respect of the line impedances and transmission losses with which the telephone set has to contend. In particular, choices have to be made concerning the most suitable sending and receiving sensitivities, regulator characteristics, sidetone characteristics and line terminal impedance for a given telephone network. These items will affect not only the levels of signals sent to line and received while listening but also the sidetone and echo performance and stability of the network. Some indication of the characteristics of present telephone sets used in the United Kingdom have been included above but the characteristics that will be desirable for future conditions require scrutiny in conjunction with proposed changes in the telephone network as a whole. Such studies can be conducted with the aid of the procedure outlined in Section 7.2 namely examination of the performance to be expected for suitably chosen hypothetical reference connections containing telephone sets of present and possible future designs.

When the desired characteristics of a telephone set intended to suit a given telephone network have been formulated and engineering design carried out, it becomes necessary to check the performance of prototype models. This can be done to a large extent by the instrumental measurements indicated above but some carefully-designed conversational tests will be necessary; Sections 3.6 and 3.7 describe suitable methods for conducting such tests. When a design has been approved, it will be necessary to determine sending and receiving loudness ratings or reference equivalents on various types of subscribers' lines; Section 3.2 explains the methods to be used.

5.2 LOUDSPEAKING TELEPHONES

Many of the problems discussed in Section 5.1 in relation to handset telephones appear also when loudspeaking telephones are considered but many of these must be treated somewhat differently. As far as possible the various features will be considered here in the same order as was done in the cases of the analogous features of handset telephones. At the present time (1970) most of the characteristics that must be specified are subject to discussion and so the information given here must be treated as provisional.

Location of microphone and loudspeaker. For specification purposes and to compare the characteristics of different loudspeaking telephones, it is necessary to define, in a rather arbitrary manner, the relative locations of the microphones and loudspeakers of loudspeaking telephone sets with respect to the mouth and ears of the user. To define quantities analogous to, and broadly comparable with, sending and receiving reference equivalents of local telephone circuits containing handset telephones, some such conventions as those described in Figure 4.31 in Section 4.4.4 must be adopted. These have not yet been standardised internationally but many suggestions are very similar to those indicated in the figure. If the loudspeaker and microphone are mounted in a single cabinet, the arrangements should locate the loudspeaker in accordance with the particulars given and the microphone should be tested without moving the cabinet.

Microphone and loudspeaker characteristics. It is not difficult to provide loudspeaking telephone sets with sensitivity/frequency characteristics rather broader in frequency bandwidth than that of handset telephones; although this cannot be taken advantage of when trunk calls are made, it is of considerable advantage for local and extension to extension PBX calls which probably include a very large proportion of loudspeaking telephone usage. There are good reasons to attenuate the lower frequency range progressively below about 1000 Hz, especially in the sending direction, otherwise the frequency responses in both directions should be smooth and approximately level.

Sending and receiving sensitivities. The results of the tests described in Section 4.4.4 have been used to define suitable sending and receiving sensitivities. These are shown in the following table together with some measured values of actual loudspeaking telephone sets which have been used in the United Kingdom; some values proposed by various organisations are also included in Table 5.10.

Table 5.10. SENDING AND RECEIVING RATINGS AND SENSITIVITIES OF PRACTICAL LOUDSPEAKING TELEPHONES

		Sending			*Receiving*		
Item	*Particulars*	*Rating rel.* ARAEN dB	'SRE' dB	*Sensitivity at 1000* Hz*	*Rating rel.* ARAEN dB	'RRE' dB	*Sensitivity at 1000* Hz†
1	As suggested in Section 4.4.4	17.5	15	+6.0	−3.8	−1	+5.5
2	United Kingdom Type A	30	27.5	−6.5	−2	+0.8	+3.7
3	United Kingdom Type B	19	16.5	+4.5	−7	−4.2	+8.7
4	United Kingdom Type C	20	17.5	+3.5	−1	+1.8	+2.7
5	Sweden	—	9‡	+12.0	—	−4.5‡	+9.0
6	Poland	—	16.4	+4.6	—	+0.9	+3.6
7	International Telephone and Telegraph Co	—	16.6	+4.4	—	−1.5	+6
8	Handset Telephone on typical average line	9	SRE=7	—	−4.5	RRE=−1	—

* Sending sensitivity is expressed here as dB relative to 1 V per N/m² with the sound pressure measured at the point to be occupied by the microphone of the LST.
† Receiving sensitivity is expressed here as dB relative to 1 N/m² per V with the sound pressure measured 500 mm from the loudspeaker.
‡ The sending and receiving ratings are defined in terms of measurement with the OBDM equipment using talking and listening distances each equal to 500 mm.

The ratings for Items 2–4 were determined primarily in terms of ARAEN as described in Section 4.4.4; conversions to 'SRE' and 'RRE' are as explained in Section 4.4.4, namely by adding, to the ratings relative to ARAEN, values of −2.5 and +2.8 which are the calculated values of sending and receiving reference equivalents of the ARAEN. Conversions to sensitivity are by allowing for the sensitivities of ARAEN at 1000 Hz. Items 6 and 7 were defined in terms of sensitivities. The handset telephone LTC quoted in Item 8 was measured subjectively relative to ARAEN giving the values quoted; the values of SRE and RRE are as determined in the CCITT Laboratory relative to SFERT.

Except for Item 2, all the loudspeaking telephone sets referred to in the table are furnished with voice switching; Item 2 is non voice-switched and it will be noticed that the sending rating is considerably greater than those for the other, voice switched, sets. The receiving rating of Item 2 is the minimum value (maximum sensitivity) which is only usable without instability under favourable conditions of line impedance. The receiving ratings of the voice switched sets are also minimum values (these sets are fitted with manual controls to reduce receiving sensitivity) but instability should not take place when these sets are used even when the line impedances are unfavourable.

It may be concluded from the results assembled in this table that the proposals embodied in Item 1 are generally in fair agreement with the values quoted for the other voice switched sets (except perhaps for Sweden, although the method of measurement may not strictly be comparable with the others). The numerical values of sending ratings ('SRE') are some 10 dB greater than those (SRE) typical of handset telephone LTCS and this is attributable, as explained in Section 4.4.4, to increased vocal levels when loudspeaking telephone sets are used and to the adoption of somewhat shorter talking distances than 500 mm as used to define 'SRE' for loudspeaking telephone sets. The numerical values of receiving ratings are very similar to those typical of handset LTCS.

Stability. The voice controlled switching arrangements must be such that sufficient loss is introduced to ensure stability when two loudspeaking telephone sets are connected directly together and when one set is connected to any line impedance likely to be encountered in actual use on telephone networks. To avoid false switching when used in rather reverberant rooms, the voice switching arrangements must include means of deferring transfer of loss after cessation of received speech until any local 'echo' has died away. This can be achieved by storing the control signal on the receive side in a capacitor/resistor so that it decays at a rate suited to the reverberation times of rooms likely to be used with loudspeaking telephone sets. Other features must be built into the control circuits to avoid other troubles analogous to those encountered by echo suppressors in a practical telephone environment. These do not seem to have been yet studied in as much detail as has been devoted to echo suppressor problems.

To ensure that the operate time for switching to the sending direction of transmission does not cause 'chopping' of the speech from the talker (except at the beginning of each utterance, when it is unavoidable but relatively harmless), the logic must prevent reversion to the receiving direction of transmission until some 300 ms has elapsed from the end of each utterance. If either local speech or received speech commences while there is no speech present in the other direction, the operate time can be made very small, say, only 10–20 ms.

Voice switching thresholds. Loudspeaking telephone sets must be capable of being used in the presence of a certain amount of continuous room noise; the received speech will also be accompanied by circuit noise. The long term r.m.s. value of speech pressure at 500 mm from a talker speaking at reference vocal level (RVL) is +65 dB relative to 20 $\mu N/m^2$; although, on average, vocal levels appreciably higher than RVL are likely to be used with loudspeaking telephones, some talkers may use levels 5–10 dB lower than RVL. The threshold for pure tones at, say, 1000 Hz ought, for satisfactory voice switching, to be set some 15 dB lower than the long term r.m.s. value of the speech and so a pure tone threshold of +40 to +45 would be desirable in the absence of any room noise. Such a low threshold may not be compatible with use in the presence of commonly occurring levels of room noise. (But see below.)

Assuming sending and receiving ratings as Item 1 in Table 5.10; and considering a maximum loss between loudspeaking telephone sets of 25 dB, the median speech voltage (SV5B) entering the LST when receiving will be about −40 dBV; putting the pure-tone threshold (at, say, 1000 Hz) 15 dB lower than this gives −55 dBV (about −53 dBm). This may be too low for use on other than rather noise free telephone connections. (But see below.)

The values of theshold level mentioned above apply when a test signal is applied to one input only (sending or receiving); in practice arrangements are usually included (as in echo suppressors) for the threshold in the send direction to be raised by an amount proportional to the signal on the receive direction and vice-versa. (Naturally, this threshold increase applies only when both input signals are higher than their respective 'single input' thresholds.) Great care is necessary to ensure that the 'differential' thresholds remain independent of the absolute levels of the input signals up to the 'overload levels' in both directions. The differential thresholds may be different for send–receive transitions from those for receive–send; i.e. some hysteresis may be incorporated. In any case a proper relationship must be observed between the absolute levels in the two directions at these thresholds; for example, a sound pressure level at the microphone (free field conditions) in the range +65 to +75 dB relative to 20 μN/m² might be equated to an electrical signal level on the line, in the receive direction, of say, −25 dBV which would seem to be appropriate when two LSTS having the sensitivities envisaged by Item 1 in Table 5.10 are connected through a line loss of about 10 dB.

Use of appropriate differential threshold arrangements will extend the range of levels of circuit noise and of room noise for which loudspeaking telephone sets are serviceable; naturally, they will always require some margin between the levels of speech and noise in each direction respectively (*see* Section 4.4.3) but this may be reduced slightly by careful choice of sensitivity/frequency characteristics of the control paths.

Miscellaneous requirements. A specification for a loudspeaking telephone set will contain various clauses common to all telephone station equipment; other matters more specifically related to loudspeaking telephone sets are as follows:

Loudspeaking telephone sets are usually furnished with a handset telephone for use when privacy is needed or to cope with line and noise conditions beyond what is possible with a distant microphone and a loudspeaker.

To avoid sending an excessive level to line (for example by talking close to the microphone), some limitation of maximum output must be provided but this must strike a reasonable compromise between excessive output and distortion of normal outputs.

A receiving sensitivity control is usually provided, giving perhaps 15 dB reduction in sensitivity from the maximum. With the control in the maximum sensitivity position, it must not be possible to hear crosstalk intelligibly; this can be ensured by arranging that a low, but sufficient, level of noise is emitted from within the equipment.

Overall conversational performances. Figure 5.16 shows some estimates of the conversational performance likely to be attainable when loudspeaking telephone connections are used. The information has been taken from Figure 4.32 in Section 4.4.4 assuming sensitivities as indicated by Item 1 in Table 5.10; it has also been assumed that the loudspeaking telephone sets are each associated with about average subscribers' lines (each having a transmission loss of 2.5 dB). The estimated values of conversational mean opinion score, Y_C, are given as functions of the transmission loss between the terminal exchanges and the results for handset telephone connections are also shown, by broken graphs, for comparison.

It may be concluded from Figure 5.16 that, under low loss conditions, when the preferred listening level can almost be achieved and when little room noise is present, the performance will be at least as good as that of a conventional handset telephone. As the line loss is increased and as the room noise becomes greater, the

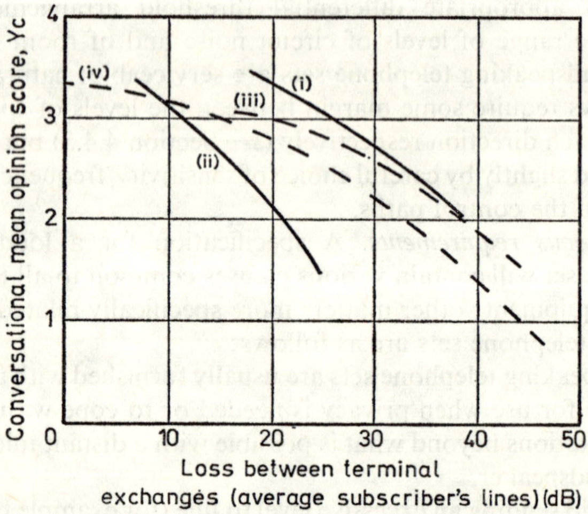

Figure 5.16. Some estimates of attainable conversational performance of loudspeaking telephone connections. (i) LSTs RNO; (ii) LSTs RN50; (iii) handsets RNO; (iv) handsets RN50.

performance of loudspeaking telephone connections will deteriorate considerably more rapidly than is the case with handset telephony. Use of loudspeaking telephones is, therefore, necessarily confined, in telephone networks designed primarily for handset telephony, to relatively low loss and low room noise situations.

The combination of a loudspeaking telephone at one end of a connection and a handset telephone at the other will be broadly acceptable when the ratings shown against Item 1 in Table 5.10 apply for the loudspeaking telephone. Some caution is needed, however, when appraising such arrangements because the user of the handset telephone will be presented with speech signals of unfamiliar quality, possibly considerably coloured by the reverberation characteristics of the room in which the loudspeaking telephone is situated; the effect is similar to introduction of severe group delay/frequency distortion. The user of the loudspeaking telephone will, of course, find the quality of speech from a handset microphone different from that from another loudspeaking telephone microphone and possibly somewhat inferior as regards nonlinear distortion, otherwise, the user of the loudspeaking telephone should experience no serious difficulty.

5.3 HEADSET TELEPHONES

A headset telephone can be used in place of an ordinary handset telephone at subscribers' stations; in this case its performance ought to conform in most respects to that required for such subscribers' sets. Headset telephones are, however, mainly used by switchboard operators and then must be suitable for bridging across circuits without introducing excessive loss as well as for use as a terminating instrument.

At present there are no recognised or generally accepted methods for determining the sensitivities (or ratings) of headset telephones and so only rather vague general statements are possible. A proper specification defining the sending sensitivity needs some conventions to be adopted regarding talking distance and vocal level; these cannot be deduced from those appropriate for handset telephones because the form of the microphone and its conditions of normal use are so different. Account also needs to be taken of the operators' habits in locating their microphones relative to their mouths and of the level of room noise found in operators' locations.

Receiving sensitivities present no greater difficulties in measurement than those associated with the measurement of ordinary handset earphones but the requirements are somewhat different. The desirable sensitivity for headset earphones may be somewhat lower than for handsets because the operator with a headset cannot react so readily to excessively loud speech from the earphone by moving it away from the ear.

Good suppression of sidetone is particularly important (but difficult to achieve, because of the very wide range of impedances that terminate an operators' telephone set); the operator will not be able to hear faint speech well if the room noise heard via the sidetone path is excessive; in addition, excessive sidetone may cause the operator to move the microphone to reduce the speech sidetone and this will degrade transmission of the operator's speech to the subscriber.

6. LINE TRANSMISSION SYSTEMS

6.1. SUBSCRIBERS' LINES AND LOCAL TELEPHONE CIRCUITS

6.1.1 General

Subscribers' lines connect subscribers' stations to their local exchanges and the network of conductors (mostly in multipair cable) is termed the local distribution network. In addition to providing the transmission medium for the speech signals, these lines conduct feeding current to the telephone sets to enable their active devices (including carbon microphones) to function; they also provide paths for signalling and dialling. A very large proportion of the total invested capital of a public telephone system is represented by the local line plant and so it is especially necessary for this to be used in the most economic manner, taking appropriate account of expected growth and changes. It must be noted, too, that the traffic density on each subscriber's line is very low.*

The radial distances of subscribers' stations from their local exchanges range, in the case of the United Kingdom, mainly up to about 3 miles (5 km); distances in dense city areas are, of course, usually much less than this and in some sparsely populated country districts they may extend much further. Particulars of the conductor pairs in cables used for subscribers' lines are shown in Tables 6.1 (for copper conductors) and 6.2 (for the newer, aluminium conductors); these tables include the more important transmission characteristics. Table 6.3 shows corresponding information for open wire lines, now very rarely used in the United Kingdom. Table 6.4 shows the characteristics of cable pairs used in various countries.

To conserve conductor material and space in underground ducts, it is necessary to use the smallest possible diameter of conductor; for a given length of subscriber's line, the minimum conductor

* To a certain extent this can be increased by sharing and by the use of concentrators, which perform a certain amount of switching outside the telephone exchange.

Table 6.1. PARTICULARS AND NOMINAL CHARACTERISTICS OF CABLE PAIRS USED FOR SUBSCRIBERS' LINES IN THE UNITED KINGDOM (Copper conductors)

Conductor size and other particulars	*Resistance* Ω		*Capac. (nF)*		*Image atten. (800/1600* Hz) (dB)		*Modulus of characteristic impedance (800/1600* Hz) Ω
	per mile	*per* km	*per* mile	*per* km	*per* mile	*per* km	
$2\frac{1}{2}$ lb (0.318 mm dia.)							
Paper core	700	435	60	37	2.82/3.99	1.75/2.48	1524/1077
New polyethylene	700	435	85	53	3.36/4.75	2.09/2.95	1280/905
For planning	700	435	75	47	3.15/4.46	1.95/2.77	1363/964
Lab. art. line	680	423	60	37	2.78/3.95	1.73/2.45	1510/1070
4 lb (0.406 mm dia.)							
PCUT	440	273	72	45	2.45/3.47	1.52/2.16	1103/780
New polyethylene	440	273	90	56	2.74/3.88	1.70/2.41	986/697
For planning	440	273	75	47	2.50/3.54	1.55/2.20	1080/764
Lab. art. line	440	273	67	42	2.36/3.34	1.47/2.08	1143/808
$6\frac{1}{2}$ lb (0.508 mm dia.)							
PCUT	270	168	75	47	1.96/2.77	1.22/1.72	846/598
Polyethylene	270	168	80	50	2.02/2.86	1.26/1.78	819/579
PCQL	270	168	69	43	1.88/2.66	1.67/1.65	882/624
New polyethylene	270	168	90	56	2.15/3.04	1.34/1.89	772/546
For planning	270	168	75	47	1.96/2.77	1.22/1.72	846/598
Lab. art. line	272	169	75	47	1.97/2.78	1.22/1.73	849/601

10 lb (0.635 mm dia)							
PCUT	176	109	83	52	1.66/2.35	1.03/1.46	650/459
PCQL	176	109	78	49	1.61/2.28	1.00/1.42	670/474
Polyethylene	176	109	73	45	1.56/2.21	0.97/1.37	693/490
New polyethylene	176	109	90	56	1.73/2.45	1.07/1.52	624/441
For planning	176	109	75	47	1.58/2.24	0.98/1.39	683/483
Lab. art. line	176	109	75	47	1.58/2.24	0.98/1.39	683/483
20 lb (0.920 mm dia.)							
PCTD	88	55	79	49	1.15/1.62	0.71/1.01	471/333
PCUT	88	55	88	55	1.21/1.71	0.75/1.06	446/315
PCQL	88	55	70	43	1.08/1.53	0.67/0.95	500/354
New polyethylene	88	55	95	59	1.26/1.78	0.75/1.11	429/304
For planning	88	55	85	53	1.19/1.68	0.74/1.04	454/321
Lab. art. line	88	55	75	47	1.12/1.58	0.70/0.98	483/342

Notes The information has been calculated in miles and converted to km (1 mile = 1.609 344 km). For planning purposes in the United Kingdom, the loss at 1600 Hz is used with the value in km rounded to the nearest 0.1 dB/km.

Polyethylene was originally used for insulation of 6½ and 10 lb cables only and, for these, average values of capacitance have been taken. Later types of polyethylene was used for all the above conductor sizes and 'maximum' average values of capacitance have been taken.

PCUT means paper core unit twin
PCQL means paper core quad local
PCTD means paper core twin distribution
Laboratory artificial lines are made up in sections not greater than ½ mile
Cables in 2½ lb conductor are available with from 100 to 4800 pairs; 4 lb ranges from 100 to 2000 pairs; 6½ lb ranges from 100 to 1200 pairs; 10 lb ranges from 100 to 800 pairs; and 20 lb cable is available only with 100 pairs.

Table 6.2. PARTICULARS AND NOMINAL CHARACTERISTICS OF CABLE PAIRS USED FOR SUBSCRIBERS' LINES IN THE UNITED KINGDOM
(Aluminium conductors)

Conductor size and other particulars	*Resis.* Ω		*Capac.* (nF)		*Image attenuation* (*800/1600* Hz) (dB)		*Modulus of characteristic impedance* (*800/1600* Hz) Ω
	per mile	*per* km	*per* mile	*per* km	*per* mile	*per* km	
0.5 mm dia	460	287	108	67	3.07/4.35	1.91/2.70	920/650
0.8 mm dia	180	112	108	67	1.92/2.72	1.20/1.70	576/407

Table 6.3. PARTICULARS AND NOMINAL CHARACTERISTICS OF OPEN WIRE PAIRS USED FOR SUBSCRIBERS' LINES (NOW UNCOMMON IN UK)
(Cadmium copper conductors)

Conductor		*Loop resis.* Ω		*Capac.* nF		*Induct.* mH		*Image attenuation* (*800/1600* Hz) dB		*Modulus of characteristic impedance* (*800/1600* Hz) Ω
dia mm	*wt. per* mile lb	*per* mile	*per* km	*per* mile	*per* km	*per* mile	*per* km	*per* mile	*per* km	
1.27	40	52	32	7.5	4.66	4.0	2.49	0.23/0.27	0.14/0.17	1216/934
1.68	70	30	19	7.9	4.91	3.8	2.36	0.16/0.18	0.10/0.11	943/782

Note. Planning values of image attenuation are shown. When dry, somewhat smaller values are obtained. Normal spacing is 9 in; 12 in is used for trunk and junction lines and this increased spacing slightly reduces the attenuation.

size is governed mainly by the transmission loss to speech signals but signalling requirements of the exchange equipment dictate a maximum permissible value of resistance for the loop, including the telephone set in its signalling condition. Usually, transmission sets the limit when the heavier gauge conductors are used but signalling restricts the maximum length of the lighter gauge conductors. In the United Kingdom, signalling and transmission limits are about equal for 6½ lb conductors. In practice, a considerable proportion of subscribers' lines are composed of a mixture of different gauges chosen as far as possible for maximum economy, usually with the lighter gauges near the exchange where duct space is particularly restricted. Some restrictions on the use of very light gauges may be set by difficulties in jointing, especially if rearrange-

Table 6.4. PARTICULARS AND NOMINAL CHARACTERISTICS OF CABLE PAIRS USED FOR SUBSCRIBERS' LINES IN VARIOUS COUNTRIES (COPPER CONDUCTORS)

Country	*Conductor size*			*Resis.* Ω		*Capac.* nF		*Image atten.* (*800/1600* Hz) dB		*Modulus of characteristic impedance* (*800/1600* Hz) Ω
	dia mm	*weight per mile* lb	*A.W.G.*	*per* mile	*per* km	*per* mile	*per* km	*per* mile	*per* km	
UK	0.318	$2\frac{1}{2}$	—	700	435	75	47	3.15/4.46	1.95/2.77	1363/964
Australia	0.318	$2\frac{1}{2}$	—	700	435	72	45	3.09/4.37	1.92/2.72	1391/983
Europe	0.32	—	—	677	420	72	45	3.04/4.30	1.89/2.67	1368/967
USA	0.321	—	28	677	420	75	47	3.10/4.39	1.93/2.73	1340/947
UK	0.406	4	—	440	273	75	47	2.50/3.54	1.55/2.20	1080/764
Australia	0.404	4	—	440	273	72	45	2.46/3.49	1.53/2.17	1100/780
Europe	0.4	—	—	434	270	81	50	2.56/3.59	1.59/2.23	1040/730
Europe	0.4	—		434	270	58	36	2.17/3.05	1.35/1.89	1220/860
USA	0.405	—	26	440	273	79	49	2.40/3.40	1.49/2.11	1050/750
UK	0.508	$6\frac{1}{2}$	—	270	168	75	47	1.96/2.77	1.22/1.72	846/598
Australia	0.508	$6\frac{1}{2}$	—	270	168	72	45	1.88/2.66	1.17/1.65	860/610
USA	0.511	—	24	273	170	72	45	1.98/2.80	1.23/1.74	870/615

UK	0.635	10	—	176	109	75	47	1.58/2.24	0.98/1.39	683/483
Australia	0.635	10	—	176	109	72	45	1.51/2.13	0.93/1.32	695/490
Europe	0.6	—	—	196	122	81	50	1.71/2.36	1.06/1.47	690/790
Europe	0.6	—	—	196	122	61	38	1.48/2.07	0.92/1.28	795/560
USA	0.644	—	22	171	106	83	52	1.64/2.32	1.02/1.44	640/455
Europe	0.8	—	—	106	66	81	50	1.22/1.69	0.76/1.05	515/360
Europe	0.8	—	—	106	66	61	38	1.08/1.48	0.67/0.92	590/415
UK	0.902	20	—	88	55	85	53	1.19/1.68	0.74/1.04	454/321
Europe	0.9	—	—	82	51	80	50	1.12/1.58	0.70/0.98	452/319
Europe	0.9	—	—	82	51	55	34	0.92/1.31	0.57/0.81	545/385
USA	0.910		19	88	55	83	52	1.18/1.66	0.73/1.03	459/325

28

ments require joints to be opened; manipulation of cable joints after installation can be reduced by extending the use of distribution cabinets and street pillars.

If the length of the subscriber's line is very long, so that large diameter conductors become necessary, the loop resistance may have to be reduced considerably below the signalling limit; the situation can be improved under such circumstances by applying loading to cable pairs chosen to meet the signalling limit with only a small margin. This procedure has been found economic in the Bell Telephone System where subscribers' lines exceeding about 3.4 miles (5.5 km) in length are loaded with as many coils as possible with consideration for appropriate spacings; Table 6.5 shows some relevant particulars.

Table 6.5. PARTICULARS AND CHARACTERISTICS OF LOADED CABLE PAIRS USED FOR SUBSCRIBERS' LINES IN THE USA WHEN THEY EXCEED 18 000 ft IN LENGTH (18 000 ft = 3.4 miles = 5.5 km)

Conductor		*Loop resis.* Ω		*Loading coils*			*Effective loss*		*Modulus of image impedance (800/1600 Hz)* Ω*
				In-duct	*Spacing*				
dia. mm	*A. W. G.*	*per* mile	*per* km	mH	mile	km	dB *per* mile	dB *per* km	
0.511	24	280	451	88	1.136	1.83	1.2	1.93	1175/1178
0.644	22	178	286	88	1.136	1.83	0.8	1.29	1039/1099
0.910	19	92	148	88	1.136	1.83	0.43	0.69	1007/1090

* Assuming ½ section termination

When planning the most economic arrangement of line plant in local telephone areas, it is necessary to have ready access to information concerning those combinations of subscriber's set, subscriber's line and exchange feeding bridge that are just acceptable from the transmission point of view. Account must also be taken of private branch exchanges (PBXS) and extension plans used at the larger subscribers' installations. As has already been explained (Section 1.1.2 and Section 7.2), the satisfaction of a customer making a particular call is a function of the totality of the connection he is using and cannot generally be related very closely with the characteristics of any one element of that connection. Nevertheless, for practical planning and design purposes, it is necessary to separate the connection into manageable parts and, for the present (local area) planning purposes it is convenient to treat 'local telephone circuits' (LTCs) as the planning unit. Naturally, when only one type of telephone set is in use and only one type of exchange feeding bridge is used, the problem is much simpler than when there is a variety of both; when only rather similar types of telephone set are in use (e.g. modern types having smooth sensitivity/frequency responses and good suppression of sidetone) the rules adopted may also be simplified. It is therefore not surprising to find that rules used at different times and in different telephone networks show considerable variety and only a few examples can be indicated. It must also be remembered that the relatively recent advances in telephone technology have yet to influence the construction of local networks; when this does take place the planning rules indicated here may well require quite substantial changes.

In the period 1905 till 1933, local line planning was performed in terms of loudness ratings which yielded equivalents between the local telephone circuit concerned and a 'standard' local telephone circuit plus (or minus) a specific length of 'standard cable'; from 1933 till 1947, the only change was to replace 'miles of standard cable' (m.s.c.) by the loss, in decibels, of a distortionless attenuator. (Further details are given in Section 7.1.) In 1947, there existed a very wide range of telephone sets in use in the United Kingdom, including the 'candlestick' telephone, handset telephones without antisidetone induction coils, antisidetone telephone sets with older, resonant earphones and those with more modern, smooth, frequency responses; several types of PBX switchboard were in use, and allowances had to be made for their effects on transmission, and much of the old, heavy gauge (20 lb), subscriber's line, was included in subscribers' lines together with some overhead construction. At least four different types of common battery exchange feeding bridges were in fairly common use and a few that required use of local battery telephone sets. Rating the relative transmission performances of such diverse compositions of local telephone circuit could not be done realistically by mere comparison of their loudness ratings and so a rather more elaborate method was devised, termed the 'transmission performance rating' method. Allowances were determined for the effects of sidetone on sending and receiving and for the differing amounts of attenuation/frequency distortion introduced and these were combined by a calculation procedure to yield ratings of all the required combinations of items in local telephone circuits. It was found that, in most practical cases, transmission was limited by the sending performance and so some further simplification was possible by determining equivalents at their respective transmission limits between each particular type of conductor pair and the loop resistance of $6\frac{1}{2}$ lb cable. The following table shows the results. The loop resistance of $6\frac{1}{2}$ lb cable that is equivalent in transmission performance rating to a given length of any other gauge is termed the TER value (transmission equivalent resistance).

Table 6.6. TRANSMISSION PERFORMANCE EQUIVALENTS OF VARIOUS GAUGES OF CABLE PAIR RELATIVE TO $6\frac{1}{2}$ lb CABLE PAIRS
(applied to 16-4T-31 set, (with regulator) on 50 V, 200+200 Ω feeding bridge)

Conductor		*Maximum length*		*Maximum loop resistance* Ω	*TER multiplying factor*	*Image attenuation at 1600* Hz dB
dia. mm	*weight* lb/mile	mile	km			
(1)	(2)	(3)	(4)	(5)	(6)	(7)
0.318	$2\frac{1}{2}$	1.90	3.1	1333*	0.75	8.5
0.406	4	2.85	4.5	1250*	0.8	10.0
0.508	$6\frac{1}{2}$	3.70	6.0	1000	1.0	10.0
0.635	10	4.73	7.6	834	1.2	10.4
0.914	20	a 7.58	12.1	667	1.5	12.9
		b 5.98	9.6	527	1.9	10.2

* Signalling limit is 1000 Ω
Note a indicates the limit originally set for 20 lb conductor; it was later reduced to the value shown by b.

The multiplying factors shown in col. 6 were obtained in 1949 from the results of transmission performance calculations based on 13-1L-18.35A and 13-1L-27 telephone sets used on the type of exchange feeding bridge in most common use at the time, namely the 50 V ballast type. The limiting length of $6\frac{1}{2}$ lb conductor for the 16-4T-31 set with regulator was chosen by conversation assessments (in about 1960) to be 3.7 miles (1000 Ω loop resistance) for the 50 V, 200+200 Ω (nonballast) feeding bridge. The same multiplying factors have been used to determine the equivalent maximum resistances for the other gauges, as shown in col. 5. The maximum lengths are obtained by dividing the maximum loop resistance values by the loop resistance per unit length as quoted in Table 6.1.

When a subscriber's line is in mixed conductor gauges, the TER multiplying factors can be used as follows to determine whether the performance of the complete LTC is within the transmission limit.

For any section of line in conductor other than $6\frac{1}{2}$ lb (0.5 mm dia), the equivalent loop resistance is determined by multiplying the actual loop resistance by the TER multiplying factor as shown in the table above; the result is the TER value of that section of the line. The several TER values of the sections of differing conductor are then added together, including the actual loop resistance of any section of $6\frac{1}{2}$ lb conductor; this sum is termed the TER value of the complete subscriber's line and must not be allowed to exceed the limiting value for $6\frac{1}{2}$ lb conductor (which happens to be 1000 Ω for the subscriber's set and feeding bridge considered in Table 6.6; different maximum TER values would apply for other combinations).

At the present time, the majority of subscribers' sets are of the 16-4T-31 type (with regulator) and almost all exchange feeding bridges are 50 V, 200+200 Ω; subscribers' sets likely to be introduced in the immediate future will have similar characteristics to the present sets. These circumstances allow some simplifications to be made and advantage is now being taken of the fact that, except for $2\frac{1}{2}$ lb conductor, the commonly used conductors (including those of aluminium), have image attenuations at 1600 Hz for limiting subscribers' lines differing little from 10.0 dB (*see* Table 6.6, col. 7). A 10 dB (1600 Hz) subscriber's line of $2\frac{1}{2}$ lb conductor would not be admissible on grounds of signalling resistance and so this exception is of no practical importance. A simplified rule has therefore been adopted of allowing the sum of the image attenuations of sections of subscriber's line of differing gauges to amount to no more than 10 dB, although, if a substantial proportion is of the lighter gauges, signalling limitations may be more restrictive than transmission.

It must be remembered that rules such as the foregoing are intended only to indicate whether or not a particular combination of items in an LTC is acceptable on grounds of transmission; the rules are not designed to enable the transmission performance of a non-limiting LTC to be determined. (This problem is discussed below in Sections 6.1.2 and 6.1.3.) Modern LTCs of limiting transmission performance but containing subscribers' lines of differing conductor gauges differ only slightly in attenuation/frequency distortion; this is because the mismatch losses, although not negligible, depend only slightly on the gauge of the conductor. The attenuation/frequency distortion does, however, vary considerably as the length of subscriber's line is changed from zero to the transmission limit. This feature is illustrated approximately in Figure 6.1 which shows the image attenuation and the insertion loss between 600 Ω terminations for a limiting (10 dB image attenuation at 1600 Hz) subscriber's line; lines less than limiting would yield results lying approximately proportionately between those shown and 0 dB.

Other telephone operating organisations have evolved somewhat different sets of rules to guide local exchange area development to suit their own circumstances. For example a fairly strict attention to maximum values of sending and receiving reference equivalents is paid in many European countries, although certain approxima-

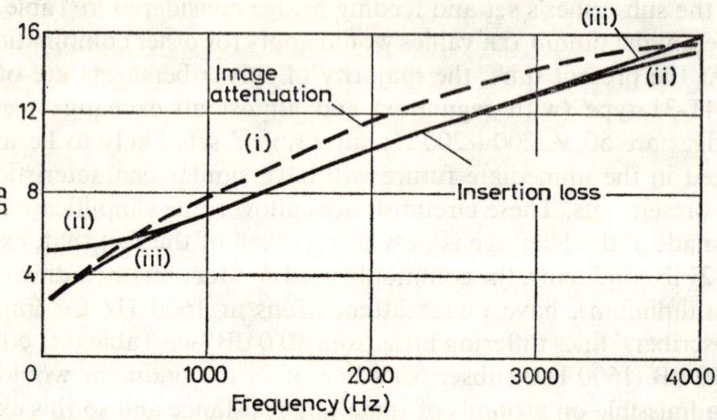

Figure 6.1. Image attenuation and insertion loss of a limiting subscriber's line (3.7 miles $6\frac{1}{2}$ lb cable).

Notes: The insertion loss between 600 Ω terminations is shown in curve ii. The insertion loss between a 16-4T-31 subscriber's set (with regulator) and a 50V, 200+200 Ω feeding bridge (600 Ω junction) is shown in curve iii

tions are usually made when combining the losses of the separate transmission elements. The Bell Telephone System has evolved a simplified method for planning which is explained by Bogan and Young (1954). With modern type telephone sets, the main feature is to select the smallest size conductor or combination of conductors of different sizes that complies with the loop resistance limit for the exchange concerned. As, mentioned above, loading is used to improve the transmission performance of the longer lines.

In the future, it may be possible to confine local cable conductors to one gauge only and to reduce the loss to the required limit by use of loading and suitable 2 wire amplifying devices. Considerable practical advantages would thereby be gained, particularly so far as manufacture and stores are concerned. Owing to the cost advantage of aluminium in comparison with copper, this is likely to be associated with cables designed to optimise the ratio of conductor to insulation volume.

The United Kingdom Post Office and many other telephone operating organisations treat local telephone areas as entities for planning purposes quite independently of their routings through junctions for access to the primary trunk switching centre (group centre). This enables the effects of all impedance mismatches within the local area to be combined in a single determination of the limiting transmission performance of the plant in that section of the telephone network. Other organisations treat subscribers' sets separately from the subscribers' lines and feeding bridges; this has the advantage that the performance can be considered in relation to the means of access to the primary trunk switching centre. However, it then becomes necessary to neglect mismatches or to assume them constant and to consider only the sum of the image attenuations of the transmission elements of the lines.

Local telephone areas must be planned to avoid excessive risk of intelligible overhearing via cable crosstalk paths. Crosstalk attenuation between pairs in a local cable can vary a great deal and the values found will depend on the type and construction of the cable. Typical values found in United Kingdom paper core unit twin local cables can be represented by the following cumulated percentages:

1% 77 dB or less; 2% 80 dB; 5% 86 dB; 10% 90 dB; 20% 96 dB; 50% 106 dB

These results, which can be represented by a Gaussian probability density distribution of mean 106 dB and standard deviation 12.5 dB, were obtained using a 1300 Hz reed hummer as source. Further information is given in Section 6.1.5.

6.1.2 Sending and receiving sensitivities of local telephone circuits

The determination of sending and receiving sensitivities of telephone sets as far as and from their line terminals is dealt with in Section 5.1.4. The present section is concerned with methods to include also the losses of feeding bridges and subscribers' lines to give the sensitivities of complete local telephone circuits up to the junction sides of their feeding bridges. When two local telephone circuits are connected together through trunk and other circuits having image impedances of 600 Ω, the sensitivities so obtained can be used directly to obtain the mouth to ear loss, L_{ME}, for the complete speech path. As shown in Section 3.2.4, this enables the overall reference equivalent to be obtained.

When a pair of identical telephone sets are directly connected together at their line terminals, L_{ME} is given by the following expression (*see* Section 5.1.4):

$$L_{ME} = L_S + L_R + 10 \log_{10} \frac{4Z_M}{Z_E} - S_M - S_E \qquad (6.1)$$

Where L_S is the transmission loss of the electrical circuit from the microphone terminals to the line terminals of the telephone set,

L_R is the transmission loss of the electrical circuit from the line terminals of the telephone set to the terminals of the earphone,

Z_M and Z_E are the respective moduli of the impedances of the microphone and the earphone,

S_M is the sensitivity of the microphone expressed in dB relative to 1 V per N/m^2, referring the sound pressure from the mouth to a point 25 mm in front of the lips and expressing the output as an open circuit e.m.f. (*see* Section 5.1.2),

S_E is the sensitivity of the earphone expressed in dB relative to 1 N/m^2 per V, measuring the input as a p.d. across the earphone terminals and referring the sound pressure output to a point at the centre of the opening of the ear canal (*see* Section 5.1.3).

The various quantities appearing in Eqn 6.1 have been defined so that, in the absence of any mismatches, they may be added together; the middle term is to take account of the impedances of microphone and earphone and of the fact that the microphone sensitivity is in terms of e.m.f. whereas the sensitivity of the earphone is in terms of p.d.

If the telephone sets connected directly together have different line terminal impedances, Z_C, there will be a reflection loss and this must be added to give L_{ME}.

Allowances must now be made for the transmission losses of feeding bridges and subscribers' lines. The general analysis of overall loss is rather complicated when several links are connected in tandem and mismatch losses are present at their interfaces and so the treatment explained here is intended only to cover a simplified selection of commonly occurring situations.

It will be assumed that two telephone sets (not necessarily identical) are interconnected via their subscribers' lines and feeding bridges and through a trunk circuit or chain of circuits which

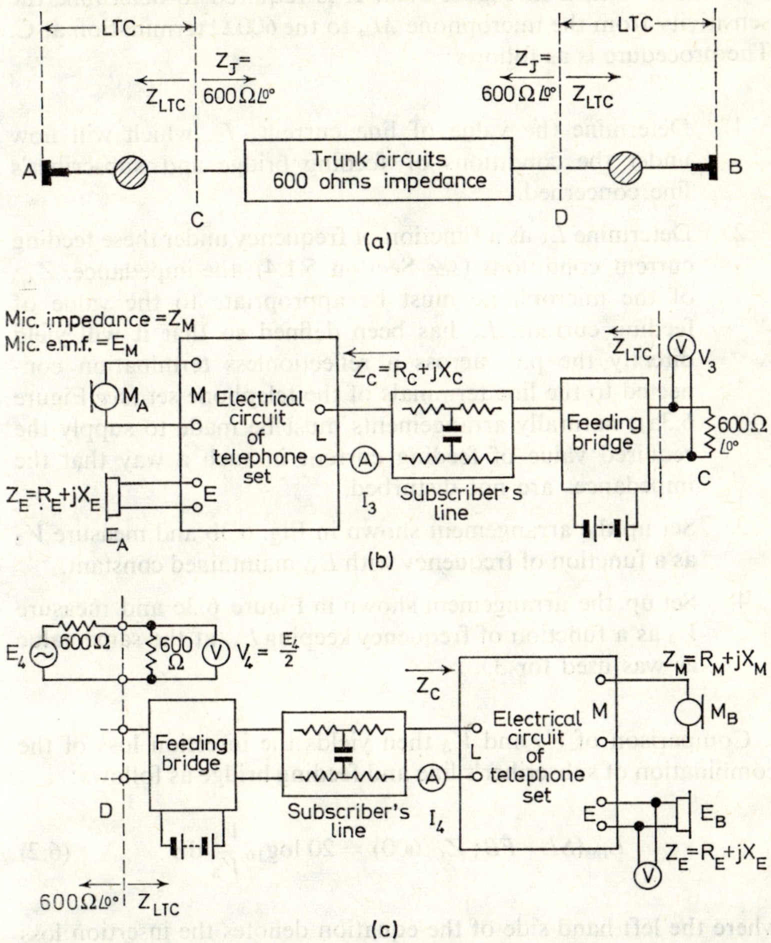

Figure 6.2. Separation of a complete connection into sending and receiving local telephone circuits. (a) Complete connection. (b) LTC *A considered for sending. (c)* LTC *B considered for receiving*

present impedances of $600/\underline{0^\circ}$ at each end at the interface with each local telephone circuit; Figure 6.2a shows the arrangement. The sending and receiving sensitivities of the local telephone circuits can be obtained as detailed below.

6.1.2.1 Sending sensitivity

Figure 6.2b shows the sending local telephone circuit (LTC) of the path from A to B in Figure 6.2a. It is required to determine the sensitivity from the microphone M_A to the 600 Ω termination at C. The procedure is as follows:

1. Determine the value of line current, I_3, which will flow under the conditions of feeding bridge and subscriber's line concerned.
2. Determine L_S as a function of frequency under these feeding current conditions (*see* Section 5.1.4); the impedance, Z_M, of the microphone must be appropriate to the value of feeding current. L_S has been defined so that it will yield directly the p.d. across a reflectionless termination connected to the line terminals of the telephone set (*see* Figure 6.3a); naturally arrangements must be made to supply the required value of feeding current in such a way that the impedances are not disturbed.
3. Set up the arrangement shown in Fig. 6.3b and measure V_2 as a function of frequency with E_M maintained constant.
4. Set up the arrangement shown in Figure 6.3c and measure V_3 as a function of frequency keeping E_M at the same value as was used for 3.

Comparison of V_2 and V_3 then yields the insertion loss of the combination of subscriber's line and feeding bridge as follows:

$$L_{\text{ins}}(SL+FB;\ Z_C/600) = 20 \log_{10} \frac{V_2}{V_3} \text{ dB} \qquad (6.2)$$

where the left hand side of the equation denotes the insertion loss of the combination between the impedances Z_C and $600\underline{/0^\circ}$ Ω.

The mismatch (or reflection) loss between Z_C and $600\underline{/0^\circ}$ is given by:

$$L_{\text{mis}}(Z_C/600) = 20 \log_{10} \frac{|Z_C+600|}{2(|Z_C \cdot 600|)^{1/2}} \text{ dB} \qquad (6.3)$$

(Values of this function can conveniently be read from graphs given in Appendix 1.)

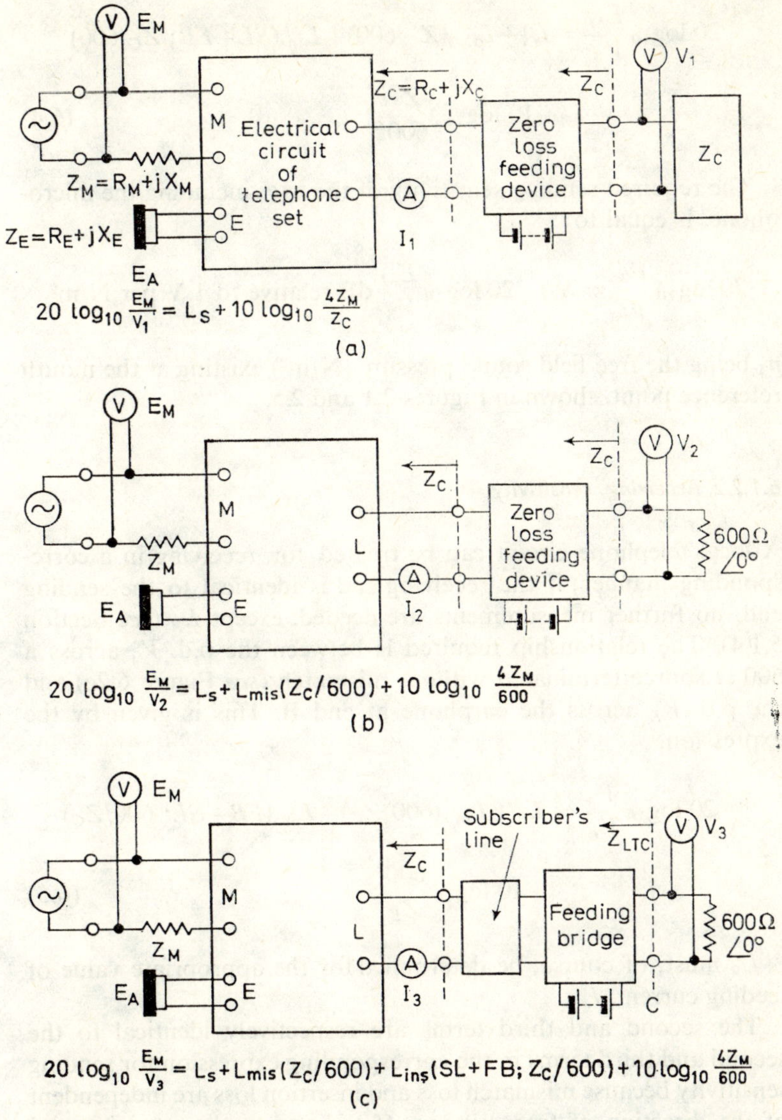

Figure 6.3. Determination of losses of electrical parts of a local telephone circuit. (a) Reflectionless termination of telephone set, (b) Telephone set terminated in 600 Ω. (c) Telephone set, subscriber's line and feeding bridge terminated in 600 Ω

The required relationship between E_M and V_3 is then given by:

$$20 \log_{10} \frac{E_M}{V_3} = L_S + L_{\text{mis}}(Z_C/600) + L_{\text{ins}}(SL+FB;\ Z_C/600)$$
$$+10 \log_{10} \frac{4Z_M}{600} \qquad (6.4)$$

The required sending sensitivity of the LTC, including the microphone, is equal to:

$$20 \log_{10} \frac{V_3}{p_1} = S_M - 20 \log_{10} \frac{E_M}{V_3} \text{ dB relative to 1 V per N/m}^2$$

p_1 being the free field sound pressure (N/m^2) existing at the mouth reference point, shown in Figures 2.1 and 2.5.

6.1.2.2 *Receiving sensitivity*

A local telephone circuit can be treated for receiving in a corresponding manner; if the receiving end is identical to the sending end, no further measurements are needed, except L_R (*see* Section 5.1.4). The relationship required is between the p.d. V_4 across a 600 Ω source terminated without mismatch (*see* Figure 6.2c) and the p.d. V_E across the earphone at end B. This is given by the expression. .

$$20 \log_{10} \frac{V_4}{V_E} = L_R + L_{\text{mis}}(600/Z_C) + L_{\text{ins}}(FB+SL;\ 600/Z_C)$$
$$+10 \log_{10} \frac{600}{Z_E} \qquad (6.5)$$

L_R must, of course, be determined for the appropriate value of feeding current, I_4.

The second and third terms are respectively identical to the second and third terms in the corresponding expression for sending sensitivity because mismatch loss and insertion loss are independent of the direction of transmission. If the LTC at the receiving end is of the same type as that at the sending end, the numerical values will also be the same; if the LTCs are of different types the mismatch and insertion losses will need to be measured separately in the manner described.

The required receiving sensitivity of the LTC, including the earphone, is then:

$$20 \log_{10} \frac{p_3}{V_4} = S_E - 20 \log_{10} \frac{V_4}{V_E} \text{ dB relative to 1 N/m}^2 \text{ per V} \quad (6.6)$$

p_3 being the sound pressure (N/m²) at the ear reference point located at the entrance to the ear canal. (*see* Figure 2.10).

The foregoing information is exemplified below by taking a typical local telephone circuit composed of the current (1970) United Kingdom telephone set with an approximately average subscriber's line. The necessary fundamental electrical parameters are assembled in Table 6.7; the electroacoustical sensitivities, S_M and S_E, of the microphone and earphone are introduced later.

6.1.2.3 Example of calculating the sending sensitivity of a LTC

The computation is shown in Table 6.8, yielding $20 \log_{10} (E_M/V_3)$ in col. 6.

Col. 3 is obtained from the formula:

$$L_{\text{mis}}(Z_1/Z_2) = 20 \log_{10} \frac{|Z_1+Z_2|}{2(|Z_1 Z_2|)^{1/2}} \text{ dB}$$

Where Z_1 and Z_2 are in the form $Z\underline{/\theta}$, charts can be used (*see* Appendix 1) which give contours of mismatch loss at convenient intervals on the plane of $|\theta_1 - \theta_2|$ and $|Z_1|/|Z_2|$.

The insertion loss of the feeding bridge and subscriber's line, in col. 4 is from Table 6.7. Col. 5 is calculated from the value of Z_M given in Table 6.7.

The remaining columns of Table 6.8 introduce the electro-acoustical sensitivity of the microphone. The sensitivity of the microphone at 47 mA is taken from the information given in Figure 5.5c; the 'envelope' of the curves shown has been used, namely the highest value of sensitivity found at each frequency. Figure 5.6 contains the information used for Z_M and the effect of changing microphone feeding current; it will be seen that an allowance of +1.0 dB is needed to adjust the sensitivity to 64 mA which is the value of current that flows in the LTC under consideration*.

* The telephone set under consideration contains no direct current shunt path across the microphone and so the microphone feeding current is the same as that flowing in the subscriber's line; this may not be so in other types of telephone set.

Table 6.7. SUMMARY OF FUNDAMENTAL ELECTRICAL PARAMETERS OF A LOCAL TELEPHONE CIRCUIT (LTC)

Telephone set	16-4T-31 with regulator
Subscriber's line	1 mile $6\frac{1}{2}$ lb cable
Feeding Bridge	50 V, 200+200 Ω, Stone type
Line current, I_3	64 mA Junction impedance, Z_J $600\underline{/0^\circ}$

Frequency Hz	L_S dB	L_R dB	Z_M Ω	Z_E Ω	Z_C Ω	Z_{SO} Ω	Z_L Ω ($Z_J = 600$)	L_{ins}(SL + FB; Z_C/600) dB	L_{ST} dB
(1)	(2)	(3)	(4)	(5)	(6)	(7)	(8)	(9)	(10)
200	6.2	9.5	67/0°	43/+50	223/+14	1588/−90	1130/−36	4.0	22.5
300	6.8	8.3	67/0°	57/+55	233/+16	1203/−90	949/−28	3.3	21.3
400	7.2	7.7	67/0°	70/+57	244/+17	1070/−90	900/−26	3.1	21.4
500	7.5	7.3	67/0°	85/+58	253/+18	1001/−81	850/−22	3.0	21.4
600	7.6	6.9	67/0°	98/+58	263/+18	925/−71	850/−23	2.9	22.2
800	7.8	7.0	67/0°	125/+55	279/+19	827/−63	806/−20	2.9	23.7
1000	8.0	7.4	67/0°	152/+48	296/+18	772/−55	782/−22	2.9	25.7
1250	7.9	8.0	67/0°	174/+39	305/+18	727/−49	759/−24	2.9	28.1
1600	7.8	7.8	67/0°	150/+42	304/+20	701/−44	720/−26	3.0	31.4
2000	8.0	7.9	67/0°	192/+47	325/+22	661/−40	679/−30	3.1	35.8
2500	8.0	9.1	67/0°	243/+36	335/+23	626/−36	629/−34	3.1	47.1
3000	8.1	8.5	67/0°	159/+45	340/+28	601/−32	582/−38	3.2	39.3
3500	8.2	8.7	67/0°	222/+56	370/+30	583/−92	540/−41	3.2	33.3
4000	8.4	9.4	67/0°	252/+56	383/+32	567/−26	505/−43	3.3	30.7

Notes. The information in cols 9 and 10 is, strictly, not fundamental; it can be derived (see text later) from the impedances given in the table but it is often more convenient to measure it directly with the aid of artificial cables, etc.

The information in cols 2, 3, 4, 6 and 7 is dependent on the value of feeding current. Col. 8 depends on the subscriber's line, feeding bridge and Z_J. Col. 9 depends on Z_C and Z_J. Col. 10 depends mainly on L_S+L_R and the return loss of Z_{SO} against Z_L (see Table 6.12)

Table 6.8. DETERMINATION OF SENDING SENSITIVITY OF A LOCAL TELEPHONE CIRCUIT (SEE TABLE 6.7 FOR PARTICULARS)
BMSP and $Z_J = 600\underline{/0°}$ have been assumed

Frequency Hz	L_S dB	L_{mis} $(Z_C/600)$ dB	$L_{ins}(SL+FB; Z_C/600)$ dB	$10\log_{10}\frac{4Z_M}{600}$ dB	$20\log_{10}\frac{E_M}{V_3}$ dB	S_M 47 mA dB	S_M 64 mA dB	$20\log_{10}\frac{V_3}{p_1}$ dB
(1)	(2)	(3)	(4)	(5)	(6)	(7)	(8)	(9)
200	6.2	+1.0	+4.0	−3.5	+7.7	−10.1	−9.1	−16.8
300	6.8	+0.9	+3.3	−3.5	+7.5	− 8.8	−7.8	−15.3
400	7.2	+0.8	+3.1	−3.5	+7.6	− 8.3	−7.3	−14.9
500	7.5	+0.7	+3.0	−3.5	+7.7	− 7.8	−6.8	−14.5
600	7.6	+0.6	+2.9	−3.5	+7.7	− 7.0	−6.0	−13.7
800	7.8	+0.5	+2.9	−3.5	+7.7	− 4.4	−3.4	−11.1
1000	8.0	+0.4	+2.9	−3.5	+7.8	− 0.6	+0.4	−7.4
1250	7.9	+0.4	+2.9	−3.5	+7.7	+ 2.3	+3.3	−4.4
1600	7.8	+0.4	+3.0	−3.5	+7.7	+ 0.4	+1.4	−6.3
2000	8.0	+0.2	+3.1	−3.5	+7.8	− 1.0	0.0	−7.8
2500	8.0	+0.2	+3.1	−3.5	+7.8	− 2.7	−1.7	−9.5
3000	8.1	+0.1	+3.2	−3.5	+7.9	− 0.3	+0.7	−7.2
3500	8.2	0.0	+3.2	−3.5	+7.9	− 4.8	−3.8	−11.7
4000	8.4	−0.1	+3.3	−3.5	+8.1	−11.8	−10.8	−18.9

Notes: 1. The units of S_M are dB relative to 1 V (e.m.f.) per N/m² (referred to a point 25 mm in front of the lips); the handset is located at BMSP.
2. The final sensitivity in col. 9 is in dB relative to 1 V (p.d. across 600 Ω) per N/m².
3. The final values of sensitivity, when allowance is made for the reference equivalent speaking position, yield a calculated value of sending reference equivalent (SRE) of 7.1 dB (*see* Table 3.8 in Section 3.2.4.3).

Col. 6 in Table 6.8 obtained by adding together the values appearing in cols 2, 3, 4 and 5 and represents the electrical loss from the microphone terminals to the junction. Col. 9 is obtained by subtracting the values in col. 6 from those in col. 8 and represents the electroacoustical sensitivity of the LTC from the mouth reference point P_1 (*see* Figures 2.5 and 3.1) to the junction. The information in col. 9 can be used to calculate the sending reference equivalent of the LTC in the manner explained in Section 3.2.4; the result is SRE = +7.1 dB. (The commonly accepted value from subjective tests is +7 dB.) It must be remembered that the sensitivity values shown in cols 7, 8 and 9 apply for the British Modal Speaking Position (BMSP) and not that required for determination of reference equivalent; the corresponding sensitivities for that purpose are obtained by increasing these values by 3.0 dB.

The sending sensitivities determined above can also be used to calculate the speech voltage produced from the LTC when it is

Table 6.9. CALCULATION OF SPEECH VOLTAGE FROM A SENDING LTC AT RVL
Sending LTC 16-4T-31 (with regulator) on 1 mile $6\frac{1}{2}$ lb cable
Talker holding the handset at BMSP

Frequency f Hz	β'_S	*Sending sensitivity* (BMSP)	*Send End Output in 1* Hz *rel. to −94* dBV	*Send End Output in 1* Hz dBV	*10* $\log_{10}$ Δf	*Level,* dBV, *represented by frequency interval* Δf	*Antilog 1/10 col. 7*
(1)	(2)	(3)	(4)	(5)	(6)	(7)	(8)
100	+55.5				+19		
200	+60.6	−16.8	+43.8	−50.2	+20	−30.2	0.00096
300	+62.0	−15.3	+46.7	−47.3	+20	−27.3	0.00187
400	+62.1	−14.9	+47.2	−46.8	+20	−26.8	0.00209
500	+61.0	−14.5	+46.5	−47.5	+20	−27.5	0.00178
600	+58.7	−13.7	+45.0	−49.0	+21.8	−27.2	0.00191
800	+53.4	−11.1	+42.3	−51.7	+23	−28.7	0.00135
1 000	+50.0	−7.4	+42.6	−51.4	+23.5	−27.9	0.00162
1 250	+47.7	−4.4	+43.3	−50.7	+24.8	−25.9	0.00257
1 600	+44.8	−6.3	+38.5	−55.5	+25.7	−29.8	0.00105
2 000	+42.4	−7.8	+34.6	−59.4	+26.5	−32.9	0.00051
2 500	+40.4	−9.5	+30.9	−63.1	+27	−36.1	0.00025
3 000	+38.5	−7.2	+31.3	−62.7	+27	−35.7	0.00027
3 500	+36.1	−11.7	+24.4	−69.6	+27	−42.6	0.00006
4 000	+33.9	−18.9	+15.0	−79.0	+28.8	−50.2	0.00001
5 000	+30.7				+30		
6 000	+29.1				+31.8		
8 000	+28.4				+33		

Sum 0.01630
Speech voltage, V = 10 $\log_{10}$ (Sum)
= −17.9 dBV

Calculated sending reference equivalent (SRE) = +7.1 dB
Therefore, speech voltage for 0 dB SRE = V+SRE = −10.8 dBV

Note 1 Speech spectrum in col. 2 was obtained from male and female talkers at their natural levels and, by rerecording, adjusted to Reference Vocal Level (RVL).
Note 2 Interposition of a pair of 4 kHz f.d.m. channel filters between the Local Telephone Circuit (LTC) and the point of measurements will reduce the reading by, perhaps, 1 dB.
Note 3 The levels of speech voltage determined above apply for the artificial condition of RVL and BMSP; under conversational conditions the level will be affected by the overall performance of the telephone connection (*see* information in Table 4.8) and by the level of sidetone (*see* information in Section 4.3.1).

spoken into at RVL/BMSP. Table 6.9 shows the speech spectrum density, β'_S, for the average of males and females speaking at their natural vocal levels but with the total vocal level standardised to that corresponding to Reference Vocal Level (RVL) (*see* Section 2.1.3.2). Col. 3 of Table 6.9 has been taken from Table 6.8. Col. 4

of Table 6.9 is the sum of the values given in cols 2 and 3 and col. 5 is 94 dB lower than col. 4 to allow for the datum used to express β'_S. Col. 6 gives the allowance, in dB, for the bandwidth represented by each frequency of the set used in the table. Col. 7 is the sum of cols 5 and 6 and the values appearing in col. 8 are in units of V^2, which can be summed to give the total V^2. The speech voltage in dB is therefore given by $10 \log_{10} \sum V^2$. It is, perhaps, interesting to note that the result shown in the table corresponds to a speech voltage output from a 0 dB SRE sending end of −10.8 dBV when RVL and BMSP are used; under conversational conditions over a telephone connection the result may be somewhat different as explained in Section 4.2.1.3.

6.1.2.4 Receiving sensitivity example

Table 6.10 shows the computation of $20 \log_{10} (V_4/V_E)$ and $20 \log_{10} (p_3/V_4)$.

Cols 3 and 4 in Table 6.10 are identical to cols 3 and 4 in Table 6.8. Col. 5 is obtained from the values of Z_E given in Table 6.7. The sensitivity of the earphone when measured on an artificial ear (BS 2042) is shown in col. 7; this requires to be corrected for the conditions that exist when a handset is held to the ear as in telephone conversations and the correction used in col. 8 has been derived from the information* that appears in Figure 5.8a in Section 5.1.3. The corrected sensitivities of the earphone are in col. 9.

Col. 6 in Table 6.10 is obtained by adding together the values appearing in cols 2, 3, 4 and 5 and represents the electrical loss from the junction to the terminals of the earphone. Col. 10 is obtained by subtracting the values in col. 6 from those in col. 9 and represents the electroacoustical sensitivity of the LTC from the junction to the ear reference point P_3 (*see* Figures 2.10 in Section 2.2.1 and 3.1 in Section 3.1.1). The information in col. 10 can be used to calculate the receiving reference equivalent of the LTC in the manner explained in Section 3.2; the result is RRE = −5.8 dB. (The commonly accepted value from subjective tests is −4 dB.)

* This correction may, under certain circumstances, yield sensitivities that are somewhat too high; more realistic corrections may be derived from threshold tests under suitable conditions. Such tests have indicated that a uniform reduction in sensitivity of about 3 dB may be appropriate.

Table 6.10. DETERMINATION OF RECEIVING SENSITIVITY OF A LOCAL TELEPHONE CIRCUIT (SEE TABLE 6.7 FOR PARTICULARS) assuming $Z_J = 600/\underline{0°}$

Frequency Hz	L_R Hz	L_{mis} *(600/Z_C)* dB	L_{ins}*(FB +SL; 600/Z_C)* dB	*10*$\log_{10}$ $\frac{600}{Z_E}$ dB	$20\log_{10}$ $\frac{V_4}{V_E}$ dB	S_E *(AE BS 2042)* dB *rel.* N/m² *per* V	*Correction* dB	S_E *(Real ear)* dB *rel.* N/m² *per* V	$20\log_{10}$ $\frac{p_3}{V_4}$ dB
(1)	(2)	(3)	(4)	(5)	(6)	(7)	(8)	(9)	(10)
200	9.5	+1.0	+4.0	+11.5	+26.0	+35.6	−8.3	+27.3	+ 1.3
300	8.3	+0.9	+3.3	+10.5	+22.7	+35.5	−1.1	+34.4	+11.7
400	7.7	+0.8	+3.1	+ 9.3	+20.9	+34.9	+1.3	+36.2	+15.3
500	7.3	+0.7	+3.0	+ 8.5	+19.5	+34.3	+2.3	+36.6	+17.1
600	6.9	+0.6	+2.9	+ 7.9	+18.3	+33.5	+2.6	+36.1	+17.8
800	7.0	+0.5	+2.9	+ 6.8	+17.2	+33.0	+2.7	+35.7	+18.5
1000	7.4	+0.4	+2.9	+ 6.0	+16.7	+32.8	+1.6	+34.4	+17.7
1250	8.0	+0.4	+2.9	+ 5.4	+16.7	+33.4	+0.8	+34.2	+17.5
1600	7.8	+0.4	+3.0	+ 6.0	+17.2	+36.1	−1.2	+34.9	+17.7
2000	7.9	+0.2	+3.1	+ 5.0	+16.2	+32.7	−4.1	+28.6	+12.4
2500	9.1	+0.2	+3.1	+ 3.9	+16.3	+29.3	−7.6	+21.7	+ 5.4
3000	8.5	+0.1	+3.2	+ 5.8	+17.6	+33.3	−6.1	+27.2	+ 9.6
3500	8.7	0.0	+3.2	+ 4.3	+16.2	+34.5	−6.3	+28.2	+12.0
4000	9.4	−0.1	+3.3	+ 3.8	+16.4	+21.8	−7.0	+14.8	− 1.6

Notes: **1. The units of S_E are dB relative to 1 N/m² (sound pressure at entrance to ear canal) per V (p.d. across earphone terminals).**
2. The final sensitivity in col. 10 is in dB relative to 1 N/m² per V (e.m.f./2 in 600 Ω source).
3. The final values of sensitivity yield a calculated value of receiving reference equivalent (RRE) of −5.8 dB (*see* Table 3.8 in Section 3.2.4.3).

6.1.2.5 Summary

Table 6.11 gives a summary of the sending and receiving sensitivities of local telephone circuits containing various lengths of 6½ lb cable subscriber's line. The sending sensitivities appropriate for determining reference equivalents are also shown.

6.1.3 Loudness ratings of local telephone circuits

For planning local telephone networks, it is necessary to use some fairly simple rule for determining the relative efficiency of different combinations of subscriber's set, subscriber's line and feeding

Table 6.11. SUMMARY OF SENDING AND RECEIVING SENSITIVITIES OF 16-4T-31 TELEPHONE SET (WITH REGULATOR) ON VARIOUS SUBSCRIBERS' LINES OF $6\frac{1}{2}$ lb CABLE

Frequency Hz	*Zero sub's. line*			*1 mile*			*3.7 miles*		
	Sending		*Receiving*	*Sending*		*Receiving*	*Sending*		*Receiving*
	BMSP	RE		BMSP	RE		BMSP	RE	
200	−13.0	−10.0	+ 3.3	−16.8	−13.8	+ 1.3	−19.7	−16.7	+ 1.1
300	−11.6	− 8.6	+13.0	−15.3	−12.3	+11.7	−17.1	−14.1	+ 9.7
400	−10.9	− 7.9	+17.1	−14.9	−11.9	+15.3	−16.4	−13.4	+13.1
500	−10.4	− 7.4	+19.1	−14.5	−11.5	+17.1	−15.6	−12.6	+14.8
600	− 9.7	− 6.7	+19.6	−13.7	−10.7	+17.8	−15.2	−12.2	+14.9
800	− 7.4	− 4.4	+20.5	−11.1	− 8.1	+18.5	−12.9	− 9.9	+15.7
1000	− 3.8	− 0.8	+19.9	− 7.4	− 4.4	+17.7	− 9.3	− 6.3	+15.2
1250	− 1.2	+ 1.8	+19.9	− 4.4	− 1.4	+17.5	− 7.3	− 4.3	+14.3
1600	− 3.3	− 0.3	+20.2	− 6.3	− 3.3	+17.7	−10.6	− 7.6	+13.1
2000	− 5.0	− 2.0	+14.6	− 7.8	− 4.8	+12.4	−12.9	− 9.9	+ 7.3
2500	− 6.9	− 3.9	+ 7.8	− 9.5	− 6.5	+ 5.4	−17.1	−14.1	− 1.9
3000	− 4.6	− 1.6	+12.1	− 7.2	− 4.2	+ 9.6	−15.3	−12.3	+ 1.9
3500	− 9.6	− 6.6	+13.9	−11.7	− 8.7	+12.0	−21.2	−18.2	+ 3.0
4000	−16.4	−13.4	+ 0.7	−18.9	−15.9	− 1.6	−29.5	−26.5	−11.6

Notes. The above sensitivities apply for $Z_J = 600/0°$
The feeding bridge is the Stone type, 50 V, 200+200 Ω with 2 μF capacitors.
The calculated sending reference equivalents are:

Zero	+3.9 dB (+5)
1 mile (1.6 km)	+7.1 dB (+7)
3.7 miles (5.9 km)	+10.9 dB (+11)

The calculated receiving reference equivalents are

Zero	−7.9 dB (−6)
1 mile (1.6 km)	−5.8 dB (−4)
3.7 miles (5.9 km)	−1.9 dB (0)

The values in brackets are taken from Fig. 6.4.

bridge. One such rule based on 'Transmission Equivalent Resistance' (TER) has been mentioned in Section 6.1.1 and this was, until recently (1970), used in the United Kingdom*. Other telephone organisations use the values of sending and receiving reference equivalent and so the relevant values are shown in Figure 6.4 for certain United Kingdom telephone sets and different gauges of subscriber's line conductor. The results shown for the 16-4T-31 set without regulator are included to show the effect of the regulation; such unregulated sets are not suitable for normal use in the network.

* It is now superseded by a rule based on the sum of the image attenuations of the sections of any particular subscriber's line; this sum must not exceed 10 dB. The effect of using this rule in the United Kingdom will differ very slightly from that of using the TER rule.

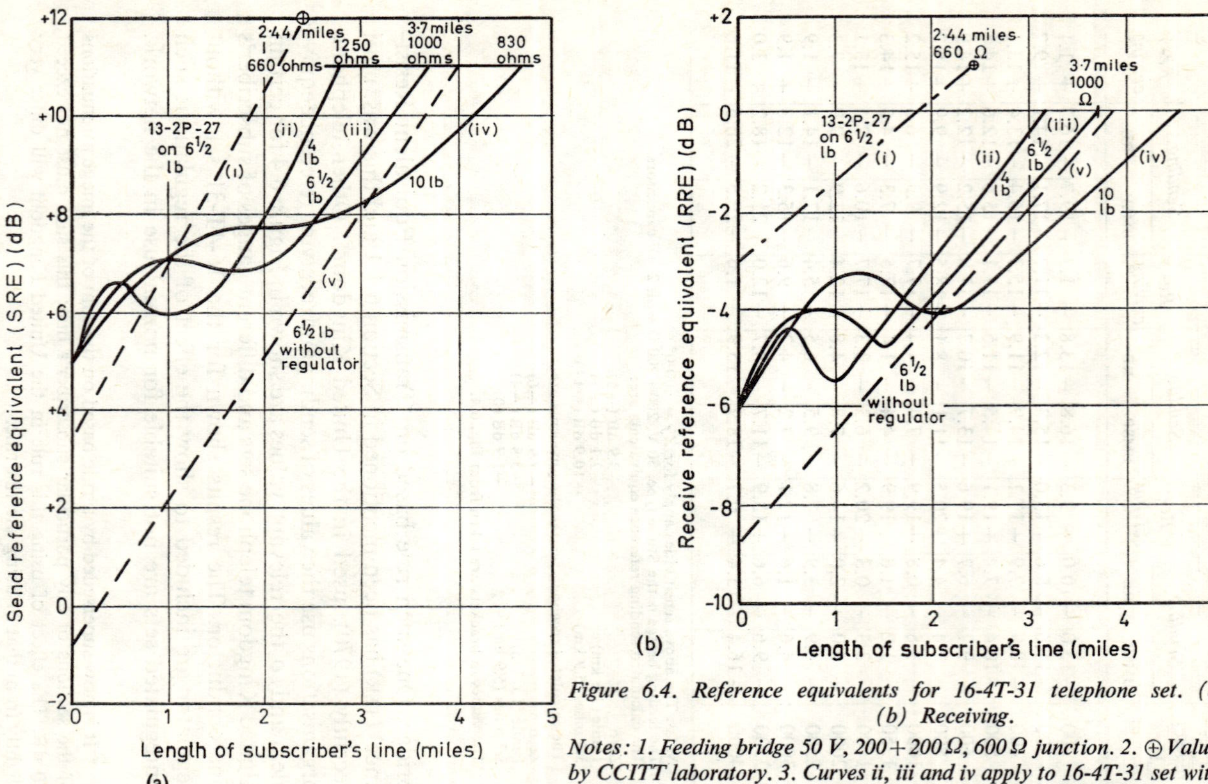

Figure 6.4. Reference equivalents for 16-4T-31 telephone set. (a) Sending. (b) Receiving.

Notes: 1. Feeding bridge 50 V, 200 + 200 Ω, 600 Ω junction. 2. ⊕ Values measured by CCITT laboratory. 3. Curves ii, iii and iv apply to 16-4T-31 set with regulator. Curve v applies to same set without regulator

When studies are being made of the satisfactoriness of a switched telephone network composed of distributed elements, it is necessary to estimate the performances of a wide variety of telephone connections. It is then hardly possible to consider the overall mouth to ear losses of the individual connections that may be set up and so it is convenient to work in terms of (among other parameters) the 'nominal overall reference equivalents'. The nominal ORE of a connection† is given by adding together the SRE and RRE of the relevant LTCS and certain nominal loss values for the junctions, trunks and interconnecting switching elements. The information given in Figure 6.4 on SRE and RRE values is used for this purpose by the United Kingdom Post Office.

6.1.4 Sidetone paths

The electrical portion of the sidetone path, namely that between the terminals of the microphone and those of the earphone, can be designated by the quantity L_{ST}. L_{ST} is governed by the values of L_S and L_R together with certain relationships between the impedances Z_C, Z_{SO} and Z_L; L_S, L_R and the impedances Z_C and Z_{SO} are characteristic of the telephone set (although they may depend on the current feeding it) but Z_L depends upon the line to which the telephone set is connected. Even when one particular telephone station is considered, Z_L will usually range rather widely according to the destination of the connection beyond the local exchange. Information on these matters is given in Section 5.1.5.2 where an expression is given for L_{ST}, the electrical transmission loss from microphone to earphone via the sidetone path. For the present purposes it is convenient to rearrange the terms as follows:

$$L_{ST} = L_S + L_R + 20\log_{10}\frac{|Z_{SO}+Z_L|}{|Z_{SO}-Z_L|} - 20\log_{10}\frac{|Z_{SO}+Z_L|}{2(|Z_{SO}Z_L|)^{1/2}}$$

$$+20\log_{10}\frac{|Z_C+Z_{SO}|}{2(|Z_CZ_{SO}|)^{1/2}} + 20\log_{10}\frac{|Z_C+Z_L|}{2(|Z_CZ_L|)^{1/2}} \quad (6.7)$$

The third term in this expression is the return loss* of Z_{SO} against

† The true overall reference equivalent (ORE) of a telephone connection can be determined from knowledge of the mouth-to-ear loss of the speech path concerned. The true ORE may differ from the nominal ORE by a few decibels, as is explained in Section 6.4.

* This can be evaluated with the aid of graphs given in Appendix 1.

Z_L; fourth, fifth and sixth terms are mismatch (or reflection) losses between the respective pairs of impedances Z_{SO}/Z_L, Z_C/Z_{SO} and Z_C/Z_L. Using the abbreviated symbols adopted in Section 6.1.2, together with a similar symbol to denote return loss, the above expression becomes:

$$L_{ST} = L_S + L_R + L_{\text{ret}}(Z_{SO}/Z_L) - L_{\text{mis}}(Z_{SO}/Z_L) + L_{\text{mis}}(Z_C/Z_{SO}) + L_{\text{mis}}(Z_C/Z_L) \quad (6.8)$$

Table 6.12 shows the magnitudes of these terms for an example of a typical LTC terminated in a 600 Ω junction. The return loss is the most important of the last four terms and therefore largely governs the difference $L_{ST} - (L_S + L_R)$. Some numerical values of this quantity are given for various line conditions in Table 5.9, which appears in Section 5.1.5.2; results are shown for the 16-4T-31 telephone set (with regulator) when it is associated with certain lengths of subscriber's line.

The sidetone path of a telephone set must be considered from two, separate, points of view; speech sidetone and room noise

Table 6.12. DETERMINATION OF L_{ST} FROM L_S, L_R, Z_{SO}, Z_L AND Z_C FOR THE LTC OF WHICH PARTICULARS ARE GIVEN IN TABLE 6.7 (16-4T-31 with regulator on 1 mile/6½ and 600 Ω junction)

Frequency Hz	L_S+L_R dB	L_{ret} (Z_{SO}/Z_L) dB	L_{mis} (Z_{SO}/Z_L) dB	L_{mis} (Z_C/Z_{SO}) dB	L_{mis} (Z_C/Z_L) dB	L_{ST} dB
(1)	(2)	(3)	(4)	(5)	(6)	(7)
200	15.7	5.3	−0.8	+2.3	+2.1	26.2 (22.5)
300	15.1	4.2	−1.2	+0.8	+1.6	22.9 (21.3)
400	14.9	3.9	−1.4	+0.1	+1.3	21.6 (21.4)
500	14.8	4.7	−1.2	−0.1	+1.2	21.8 (21.4)
600	14.5	7.0	−0.8	−0.2	+1.0	23.1 (22.2)
800	14.8	8.1	−0.6	−0.5	+0.8	23.8 (23.7)
1000	15.4	10.6	−0.4	−0.5	+0.6	26.5 (25.7)
1250	15.9	13.0	−0.2	−0.5	+0.4	29.0 (28.1)
1600	15.6	15.9	−0.1	−0.5	+0.2	31.3 (31.4)
2000	15.9	21	−0.1	−0.6	−0.3	36 (35.8)
2500	17.1	35	0.0	−0.7	−0.6	51 (47.1)
3000	16.6	25	0.0	−0.8	−1.1	40 (39.3)
3500	16.9	18.8	0.0	−0.9	−1.6	33.2 (33.3)
4000	17.8	15.9	−0.1	−1.0	−1.9	30.9 (30.7)

Note. Col. 7 = col. 2+col. 3−col. 4+col. 5+col. 6
The figures in brackets in col. 7 are those obtained by direct measurement.

sidetone. This has been explained in Section 3.1. The electrical portion, characterised by L_{ST}, and the earphone operate identically for speech and for room noise but the operation of the microphone is different for the two cases. For speech sidetone, the sensitivity of the microphone is the same as for transmitting the speech to the line and the same value of S_M is appropriate. When transmitting room noise, however, the form of the applied sound field is different and, with carbon microphones at the rather low sound pressures of room noise ordinarily experienced in telephone locations, the sensitivity may differ from that at speech sound pressures. (Room noise rarely exceeds about 70 dB above the conventional datum of 20 $\mu N/m^2$ as compared with about 90 dB for speech at a handset microphone.) The speech sidetone case will be treated first.

The mouth to ear transmission loss of the speech sidetone path will be denoted by L_{MEST}.

$$L_{MEST} = L_{ST} + 10 \log_{10} \frac{4Z_M}{Z_E} - S_M - S_E \qquad (6.9)$$

Table 6.13 shows the computation of L_{MEST} for the case of which particulars are given in Table 6.7. In this table col. 4 is the sum of cols 2 and 3 and represents the ratio of transmitter e.m.f. to earphone p.d. Cols 5 and 6 are as used in Tables 6.8 and 6.10 respectively. Col. 7 is equal to col. 4 minus the sum of cols 5 and 6. This set of values of L_{MEST} can then be used to determine the sidetone reference equivalent for speech, in the manner described in Section 3.2; first, however, a correction must be made to allow for the different speaking position specified for determining speech sidetone reference equivalents according to CCITT Recommendation P. 73. The required values of L_{MEST} for calculating sidetone reference equivalent are shown in col. 8, the resultant value of STRE being +3.8 dB. STRE determined in the same manner for zero subscriber's line and 3.7 miles $6\frac{1}{2}$ lb cable are −0.1 dB and +8.2 dB respectively.

The case of room noise sidetone is conveniently treated by determining the equivalent transmission loss of the sidetone path to ambient noise. The spectrum density of room noise having the Hoth spectrum at a level of 50 dBA is shown in col. 2 of Table 6.14 using the normal datum of 20 $\mu N/m^2$; these are converted to the datum of 1 N/m^2 as shown in col. 3. Col. 4 shows the electrical spectrum density measured at the output of the microphone and expressed, in terms of e.m.f., in dBV; for this measurement the

Table 6.13. DETERMINATION OF MOUTH TO EAR TRANSMISSION LOSS OF A SPEECH SIDETONE PATH (L_{MEST}); PARTICULARS OF LTC IN TABLE 6.7

Frequency Hz	L_{ST} dB	$10\log_{10}\frac{4Z_M}{Z_E}$ dB	$20\log_{10}\frac{E_M}{V_E}$ dB	S_M *64* mA dB (BMSP)	S_E *(Real ear)* dB	*20* $\log_{10} p_1/p_3 = L_{MEST}$ (BMSP) dB	L_{MEST} *(RE) for calc. of* STRE	*As col. 8 but zero sub.'s line*	*As col. 8 but 3.7 miles* $6\frac{1}{2}$ lb *cable sub.'s line*
(1)	(2)	(3)	(4)	(5)	(6)	(7)	(8)	(9)	(10)
200	22.5	+8.0	+30.5	−9.1	+27.3	+12.3	+9.3	+9.9	+6.7
300	21.3	+6.7	+28.0	−7.8	+34.4	+1.4	−1.6	−2.0	0
400	21.4	+5.8	+27.2	−7.3	+36.2	−1.7	−4.7	−5.8	−1.4
500	21.4	+5.0	+26.4	−6.8	+36.6	−3.4	−6.4	−8.2	−1.1
600	22.2	+4.4	+26.6	−6.0	+36.1	−3.5	−6.5	−8.7	−0.2
800	23.7	+3.3	+27.0	−3.4	+35.7	−5.3	−8.3	−11.4	−1.2
1000	25.7	+2.5	+28.2	+0.4	+34.4	−6.6	−9.6	−13.4	−2.1
1250	28.1	+1.9	+30.0	+3.3	+34.2	−7.5	−10.5	−14.9	−1.4
1600	31.4	+2.5	+33.9	+1.4	+34.9	−2.4	−5.4	−11.4	+2.6
2000	35.8	+1.5	+37.3	0.0	+28.6	+8.7	+5.7	−3.0	+7.5
2500	47.1	+0.4	+47.5	−1.7	+21 7	+27.5	+24.5	+7.2	+14.1
3000	39.3	+2.3	+41.6	+0.7	+27.2	+13.7	+10.7	+1.1	+6.5
3500	33.3	+0.8	+34.1	−3.8	+28.2	+9.7	+6.7	+3.4	+6.9
4000	30.7	+0.3	+31.0	−10.8	+14.8	+27.0	+24.0	+23.6	+26.4

Notes. The values of L_{MEST} shown in cols. 8, 9 and 10 yield calculated values of STRE as follows:

Zero subscriber's line (600 Ω junction), −0.1 dB
1 mile, $6\frac{1}{2}$ cable (600 Ω junction), +3.8 dB
3.7 miles, $6\frac{1}{2}$ lb cable (600 Ω junction), +8.2 dB
The calculated values of SRE and RRE are given under Table 6.11.

handset was held in the hand and the measurements were made after the microphone had been spoken into, and the appropriate value of feeding current flowed in the microphone. The electrical spectrum was measured with a set of $\frac{1}{2}$ octave filters and the values at the required set of frequencies were obtained by interpolation. The required sensitivity of the microphone when stimulated by the room noise is given by the difference col. 4–col. 3 which is entered in col. 5. For comparison, col. 6 shows the sensitivity of the microphone at the same feeding current but when stimulated by speech from a speech source with mouth reference point, P_1, located 25 mm in front of the lips; col. 7 shows the difference. About 4–5 dB of this difference can be accounted for by purely linear effects attributable to differences in the form of the sound field and relative locations of the points at which the reference sound pressure is

Table 6.14. DETERMINATION OF SENSITIVITY, S_M, OF A TELEPHONE MICROPHONE WHEN STIMULATED BY ROOM NOISE

The microphone is the Post Office Transmitter inset No. 16 with 64 mA feeding current.

Frequency Hz	*RN50 Spectrum density* dB *rel.* *20* μN/m²	*RN50 Spectrum density* dB *rel.* *1* N/m² *See Note*	*Spectrum density of* E_M *in* *50* dBA *room noise* dBV	S_M *for room noise* dB *rel.* *1* V *per* N/m²	S_M *for speech* BMSP dB *rel.* *1* V *per* N/m²	*Difference col. 5– col. 6*	*As col. 7 but for a linear microphone in the same type of handset*
(1)	(2)	(3)	(4)	(5)	(6)	(7)	(8)
200	27.6	−66.4	−97.2	−30.8	−9.1	−21.7	−6.6
300	24.7	−69.3	−93.6	−24.3	−7.8	−16.5	−7.0
400	22.7	−71.3	−92.8	−21.5	−7.3	−14.2	−6.6
500	21.1	−72.9	−91.3	−18.4	−6.8	−11.6	−4.8
600	19.9	−74.1	−92.0	−17.9	−6.0	−11.9	−6.3
800	17.8	−76.2	−93.3	−17.1	−3.4	−13.7	−8.4
1000	16.2	−77.8	−92.7	−14.9	+0.4	−15.3	−8.9
1250	14.6	−79.4	−90.8	−11.4	+3.3	−14.7	−11.0
1600	12.9	−81.1	−89.2	−8.1	+1.4	−9.5	−13.3
2000	11.3	−82.7	−88.3	−5.6	0.0	−5.6	−14.7
2500	9.6	−84.4	−89.9	−5.5	−1.7	−3.8	−10.8
3000	8.2	−85.8	−92.5	−6.7	+0.7	−7.4	−11.2
3500	6.7	−87.3	−94.5	−7.2	−3.8	−3.4	−17.0
4000	5.4	−88.6	−97.8	−9.2	−10.8	+1.6	−12.1
5000	2.6	−91.4	—	—	—	—	−4.2

Note. Change in datum from col. 2 requires subtraction of 94 dB

$$= \left[20 \log_{10}\left(\frac{1\ \text{N/m}^2}{20\ \mu\text{N/m}^2}\right)\right]$$

measured. The remaining difference, attributable to nonlinear characteristics of the microphone is mainly confined to the lower frequency region where, as shown later, room noise sidetone is usually masked by leakage of room noise past the earcap. Corresponding information for a linear handset microphone is given in col. 8.

The transmission loss of the sidetone path for room noise can be determined in the same manner as that for speech but use must be made of the appropriate values of microphone sensitivity, namely those appearing in col. 5 of Table 6.14. The determination is otherwise exactly the same as that in Table 6.13. Table 6.15 shows the determination and the results for the same local telephone

Table 6.15. DETERMINATION OF TRANSMISSION LOSS FOR ROOM NOISE OF THE SIDETONE PATH (L_{RNST}) OF A TELEPHONE SET 16-4T-31 ON 1 MILE $6\frac{1}{2}$ lb CABLE AND 600 Ω JUNCTION

Frequency Hz	L_{ST} dB	$10 \log_{10} \frac{4Z_M}{Z_E}$ dB	$20 \log_{10} \frac{E_M}{V_E}$ dB	S_M for room noise dB rel. 1 V per N/m² (Table 6.14)	S_E (real ear) dB rel. 1 N/m² per V	$20 \log_{10} \frac{p_{RN}}{p_3} = L_{RNST}$ dB (see Note)	Loss of earcap leakage path dB
(1)	(2)	(3)	(4)	(5)	(6)	(7)	(8)
200	22.5	+8.0	30.5	−30.8	+27.3	34.0	3.6
300	21.3	+6.7	28.0	−24.3	+34.4	17.9	6.2
400	21.4	+5.8	27.2	−21.5	+36.2	12.5	8.7
500	21.4	+5.0	26.4	−18.4	+36.6	8.2	10.6
600	22.2	+4.4	26.6	−17.9	+36.1	8.4	12.4
800	23.7	+3.3	27.0	−17.1	+35.7	8.4	16.8
1000	25.7	+2.5	28.2	−14.9	+34.4	8.7	20.2
1250	28.1	+1.9	30.0	−11.4	+34.2	7.2	23.1
1600	31.4	+2.5	33.9	−8.1	+34.6	7.4	24.4
2000	35.8	+1.5	37.3	−5.6	+28.6	14.3	23.3
2500	47.1	+0.4	47.5	−5.5	+21.7	31.3	20.6
3000	39.3	+2.3	41.6	−6.7	+27.2	21.1	19.2
3500	33.3	+0.8	34.1	−7.2	+28.2	13.1	18.2
4000	30.7	+0.3	31.0	−9.2	+14.8	25.4	18.4

Note. p_{RN} denotes the free field sound pressure (due to a relatively remote source of room noise) at the point to be occupied by the telephone microphone.

set as that used in Table 6.13; for comparison, col. 8 shows the loss of the earcap leakage path and it will be seen that noise heard via that path will mask that via the sidetone path at frequencies lower than 500 Hz. So far as room noise sidetone is concerned, it is therefore less important to secure good sidetone suppression at low frequencies than in the range 600–1600 Hz (suppression in the range from 2000 Hz upwards is already sufficient for the example taken).

6.1.5 Crosstalk paths

Crosstalk in the local distribution network (especially near-end) is a particularly difficult problem to deal with for the following reasons:

1. In the local network, close to the telephone set, the greatest possible difference in transmission level exists and so the necessary crosstalk attenuation must be higher than elsewhere, where smaller differences exist.
2. The persons affected by near-end crosstalk in local lines may well know each other, especially if they are served by the same distribution point, and so the consequences of intelligible overhearing are likely to be regarded as more serious than that between two persons who are never likely to be mutually affected again and who do not know each other.

Local distribution cables are designed to very severe limits of crosstalk attenuation (mainly caused by capacitance unbalance) and the values found in practice in the United Kingdom have a median value at 1300 Hz of about 106 dB, with standard deviation about 12.5 dB. Equipment connected to the subscriber's line must also be well balanced if it is not seriously to degrade the crosstalk performance of the cable; it is reasonable, for planning, to reduce the above median value by about 3 dB to allow for practical amounts of such unbalances.

Crosstalk attenuation at audio frequencies is usually measured with a $600\underline{/0°}$ Ω source and with a $600\underline{/0°}$ Ω measuring device; the opposite ends of both the disturbing and disturbed pairs are also closed with $600\underline{/0°}$ Ω terminations. The values of crosstalk attenuation quoted are therefore insertion losses between 600 Ω terminations.

Ideally, of course, crosstalk attenuation ought to be measured as a function of frequency covering the complete telephony range but such extensive results are rarely available, measurements commonly being made at a single frequency, e.g. 1000, 1300 or 1600 Hz. Table 6.16 shows approximately the variation in crosstalk attenuation with frequency to be expected in local cables. These results were obtained from the mean values for 134 measurements on spare pairs in working cables (600 Ω terminations were used).

Examination of the results for different lengths of subscriber's cable shows no systematic differences in crosstalk attenuation for different lengths, at least down to 0.5 miles.

Studies of the crosstalk performance of telephone networks can conveniently be performed in terms of the nominal overall reference equivalent of the crosstalk path. This can be determined by adding the appropriate SRE and RRE values of the disturbing and disturbed paths, up to and from the points where the crosstalk attenuation

Table 6.16. VARIATION IN CROSSTALK ATTENUATION OF LOCAL CABLE WITH FREQUENCY

Freq. Hz	*Relative attenuation* dB	*Freq.* Hz	*Relative attenuation* dB	*Freq.* Hz	*Relative attenuation* dB
200	+5	800	+1.6	2500	− 6.3
300	+4.7	1000	0	3000	− 7.3
400	+4.2	1250	−1.3	3500	− 8.7
500	+3.7	1600	−3.3	4000	−10.3
600	+3.1	2000	−5.1		

values apply, and adding a single 'equivalent crosstalk attenuation' value determined at a particular frequency or averaged in some manner to represent the loudness equivalent attenuation of the electrical crosstalk path concerned. Using the variation in crosstalk attenuation given in the above table, the calculated reference equivalent attenuation is equal to the single frequency crosstalk attenuation at approximately 1000 Hz.

A detailed study of the effects of changing the terminating impedances and of other matters concerned with crosstalk is given by Hayes, Seymore and Bray (1938).

6.2 JUNCTIONS

The circuits that interconnect local exchanges directly with each other in a multi-exchange area are called local junctions and, in the London Region, they average about 5 miles. In large multi-exchange areas, like London, local exchanges are also interconnected through tandem exchanges; circuits between a local exchange and a tandem are called tandem junctions which, in London average about 10 miles in length. Local exchanges have access to their primary (Group) switching centre, and so to the trunk network, through trunk junctions which average about 6 miles in length. Neither local, tandem nor trunk junctions exceed about 20 miles in length. The nominal transmission losses of junctions are shown in Figure 7.3 for the United Kingdom; these nominal values are the limits for new work but older plant exists which may have been installed to worse limits and which it has not proved economic to improve; with the increasing installation of pcm systems in junctions, however, it becomes possible to reduce the loss of each junction to only about 2–3 dB at no extra cost and where there is no disadvantage in doing so this will be done.

Until recently, such comparatively short circuits have been provided most economically by unamplified (but possibly loaded) audio cable pairs. The characteristics of typical pairs are shown in Table 6.17. The longer circuits will be amplified using 2 wire negative impedance repeaters but 4 wire audio amplification does not become economic for circuits less than about 20 miles in length.

Certain designs of FDM system have been developed for short distance routes using the same type of cable as for audio circuits but it is difficult to justify the cost of the terminal equipment over such short distances. The advent of TDM/PCM junction systems, which also use deloaded audio cables, has, however, reduced to about 10 miles the route length over which multiplex transmission can be justified economically, PCM systems are therefore likely to be used in the future for a very large proportion of junction routes requiring a capacity of 24 circuits* or more. In the London Director Area, the number of circuits per local junction route averages about 10; the averages for tandem junction routes is about 26 and trunk junctions 32.

Because of their relatively short length, unamplified audio junction circuits will degrade transmission mainly by introducing loss, although some attenuation/frequency distortion will be introduced and the influence on return losses and sidetone of the image impedance of these lines (*see* Table 6.17) must be considered. Impedance matching tends to be rather expensive and has not been used on a wide scale in the United Kingdom; some telephone operating organisations do, however, consider it worth while to use impedance correcting arrangements to improve return losses and so permit transmission losses to be reduced.

When amplification is introduced, care must be taken to provide a large enough power handling capacity and to avoid introducing appreciable circuit noise. The more sophisticated, multiplex systems (FDM or TDM/PCM) must, of course, be properly designed to avoid these troubles and also with proper regard to suitable limits for attenuation/frequency distortion, nonlinear distortion and crosstalk.

* At present (1970), 24 channel PCM systems are used in the United Kingdom and other countries. So far as Europe is concerned, it is likely that 30 channel systems, which are now being standardised by the CCITT for international circuits, will be used in the future.

Table 6.17. NOMINAL CHARACTERISTICS OF CABLE PAIRS USED AT AUDIO FREQUENCIES FOR JUNCTIONS AND SHORT DISTANCE TRUNK CIRCUITS

Conductor size and other particulars (copper)	*Resis.* Ω		*Image attenuation,* dB/mile						*Modulus of image impedance (800/1600 Hz)* Ω
	per mile	*per* km	*500* Hz	*800* Hz	*1000* Hz	*1600* Hz	*2000* Hz	*3000* Hz	
10 lb PCQ (0.635 mm dia) unloaded	176	109	1.16	1.45 (1.56)	1.62	2.05	2.28	2.78	686/485
20 lb PCQT (0.902 mm dia) unloaded	88	55	0.78	0.98 (1.01)	1.08	1.35	1.49	1.75	506/358
40 lb PCQT (1.270 mm dia) unloaded	44	27	0.60	0.74 (0.70)	0.81	0.97	1.06	1.21	361/257
20 lb PCQT (0.902 mm dia) loaded 88 mH at 1.136 mile	91.3	56.7	0.340	0.345 (0.37)	0.347	0.353	0.357	0.385	1120/1180* 1080/992† (see Note 3)

Notes. 1. The values shown were obtained from measurements on actual cables. Values shown in brackets are used for planning purposes.

2. 6½ lb conductor (0.5 mm dia.) is also used for short lengths (planning value of loss, 1.94 dB/mile

3. * ½ section termination
† ½ coil termination

4. PCQ means paper core quad; PCQT means paper core quad trunk.

Because of the increasing importance of PCM as a junction transmission method, the remainder of this section is devoted to considering the transmission performance problems associated with such systems.

In the future, it may prove economic to use TDM systems (PCM or deltamodulation) on specially designed cables rather than using existing audiofrequency cable pairs. Small diameter coaxial or balanced pair arrangements may be used. Naturally, such systems become increasingly attractive as the density of the traffic between the exchanges and switching centres increases; 60 or 120 channel systems may be used.

At present, much of the attractiveness of PCM for junction circuits lies in the fact that it exploits more economically existing underground plant originally designed only for audio transmission; this enables the number of circuits between exchanges or switching points to be multiplied substantially at the cost of terminal equipment and repeaters, only. Each pair of a multi-pair cable will, when suitably repeatered, transmit at 1.5 or 2 Mbit/s to carry 24 or 30 channels so enabling two pairs to carry as many conversations as 24 or 30 2 wire audio pairs. As the number of pairs devoted to PCM in a cable is increased, the mutual interference by crosstalk at the very high frequencies concerned increases and the maximum spacing at which digital repeaters (regenerators) must be located becomes less. In the United Kingdom, much audio junction cable is loaded at intervals of 2000 yd, the loading coils being installed in accessible chambers or footway boxes; when pairs in such a cable are deloaded for PCM transmission, the loading coil chambers therefore provide convenient sites for the repeaters, provided that the same spacing of 2000 yd is suitable from the interference point of view. The PCM junction systems now being installed (and those contemplated for the future) make use of this arrangement the interference being kept to an acceptable level by limitation, where necessary, of the proportion of total cable capacity devoted to PCM. If a single cable is used for both directions of transmission, the crosstalk interference problem is, naturally, more onerous than when two cables are available on the route and one can be used for each direction. Experience with single cable schemes is showing that, by suitable choice of pairs in different layers of the cable, a substantial proportion of the total cable capacity can be used without excessive risk of digital errors.

The following information is concerned with the problem of ensuring adequate performance of PCM systems between audio input and output ports.

In order to assess properly the transmission performance adequacy of an element such as a PCM system, consideration must be given to suitable hypothetical reference connections of which it could form part. One such HRC is shown in Figure 6.5; this has been used to study suitable characteristics for junction (and other short haul) PCM systems. Reference should be made to Chapter 7 for the principles on which such studies ought to be conducted; in accordance with the method described there, the present example shows how the percentage of customers likely to experience difficulty can be estimated. Each of the four PCM links present in the HRC shown in Figure 6.5 will be assumed to have the following characteristics and the switching between PCM links will be assumed to take place by analogue methods, i.e. there will be non-digital interfaces between adjacent PCM systems.

1. Sampling rate, 8000 per second
2. Number of quantised amplitudes each side of zero, $B = 64$; this permits encoding the speech samples in 7 binary digits
3. Load capacity, $T_{max} = +2$ dBmO
4. Companding advantage, $G = 24.1$ dB
5. Compression characteristic, 13 segments arranged so that the interval between adjacent decision amplitudes increases in the ratio 2 : 1 on moving into the next higher amplitude segment

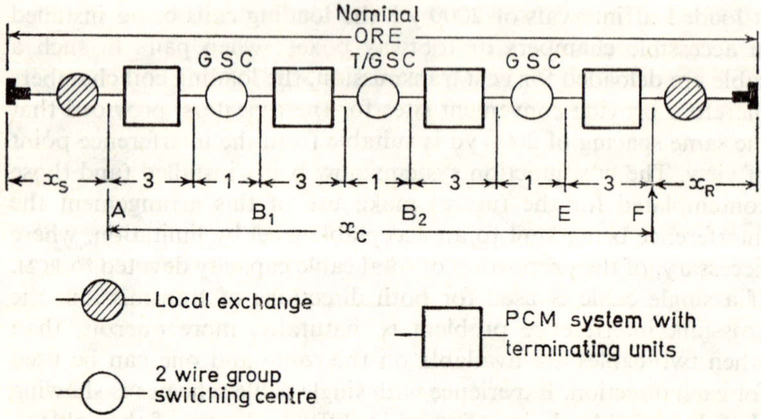

Figure 6.5. A hypothetical reference connection containing four p.c.m. systems in tandem and representing national telephone connections through three GSCs with analogue switching.

Notes: 1. The nominal overall reference equivalent is 28 dB when the subscribers' lines are limiting. 2. The SRE, *x_S, is 12 dB for limiting subscribers' lines and 7 dB for average lines. 3. The* RRE, *x_R, is 1 dB for limiting subscribers' lines and −1 dB for average lines. 4. This* HRC *represents the same connections as Figure 7.5d (which appears in Section 7.3.) but the junctions and trunks are provided by* PCM *systems instead of analogue transmission systems;* PCM *systems avoid mismatch losses at* GSCs *and their actual transmission losses do not depart appreciably from their nominal values*

Theoretically, the quantising distortion produced by a single such system can be described quantitatively in terms of the quantity R as a function of the level (s, dBm) of a Gaussian input test signal, together with the quantity L which represents the effective loss (also as a function of the level of the input test signal). R is the ratio, expressed in dB, of signal power to power of the quantising distortion products; R can be calculated by determining the r.m.s. value of the departure of output amplitudes from the linear regression line of output amplitude on amplitude of corresponding sample of the input amplitude. The slope of this regression line yields the effective gain of the system, from which L (in dB) can be found. Real p.c.m. codecs* can be tested using a Gaussian test signal of restricted bandwidth and measuring the power of the

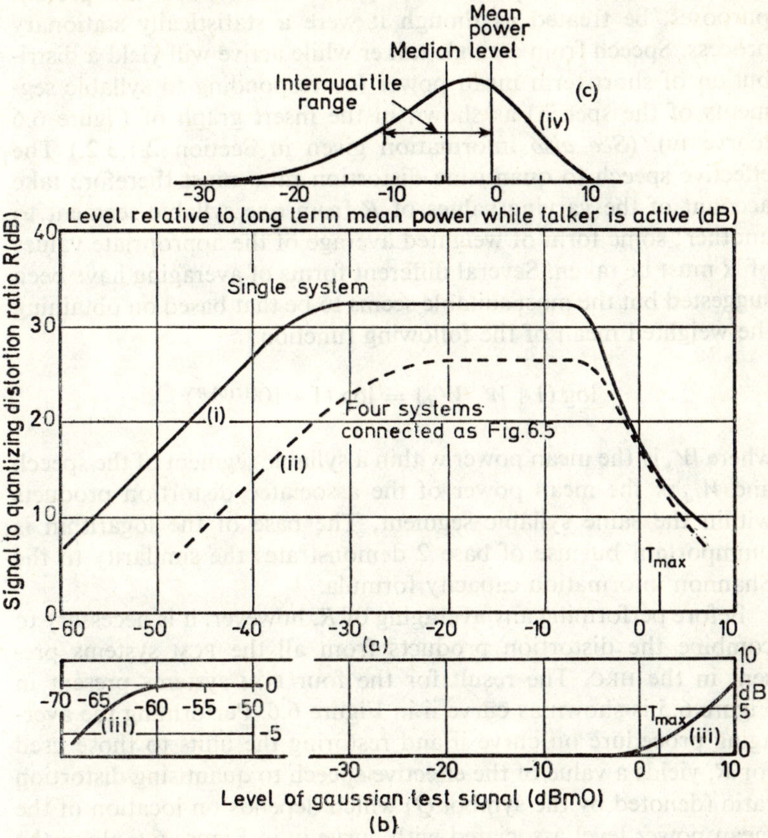

Figure 6.6. Effects of quantising by PCM *systems. Performance when measured with Gaussian test signals. (a) Signal/quantising distortion ratio, R. (b) Effective loss, L. (c) Probability density of levels of syllabic segments emitted by a given talker*

distortion products outside this frequency band, but within the nominal frequency range (300–3400 Hz) of the system. Figure 6.6 shows the results of such calculations for the PCM system defined

* Codec is the term used to denote any encoder/decoder pair in a digital transmission system; two codecs, one at each end of the circuit, provide a channel in each direction.

above. Over a considerable range of input level, R remains constant; outside this range, however, the performance deteriorates as R decreases.

The statistical properties of speech differ from those of a Gaussian test signal; in particular, speech cannot, for the present purposes, be treated as though it were a statistically stationary process. Speech from a single talker while active will yield a distribution of short term mean power (corresponding to syllabic segments of the speech) as shown in the insert graph of Figure 6.6 (curve iv). (*See also* information given in Section 2.1.3.2.) The effective speech to quantising distortion ratio must therefore take account of the varying values of R from one syllabic segment to another; some form of weighted average of the appropriate values of R must be taken. Several different forms of averaging have been suggested but the most suitable seems to be that based on obtaining the weighted mean of the following function:

$$\log(1+W_s/W_d) = \log(1+10^{(1/10)R})$$

where W_s is the mean power within a syllabic segment of the speech and W_d is the mean power of the associated distortion products within the same syllabic segment. The base of the logarithm is unimportant but use of base 2 demonstrates the similarity to the Shannon information capacity formula.

Before performing any averaging of R, however, it is necessary to combine the distortion products from all the PCM systems present in the HRC. The result for the four PCM systems present in Figure 6.5 is shown as curve ii in Figure 6.6. Performing the averaging procedure on curve ii and restoring the units to those used for R, yields a value of the effective speech to quantising distortion ratio (denoted by the symbol Q) which depends on location of the mean power level associated with curve iv in Figure 6.6 along the scale of s. Q is thus a function of the level of the long term mean power of the speech signal while active (denoted by the symbol S, dBm); this is illustrated in Figure 6.7 but, for convenience, speech level has been expressed in terms of V_A (dBV), the reading on a Speech Voltmeter Type 5B connected at point A in Figure 6.5. Q may be thought of as that value of R, independent of s, that would be equivalent subjectively to the actual R versus s function; defined in this way, it will be noticed that Q can be identified with a setting of the Modulated Noise Reference Unit (MNRU) described in Section 4.2.2.1. The information given there on the performance of speech links degraded by various settings of the MNRU can thus be used in studying the present problem.

The values of Q shown in curve i of Figure 6.7 apply for four ideal PCM systems in the HRC shown in Figure 6.5; practical systems are likely to contain imperfections with the result that the values of R, and consequently Q, will be somewhat lower than as shown

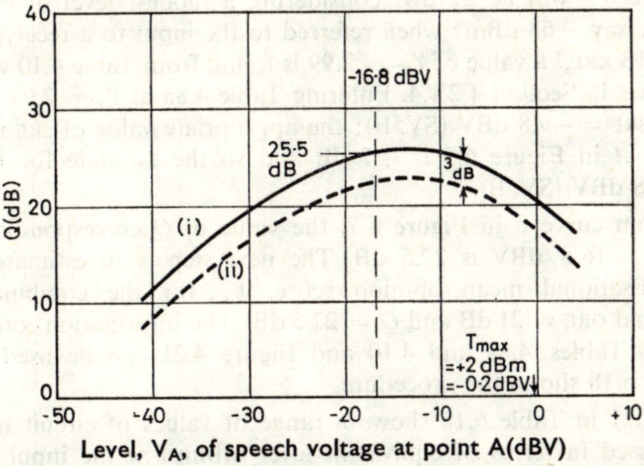

Figure 6.7. Q for the complete connection shown in Figure 6.5. as a function of the level of speech voltage at the input to the first PCM *system in the connection (Point A).*

Notes: 1. The speech voltage indicated by the broken line applies for a median talker on an average subscriber's line; measurement by SV5B. 2. Curve i applies for ideal PCM *systems; curve ii includes allowances for imperfections in practical* PCM *systems. 3. The load capacity,* T_{max}, *shown is the level of the maximum sinusoidal test tone that can be applied to the first* PCM *system without any 'peak-limiting'. 4. S (in dBm) =* V_A *(in dBV) +2.2. dB*

for ideal systems. Allowing a margin of 3 dB, independent of s, will result in the values of Q shown in curve i of Figure 6.7 being reduced by this amount to those shown in Curve ii.

In this example the send and receive reference equivalents x_S and x_R of the LTCS will be taken respectively as +7 dB and −1 dB which are approximately average values in the telephone network of the United Kingdom Post Office. The nominal ORE then becomes 21 dB for the HRC shown in Figure 6.5. To determine from Figure 6.7 the value of Q to be used, it is necessary to estimate the level of speech voltage appearing under conversation conditions at the input to the first PCM link of the connection (Point A in Figure

6.5). The standardised level of speech voltage, V_C+SRE, is shown in Table 4.8a (which appears in Section 4.2.1.3) as a function of Y_C the conversational mean opinion score. At this stage of the present discussion, Y_C is not known exactly but a sufficiently good estimate can be made to continue the argument; if necessary a second cycle of estimation can be performed. For average subscribers' lines, the nominal overall reference equivalent of the HRC shown in Figure 6.5 will be 21 dB; considering a modest level of circuit noise, say −68 dBm* when referred to the input to a receive end of 0 dB RRE,† a value of $Y_C = 2.99$ is found from Table 4.10 which appears in Section 4.2.1.4. Entering Table 4.8a at $Y_C = 2.99$ gives V_C+SRE = −9.8 dBV (SV5B); the appropriate value of SRE up to Point A in Figure 6.5 is +7 dB and so the estimate for V_C is −16.8 dBV (SV5B).

From curve ii in Figure 6.7, the value of Q corresponding to $V_A = -16.8$ dBV is 22.5 dB. The next step is to estimate the conversational mean opinion score, Y_C, for the combination nominal ORE = 21 dB and Q = 22.5 dB. The information contained in Tables 4.8a and 4.10 and Figure 4.21 can be used and Table 6.18 shows the procedure.

Col. 1 in Table 6.18 shows a range of values of circuit noise, expressed in terms of equivalent level (dBmp) at the input to a receive end of 0 dB RRE; col. 2 shows the same values unweighted for a restricted bandwidth from 300 to 3400 Hz (+2.5 dB) and in units of voltage across 600 Ω instead of power (−2.2 dB). Col. 3 shows the values of Y_C obtained from Table 4.10 for the given combination of nominal overall reference equivalent (21 dB) and level of circuit noise (col. 1). Col. 4 gives the standardised level of speech voltage for the values of Y_C appearing in col. 3; Table 4.8a was used. Col. 5 gives the level of speech voltage at the input to a (hypothetical) receive end of 0 dB RRE, i.e. referred to the same point as applies for the levels of circuit noise given in cols 1 and 2; col. 5 = col. 4 minus the nominal overall reference equivalent. (Note that, as shown in Figure 6.5, nominal ORE = SRE+x_C+RRE.) The quantity G_n appearing in col. 6 is the ratio, expressed in dB, of the power of the speech signals to the unweighted (but restricted to the

* This level is equal to that attributable to idle channel noise, see later.

† The actual level of circuit noise found at a particular point in a telephone connection, e.g. Point F in Figure 6.5, can be standardised by subtracting from it the value of RRE applicable to that portion of the connection between the point in question and the listener.

Table 6.18. DETERMINATION OF Y_C FOR A GIVEN COMBINATION OF NOMINAL ORE AND Q

N dBmp	N dBV *unweighted**	Y_C (*Nom.* ORE=*21* dB)	V_C+SRE dBV	V_C-x_C −RRE dBV	G_n dB	Q dB
(1)	(2)	(3)	(4)	(5)	(6)	(7)
None	None	3.02	−9.9	−30.9	∞	∞
−70	−69.7	2.99	−9.8	−30.8	38.9	22.8
−68	−67.7	2.99	−9.8	−30.8	36.9	22.1
−66	−65.7	2.98	−9.8	−30.8	34.9	21.4
−64	−63.7	2.97	−9.7	−30.7	33.0	20.8
−62	−61.7	2.96	−9.7	−30.7	31.0	20.1
−60	−59.7	2.94	−9.7	−30.7	29.0	19.3
−58	−57.7	2.92	−9.6	−30.6	27.1	18.6
−56	−55.7	2.90	−9.6	−30.6	25.1	17.8
−54	−53.7	2.87	−9.5	−30.5	23.2	17.0
−52	−51.7	2.83	−9.4	−30.4	21.3	16.2
−50	−49.7	2.78	−9.3	−30.3	19.4	15.3
−48	−47.7	2.72	−9.1	−30.1	17.6	14.4

* Measured 'flat' in the frequency band 300–3400 Hz

telephone channel bandwidth) power of Gaussian circuit noise uncorrelated with the speech; G_n is therefore given by subtracting the entries in col. 2 from those in col. 5. The values of Q shown in col. 7 are derived from G_n (col. 6) by the relationship shown in Figure 4.21.

From Table 6.18, the value of Y_C when nom. ORE = 21 dB and Q = 22.5 dB is 2.99; this can be converted to the percentage of participants likely to experience difficulty in conversation by use of Table 3.18 which appears in Section 3.7.1 and the result is 7.6%.

Transmission loss and quantising distortion are, of course, not the only degradations introduced into a telephone connection by PCM systems; allowances must also be made for idle channel (speech off) noise and attenuation/frequency distortion attributable to the input and output channel filters (nominally 300–3400 Hz passband).

The level of circuit noise from a channel involving a single pair of PCM codecs when there is no input signal and the central decision amplitude is at zero will, in theory, be equal to:

$$N_{MT} = T_{\max}+3-6.02(1+\log_2 B)-G-3.6 \text{ dBmOp}.$$

For the characteristics being considered, $N_{MT} = -64.8$ dBmOp;

in practice, such adverse circumstances are likely to be rare and a reasonable estimate is a value 3 dB lower than this, namely −67.8 dBmOp. The total level of speech off noise reaching the listener when there are several PCM systems in tandem is difficult to estimate precisely because it is influenced by noise from other items in the connection, by sources of noise within the codecs and by the extremely nonlinear transmission characteristics of each PCM system through which any noise is transmitted. A simple approximation is to allow for the losses shown in Figure 6.5 between input points of the successive PCM systems and to assume that these apply also for the noise produced separately by each source. This, of course, ignores the nonlinearity at low input levels. The result is as follows, all noise components being referred to the input to a receive end of 0 dB RRE:

From 1st PCM system,	−67.8−14 dB = −81.8 dBmp;	equivalent to 6.6 pWp
2nd	−67.8−10 dB = −77.8 dBmp;	16.6 pWp
3rd	−67.8− 6 dB = −73.8 dBmp;	41.7 pWp
4th	−67.8− 2 dB = −69.8 dBmp	104.8 pWp
	Total	169.7 pWp
	or	−67.7 dBmp

In practice, assuming well designed and engineered PCM systems, the total idle channel noise is likely to be somewhat less than this computation indicates. If only the last PCM system in the chain is considered, the value becomes −69.8 dBmp. For the present example a value of −68 dBmp would seem to be a reasonable choice. The value of Y_C for this level of circuit noise and a nominal ORE of 21 dB is (from Table 4.10) 2.99; the percentage of customers likely to experience difficulty on account of the transmission loss and the idle channel noise (speech off noise) is therefore (from Table 3.18) 7.6%.

The effects of attenuation/frequency distortion introduced by the input and output channel filters of the four systems present in the HRC under consideration can be estimated approximately as follows. Typical specification limits for attenuation/frequency distortion at 4 wire ports of a single PCM system are as shown in Table 6.19.

Table 6.19. TYPICAL SPECIFICATION LIMITS FOR ATTENUATION/FREQUENCY DISTORTION ATTRIBUTABLE TO A PAIR OF PCM CHANNEL FILTERS

Frequency Hz	*Maximum and minimum loss relative to that at 800* Hz
300– 400	+0.5 to −0.5 dB
400– 500	+0.3 to −0.5 dB
500– 800	+0.2 to −0.5 dB
800–2400	+0.5 to −0.5 dB
2400–3000	+1.0 to −0.5 dB
3000–3400	+2.0 to −0.5 dB

Attenuation/frequency distortion will also be introduced by the eight 2 wire/4 wire terminating units and the signalling relay sets which would be present in real connections represented by the HRC shown in Figure 6.5 and so the total may well reach an amount that would attract an impairment value (defined in Section 4.2.1.2 as $I_{2.5}$) of 1 dB. For comparison the attenuation/frequency distortion shown by curve i in Figure 4.17 attracts an impairment value (defined in the same way) of 2 dB.

Taking the impairment value as 1 dB, in association with the nominal ORE of 21 dB, yields a value of Y_C attributable to loss and attenuation/frequency distortion only of 2.96 (from Table 4.11 in Section 4.2.1.4); the corresponding percentage 'difficult' (from Table 3.18) is 8.4%.

The foregoing results may be summarised as shown in Table 6.20.

Table 6.20. SUMMARY OF 'PERCENTAGE DIFFICULT' ATTRIBUTABLE TO THE VARIOUS LISTENING FACTORS AND DETERMINATION OF THE TOTAL 'PERCENTAGE DIFFICULT'

Percentage attributable to 'loss', 'loss and quantising', 'loss and circuit noise' and 'loss and linear distortion'	*Proportion attributable to 'loss', 'quantising', 'circuit noise' and 'linear distortion' separately*	*Proportion attributable to all four factors simultaneously*
(1)	(2)	(3)
p_l = 7.0%	p_l = 0.0700	
P_{lq} = 7.6%	p_q = 0.0064	P_{lqnd} = 0.0934 (9.3%)
P_{ln} = 7.6%	p_n = 0.0064	
P_{ld} = 8.4%	p_d = 0.0150	

$p_l = 7.0\%$ in col. 1 applies for the hypothetical case when no quantising distortion, no speech off noise and no attenuation/frequency distortion are present and the connection is degraded only by loss (nominal ORE = 21 dB); the value of Y_C for this case would be 3.02, corresponding to 7.0% 'difficult'. The remaining entries in col. 1 are from the text above using the symbols explained in Section 4.2.4 which also gives the expressions used to derive the values entered in col. 2 and that used to estimate the proportion of 'difficult' opinions to be expected when all four listening factors are present together, *see* col. 3.

Subject to acceptance of the assumptions made in arriving at these estimates (explained in Section 4.2.4), col. 2 shows that, in this example, 'loss' is by far the most serious cause of difficulty and that 'linear distortion' (attenuation/frequency distortion) is more serious than either 'quantising' or 'circuit noise'.

6.3 TRUNK CIRCUITS

Until the advent of amplification (in the early 1920s), transmission problems in providing trunk circuits were little different in nature from those encountered with subscribers' lines and junctions. Heavier gauge conductors were, of course, needed to keep transmission losses to acceptable limits. The types of line plant then in use for trunk circuits are shown in Table 6.21 which includes values of relative loss per unit distance.

With aerial lines, trunk routes up to 500 miles or more were feasible. The velocity of propagation over such lines is fairly high and so, because the transmission losses were also fairly large, no trouble was caused by echoes. Audio amplification enabled lightly loaded cables to be used over long distances and these circuits suffered from reduced velocity of propagation and so minimum values of loss had to be set to avoid unacceptable echoes. The minimum permissible loss had to be made larger as the propagation time increased so that the full value of the amplification could not be used. This problem was solved by development of voice switched echo suppressing devices but these complicated the operation of trunk circuits as well as adding terminal equipment cost.

High velocity wideband frequency division multiplex transmission systems, over coaxial cables or microwave links, have rendered unnecessary any special measures of echo control unless route distances exceed about 1000 miles (one way propagation

Table 6.21. SPEECH TRANSMISSION EQUIVALENTS OF TRUNK LINE PLANT USED IN 1906 (FROM J. G. HILL)

Conductor weight per mile, lb *(see Note)*	*Speech transmission equivalent, relative to Standard Cable**		
	Aerial lines	*Air space cable*	*Guttapercha cable*
800	0.022		
600	0.027		
400	0.038		
300	0.048		
200	0.068	0.29	
150	0.086	0.34	
100	0.120	0.41	
70	(bronze) 0.26	0.55	
40	(bronze) 0.35	0.68	1.1
20		1.00*	
10		1.64	

* Standard Cable is equivalent to 20 lb air space cable and has an image attenuation at 800 Hz of 0.922 dB/mile.

Note. Conductors are copper unless otherwise stated.

time 10 ms): minimum transmission loss is therefore now set by stability requirements rather than echo.

Most of the trunk circuits of the United Kingdom are provided on a frequency division multiplex (FDM) basis. Twenty-four channel amplitude modulated transmission over pair-type cable can be used economically on routes at least 20 miles in length but such systems are not now being installed in the United Kingdom. Systems with larger numbers of channels are carried mainly on coaxial tubes 0.375 in diameter which transmit bandwidths up to 12 MHz. (Some systems use only up to 4 MHz.) Fairly recently small diameter coaxial tubes 0.174 in diameter have been introduced and, with transistorised equipment, these will also transmit up to 12 MHz. Microwave radio relay systems are also used working in the 2, 4 and 6 GHz bands with frequency modulated carriers bearing baseband blocks of channels assembled in exactly the same manner as the amplitude modulated, single sideband, suppressed carrier FDM systems mentioned above. Such radio relay systems will carry up to 2700 4 kHz telephony channels (or the equivalent capacity in television channels, etc.).

Telephony channels, spaced at 4 kHz, are assembled in FDM systems according to a hierarchy of groups (12 channels), supergroups (5 groups, 60 channels), mastergroups (5 supergroups, 300 channels) and supermastergroups (15 supergroups, 900 channels). Larger capacity systems are then composed of various numbers of supermastergroups. Table 6.22 shows the frequency ranges of these basic 'building blocks' of channels.

Table 6.22. EXAMPLES OF BASIC BLOCKS OF 4 kHz SPACED CHANNELS STANDARDISED BY THE CCITT FOR FDM SYSTEMS

Number of channels	*Title of block*	*Minimum and maximum frequencies* kHz	*Frequency bandwidth*
12	Basic group B	60–108	48 kHz
60	Basic supergroup	312–552	240 kHz
300	Basic mastergroup	812–2 044	1.2 MHz
900	Supermastergroup	8516–12 388	3.8 MHz

When FDM transmission systems are designed, the following factors are important and must be taken into account:

1. Power handling capacity
2. Circuit noise
3. Attenuation/frequency and group delay/frequency distortion
4. Audio crosstalk between channels

The levels of speech signals arriving at the input to a trunk circuit depend upon the following:

1. The person who is talking
2. The sending reference equivalent from the subscriber's station up to the input to the trunk circuit
3. The sidetone reference equivalent of the talker's telephone set
4. The overall transmission performance of the connection being used
5. The topic of the conversation.

The speech level cannot be treated on an individual call basis and so consideration must be given to statistical descriptions of aggregates of calls. In Table 4.8, Section 4.2.1.3, it can be seen that the median value of speech voltage, obtained from a sending end of 0 dB SRE when the participants are conversing over a telephone connection of such overall transmission performance that they would yield a conversational mean opinion score of 2.8, is −9.3 dBV for 13-2P-27 sets and −11 dBV for 16-4T-31 sets with regulators. The mean opinion score of 2.8 corresponds to about 13 per cent of participants experiencing 'difficulty', which is fairly typical of the transmission performance of long international connections. The difference between the speech voltages from the two types of telephone set is probably attributable to the lower sidetone reference equivalent of the latter set; taking a compromise value and converting to power level relative to 1 mW in 600 Ω yields a median value of about −8 dBm. Under constant physical conditions, talkers will yield speech levels dispersed approximately in accordance with a Gaussian distribution of 4.1 dB standard deviation.

On average, the sending reference equivalent (SRE) up to and including the local exchange is about 7 dB with a certain dispersion (see Section 8.2); from the local exchange to a 0 dBr point at the input to the first trunk circuit, the transmission loss may (nominally) lie between 0 and 4.5 dB (*see* Section 7.2). (Trunk junctions in the London Director Area have an average loss of 3.9 dB with standard deviation 2.05 dB (*see* Section 8.2). Users of international circuits may, on average, be more favourably located relative to the international trunk switching centre and so the average SRE may be somewhat less than the value of 10.9 dB indicated by the foregoing; assuming 8 dB with standard deviation (subscribers' lines and junctions taken together) 3 dB, the speech level entering a trunk circuit becomes about −16 dBmO (median) with standard deviation about 5 dB. Speech level measurements at the input to transatlantic submarine cable circuits have yielded the results shown in Table 6.23 and these are in good agreement with the above estimates.

The results were obtained by measuring the level, S (dBm), of the mean power, W (mW), produced by each talker while he was actively emitting speech ($S = 10 \log_{10} W$). The values in cols 3 and 4 result from calculating the mean and standard deviation of S.

To determine the long term mean power loading per channel, averaged over all talkers while active, W must be averaged to obtain $\overline{W}$, from which the level of the long term mean power while active is given by $10 \log_{10} \overline{W}$, which is not the same as $\overline{S}$, because

Table 6.23. LEVEL OF MEAN POWER OF TALKERS WHILE ACTIVE (TRANSATLANTIC CABLE CIRCUITS)

Date	*Talker's location*	*Median level* $\bar{S}$	*Standard deviation* σ_S	*Proportion of traffic*
(1)	*(2)*	*(3)*	*(4)*	*(5)*
		dBmO	dB	
1957/58	London Region	−10.0	5.3	
	Outside London	−12.2	6.8	
1965	a. Inner London (City and Central Exchange Areas)	−14.9	5.2	60%
	b. Outer London (Remainder of Director Area)	−18.3	5.1	35% (including d and e)
	c. Large provincial Director Areas	−13.9	6.1	5%
	d. Outer provincial	−18.1	6.1	see above
	e. Unit Automatic Exchanges	−19.3	6.4	see above
	Traffic weighted mean	−16.0	5.8	100%

W is distributed in an asymmetric manner. As an approximation*, $10 \log_{10} \bar{W} = \bar{S}+0.115(\sigma_S)^2$.

Taking the traffic weighted mean (1965), the level of mean power averaged over all talkers while active is $-16.0+0.115(5.8)^2 = -12.1$ dBmO. Allowing an assumed activity factor of 0.25 (proportion of time that a channel is transmitting active speech, including within conversation activity and circuit occupancy factors†), the level of mean power loading per channel, averaged

* This assumes the distribution of S to be Gaussian, i.e. W is lognormal.

† Measurements made in association with the 1965 speech level measurements quoted in Table 6.23 yielded the following:

Conversational activity factor,	0.41
Circuit occupancy factor,	0.66
Total activity factor, 0.41×0.66 =	0.27

These results apply during the 'busy hour'.

over inactive as well as active time, becomes $-12.1+10 \log_{10} 0.25 = -18.1$ dBmO. (This is equivalent to 15.5 μW per channel, which is slightly less than the allowance of 22 μW per channel conventionally allowed for speech signals, telegraph signals and leakage of carriers.)

Estimation of the power handling capacity of the wideband amplifiers which transmit the total power from the blocks of channels is complicated by the fact that the probability density distribution of this multiplex signal depends on the number of channels and other particulars; for very large blocks of channels (say 240 or more) the multiplex signal may be assumed to be Gaussian, but for lesser numbers the peak to r.m.s. ratio is greater than for such a signal.

Circuit noise is produced in FDM transmission systems from thermal sources, from intermodulation produced by nonlinearity of the transmission path of the multiplex signal and in the modems* of the frequency translating equipment at the line terminals. According to CCITT Recommendations (see Section 2 of CCITT White Book, 1968, Volume III) such systems ought to be designed to noise limits of 3 pW per km of circuit length for thermal and intermodulation noises (which may be expected to be proportional to length) plus 2500 pW allocated for all the frequency translating equipment likely to be present in a circuit 2500 km long. Similar standards may be applied to inland national circuits (except very short FDM circuits which are allowed a fixed amount of noise of 500 pW, independent of their length). Submarine cable circuits are usually designed to 1 pW/km but satellite systems are allowed 10 000 pW for a single hop irrespective of the relative locations of the earth stations. Radio relay systems are permitted certain concessions for very small proportions of their operating time to allow for their fading characteristics.

Attenuation/frequency and group delay/frequency distortion arises in FDM carrier systems mainly from the cut off characteristics of the input and output channel filters of the modems; edge channels in groups may also suffer from effects due to group filters of translating equipment. Within the boundaries of the 4 kHz space between carrier frequencies, it is possible to provide a nominally level frequency response channel from 300 to 3400 Hz

* Modem is the term used to denote a pair of equipments for modulating the signals in one direction and demodulating those in the other. Correspondingly, the term codec is used to denote an encoder/decoder pair in a digital transmission system.

but the tolerances at these limit frequencies and the rate of increase in loss outside the nominal passband depends upon the design of the filters and this is influenced by cost. Table 6.24 shows the limits of transmission loss permitted according to CCITT Recommendation G.232 together with the limits appearing in former and current United Kingdom Post Office specifications for 12 channel terminal equipments.

Table 6.24. LIMITS OF LOSS, RELATIVE TO THAT AT 800 Hz, FOR A PAIR OF 12 CHANNEL TERMINAL EQUIPMENT FILTERS

Item	*Frequency*, Hz					
	200–300 dB	300–400 dB	400–600 dB	600–2400 dB	2400–3000 dB	3000–3400 dB
CCITT G.232 (Mean for the 12 pairs of a group)	—	+1.7 −0.4	+0.9 −0.4	+0.4 −0.4	+0.9 −0.4	+1.7 −0.4
CCITT G.232 (Worst pair of a group)	—	+3.0 −0.9	+1.7 −0.9	+0.9 −0.9	+1.7 −0.9	+3.0 −0.9
UKPO Specification (Former type equipment)	—	+4.0 −1.0	+2.0 −1.0	+1.0 −1.0	+2.0 −1.0	+4.0 −1.0
UKPO Specification (High grade type equipment)	+1.0 −1.0	+0.5 −0.5	+0.5 −0.5	+0.5 −0.5	+0.5 −0.5	+1.0 −0.5

Note: The UKPO Specifications give separate (but equal) limits for send and receive channels. The values quoted are the sums of these limits; on average the equipment will, course, be appreciably better.

The delay/frequency distortion permitted for the present (high grade) type United Kingdom Post Office equipment is as follows; the delays are relative to that at 800 Hz:

at 250 Hz and at 3400 Hz, 1 ms
in the range 565–2550 Hz, 0.5 ms

These limits are required primarily for data transmission rather than for speech.

Typical characteristics of equipment designed to these specifications are shown in Figure 4.16 (attenuation/frequency distortion) and 4.24 (group delay/frequency distortion) which appear in Section 4.2.1.2 and 4.2.3; these figures also include information relating to 3 kHz channelling equipment (16 channel groups) which is used in submarine cable systems.

Crosstalk between audio channels can be produced by FDM modems and PCM codecs and this must be controlled to suitable limits. In FDM modulators and demodulators this requires proper attention to decoupling of power and carrier supplies and to the amount of suppression provided in the stop range of channel filters. In some cases the crosstalk is unintelligible and then less severe limits may be allowed. In PCM systems, crosstalk can be a very severe problem on account of the nonlinear transmission paths that may be involved (*see* Section 4.5.3).

In order to achieve satisfactory performance, it may be necessary to associate certain types of audio frequency equipment with trunk circuits; examples are echo suppressors, needed when the mean one way propagation time exceeds about 15 ms (*see* Section 4.3.2), and compandors, used to reduce the subjective effect of circuit noise (*see* Section 6.5). Information on these items will be found CCITT Recommendations G.143, G.161 and G.162.

6.4 OVERALL TRANSMISSION LOSS OF A CONNECTION

To estimate the transmission performance of a given speech link without actually conducting any subjective tests upon it, the overall air to air transmission losses must be known for each of its constituent paths (*see* Figure 3.2 and accompanying text appearing in Section 3.1). Sidetone paths (c and d in Figure 3.2), which involve both speech and room noise, have been treated in Section 6.1.4 and the crosstalk paths (i and j in Figure 3.2) are considered in Section 6.1.5. The present section is concerned with the principal speech paths (a and b in Figure 3.2) from the mouth of the participant at one end to the ear of the participant at the other.

The mouth to ear loss, L_{ME}, of the speech path in question could, in certain circumstances, be measured directly using suitable artificial mouth and artificial ear test equipment but this would be possible only if the speech link actually existed. It is more commonly the case that the information is required for a hypothetical connection, or one having no permanent existence as is so in a switched telephone network. In such cases the values of L_{ME} (L_{ME} is, of course, a function of frequency) must be built up from knowledge of the characteristic parameters of the constituent parts of the connection.

Any telephone connection contains at least a pair of local telephone circuits (LTCS) and one method for building up their sending and receiving sensitivities is described in Section 6.1.2. That method makes use of the sensitivities of microphone and earphone, determined as described in Section 5.1.2 and 5.1.3, the electrical losses of the telephone sets, certain mismatch losses and suitably determined insertion losses of subscribers' lines and feeding bridges. The measurements of the various losses must be performed on actual items or laboratory models of these items; if such measurements are not possible or the requisite models are not available, the information can be obtained analytically but rigorous treatment involves rather complicated calculations. Fairly good approximation is possible by use of insertion loss data with arbitrary (say 600 Ω resistance) terminations.

For the special case of a local call over a connection consisting of two identical telephone sets joined together at their line terminals through a lossless feeding bridge, i.e. without any intervening subscribers' lines, junctions or trunk circuits, the mouth to ear loss is given by:

$$L_{ME} = L_S + L_R + 10 \log_{10} \frac{4Z_M}{Z_E} - S_M - S_E \qquad (6.10)$$

Allowance for intervening subscribers' lines and a feeding bridge can be made easily if the insertion loss of the appropriate combination of these items can be specially measured between the appropriate values of Z_C. This procedure will yield appropriate values of L_{ME} for local call connections; cases of asymmetrical local connections especially are, perhaps, most easily treated by special end to end measurements of the electrical losses.

When the telephone connection under consideration contains a section of, say, 600 Ω image impedances (*see* Figure 6.2a) and this has substantial transmission loss (so that each end presents practically its image impedance) and there are no junctions, use can be made directly of the sending and receiving sensitivities derived for an LTC in Section 6.1.2; in fact adding them directly together and reversing the sign (to change sensitivity to transmission loss) yields L_{ME}.

Thus: $$L_{ME} = 20 \log_{10}(E_M/V_3) - S_M + x_{CD} + 20 \log_{10}(V_4/V_E) - S_E \qquad (6.11)$$

Provided the transmission loss x_{CD} of the 600 Ω section (representing the trunk circuits) is sufficient to render input and output impedances independent of the impedance connected to the remote end, the transmission losses ascribable to the trunk circuits can be introduced by direct addition. Where this is not so, some rather complicated allowances ought to be made, the sensitivities of the LTCS having been defined specifically in terms of 600 Ω terminations.

Where any junctions are present between an LTC and the 600 Ω representing the image impedance of the trunk circuits, and these junctions are not of 600 Ω image impedance (e.g. if they are of unloaded or loaded cable), their insertion losses at points C and D in Figure 6.2a must be determined. Such insertion losses can readily be measured if a physical model of the connection is available but calculations alone are rather laborious, depending as they do, upon the specific values of the impedances Z_{LTC} and those of the junctions and the trunk circuits. The insertion losses of junctions cannot, therefore, be generalised.

In view of the foregoing complications in treating strictly the overall losses of telephone connections and remembering that all these losses are functions of frequency, it is not surprising to find that, in practice, considerable simplifications are usually adopted and the strict determinations confined to laboratory situations.

The first simplification is to avoid treating transmission losses as functions of frequency and to consider only 'averaged' or 'equivalent' losses. In principle, these are intended to be the hypothetical values of loss of each element, independent of frequency, that could be used to replace the actual, frequency dependent, loss without any change in the subjective assessment of the connection. Although equivalent losses could be based on other subjective criteria, it is usual in telephony to consider loudness equivalent losses, i.e. the frequency independent loss that would leave unchanged the loudness of the received speech. The system of reference equivalents is based on this principle (*see* Section 3.2 for details).

Determinations of reference equivalent according to current CCITT (Recommendations P.42 and P.72) are laborious and strict interpretation of the principle is usually confined to the determination of sending and receiving reference equivalents of a few examples of LTCS. When suitable sensitivity values are available it is possible to calculate these reference equivalents as explained in Section 3.2.4. Allowances in reference equivalent for junctions and trunk circuits are usually made on the basis of very simple rules, such as those indicated below. The effects on the sending and receiving reference equivalents of LTCS of certain changes in subscriber's line (changes in length and in conductor) are also commonly made by use of simple arbitrary rules. These rules, which yield single numbers for the equivalent loudness losses of the electrical elements are then added together and combined with the sending and receiving reference equivalents of the relevant LTCs to produce what is termed the 'nominal overall reference equivalent' of the connection. (This is also explained in CCITT Recommendation P.11, *see* Section 3.2.3.)

The following rules are used in the United Kingdom:

1. Unloaded cable in the local network or in junctions is treated by taking the equivalent loss to be equal to its image attenuation at 1600 Hz. Mismatches at changes in conductor or in type of line are ignored.
2. Loaded cable in junctions is treated by taking the image atenuation at 800 or 1000 Hz.
3. Sections nominally of 600 Ω image impedance are treated by taking their nominal losses at the line up frequency of 800 or 1000 Hz.

The manner in which sending and receiving reference equivalents are specified (in particular, use of a relatively wide frequency band receive end for determining sending reference equivalent and a similarly wide frequency band send end for determining receiving reference equivalent) results in some interaction effects in consequence of which the overall reference equivalent of a telephone connection may differ appreciably from the sum of sending and receiving reference equivalents of the LTCs even when strictly correct allowance has been made for intervening losses between the LTCs. Some examples of the magnitude of this interaction effect are given in Section 3.2.4; the overall reference equivalent may be 2–3 dB less than the sum of SRE and RRE with correct allowance for inter LTC loss. Thus the nominal overall reference equivalent may on this account somewhat exceed the true overall reference equivalent.

The use of rules such as those indicated above, however, can underestimate the true overall loss because mismatch losses are largely ignored and these can commonly be positive in their total effect. The rules can also result in underestimation of the true loudness equivalent because attenuation/frequency distortion is also ignored. Under rather adverse conditions (e.g. a connection representing a very unfavourable international connection) these discrepancies seem approximately to cancel each other as indicated in CCITT Recommendation P.11 but, under more commonly occurring national connection conditions, the true overall reference equivalent is likely to be a few dB less than the nominal overall reference equivalent.

When estimating the overall transmission loss of a connection as it is experienced by the customer, care must be taken to ensure that all the transmission elements present have been properly considered. For example, a switched connection usually contains, in it transmission path, switching elements, 2 wire/4 wire terminating units, signalling sets and other items necessary for setting up, controlling and supervising the connection; it may contain, in addition, echo suppressors and compandors. Many such items are located in parts of the connection that escape measurement when circuits are lined up; even when included in the lining up procedure, this will be done at a single test frequency and so they may introduce attenuation/frequency distortion not otherwise taken into account. When the composition of these additional transmission elements is known, measurements of their losses can be made; it is usually sufficient to measure them between 600 Ω terminations and to take the mean of the insertion losses (in dB) at the four frequencies 500, 1000, 2000 and 3000 Hz. If any item introduces serious attenuation/frequency distortion, more elaborate treatment is required.

6.5. ECONOMY SYSTEMS

An orthodox linear speech channel designed to cater for the range for amplitude of speech sounds emitted from the mouth and to have a negligibly low level of circuit noise and distortion will require a channel capacity of almost 100 000 bit/s (say, 8000 Hz bandwidth and 12 bit uniform p.c.m. encoding or 70 dB signal/noise ratio). In ordinary telephony, a bandwidth of 3100 Hz suffices and 30 dB signal/noise ratio is acceptable (30 000 bit/s) provided the channel can be adjusted for each talker individually; a signal/noise ratio of about 60 dB is necessary in practical telephony networks to cater for the range of talkers and of transmission losses between the talker and the part of the telephone connection at which the most severe channel restriction occurs.

The channel capacity that is necessary can, however, be reduced by taking advantage of redundancies in speech waveforms and of the limitations in information processing capacity of the ear/brain mechanism. The simpler methods of exploiting these characteristics of speech, hearing and conversational behaviour are relatively crude; for example, a frequency bandwidth from 250 to 2850 Hz (2600 Hz in width) is practically as acceptable as one restricted severely to the range 300–3400 Hz (3100 Hz in width), as shown in Section 4.2.1.2. Another simple method is that of instantaneous 'companding', used in PCM transmission systems; this exploits, rather crudely, the fact that high amplitude components of the speech waveform do not need to be transmitted with the same absolute precision as low level (but still audible) components.

The principles on which channel capacity economy systems have been based can be classified as follows:

1. Amplitude compression or limiting
2. Time compression
3. Frequency compression
4. Analysis/synthesis

This classification is not intended to be rigorous but only to provide a convenient basis for description and explanation. Class 1 includes not only the rather crude methods indicated above but syllabic companding and those systems designed to transmit separately the amplitude and frequency information of the speech; these exploit the fact that speech waveforms can be treated as quasi-stationary statistical processes which change their parameters only rather slowly (because of the manner in which speech is produced).

Class 2 is represented by those systems which attempt to exploit silences on telephone channels or which sample waveforms in blocks of about the larynx period and only transmit one in, say, two, the missing blocks being filled at the receiving end by repetition of the preceding block. Greater efficiency in exploiting actual silences can be achieved by using a group of several talkers in different conversations which, by suitable switching in accordance with the speech 'bursts', can be transmitted over perhaps only half that number of real channels; in this way incomplete circuit occupancy can be exploited as well as the presence of silences in the speech of a given talker (partly because he is listening to his respondent).

In its simpler forms, frequency compression (Class 3) consists of the removal, by filtration, of parts of the frequency band occupied by speech; besides judicious removal from top and bottom, some advantage can be obtained by removal of all except perhaps three bands 500 Hz wide located with mid-frequencies at 500, 1500 and 2500 Hz. More sophisticated procedures involve analytic separation of amplitude and frequency followed by halving, etc, the frequencies in the latter component.

The aim of analysis/synthesis methods (Class 4) is to extract from the speech waveforms the essential, information bearing, elements so that as much as possible of the redundancy is removed (*see* Section 2.1.4). The ultimate success of such methods would be achieved if it ever became possible to distinguish speech elements on a phonemic basis, to code these into perhaps 40 distinguishable signals for transmission and reconstitute speech at the receiving end by switching on appropriate stereotypes to represent each phoneme.

It must be noted that any economy of information capacity for transmitting speech renders the channel less versatile and, consequently, transmission of other forms of signal, e.g. multi-frequency data or telegraphy, may suffer certain restrictions. Furthermore, the restriction in versatility may also increase difficulty in conversation even though all the speech sounds can be understood quite clearly by both participants; listeners may recognise the speech sounds but fail to note certain peculiarities they should possess. In communicating, talkers engaged in conversation impress upon their phonemic emissions certain 'extra messages', conveyed by such features as unusual changes in pitch or duration, sometimes intended to indicate some emotional content of the message.

7. PLANNING OF TELEPHONE NETWORKS

7.1 A LOOK BACKWARDS

Engineers and others concerned with different parts of a telephone network will, naturally, adopt different attitudes towards 'planning'; those occupying different viewpoints will attach correspondingly different meanings to the term. Those who deal with the longer term aspects will attribute a 'strategic' significance to planning, whilst those whose job it is to ensure the most economic distribution of plant in local areas will think in 'tactical' terms; decisions on the installation of longer distance transmission equipment perhaps requires an attitude intermediate between those mentioned above. The various facets of the network planning process are reviewed later (Section 7.2) but it is instructive first to glance at the methods used in the past to guide the development of the telephone network in the United Kingdom.

Reference to the information in Section 1.1* reveals certain trends and it is worth noting their developments under different headings, for example the following: First, 'transmission standards' are frequently mentioned and these undergo change from time to time; secondly, assessment of the performances of individual telephone connections (and parts of connections) is performed in accordance with evolving, contemporarily favoured, criteria and methods; thirdly, 'the transmission standard' has been interpreted somewhat differently at different times, for example either as an 'overall' standard or as separate parts used independently for planning local telephone areas and the longer distance, junction and trunk, system; fourthly, there has often been confusion between those who treated the transmission standard as a proper representation of the efficiency of a defined class of telephone connections and those who treated it merely as a basis for formulating installation instructions concerned only with those items (e.g. losses of line plant) that were under planning control.

* *See also* Section 7.1 of the Bibliography

The term 'transmission standard' has been applied to the assembly of a pair of identical local telephone circuits (LTCS) joined together through a length (either adjustable or having certain defined values) of 'standard cable' or, since 1933, through a 600 Ω attenuator (either capable of adjustment or having one or more defined values).

In the early stages of development of telephony, telephone plant had been installed in the United Kingdom (*see* Section 1.1) so as to provide an economic service acceptable to the majority of those who paid for calls, rather than to provide a good public service for the whole country. This had been achieved by efficient planning within rather restricted regions, wherein most of the traffic circulated, together with some attention to the needs (mainly business) for satisfactory communication between pairs of large cities in which most of the telephone users were situated rather close to the terminals of the main long distance circuits. It was certainly possible, at least from 1896, for connections to be established between even remote areas of the country but there is no doubt that such 'fringe to fringe' connections afforded performances very much inferior to those of local calls; this was quite justified until the advent of practicable amplification (in the 1920s) because the revenue earned by calls over such 'unfavourable' routings could not possibly justify the heavy cost of making substantial improvements in their performances. The need for a more uniform standard of performance was, however, recognised, even as early as 1891 (see Section 1.1) although the economic 'facts of life' were then dictating a very inferior performance for the unfavourable and rarely required connections.

A transmission standard applicable for the whole of the United Kingdom was first defined in 1905 when an agreement was signed between the Post Office, who then operated the long distance lines, and the National Telephone Company, who, at that time, operated most of the local systems. This agreement was formulated in terms of a physical representation of a telephone connection consisting of standard subscribers' sets, selected from current supplies at the time and average items taken, feeding bridges representing 22V CB No. 1 Exchanges, 300 Ω resistors to represent subscribers' lines, and certain specified lengths of standard cable to represent the junctions and trunk circuits between the terminal exchanges. The composition of this representation of a Transmission Standard*

* In 1905 this assembly of equipment was referred to as the 'transmission test circuit'; the term 'transmission standard' strictly applied to the performance represented by a particular setting of the 'standard cable'.

and its successors are shown in Figure 7.1; certain important transmission characteristics of these Transmission Standards are tabulated in Table 7.1. The agreement provided for an upper (worst) limit for the most unfavourably routed trunk connections represented by a setting of 46 miles of standard cable (46 m.s.c.) and certain lesser values for connexions between subscribers' stations located not more than 50 miles apart (*see* Table 7.1).

Any particular telephone connection was compared with the transmission standard by determination of the equivalent setting of the standard cable using the loudness of received speech as the criterion of equality. Naturally only representations of actual, particular connections could be compared directly and so comparisons were made mainly by using information derived from series of carefully controlled laboratory tests.

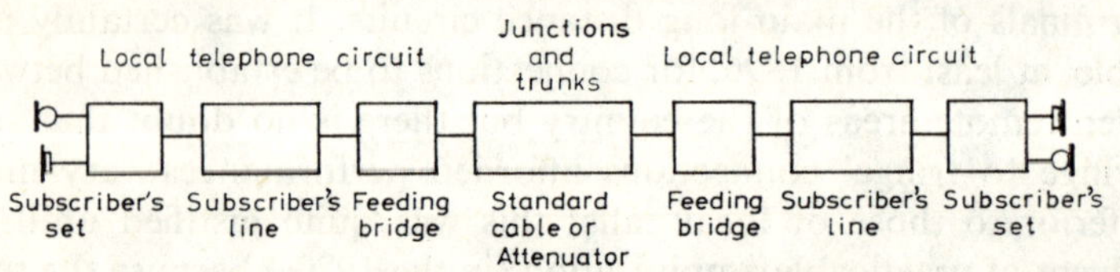

Figure 7.1. Transmission standards and representative limiting connections used for planning in UK

The 1905 Transmission Standard was treated as an overall limit; that is, local areas that had direct access to the long distance trunk circuits could take advantage of this by allocating larger equivalents to their subscribers' line plant than in areas accessible only through additional switching points. Similarly, the equivalent of the 300 Ω resistor, representing the subscriber's line, could be exceeded when the loss in the trunk junctions was less than the nominal limit.

In 1915, the 46 m.s.c. standard was recognized as excessive and an objective of 35 m.s.c. was set, with correspondingly smaller values for the more commonly used connections (*see* Table 7.1). This standard was treated in the same manner as that of 1905 (i.e. as an overall limit) and assessment remained, fundamentally, in terms of loudness of received speech.

By 1922 even the objective of 35 m.s.c. was considered inadequate and the representation of the transmission standard was upgraded by removal of the 300 Ω resistors that had represented the subscribers' lines at each end. This effectively reduced the transmission equivalent by about 12 m.s.c., the equivalent of the subscribers' lines now being considered as part of the 35 m.s.c. which was

Table 7.1. SOME TRANSMISSION PARTICULARS OF TRANSMISSION STANDARDS AND REPRESENTATIVE LIMITING CONNECTIONS

(a) Composition (see Fig. 7.1)

Date of formulation	*Particulars of local telephone circuit at each end*			*Standard cable or attenuator setting*
	Type of subscriber's set	*Subscriber's line*	*Feeding bridge*	
1905	CB1-1A-3 (Telephone No 2, 'candlestick' type)	300 Ω nonreactive resistance	22 V Repeating coil type	maximum 46 m.s.c.
1915	ditto	ditto	ditto	maximum 35 m.s.c.
1922	ditto	zero	ditto	ditto
1933	10-1L-14. 35A. (Telephone No 162, handset type)	300 Ω nonreactive resistance	ditto	15 dB
1947 (Figure 4.2a)	13-1L-18.35A (Telephone No 162)	450 Ω 10 lb/mile cable	50 V Stone type 200+200 Ω	maximum 27 dB
1957	13-2P-27	660 Ω $6\frac{1}{2}$ lb/mile	ditto	ditto
1960 (Figure 4.1b)	ditto	ditto	ditto	20 dB
1969 (Figure 7.4)	16-4T-31 (with regulator) (Telephone No 706)	1000 Ω $6\frac{1}{2}$ lb/mile	ditto	ditto

(b) Transmission particulars

Date of formulation	LTCS		*Standard Cable or Attenuator Setting (Note 1)*	*Nominal* ORE dB *(nearest 1 dB)*	*Remarks*
	SRE dB	RRE dB			
1905	+12.2 Note 2	+5.5 Note 2	46 m.s.c.	60	Extreme limit
			30 m.s.c.	45	Stations separated by ⩽ 50 miles
			20 m.s.c.	36	Stations separated by ⩽ 10 miles

Table 7.1. *(cont.)*

Date of formulation	LTCS SRE dB	LTCS RRE dB	*Standard Cable or Attenuator Setting (Note 1)*	*Nominal* ORE dB *(nearest 1 dB)*	*Remarks*
1915	+12.2 Note 2	+5.5 Note 2	35 m.s.c.	50	Extreme limit
			30 m.s.c.	45	Stations separated by ⩽ 300 miles
			25 m.s.c.	41	Stations separated by ⩽ 200 miles
			20 m.s.c.	36	Stations in same urban area
1922	+2.9 Note 2	+2.8 Note 2	35* m.s.c.	38*	Objective for worst cases*
			30 m.s.c.	33	Stations in same local area (10, 7 or 5 miles radius of a central point)
1933	+10.8 Note 3	+4.5 Note 3	15* dB	30*	Objective for all connections*
1947 (Figure 4.2a)	+10 Notes 3, 7	+4 Notes 4, 7	27 dB Note 6	41	Limit for fringe to fringe connections
1957	+12 Note 5, 7	+1 Notes 5, 7	27 dB	40	Limit for fringe to fringe connections
1960 (Figure 4.1b)	+12 Notes 5, 7	+1 Note 5, 7	20 dB	33	Objective for calls over new transit network via most unfavourable routing
1969 (Figure 7.4)	+11 Note 7	0 Note 7	20 dB	31	Objective for calls over most unfavourable routings

* Proved unrealistic and was not achieved at that time.

Note 1 Originally the standard cable was identical in characteristics to that used for similar purposes in the United States of America by the American Telephone and Telegraph Company. The primary constants were as follows:

Loop resistance per mile of cable, 88 Ω per mile
Capacitance per mile of cable, 54 nF per mile

At a later date, when actual rather than artificial cable was used by the National Telephone Company (1905) and the Post Office from 1912, the primary constants also included:

Inductance per mile of cable, 1 mH
Leakance per mile of cable, 1 μS

(Naturally, leakance is a somewhat variable quantity; 5 μS was quoted by Cohen in 1916).

When the use of 'miles of standard (or test) cable' was superseded by that of transmission units in dB or Np, during the late 1920s the following relationship was established:

1 m.s.c. (British, with inductance and leakance) = 0.92209 dB
0.10616 Np

1 m.s.c. (USA, zero inductance and zero leakance) = 0.94670 dB
0.10899 Np

These relationships apply for the frequency $5000/2\pi$ Hz (approx. 800 Hz) which may be taken to correspond, for loudness, to the equivalent frequency of speech signals from and heard via the telephone sets current from 1904 to 1930.

Note 2 These values of reference equivalent are quoted from Research Branch Handbook of Technical Information for Communications Engineers (1941) with a correction of +4 dB added to the sending reference equivalents because they are quoted therein in terms of the 'P.O. Reference System' which was then thought to be identical to SFERT but which was later found to differ because account had not been taken of certain changes that had been made in the real SFERT in Paris. The most serious change consisted of fitting a rather large heating device round the capacitor microphone to obviate trouble from condensation. This change was never properly reported in published descriptions of SFERT.

These are 'candlestick' telephones and the values of SRE quoted apply for a talking distance o ¼ in (6 mm).

Note 3 The derivation of these reference equivalents is the same as indicated in Note 2. Various slightly different values are on record, for example Stratton and Luxton (1933) and CCIF Book, Budapest (1934) which quote +7.9 dB for sending (+11.9 dB after correction) and +5.0 dB for receiving.

Note 4 The values quoted were decided upon as most representative in 1957 on the occasion of a review of available information on the subject. Values of +6.2 (+10.2 after correction, see Note 1) for sending and +4.5 dB for receiving are given in the Research Branch Handbook (1941) and one determination in 1954 (believed to be reliable) by the CCITT Laboratory yielded +10.0 and +4.1 dB respectively for Sending and Receiving (Doc 51, Period 1952–55, also quoted in CCITT, 12th CE, Contribution No. 41, Period 1957–60).

Note 5 The reference equivalents quoted above for the 13–2P–27 set on 660 Ω 6½ lb cable and 50 V 200+200 Ω feeding bridge resulted from the review, in 1957, of available evidence (which was then somewhat conflicting); the values chosen agree reasonably well with a determination at the CCITT Laboratory in 1954 (which also contained the 13–1L–18.35A LLTC referred to above in Note 3) yielding +12.6 dB for Sending and +1.0 dB for Receiving. Subsequent evidence supports the present choice of +12 and +1 dB.

Note 6 The change from 15 dB to 27 did not involve any change in actual installation practices and so there was no worsening of performance to the customers. The change was to render the value realistic by taking into proper account switching and mismatch losses, departures of losses of line plant from their nominal values and certain concessions which had been found necessary since 1933. The value can be compared realistically with values quoted for later dates because it is now the practice to recognise the presence of the transmission features mentioned.

Note 7 The method used from 1947 for comparing local telephone circuits and establishing limiting subscribers' lines of various gauges (including mixtures) involved a series of calculations taking all the fundamental transmission factors into account including the effects of sidetone on sending and on receiving in the presence of some room noise. In 1949 a simplified version was developed which permitted Transmission Equivalent Resistances to be determined very simply. Transmission Equivalent Resistance is the loop resistance of a subscriber's line of 6½ lb/mile cable (0.5 mm copper conductors) that gives the same transmission performance (for sending in most cases) as the line in question. With the most recent types of telephone set it has been found possible to simplify the procedure still further by relying only on the sum of the image attenuations at 1600 Hz of the individual homogeneous sections of the line; this value must not exceed 10 dB for the 706 telephone set. Naturally, any such method will result in slight anomalies when the corresponding values of reference equivalent are compared but these are unlikely to exceed a few dB in any practical cases.

retained as the length of standard cable to be inserted between the new local telephone circuits (having zero lines). This standard was also considered to be an objective, it being realised at the time that it would be very difficult to achieve generally, and especially on the more unfavourable connections, until certain advances in amplifying techniques came to maturity. This transmission standard was interpreted as to apply for the most unfavourable routings and so the limits set for individual trunk circuits were functions of their length measured from the Zone Centre; these limits took into account the types of line plant then practicable for routes of different lengths. In the event, it proved difficult to comply with all the requirements, particularly so far as interconnection of all pairs of Zone Centres was concerned, and the cable equivalents were often greater than intended because circuits had to be used in tandem.

Amplification of long distance circuits had become general by 1933 and it was believed that substantially zero loss circuits would become quite practicable. A new type of telephone set was also being introduced which was appreciably better in performance than the 'candlestick' variety which embodied the 'solid back' microphone and 'Bell' earphone which had remained standard for CB and Automatic exchanges* since about 1905. Accordingly the transmission standard was revised to that shown for 1933 in Table 7.1. It will be noticed that the standard cable has been replaced by a non-reactive (distortionless) attenuator; this was more appropriate to the use of amplified circuits on lightly loaded audio cables. The original 22 V repeating coil feeding bridge was retained and the 300 Ω resistors representing the subscribers' lines were restored.

The 1933 Transmission Standard was intended to represent the worst transmission performance likely to be experienced even over the most unfavourable routings and the attenuator setting chosen (15 dB) to mark 'standard transmission' corresponded to quite a high grade of performance†. Unfortunately, as had happened in the past, the objective was not achieved generally. The reasons for this failure were mainly the treating of the 15 dB loss as all available for the sum of the image attenuations of the lines only, no allowances being made for switching and mismatch losses; it was also found impossible to work Zone to Zone circuits at 0 dB loss with 2 wire switching; also tandem connected Zone to Zone circuits could not be eliminated. A later (about 1947) assessment of the state of the junction and trunk plant revealed the performance of connections over the most unfavourable routings to be equivalent to about a 27 dB setting of the attenuator; this includes reasonable allowances for switching and mismatch losses and for the effects of attenuation/frequency distortion in unloaded cable circuits.

* Until the introduction of the new types in about 1933, Magneto and CBS exchanges used a different type of telephone set.

† No more than about 5% or users would be likely to regard such connections as 'Poor or Bad' and only about 20% would experience any difficulty in conversing.

As explained, until 1933, the transmission standard was treated as an overall representation of performance; after this date, however, the local telephone circuits were treated as independent representations of the limit permissible in the local exchange area. Where the local telephone circuit differed, in respect of the telephone set, the gauge of conductor and/or the feeding bridge, from the representation in the transmission standard, a set of local line allowances was established on the basis of loudness balancing against the transmission standard. Sending and receiving allowances were necessary and the limit was set by whichever was the worse.

Until the 1930s, transmission loss constituted practically the only feature that could, or indeed required to, be controlled in transmission planning. Noise was kept as small as possible and attenuation/frequency distortion was dominated by that present in the microphones and earphones then available. For these reasons the criterion of received impression of loudness was quite sufficient for comparing telephone connections with the representation of the transmission standard. When the modern variety of handset telephones appeared with moulded plastic handset, microphone with immersed electrodes and much better earphone, these had better frequency responses and gave greater suppression of sidetone and so it became necessary to develop assessment methods to enable proper account to be taken of these features in planning. At the same time long distance plant was posing its own problems of stability, echo and circuit noise and these also needed to be properly assessed for planning purposes. The first step in modernising the assessment methods was taken in 1947 when the criterion for equality of transmission performance was changed from the previous one of loudness to one based on articulation ratings with certain allowances, derived from conversation tests, for features like speech sidetone which cannot be assessed by articulation testing because a fixed talking level is used in such tests.

The new assessment method was first termed 'effective transmission' but that term was subsequently reserved for assessments made by the repetition rate method; the term 'transmission performance rating' was therefore used for the new method adopted in 1947.

At first the change in criterion affected mainly the planning of local exchange areas and to facilitate its application the representation of the local telephone circuit in the transmission standard was made more realistic. Not only was the new telephone set incorporated but also the subscriber's line was represented by artificial cable instead of by a pure resistor; a 50 V Automatic exchange feeding bridge was also used. The new arrangement is shown in Table 7.1. It will be noticed that the values of SRE and RRE of the transmission standard were practically unchanged by the alterations of subscriber's line and feeding bridge. In planning local exchange areas, the new assessment method yielded limiting loop resistances for subscribers' lines that, for any given feeding bridge, were greater for lighter gauge cables than for those of heavier gauges; previously the loudness criterion had indicated limiting loop resistances practically independent of conductor gauge. The increased use of lighter gauge conductors was therefore encouraged although account had also to be taken of signalling resistance.

The new assessment criterion of 1947 did not appreciably affect practices in the longer distance parts of the system, except to substitute 1600 Hz as the equivalent single frequency to determine the speech transmission losses of unloaded junction cables instead of 800 Hz which had been customary with the loudness criterion.

Under the influence of the more appropriate assessment technique, the use of improved telephone sets was encouraged and by 1957 the 13-1L-18.35A set introduced in 1933 had not only become obsolete but had been overtaken numerically in service by the anti-sidetone set 13-2P-27 (first introduced in 1949) which included also a better earphone. The transmission limit for this set was greater, in terms of subscriber's line length or resistance than the 1933 set and the most common conductor gauge had become $6\frac{1}{2}$ lb/mile. It was therefore decided in 1957 to change the representation of the local telephone circuit in the transmission standard; this local telephone circuit also represented the transmission limit for local planning. The representation adopted in 1957 was, according to the transmission performance rating criterion equal in sending rating to the 1933 representation of the local telephone circuit and about 3 dB better in receiving rating, the overall performance of the whole transmission standard being thereby improved by 3 dB no change being made in the value of the attenuator connecting the two local telephone circuits together. It will be seen in Table 7.1 that the sending reference equivalent (loudness rating) for the 1957 limiting local telephone circuit is 2 dB greater (which conflicts with the transmission performance rating for sending) but the receiving reference equivalent is 3 dB less (which happens to agree with the transmission performance rating for receiving); such discrepancies are, of course, inevitable when assessment criteria are changed but are likely to become negligible in magnitude as telephone sets and line plant approach stability in evolution of their characteristics.

From about 1952, further improvements in assessment techniques, particularly conversational opinion methods and application of statistics to the estimation of customer satisfaction, were developed and these enabled realistic studies to be made of the effects of introducing 4 wire switching and other features into the longer distance parts of the telephone network. These studies resulted in 1960 in the formulation of a plan for a new transit network for the inland trunk network of the United Kingdom. It was found possible to upgrade the transmission standard by 7 dB as is shown in Table 7.1.

The modern methods described in Chapter 8 yield estimates of the degrees of satisfaction likely to be achieved not only for the most unfavourable routings but also several other classes of much more commonly occurring connections. It is therefore no longer appropriate to refer to the connection that represents only the most unfavourable routings as the 'transmission standard'; many will believe 'standard transmission' to be all that is really necessary and any improvement beyond this to be sheer luxury; this is certainly not true because customers will judge mainly by their general experience which must therefore be made considerably superior to that which is just marginal. It is therefore proposed to term the connection that represents the 'just marginal' performance appropriate to the most unfavourable routings the 'representative limiting connection'; this term will be used henceforth in the present chapter and in Chapter 8.

To bring this story up to date it is necessary only to record that in 1969 it was decided to revise the representation of the limiting local telephone circuit by substituting the now commonest type of telephone set in use in the United Kingdom in association with the limiting subscriber's line for which it is suitable. Table 7.1 gives the particulars. It is worth mention again that the modern criteria of assessment do, under certain circumstances, result in local telephone circuits being considered as equivalent although their sending, and receiving, reference equivalents may differ slightly; for this reason, it is intended to retain values of reference equivalent for typical limiting local telephone circuits in the United Kingdom as +12 dB for sending and +1 dB for receiving in spite of the slightly lower values applicable for the 1969 representation shown in Table 7.1. In the United Kingdom values of reference equivalent are used mainly for certain international planning purposes; naturally, even in international planning, the need is commonly found for more general assessment criteria and the reference equivalents then have to be supplemented by suitable additional information.

7.2 THE NETWORK PLANNING PROCESS

Telephone networks can hardly ever be designed from first principles; their present states are the result of a process of evolution in which planning provides a guiding hand rather than imposes a grand design. Two features are conspicuous; the repetitive nature of the constituents of the network; and the intractability of an established network to change. Planning must seek to introduce new, and eventually more economic, technologies without excessive short term cost and without any degradation of the transmission quality of the connections. New demands from customers must also be provided for in the most economic manner.

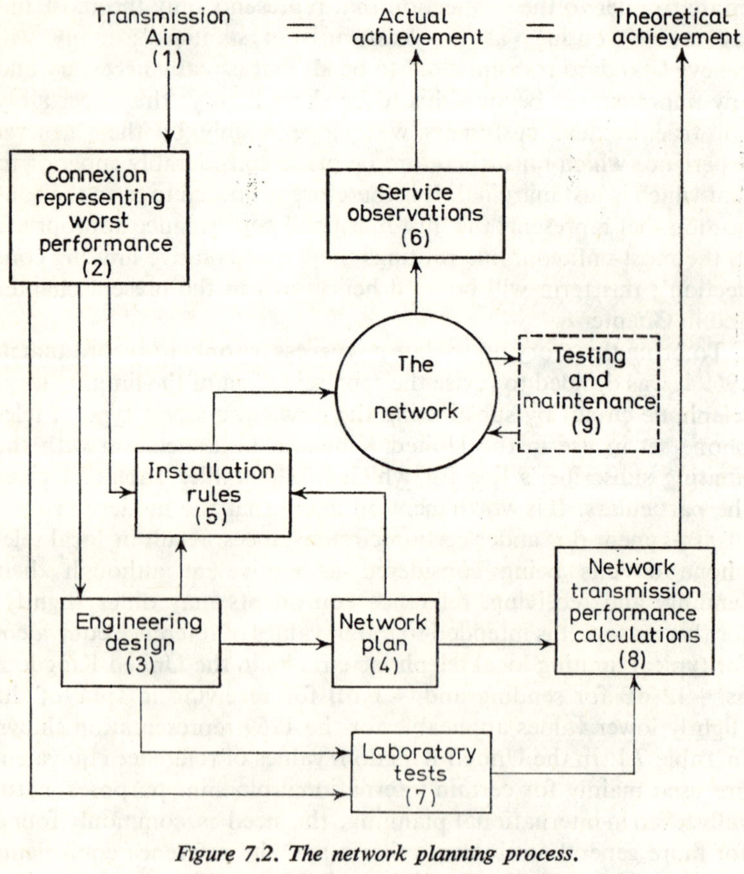

Figure 7.2. The network planning process.
Note: Numbers correspond to list in text.

When making far reaching changes in an established network to admit, for example, exploitation of entirely new engineering designs, the problems posed require for the solutions, application of a process of successive approximation and the interdependencies can be illustrated in the form of a flow diagram such as is shown in Figure 7.2.

Logically, the process begins at 1 by deciding on the *transmission aim*, which must be defined in terms related to the purpose for which the telephone network is being provided. For the present argument, the transmission aim may, initially, be expressed in rather vague terms such as: 'To ensure that almost all possible connections will be reasonably satisfactory for conversing under conditions likely to be present during telephone calls'.

Historically, however, network plans were being used long before any well defined transmission aim (in the above sense) was recognised. In the early stages of development of telephony, planning was correspondingly primitive and it was sufficient to consider only a set of 'limits' applying first to loss only and, later, also to circuit noise and attenuation/frequency distortion. This procedure was suitable so long as loss remained, for economic reasons, overwhelmingly the most serious degrading factor; the remaining degradations could be treated as of lesser importance in the planning procedure and could be considered on rather arbitrary bases, taking account of what was technically feasible. At that stage, making telephone calls was recognised by the users as being a rather difficult business and so no very high standard of satisfaction (in the modern sense) was expected. When it became eonomically possible to make substantial reductions in overall losses of telephone connections, an intermediate stage in the development of methods of planning telephone networks was reached and the various degradations were treated as separate statistical variables and what might be termed 'planning objectives' were established. These objectives were formulated in terms of the proportion of connections that suffered overall losses greater than x dB; and the proportion that produced noise levels at the listener greater than y dBmp. This form of statement does not constitute a transmission aim in the current sense because it is not related to the purpose for which the network is expected to be used. This method is now inappropriate because the variety of degrading factors is much wider and, because the effects of loss are now much less, the effects of all the other factors to be considered are much more nearly of comparable importance. In the present method of planning, illustrated in Figure 7.2, the effects of all the degrading factors must be treated together on a common basis related directly to the fitness for purpose of the complete telephone network; clearly this requires proper understanding of the manner in which the effects of the different factors interact with each other.

The network planning process, under present conditions, will now be explained in terms of Figure 7.2 by starting with 4 the *network plan.* A network plan consists of the description of a suitable arrangement of the constituent elements which have undergone basic

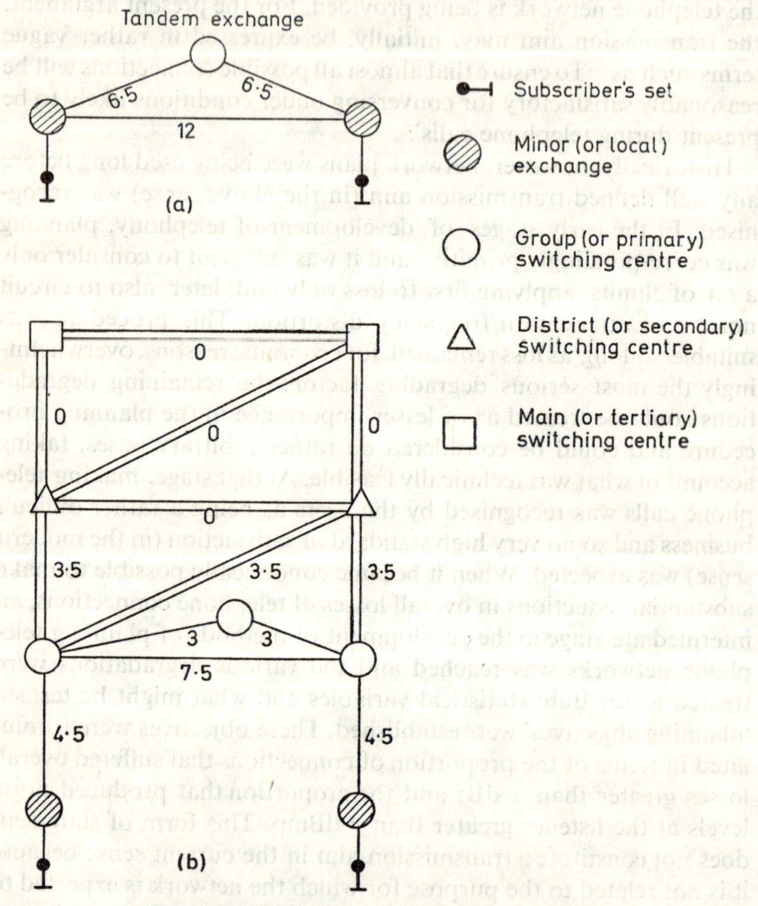

Figure 7.3. Switching, transmission and routing plan (1960) for United Kingdom. (a) Local traffic, (b) Longer distance traffic.

Notes: 1. Subscriber's lines and junctions are 2 wire (except where p.c.m. junction systems are used). 2. Trunk circuits are (with few exceptions) 4 wire. 3. Switching at minor exchanges and group switching centres is 2 wire. 4. Switching at district and main switching centres is 4 wire. 5. Where economically feasible, smaller losses are used

engineering design. The network plan must be chosen to be consistent with the transmission aim; although an ultimate aim of perfect satisfaction under all conditions would be desirable, this can be approached only gradually and so the aim practicable at a given time must be limited by the economic possibilities of the contemporary technologies; this is one reason why successive approximation is necessary in the planning process.

Figure 7.3 shows a switching, transmission and routing plan such as is applicable in the United Kingdom; this is an important part of the complete planning intentions for the network which must include also particulars of allowable circuit noise and distortions as well as all specifiable transmission characteristics, of the telephone sets and subscribers' lines. Much of the information required is to be found in the performance specifications of the individual engineering items.

The network plan is implemented and plant is put into service by use of suitably conceived *installation rules* 5. These rules and the associated procedures consist of engineering instructions designed to ensure that the equipment installed will provide the distribution intended by the network plan in the most economic manner. To a large extent these installation rules are based on a *representative limiting connection* 2 which consists of a pair of limiting local telephone circuits* connected together through an attenuator. This attenuator represents the nominal maximum total loss of the intervening junctions and trunk circuits plus any switching losses and any systematic departures from the nominal losses together with suitable allowances for reflection losses at impedance mismatches.

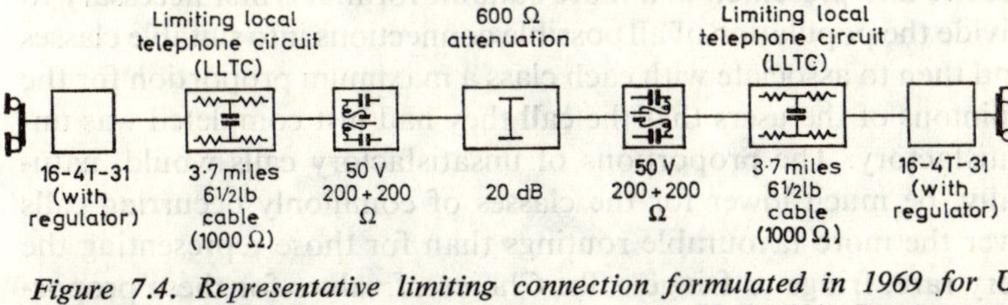

Figure 7.4. Representative limiting connection formulated in 1969 for UK telephone network. Send RE *of above* LTCs *= +11 dB (+12), Receive* RE *of above* LTCs *= 0 dB (+1). Therefore nominal overall* RE *of above* RLC *= 31 dB (33). For certain planning purposes, the values shown in brackets are used to allow a margin for tolerances*

* The present practice is to plan local networks independently of their mode of access to the trunk network; other practices may apply elsewhere and then the representative limiting connection may not be divisible into limiting local telephone circuits.

Figure 7.4 shows the representative limiting connection (RLC) used for this purpose at present in the United Kingdom; it represents the total transmission loss of the most unfavourable physical connection possible in the network according to the network plan (Figure 7.3). The items that are assembled to form the representative limiting connection must be of practical telephone equipment properly representative of their counterparts in the real network. The basic *engineering design* 3 must aim at suitability at least for incorporation as a replacement for part of, or an addition to, the representative limiting connection. More favourable representative connections than such 'fringe to fringe' ones are also needed and can be derived from the representative limiting connection in Figure 7.4 by modifiying the values of one or more of its components, such as the loss of the attenuator and the lengths of the subscribers' lines. These representative connections have several other uses in the planning process shown in Figure 7.2; they must be properly related to the transmission aim 1 and they are required for use in laboratory tests 7.

When the installation rules and procedures have been properly devised and they have been diligently applied, the plant in service should comply with the network plan. *Service observations* 6 can then be made to assess the degree to which the actual network achieves its purpose as a working system; the results which yield a measure of *actual achievement* can be compared with the transmission aim 1 thus completing one loop of the diagram in Fig 7.2.

Clearly the transmission aim must be expressed in terms that can be compared with the results of feasible service observations. The vague form of statement given earlier must therefore be made more precise and presented in a more suitable form. It is first necessary to divide the population of all possible connections into suitable classes and then to associate with each class a maximum proportion for the opinions of the users that the call they had just completed was unsatisfactory. The proportions of unsatisfactory calls would, naturally, be much lower for the classes of commonly occurring calls over the more favourable routings than for those representing the very rare, fringe to fringe calls. Choice of values for these proportions must be guided by the need to achieve a proper economic balance between the desire to provide the best possible transmission service and the willingness of customers to pay for a very high standard. The actual values will also depend on the definition of 'unsatisfactory'; for the present purpose, an opinion will be considered as contributing to the proportion of 'unsatisfactory' calls when the user would, after he has completed the call, reply 'yes' to the question:

'Did you experience any difficulty in conversing with your partner?'. According to this criterion and assuming an average level of room noise at each end the proportion of fringe to fringe calls likely to cause difficulty in the present United Kingdom network is about 30–40% depending on the level of circuit noise present but it must be noted that such a class of connections will carry only a few calls per week.

The foregoing description has covered the loop in Figure 7.2 that includes 1, 2, 3, 4, 5 and 6. The effects of engineering developments and of changes in the network plan take a very long time to affect the results of any service observations and, by the time they have become manifest, it may be too late to restore the desired performance by any practicable change in network plan. It is therefore necessary to examine in advance the likely consequences of implementing any proposed network plan and this can be done by performing *network transmission performance calculations* 8 which yield measures of theoretical achievement in the same terms as those used for expressing the transmission aim with which they can then be compared. These network transmission performance calculations require basic information, derived mainly from *laboratory tests* 7, which show the relationships between the various fundamental factors like overall loss, circuit noise reaching the listeners, amount of distortion of various types, propagation time, losses of sidetone and echo paths, etc. The results of these laboratory tests must be properly related to the characteristics of actual engineering plant that has been or might be, developed; the results of engineering design 3 must be expressed quantitatively in terms compatible with definitions of variables used in the laboratory tests. It is therefore usually desirable to include the representative limiting connection 2, or some similar representative connection derived from it, in at least some of the laboratory tests; this will help to ensure that the information assembled has been properly checked against at least some of the clauses in which the transmission aim 1 is expressed.

The loop in Figure 7.2 embracing 1, 2, 3, 4, 7 and 8 therefore provides the means for examining the relative merits of different proposed network plans or of including in one plan different realisations of transmission equipment.

The eight aspects of the network planning process shown in Figure 7.2 can now be summarised as follows:

1. Deciding on the transmission aim, e.g. 'Almost all possible connections should be reasonably satisfactory for conversing

by speech under conditions likely to be present during telephone calls'.

2. Translating 1 into terms suitable for engineering design; this is usually accomplished by setting up a transmission standard or representative limiting connection.

3. Basic engineering design of the constituent elements.

4. Deciding on the distribution of elements 3 that will satisfy the representative limiting connection (or transmission standard) 2; this consists of choosing a switching and transmission plan.

5. Devising installation procedures to ensure that the equipment installed provides the distribution decided by 4 in the most economic manner; this is accomplished by means of engineering instructions, etc.

6. Devising and applying methods of assessing the degree of satisfaction with which the network performs as a working system; this is to ascertain whether 1 and 2 are being achieved; this can be done by transmission surveys, special observations, noting complaints, etc.

7. Conducting laboratory and other tests to check that the engineering design 3 will satisfy at least the requirements defined by the representative limiting connection 2 and to provide data on which 8 can be based.

8. Estimating the degree of satisfaction likely to be achieved by implementing the network plan 4; this part of the process yields a theoretical estimate of achievement of the transmission aim 1 by making use of a suitable set of hypothetical reference connections.

Although not part of the planning process, *testing and maintenance* are important short term means of ensuring that the condition of the network remains as intended by the installation rules; this feature of the operation of a telephone network is indicated in Figure 7.2 by Item 9. The measurements used in such testing are very simple and are intended merely to check losses, noise levels and other physical particulars of individual transmission elements, unlike the service observations which are highly sophisticated and are applied to the users rather than the connections they use.

A general explanation of the philosophy of the network planning process has been given above and some notes follow dealing with certain practical aspects of each of the Items 1–8 listed above; these are necessarily very brief and, for further information, reference should be made to the sources indicated in Sections 7.2 and 7.3 of the Bibliography.

7.3 NOTES ON SPECIFIC ASPECTS OF THE NETWORK PLANNING PROCESS

Item 1 Transmission aim. Practical statement of transmission aim in terms suitable for the process illustrated in Figure 7.2 requires first subdividing the possible connections into suitable classes and setting a maximum proportion for the opinions of calls over connections within each class that are unfavourable. The proportions must naturally, be lower for the commonly occurring calls over the more favourable routings than for the rare, fringe to fringe connections. Customers' expectations must also be taken into account; those connections that are easily set up will be expected to have very good transmission performance. If the customer finds difficulty in getting connected to the number he is calling, he may not reject the connection when it has a poor transmission efficiency although he is not satisfied; as connections become easier to obtain, customers may well reject more readily those found unsatisfactory for transmission.

At present (1970) the United Kingdom Post Office is studying the consequences of setting the transmission aim in terms of maximum percentage of customers expected to experience difficulty in conversing over connections established for certain specific classes of traffic defined according to its routing. For the most unfavourable, fringe to fringe routings the maximum percentage envisaged is about 25%; for commonly occurring calls, the percentage will be about 5%.

In principle, the transmission aim, defined, for example, in the above terms, ought to take into account the following three classes of factors which affect the degree of satisfaction of telephone connections to the users: *X* factors, which are under the control of the telephone operating organisation. These are the physical particulars of the telephone equipment, like loss, circuit noise, etc. which are the responsibility of the telephone network planning authority.

Y factors, which are those physical features of the telephone environment that are outside the planning control of the telephone operating organisation; these include the level and other particulars of any room noise present at the telephone locations and the manner in which the users hold their handsets.

Z factors, which are physiological and psychological and concerned with the tolerances of the telephone users. Vocal level, threshold of audibility and telephonic ability are the most important examples of *Z* factors.

Most of the forms of degradation treated in Chapter 4 are examples of *X* factors and no further comment on them is needed here. The telephone network planner must take into account a reasonable range of *Y* and *Z* factors although he cannot control their intensity. Surveys of the levels and character of room noise found at telephone locations have been made (see references in Section 4.2.1 of the Bibliography) and for most planning purposes it is reasonable to assume the presence of room noise having the so-called 'Hoth' spectrum at a level of 50 dBA; sufficient margin should be available, however, to ensure that excessive difficulty is not experienced in the presence of room noise at 60 dBA. Variations in users' methods of holding telephone handsets and other behaviour are catered for by basing all the fundamental planning data on the results of conversation tests in which the subjects behave quite freely under conditions simulating those of actual telephone calls. Such conversation tests will also properly take into account the *Z* factors. Under certain circumstances non-conversation tests may have to be used and then special allowances may have to be made for appropriate dispersions attributable to certain *Z* factors. Conversation tests conducted as explained in Section 3.6 include appropriate attention to simulating the usual manner of setting up the connection and giving the participants suitable typical tasks to perform over the test connections.

Item 2 Representative Limiting Connection. Unlike reference systems, RLCS must be composed of practical telephone equipment suitably representing corresponding items in the real network. Consequently, the RLC must periodically be revised as telephone sets become obsolete and no longer representative of the current practice; revision is also necessary as the transmission aim is improved. Section 8.1 illustrates the extent of this improvement by considering hypothetical reference connections having the same performance as the RLC currently used at various dates from 1905.

In the future, it is to be hoped that instrumental measurements of the characteristics of telephone sets will have advanced to the stage that RLCs will no longer be required for defining the transmission aim; some such assembly of practical equipment will, however, still need to be included in certain laboratory experiments to ensure that realistic assessments are being made.

Telephone connections representative of transmission conditions less unfavourable than those of fringe to fringe connections can be derived from the RLC by suitable modifications of the subscribers' lines and the value of the attenuator connecting the LTCS together.

In the past, much transmission performance information has been presented in terms of ratings relative to the 'transmission standard'. This term is synonymous with, in present parlance, 'representative limiting connection' (RLC); use of the word 'standard' in this sense is now considered undesirable because it has led to misunderstandings on account of common usage in other contexts to denote some specially high grade item used in standardising laboratories.

Table 7.2 has been prepared to assist in interpreting the older published information; it shows assessment scores, according to

Table 7.2. ESTIMATES OF ASSESSMENT SCORES FOR SOME REPRESENTATIVE LIMITING CONNECTIONS (RLCS)

Representative Limiting Connection	Nom ORE dB	Level *of* RN dBA	MOS Y_C *(OS9)*	% *F+P+B*	% *P+B*	% *B*	% *Not usable*	% *Difficult*
1947 13-1L-18.35A 2.56 miles 10 lb 27 dB (Figure 4.2a)	about 41	50	1.66	86	44	8.8	—	68
		60†	1.11	95	72	28	54*	90
1957 13-2P-27 2.44 miles 6½ lb 27 dB	40	50	1.91	78	30	4.3	—	55
		60†	1.40	91	58	16	—	80
1960 13-2P-27 2.44 miles 6½ lb 20 dB (Figure 4.2b)	33	50	2.41	54	9.6	0.6	—	28
		60†	1.99	75	26	3.2	—	50
1969 16-4T-31 (with regulator) 3.7 miles 6½ lb 20 dB (Figure 7.4)	31	50	2.53	48	6.5	0.3	—	23
		60†	2.14	68	19	1.8	—	42

* From a test conducted in 1950 using the scale 'usable/not usable'

† Except for the 'not usable' criterion, the results at 60 dBA room noise have been estimated from information at 50 dBA level of room noise by adjustment for the effect of the increase in level of room noise.

present assessment criteria in use, for certain representative limiting connections (transmission standards) referred to elsewhere. Naturally the RLC, in itself, does not define the transmission performance assessment of a connection because the room noise present at the two terminals must also be stated; accordingly information relating to use in 50 and 60 dBA room noise is given. In some literature, use of the RLC in 60 dBA room noise has been treated as a 'transmission performance datum' and this has been used for rating the performances of other connections relative to it.

Item 3 Basic engineering design. Engineering design of telephone sets is discussed in some detail in Chapter 5 and references are made in Chapter 6 to the engineering design of line and other transmission equipment. In all cases, the item under consideration must be capable of satisfactory operation when associated with the appropriate parts of the RLC. This is necessary but not always sufficient to ensure satisfactory performance under all possible conditions of use and so other representative connections usually have to be considered as well. The results of such studies are embodied into the performance specification of each item of engineering equipment. Although basic engineering design is a separate process from network planning, it must proceed closely in step with it. When assessing the suitability of any proposed item that has been designed, not only must a prototype be constructed and tested but also attention must be given to likely variations in performance of manufactured items in actual service from that obtained with the prototype.

Item 4 Network plan. Until the present time, loss has been the dominant form of degradation experienced by telephone users and so network plans are commonly described in terms of a 'routing, switching and transmission' diagram such as that shown in Figure 7.3. The other degradations, such as circuit noise, attenuation/frequency distortion, nonlinear distortion, propagation time, sidetone, echo and crosstalk, must, however, also be controlled in planning; overall limits must be set for these and appropriate allocations made for each degradation to every element present in any connection. Such distributions are very difficult to decide because account must be taken of how the transmission items are likely to be combined in various hypothetical reference connections and the economic consequences of making particular demands in the various transmission elements. Appropriate hypothetical reference connections must therefore be employed for such studies and reliable estimates must be made of the degree of satisfaction to be expected; this is the function of Item 8 (Network Transmission Performance Calculations) in the planning process.

Item 5 Installation rules. In the United Kingdom, local telephone networks are planned independently of the junctions and other plant through which customers have access to long distance circuits. Planning of local telephone networks can therefore be based on a section of the complete representative limiting connection, namely the *limiting local telephone circuit*. The permitted combinations of different kinds of local cable, etc. used in subscribers' lines, in association with the different types of subscribers' sets and exchange feeding bridges, can be embodied in simple planning rules which take into account the sending, receiving and sidetone performances of the local telephone circuits. The form of such rules has changed with circumstances and some examples are given in Section 6.1. At present, the transmission criterion is based on the sum of the image attenuation values of the sections of the subscriber's line, sidetone being controlled separately by including suitable clauses in the specifications of the subscribers' sets. Allowances are necessary for any PBXS and extension plans included.

The rules for installing junction and trunk circuits often distinguish between those used for terminating traffic and those for transit calls. The values of nominal loss, shown in the transmission plan (e.g. Figure 7.3), are supplemented by suitable limits for circuit noise, attenuation/frequency distortion, etc., In the case of multiplex transmission systems, loading is particularly important and the performance may depend on signals present in channels in addition to that under consideration. Further information is given in Chapter 6 which also indicates sources of other relevant information.

Item 6 Service observations. At one time service observations were very highly regarded and it was considered that much of the transmission planning data could be derived from such measurements; it is now generally admitted that service observing is a rather crude measuring instrument which yields results averaged over many of the variables that must be treated separately in planning. Service observing does, however, provide a measure of the overall adequacy of the telephone network as a working system. Naturally, observations that can be related to individual calls are much more informative than more generalised polling but they are more costly and very time consuming. Marketing surveys are related to, but different from, service observations; they are conducted to explore the prospects for some facility that may not exist and which the telephone customers questioned cannot have experienced as a service.

Service observations must, for the present purpose, yield results in the same terms as are used to express the transmission aim. An example of observations that have been used successfully is the method in which individual customers are called back, on a random sample basis, after they have completed an identifiable, paid for, call; if the questions asked are suitably framed, the results can be compared with estimates based on appropriate forms of laboratory assessment.

Item 7 Laboratory tests. The various methods used for conducting laboratory tests are described in Chapter 3 and Chapter 4 is devoted to assembling the required fundamental planning data. Conversational tests are usually needed and the results derived from them can be compatible with the forms used to express the transmission aim, i.e. in terms of proportions of unfavourable opinions expressed by participants after conversations over the telephone connections. Representative limiting connections, reference telephone connections and reference devices to simulate certain types of distortion are required: models of engineering equipment may also be included in certain experiments.

In principle, the laboratory tests should include all possible combinations of the physical features of the connections that can be set up according to the plan, associated with the appropriate distributions of the room noise conditions likely to be present during real calls and with proper account taken of the whole distribution of telephonic ability of the actual customers. This cannot be achieved by direct inclusion of every possible combination of all these factors and so the amount of actual conversation testing must be strictly limited and interpolations, made by suitable simplified methods, used to fill in the remainder indirectly.

Laboratory tests are used mainly to build up fundamental information relating assessment scores in the required units to certain quantities that can be used to define the physical characteristics of speech links in general; occasionally, it becomes necessary to conduct special tests for a particular study. The quantities concerned include overall (mouth to ear) losses, expressed as loudness ratings, circuit noise reaching the listener, expressed in terms of the level of circuit noise referred to the input to a receive end of defined sensitivity, and distortions defined in suitable ways.

At present loudness ratings require, for their determination, certain laboratory tests using reference systems such as ARAEN and NOSFER; it is expected that these measurements using subjects will be replaced by suitable instrumental determinations and calculations which are now being defined.

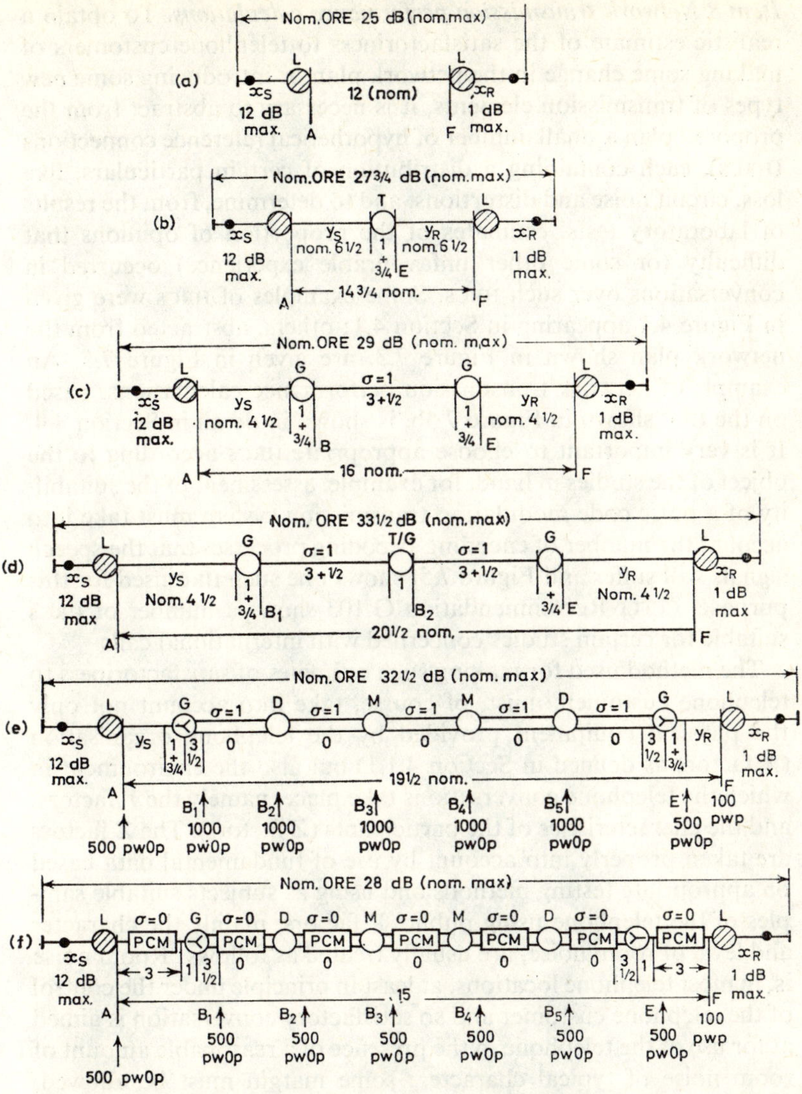

Figure 7.5. Some hypothetical reference connections based on the UK plan (Figure 7.3.). Diagrams apply for transmission left to right.

Notes: 1. Appropriate levels of circuit noise must be assumed to originate at points A, B, E and F. 2. Appropriate allowances must be made for attenuation/frequency distortion. 3. In e, y_S and y_R are nominally $4\frac{1}{2}$ dB but may, for certain purposes, be taken as having mean 2.9 dB and s.d. 1.0 dB corresponding to 95% within $4\frac{1}{2}$ dB. 4. Noise sources shown in f represent background noise assumed power additive. Quantising distortion is allowed for separately

Item 8 Network transmission performance calculations. To obtain a realistic estimate of the satisfactoriness to telephone customers of making some change in the network plan or introducing some new types of transmission elements, it is necessary to abstract from the proposed plan a small number of hypothetical reference connections (HRCS), each containing a distribution of certain particulars, like loss, circuit noise and distortions, and to determine, from the results of laboratory tests, estimates of the proportion of opinions that difficulty (or some other unfavourable experience) occurred in conversations over such HRCS. Some examples of HRCS were given in Figure 4.1 appearing in Section 4.1; others, abstracted from the network plan shown in Figure 7.3, are given in Figure 7.5. An example of network transmission performance calculations, based on the HRC shown in Figure 7.5b, is shown in detail in Section 8.4. It is very important to choose appropriate HRCS according to the object of the studies in hand, for example, assessment of the suitability of a pulse code modulation transmission system must take into account the number of encoding/decoding processes that the speech signals will suffer and Figure 7.5f shows one such HRC used for this purpose. CCITT Recommendation G.103 shows a number of HRCS suitable for certain studies concerned with international calls.

The method used for making such estimates of satisfactoriness to telephone customers must, of course, take into account not only the physical equipment provided by the telephone organisation (*X* factors as defined in Section 4.1.1) but also the environment in which the telephone conversations take place, namely the *Y* factors, and the characteristics of the participants (*Z* factors). The *Z* factors are taken properly into account by use of fundamental data based on appropriate testing methods and using as subjects suitable samples of the telephone using public. *Y* factors, mainly the character and level of room noise, are usually treated as follows: Room noise is, in most telephone locations, at least in principle under the control of the telephone customer and so satisfactory conversation is aimed at for use of the telephone in the presence of a reasonable amount of room noise of typical character; some margin must be allowed, however, so that conversations can still be carried out when the room noise is somewhat worse than average, for example in a call office where the user cannot control the environment.

8. TRANSMISSION PERFORMANCE OF TELEPHONE NETWORKS

8.1 SOME ESTIMATES OF THE TELEPHONIC PERFORMANCE OF THE TELEPHONE NETWORK OF THE UNITED KINGDOM SINCE 1905

As explained in Section 7.2, the telephonic transmission performance of a telephone network may be expressed in terms of the proportion of calls that give satisfaction, or are expected to give satisfaction, to customers. Section 8.4 of the present chapter is devoted to the description of a method for estimating the percentage of customers likely to be unsatisfied if a certain transmission plan were adopted, a certain distribution of plant were installed and traffic were distributed between different classes of connection in a certain manner. Some indications of the statistical distributions of transmission losses in actual networks are given in Section 8.2 and of traffic incidence in Section 8.3. The present section provides an introduction to these topics by illustrating the evolution of transmission performance of the United Kingdom telephone network in terms of some estimates of the telephonic communication efficiency of the contemporary plant at various times since 1905*.

Table 7.1 in Section 7.1 shows the currently declared 'standards of transmission' at various periods and this information, together with known characteristics of the contemporary telephone sets and line plant, enable the physical transmission characteristics of representative (although hypothetical) connections at various times to be chosen. It must be remembered that these 'transmission standards' were intended to define the performance objective for 'fringe to fringe' connections† and not for the commonly experienced

* This date has been chosen because a complete assessment of the condition of the telephone plant of the United Kingdom was made from this date in preparation for the taking over by the Post Office of practically all the local networks in 1912. (*see* Section 1.1).

† Fringe to fringe connections are those between customers both of whom are on limiting subscribers' lines and are being connected together through the most unfavourable combination of junctions and trunks.

Table 8.1. SOME ESTIMATES OF THE TRANSMISSION PERFORMANCE OF HYPOTHETICAL FRINGE TO FRINGE CONNECTIONS AT VARIOUS PERIODS (1905–70)

Period	*Hypothetical Reference Connection*			*Nom.* ORE dB*	*Estimated Transmission Performance* §		
	Local telephone circuits (LTCS)	*Junctions (each end)*	*Trunk circuits*		*Listening* MOS	*Conversation* MOS	*Percentage Difficult*
1905–1914	CB1-1A-3 on 450 Ω 10 lb cable SRE = +10.5 dB RRE = +7.5 dB	Unloaded cable; $4\frac{1}{2}$ dB at 800 Hz	Overhead open wire equiv. to 36 m.s.c. (33 dB RE)	60	1.00	1.11	>85
1915–1921	Ditto	Ditto	Open wire equiv. to 25 m.s.c. (23 dB RE)	50	1.24	1.34	>80
1922–1932	Ditto	Ditto	4 amplif. tk. ccts‡ (13 dB at 800 Hz) +5 dB	45	1.42	1.51	75
1933–1946	13-1L-18.35A on 450 Ω 10 lb cable SRE = +10 dB RRE = +4 dB	Ditto	Ditto	41	1.92 (1.45)	1.93 (1.54)	>50 (>70)
1947–1956	Ditto	Unloaded cable; $4\frac{1}{2}$ dB at 1600 Hz	Same but carrier circuits ‡	41†	1.99 (1.53)	1.98 (1.62)	>50 (70)

33

1957–1968 (non-transit)	13-2P-27 on 660 Ω $6\frac{1}{2}$ lb SRE = +12 dB RRE +1 dB	Ditto	Ditto	40†	2.09 (1.62)	2.06 (1.69)	47 (65)
1960 Transit network§§ plan	16–4T–31 (with regulator) on 1000 Ω $6\frac{1}{2}$ lb cable SRE = +11 dB RRE = 0 dB	Unloaded cable; $4\frac{1}{2}$ dB at 1600 Hz $+\frac{3}{4}$ dB	9 dB††	$30\frac{1}{2}$†	3.22 (2.72)	2.90 (2.54)	10.0 (22)
1969–***	Ditto	$4\frac{1}{2}$ dB** $+\frac{3}{4}$ dB‡‡	10 dB	$31\frac{1}{2}$	3.02 (2.57)	2.76 (2.43)	14 (28)

* Nominal ORE is equal to the sum of the following:

SRE of the LTC at one end
RRE of the LTC at the other end
Loss of unloaded cable junctions at 800 Hz
Loss of trunk circuits at 800 Hz

The true ORE may differ somewhat from the nominal value (see Section 6.4).

† In these cases, the current rule in the UK of using the loss at 1600 Hz (instead of 800) has been adopted for calculating the reference equivalents of unloaded junctions.

‡ Four amplified trunk circuits, or four early type carrier circuits, in tandem were assumed to have transmission losses as follows:

200 Hz	300	800	1000	1600	2500	3000
39 dB	29.2	13.0	11.4	12.2	18.5	23.3

These losses include those introduced by switching equipment. The shape of this loss/frequency characteristic is taken from curve ii in Figure 4.17 (*see* Section 4.2.1.2).

§ The estimates of transmission performance shown without brackets are based on the assumptions that the connections are free from circuit noise and that the telephone locations are quiet. The values in brackets include an allowance corresponding approximately to typical values of circuit and room noise. (15 dB masking of the listener's threshold of audibility was taken; this is slightly less onerous than the effects of 50 dBA room noise and −65 dBm circuit noise). The estimates are also based on assessment scores by telephone users conditioned to normal telephone experience in about 1957; customers conditioned to worse performance (i.e. in earlier periods) might well have given better scores; conversely, present day customers show evidence of yielding relatively worse scores.

** It is assumed that a substantial proportion of junctions in loaded cable will be included in these connections; loss independent of frequency is assumed.

†† This reduced value of loss is attributable largely to the introduction of 4 wire switching; at the same time it has been assumed here that the attenuation/frequency distortions present with earlier transmission and switching equipment are now absent and that the full frequency range of the telephone sets can be transmitted.

‡‡A mismatch loss of $\frac{3}{4}$ dB has been assumed between the image impedance of the junction and 600 Ω port of the trunk network.

§§ Fig. 7.5e refers (*see* Section 7.3) but SRE and RRE are taken as for 16-4T-31 (with regulator) sets.

*** Fig. 7.5d refers (*see* Section 7.3) but SRE and RRE are taken as for 16-4T-31 (with regulator) sets.

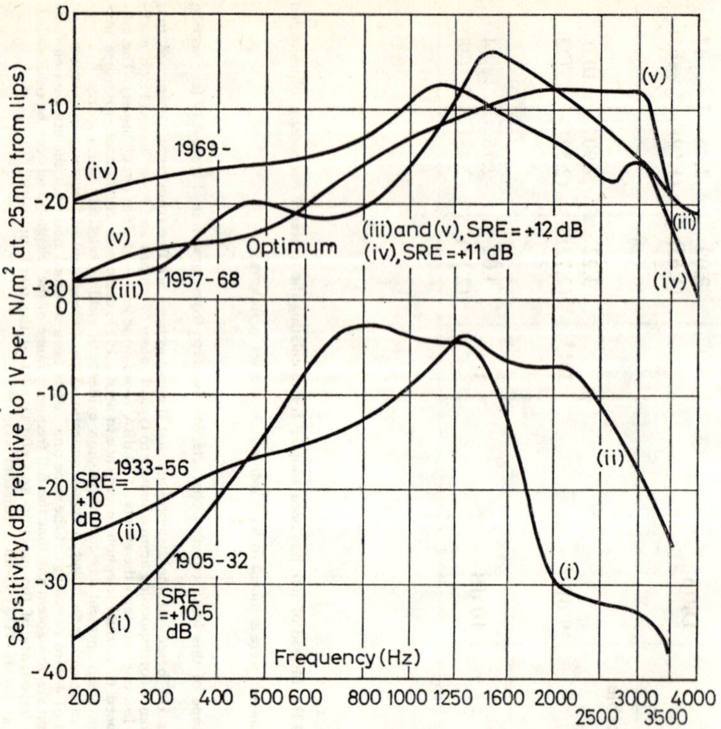

Figure 8.1. Sending sensitivities of limiting local telephone circuits

classes of connections; nevertheless, for the present illustrations, the performances of such 'maximally adverse' connections are quite informative.

Figures 8.1 and 8.2 show sensitivity/frequency response curves for the various limiting local telephone circuits considered and Table 8.1 shows the particulars of hypothetical reference connections chosen to represent the class of fringe to fringe connections during each of the periods chosen*. It should be noted that the local telephone circuits chosen are typical, and representative, of the limiting local transmission conditions and not necessarily identical to those forming parts of the transmission standards in force at the time; the sending and receiving reference equivalents of these LTCS are, however, approximately equal to those of the transmission standards. Suitably representative contemporary junctions

* The starting date of each period coincides with the promulgation of some new transmission plan or some change in transmission standard.

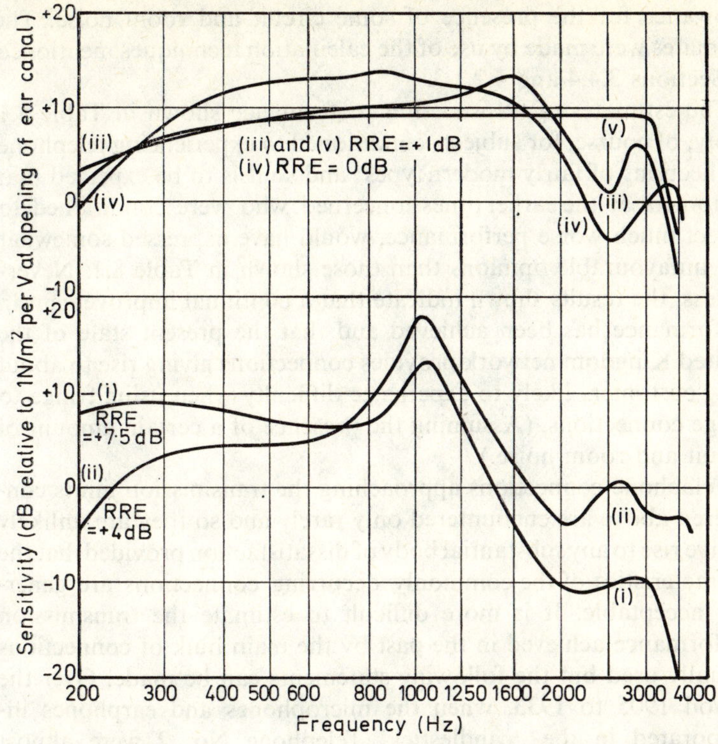

Figure 8.2. Receiving sensitivities of limiting local telephone circuits

and trunk circuits have also been chosen, as shown in the table. In some cases, the combined losses of the trunks and junctions in Table 8.1 are somewhat greater than the loss allocated to these items in the respective 'transmission standard' or 'representative limiting connection' defined in Figure 7.1 and Table 7.1 (*see* Section 7.1). This is because these transmission standards were set up as design objectives, often not achievable until certain anticipated technological advances had borne fruit and become established practices. Sometimes the advances in technology did not materialise and the transmission objective of the period then proved over optimistic even for fringe to fringe connections.

The foregoing information has been used to estimate the transmission performance obtained from each HRC and the results are shown in Table 8.1; in each case, the estimate applies for favourable noise conditions (no circuit noise and no room noise) but the later examples include also (values in brackets) estimates that include allowance for the presence of some circuit and room noise. The estimates were made by use of the calculation techniques mentioned in Sections 3.4.4 and 4.2.1.3.

The estimates of transmission performance shown in Table 8.1 apply, of course, for subjects conditioned to experiencing telephone connections of fairly modern types* and so it is to be expected that customers at the earlier times concerned, who were conditioned to expect much worse performance, would have expressed somewhat less unfavourable opinions than those shown in Table 8.1. Nevertheless, the results shown indicate that a continual improvement in performance has been achieved and that the present state of the United Kingdom network provides connections giving rise to about 25% customers likely to experience difficulty when using fringe to fringe connections. (Assuming the presence of a certain amount of circuit and room noise.)

Telephone connections approaching the transmission limits considered above are encountered only rarely and so they are unlikely to give rise to any substantial body of dissatisfaction provided that the characteristics of the commonly occurring connections are generally acceptable. It is more difficult to estimate the transmission performance achieved in the past by the main bulk of connections actually used but the following statements can be made. Over the period 1905 to 1933, when the microphones and earphones incorporated in the 'candlestick' Telephone No. 2 were almost universal, the transmission losses of the most common classes of call were being gradually reduced but, even at the end of this period, it is unlikely that many users felt that they were being provided with completely adequate levels of speech, even accepting the distortions. Probably never more than about 80% of participants completed calls without experiencing some difficulty. After 1933, the handset telephones provided a somewhat broader frequency response enabling some benefit to be obtained from speech components at frequencies up to 3000 Hz.† At the same time, transmission losses were reduced under the more commonly occurring circumstances so that most conversations could then take place with reasonably adequate listening levels, although the attenuation/fre-

* The experimental data on which these estimates are based was obtained in about 1957.

† This extension in bandwidth of telephone sets was accompanied by successive changes in loading procedure in audio cable pairs and in the design of f.d.m. carrier systems to permit frequencies up to 2400 Hz (1930), 2600 Hz (1934) and eventually 3400 Hz (1938) to be transmitted.

quency distortion remained fairly severe by modern standards. Generally fewer than 10% of users would have experienced any difficulty except under unfavourable circumstances, e.g. when room noise was loud.

From 1947, when the first equalised earphones (Receiver inset No. 2P) came into common use in the United Kingdom, the attenuation/frequency distortion of the telephone sets in use began to be reduced from its previous severity and sidetone performance was much improved.* With the advent in 1956 of even better equalised earphones (Receiver inset No. 4T) and of equalised carbon granule microphones in 1960 (Transmitter inset No. 16), modern telephone sets (such as the Post Office telephone No. 706) can be regarded as having fair approximations to optimum frequency responses although some appreciable nonlinear distortion remains from the action of the carbon granule microphone.

As the distortions in subscribers' telephone sets have been gradually reduced, the proportion of users able to converse without difficulty has increased. Under the most commonly occurring conditions the proportion who may, according to estimates based on present plans, be expected to experience any transmission difficulty amounts to, perhaps, 5% or less; under relatively unfavourable routing conditions which occur only rarely, the proportion may reach about 25%. This potential degree of excellence may, however, not always be realised in practice because a great deal of plant remains in service which was installed to inferior standards of transmission performance than those now in force. A tendency towards erosion of the present potential performance must also be resisted; this arises partly from the introduction of new transmission systems which, broadly, are beneficial both economically and so far as their transmission performance is concerned but may introduce degradations not previously experienced by customers.

* Even before the 1930s, it had been recognised that transmission factors in addition to overall loss, which was assessed on a loudness basis, were of importance; it is, however, comparatively recently that a wide range of factors could be taken into account in planning. The first steps were taken in 1947 when the previous reliance upon loudness of received speech as sole assessment criterion was discontinued; the new method was first termed 'effective transmission' but later called 'transmission performance rating' and was devised to take into account features of telephone sets and local lines previously ignored. For example, the effects of sidetone were properly assessed and the equivalent losses of unloaded cables were taken as at 1600 Hz instead of at 800 Hz which was appropriate for the older type telephone sets assessed on the old loudness basis Modern methods of assessment are described in Chapter 3.

Examples of such transmission systems are satellite communication systems, which increase transmission delay, and PCM systems, which introduce quantising distortion and other nonlinear effects. There is also a tendency towards increasing the number of transmission links in telephone connections and, when this increases the number of pairs of modems or codecs in the connection, the total attenuation/frequency distortion introduced by the channel filters may become a very serious consideration indeed especially under the low loss conditions likely to prevail in the future.

8.2 DISTRIBUTION OF TRANSMISSION LOSSES

Examination of a switching, transmission and routing plan, such as Figure 7.3 in Section 7.2, will enable the total value of the nominal maximum losses of the lines to be determined for any given routing between local exchanges. To obtain the nominal maximum value of overall reference equivalent for a routing, it is necessary to add also the following:

1. nominal maximum values of sending and receiving reference equivalent of the local telephone circuits (customer up to local exchange),
2. appropriate allowances for the transmission losses introduced at each switching point (say, 1 dB) (except at the local exchanges whose losses are already included in 1),
3. suitable values of mismatch loss to allow for impedance mismatch between local telephone circuit and junction and between junction and the trunk network (a compromise value of $\frac{3}{4}$ dB has been used),
4. allowances for any systematic departure of line losses from their nominal values. In the past, it has sometimes been found that the planned values of line loss could not be adhered to; under conditions of 2 wire switching of amplified circuits, nominally 0 dB lines have been found in practice to average about $1\frac{1}{2}$ dB.

The hypothetical reference connections shown in Figure 7.5 show examples derived in this manner from Figure 7.3.

It must, however, be remembered that such nominal maximum values of total transmission loss of telephone connections will be experienced only rather infrequently over each routing. For any given routing, the nominal overall reference equivalent will be considerably dispersed and the average value may be, perhaps, 8–10 dB less than the nominal maximum. Of course, in extremely rare cases even the nominal maximum value will be exceeded but this will be so infrequent that it will have negligible effect on customers' opinions.

The dispersion of transmission losses can be obtained by combining distributions for the losses in the various parts of the connections under consideration; it is convenient to treat separately the local telephone circuits (sending and receiving), junctions and trunk circuits.

Local telephone circuits. Table 8.2 shows the results of two surveys of subscribers' line plant. The first was conducted in 1953 when sectors of 62 automatic exchange areas* were studied, totalling about 65 000 subscribers' lines; the areas were classified as follows and the percentages of then existing (March 1953) exchange connections in automatic areas are shown:

Table 8.2. DISTRIBUTION OF TRANSMISSION EQUIVALENT RESISTANCE (TER) OF SUBSCRIBERS' LINES

TER class interval Ω	*Category of exchange area; 1953 survey* A	B	C	D	E	*Weighted UK mean (lines)* %	*Category of exchange area; 1964/5 survey* F	G	*Weighted UK mean*†
(1)	(2)	(3)	(4)	(5)	(6)	(7)	(8)	(9)	(10)
0–100	28	12	29	15	33	19.6	9.2	16.2	14.3
100–200	50	32	35	20	24	27.4	30.1	27.2	28.0
200–300	18	35	27	20	13	22.9	25.1	18.9	20.7
300–400	4	18	7	15	9	13.2	18.8	15.7	16.5
400–500		3	2	19	9	10.2	8.5	11.0	10.0
500–600				10	8	5.5	6.0	6.4	6.2
600–700				1	3	1.0	1.5	2.9	2.8
700–800					1	0.2	0.7	1.2	1.0
800–900							0.1	0.4	0.4
900–1000								0.1	0.1

† This mean is weighted in accordance with the incidence of line plant and cannot be derived directly from *F* and *G*.

* At the date automatic exchange connections (lines) represented about 70% of all exchange connections in the United Kingdom; now there are very few non-automatic areas remaining.

A. Inner London (7.6%); exchange areas in the City and Centre Telephone Areas of the London Telecommunication Region (LTR).

B. Outer London (26.2%); remainder of the automatic exchange areas of the LTR, except Unit Automatic Exchanges (UAXS).

C. Inner Provincial (6.7%); exchanges mostly in director areas, but some in non-director areas, near the centres of heavy industrial and commercial locations (e.g. Birmingham, Manchester, Liverpool, Glasgow, Leeds, Newcastle, central exchanges).

D. Outer Provincial (40.7%); remainder of automatic exchanges in director, non-director and satellite areas outside London and excluding UAXS.

E. Unit Automatic Exchanges (18.8%); all types and sizes of UAXS, which serve exchange areas of up to about 2000 lines, mostly in country locations.

In Table 8.2, col. 7 shows the mean value averaged over Classes A to E with weighting according to the percentages shown above; this can be regarded as the national average for the United Kingdom at that date.

The second survey was conducted in 1964/65 when the following populations in the United Kingdom were sampled in representative manners:

F. London Telecommunication Region telephone exchange areas.

G. Provincial telephone exchange areas.

The results for these populations are included in Table 8.2 together with the results for a sample drawn in a representative manner from the whole of the United Kingdom; this cannot be derived directly from the results for the LTR and Provincial samples.

The information presented in Table 8.2 is in the form of relative frequencies of occurrence within class intervals of 100 Ω transmission equivalent resistance (TER). TER is the loop resistance of a subscriber's line wholly of $6\frac{1}{2}$ lb/mile copper conductor which could replace the actual line (commonly of mixed gauges) without changing the transmission performance; this is explained in Section 6.1.1 (*see* Table 6.6). Additional information was collected during these surveys, relating to d.c. resistance and line length. Mean values for

Table 8.3. SUMMARY OF MEAN VALUES OF TRANSMISSION EQUIVALENT RESISTANCE (TER), DC RESISTANCE AND ROUTE LENGTH OF VARIOUS CLASSES OF SUBSCRIBERS' LINES IN THE UNITED KINGDOM

Item	*1953 Survey*						*1964/65 Survey*		
	Class A	*B*	*C*	*D*	*E*	*UK average*	*F*	*G*	*UK average*
(1) Mean TER Ω	148	218	168	287	228	235	265	268	268
(2) Mean d.c. resistance Ω	153	207	153	265	206	222	264	266	266
(3) Average TER-factor	0.97	1.06	1.10	1.08	1.11	1.06	1.00	1.01	1.01
(4) Average length per 100 ohms DC Resis. (miles)	0.34	0.42	0.46	0.44	0.47	0.42	0.37	0.38	0.38
(5) Approx. length of average line (miles)	0.52	0.87	0.70	1.16	0.97	0.93	0.98 (0.68)	1.01 (0.81)	1.01 (0.74)

Note 1. Classes of automatic telephone local area are as for Table 8.2.
2. Values in brackets are radial distances of subscribers' stations from their local exchanges.

each class are shown in Table 8.3; row 1 gives the mean values of TER obtained from the distributions in Table 8.2 and row 2 shows the mean values of d.c. resistance. It is instructive to assemble the ratios of TER/d.c. resistance which can be treated as averaged values of TER multiplying factor and these are shown in row 3. Where the average TER factor is less than 1, the average conductor gauge is lighter than 6½ lb/mile and this is seen to apply for inner London areas (Class A). Relatively heavy gauge conductors tend to be used in Outer Provincial and UAX areas (Classes C and E) and this is shown up by the relatively high average TER factors. The average conductor gauge has also become lighter for the country as a whole between 1953 and 1964/65. The TER multiplying factors shown in Table 6.6 for the different gauges of conductor are approximately linearly related to the ratio line length/100 Ω d.c. resistance, the slope being about 1.1 when the length is expressed in miles; this fact enables the average line length to be estimated from the mean d.c.

resistance (row 2) and the average TER factor (row 3); the results are shown in row 5, with row 4 giving the estimated average length per 100 Ω d.c. resistance using the above relationship in the form of the following expression:

$$\text{Length (miles)/100 }\Omega\text{ d.c. resistance} = 0.37 + \frac{(\text{TER-factor}) - 1}{1.1}$$

The estimates given in row 5 are in fairly good agreement with the mean values obtained direct from the samples. Values of radial distance of the subscribers' stations from their exchange were also available from the second survey and these averages are included in Table 8.3.

In 1953, most of the local line plant in automatic areas had been installed to a transmission limit of 660 Ω TER but, by 1964/65 a large proportion of the plant had been exposed to a transmission limit of 1000 Ω TER made possible by the introduction of the 16-4T-31 (with regulator) subscriber's set and the use of lighter gauge conductors (4 and $2\frac{1}{2}$ lb) had become extensive. Nevertheless, the distributions shown in Table 8.2 show remarkably little change when col. 7 is compared with col. 10.

Transmission performance studies usually require distributions of subscribers' lines to be associated with particular types of subscribers' sets and the information to be presented in the form of distributions of transmission loss attributable to the local telephone circuits. It is, at present, convenient to quantify losses in local telephone circuits as sending and receiving reference equivalents (SRE and RRE); the distributions with TER as variate can be converted to those with SRE and RRE as variate by use of relationships such as are shown in Figure 6.4 in Section 6.1.3. (The graphs applicable to $6\frac{1}{2}$ lb/mile conductor should be used and the abscissa scale in miles converted to TER by the relationship 270 Ω/mile.)

The results for the 13-2P-27 subscriber's set, which is unregulated, can be represented by the following linear equations:

$$\text{SRE} = 3.54 + 0.01232\ (\text{TER})$$

$$\text{RRE} = 3.00 + 0.00606\ (\text{TER})$$

Results for regulated telephone sets usually require nonlinear equations and the extra complication of curve fitting would rarely be justified; values should be read directly from the graphs.

The following approximate numerical procedure can be used to determine probability density histograms with SRE and RRE as variates; Table 8.4 illustrates the procedure for SRE with the 13-2P-27 subscriber's set.

Table 8.4. DETERMINATION OF PROBABILITY DENSITY HISTOGRAM WITH SENDING REFERENCE EQUIVALENT (SRE) AS VARIATE FOR THE 13-2P-27 SUBSCRIBERS' SET

TER			*Corresponding limits of* SRE	*Subdivision of* SRE	*Corresponding subdiv. of freq.*	SRE	
Class Mark	*Class Interval*	*Frequency %*				*Class Interval*	*Frequency %*
(1)	(2)	(3)	(4)	(5)	(6)	(7)	(8)
50	0–100	19.6	3.5–4.8	3.5–4.0	7.5	3–4	7.5
				4.0–4.8	12.1	4–5	16.3
150	100–200	27.4	4.8–6.1	4.8–5.0	4.2		
				5.0–6.0	21.1	5–6	21.1
				6.0–6.1	2.1	6–7	18.0
250	200–300	22.9	6.1–7.4	6.1–7.0	15.9		
				7.0–7.4	7.0	7–8	13.1
350	300–400	13.2	7.4–8.7	7.4–8.0	6.1		
				8.0–8.7	7.1	8–9	9.5
450	400–500	10.2	8.7–10.0	8.7–9.0	2.4		
				9.0–10.0	7.8	9–10	7.8
550	500–600	5.5	10.0–11.2	10.0–11.0	4.6	10–11	4.6
				11.0–11.2	0.9	11–12	1.5
650	600–700	1.0	11.2–12.5	11.2–12.0	0.6		
				12.0–12.5	0.4	12–13	0.5
750	700–800	0.2	12.5–13.8	12.5–13.0	0.1		
				13.0–13.8	0.1	13–14	0.1
	Sum	100.0	—	—	—	—	100.0

Cols 1, 2 and 3 show class marks, class intervals and probability density (frequency) from cols 1 and 7 of Table 8.2. Col. 4 shows the values of SRE corresponding to the limits of TER in col. 2, using the equation above; col. 5 shows the subdivision of the ranges of SRE so that limits at 3, 4, 5, etc. dB can be formed and col. 6 shows the proportionate subdivision of the frequencies given in col. 3. Col. 7 shows the desired class intervals of SRE and the corresponding frequencies in col. 8 are formed by suitable combinations of frequencies in col. 6.

A similar table must be constructed to arrange the frequencies in suitable intervals of RRE, but this is not shown here. Table 8.5

Table 8.5. PROBABILITY DENSITY HISTOGRAMS OF SRE AND RRE FOR 13-2P-27 AND 16-4T-31 (WITH REGULATOR) SUBSCRIBERS' SETS

Class mark dB	*Class Interval* dB	*13-2P-27* SRE %	RRE %	*16-4T-31 (with reg)* SRE %	RRE %
(1)	(2)	(3)	(4)	(5)	(6)
−5.5	−6 to −5				9.5
−4.5	−5 to −4				81.1
−3.5	−4 to −3				5.8
−2.5	−3 to −2		37.9		2.4
−1.5	−2 to −1		36.4		1.0
−0.5	−1 to 0		19.0		0.2
+0.5	0 to +1		6.2		
+1.5	+1 to +2		0.5		
+2.5	+2 to +3				
+3.5	+3 to +4	7.5			
+4.5	+4 to +5	16.3			
+5.5	+5 to +6	21.1			
+6.5	+6 to +7	18.0		13.7	
+7.5	+7 to +8	13.1		78.7	
+8.5	+8 to +9	9.5		4.7	
+9.5	+9 to +10	7.8		1.8	
+10.5	+10 to +11	4.6		0.9	
+11.5	+11 to +12	1.5		0.2	
+12.5	+12 to +13	0.5			
+13.5	+13 to +14	0.1			
	Sum, %	100.0	100.0	100.0	100.0
	Mean, dB	6.6	−1.6	6.5	−4.5
	Standard deviation, dB	2.02 (2.00)	0.93 (0.88)	0.61 (0.53)	0.61 (0.53)

Note. Standard deviations shown in brackets include application of Sheppard's correction.

summarises the results for SRE and RRE for 13-2P-27 and 16-4T-31 (with regulator); col. 10 of Table 8.2 has been used for the 16-4T-31 (with regulator) subscriber's set, instead of col. 7, because it is more appropriate for the more modern type of set.

Junctions. Three types of junctions must be distinguished, namely as follows:

Local junctions, which are used to interconnect local exchanges directly and, in the United Kingdom, these are allowed a maximum nominal value of loss of 12 dB (*see* Figure 7.3 in Section 7.2). The nominal loss of a junction is taken to be equal to the image attenuation at 1600 Hz if unloaded, the image attenuation at 800 or 1000 Hz if unamplified but loaded or at the line up frequency (800 or 1000 Hz) if amplified (see Section 6.4).

Tandem junctions, which connect local exchanges to tandem exchanges through which the local exchanges have access to each other. These are allowed maximum losses of 6.5 dB (see Figure 7.3). Tandem exchanges are found in the United Kingdom only in London.

Trunk junctions, which provide the means of access for local exchanges to the trunk network. (There will be no trunk junction needed when a local exchange is co-sited with the group switching centre (GSC)). Trunk junctions are, in the United Kingdom, allowed a maximum loss of 4.5 dB although, until recently, a special concession was given in London permitting losses up to 6.5 dB.

Distributions of nominal losses for these three types of junction in London are shown in Table 8.6; the table also shows some distributions for trunk junctions outside London.

Most of the junction plant included in the foregoing statistics was unamplified (the survey was made in 1967) but, in the future PCM junction systems will be extensively used and, because amplified junctions are lined up to 3 dB, this may be expected progressively to reduce mean losses. It is worth remarking that the information given in Table 8.6 reveals a considerable proportion of junctions with losses greater than the planning maximum; this is partly accounted for by plant remaining unchanged after former planning rules had been revised, e.g. before about 1950, the nominal loss of unloaded cable was calculated at 800 Hz instead of the current practice of using 1600 Hz. It should also be noted that many junction circuits are composed of mixtures of unloaded and loaded sections and this will result in mismatch losses which will render the true transmission losses somewhat greater than the nominal values which, in such cases, have been taken as the sum of the image attenuations.

Trunk circuits. Modern trunk plant is almost entirely of FDM carrier systems which have the losses of groups stabilized by pilots;

Table 8.6. PROBABILITY DENSITY HISTOGRAMS OF NOMINAL TRANSMISSION LOSSES OF JUNCTIONS (ON A BASIS OF CIRCUITS)

Nominal loss		*London (Director)*			*Birmingham Director Area*	*Stoke-on-Trent and Bradford*
Class mark dB	*Class Interval* dB	*Local junctions* %	*Tandem junctions* %	*Trunk junctions* %	*Trunk junctions* %	*Trunk junctions* %
0.5	0–1	1.5	0.8	4.5	8.0	5
1.5	1–2	7.6	6.9	17.1	23.1	24
2.5	2–3	10.4	14.5	18.9	23.4	33
3.5	3–4	10.8	18.2	12.9	22.0	25
4.5	4–5	10.9	16.2	12.1	12.1	10
5.5	5–6	10.6	17.5	13.1	5.5	3
6.5	6–7	10.3	16.3	14.9	2.2	
7.5	7–8	9.1	5.9	5.8	2.1	
8.5	8–9	8.2	2.2	0.6	1.0	
9.5	9–10	6.9	1.0	0.1	0.4	
10.5	10–11	5.4	0.1		0.2	
11.5	11–12	3.8	0.4			
12.5	12–13	2.2				
13.5	13–14	1.0				
14.5	14–15	0.6				
15.5	15–16	0.4				
16.5	16–17	0.2				
17.5	17–18	0.1				
	Sum	100.0	100.0	100.0	100.0	100
Mean loss, dB		6.15	4.64	3.93	3.03	2.70
Standard deviation of loss, dB		3.27	1.92	2.05	1.74	1.15

consequently the losses of channels depart very little from their nominal values to which they are lined up. To cater for such residual fluctuations with time and between channels on a route, it is customary to assume a standard deviation of 1 dB per circuit. In the future, digital transmission systems are likely to extend their application from junctions (as at present) to trunk circuits and this may be expected to reduce fluctuations still further, perhaps to a standard deviation of 0.5 dB per pair of codecs. Any reduction in variation of circuit losses has repercussions on the minimum loss needed to ensure stability in normal operation and so it may become possible to reduce the nominal losses below those now considered necessary.

8.3 FREQUENCIES OF CALLS OVER VARIOUS CLASSES OF CONNECTION

When estimating the merit of a particular transmission plan, it is convenient to distinguish different classes of connection in terms of their routings through the switching stages and transmission links. For any proposed new plan, the relative frequencies with which calls will be made over different classes of connection can only be estimated; such estimates are, however, helped by knowledge of statistics applicable to current or past conditions.

In March 1969, there were in the United Kingdom 7 868 000 exchange connections which served 12 805 000 subscribers' stations connected to public exchanges. The exchange connections can be subdivided as follows:

Business, exclusive lines	2 341 000
shared service	62 000
Residence, exclusive lines	3 918 000
shared service	1 380 000
Call offices	75 000
Post Office service lines	92 000

All but 158 000 exchange connections were to automatic exchanges.

During the year ending March 1969, about 8 644 million calls involving United Kingdom subscribers were made; these can be subdivided as follows:

Inland, local (same or adjacent group charging areas; excluding Enquiry calls)	7 086 million
Enquiry calls	334
trunk, subscriber dialled	810
manually connected	388
International, Continental, outward	11.966 million
inward	11.464

Intercontinental,	outward	1.229
	inward	1.200
Short range radiotelephone ship to shore,		
	outward	28 thousand
	inward	203

Periodically, an analysis has been made of the actual routings of all international calls and samples of inland calls are taken to determine the traffic patterns; some of the results (applicable at about 1969) are presented in Figure 8.3. The pattern of inland calls

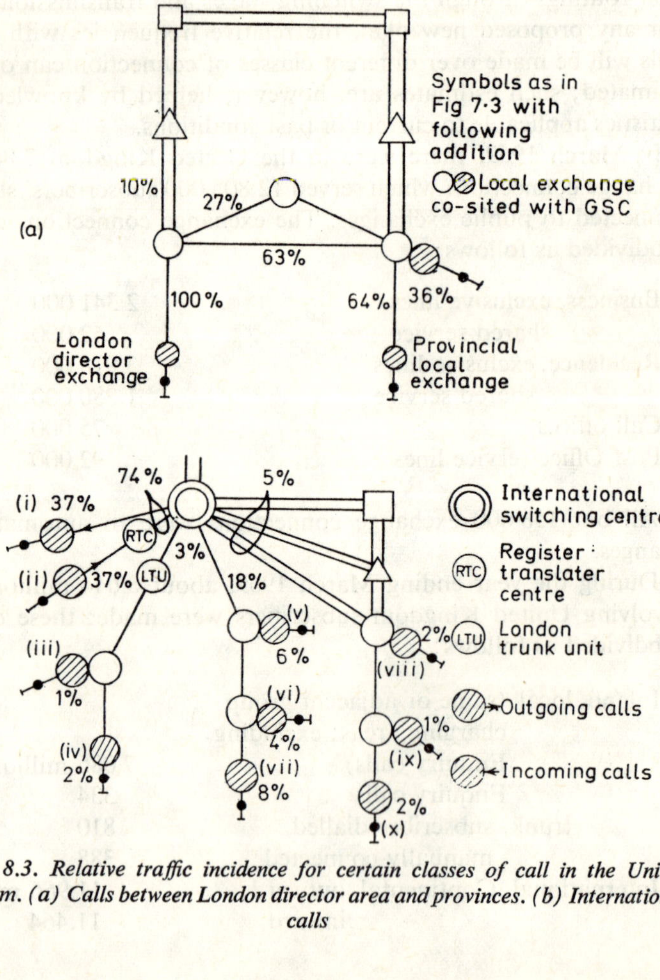

Figure 8.3. Relative traffic incidence for certain classes of call in the United Kingdom. (a) Calls between London director area and provinces. (b) International calls

is illustrated for those between the London Director Area and exchanges in the provinces; the percentage of such calls following each possible routing has been marked. This part (a) of the figure is self-explanatory. Figure 8.3b shows the pattern for international calls which pass through the international switching centre (ISC) to differently located terminal exchanges as follows:

i. incoming to local exchanges in the London Director Area.
ii. outgoing from local exchanges in the London Director Area; these are routed through an additional switching centre containing the register translaters needed for outgoing calls.
iii. to and from Group Switching Centres (GSCs) via a London Trunk Unit and thence to a local exchange co-sited with the GSC.
iv. as iii but to a local exchange via a trunk junction from the GSC.
v. to and from GSCs connected directly to the ISC and thence to a local exchange co-sited with the GSC.
vi. to and from GSCs via a tandem GSC and thence to a local exchange co-sited with the GSC.
vii. as vi but to a local exchange via a trunk junction from the GSC.
viii. through the transit network to and from a provincial switching centre having a co-sited local exchange.
ix. through the transit network to a GSC via an additional switching centre and thence to a local exchange co-sited with the GSC.
x. as ix but to a local exchange via a trunk junction from the GSC.

The routings for international calls are shown in more detail than those for inland calls because the information could be obtained more readily. At the time of the survey international traffic was handled manually and so the statistics could be assembled from examination of the call tickets whereas inland calls were largely subscriber dialled so that special steps were required to obtain records of their incidence.

Within the areas of local exchanges no information is available on the relationship between traffic density and length (or transmission loss) of subscribers' lines.

8.4 ESTIMATION OF FUTURE NETWORK TRANSMISSION PERFORMANCE

The network planning process has been described in principle in Sections 7.2 and 7.3 and the purpose of the present section is to provide a detailed example of Item 8 in that process (network transmission performance calculations). The network plan considered here is that shown in Figure 7.3 (Section 7.2) and attention will be concentrated for this example on calls within the London area via tandem exchanges; the appropriate hypothetical reference connection (HRC) is that illustrated in Figure 7.5b. The problem to be considered is to estimate the improvement in transmission performance to be achieved by extensive introduction of PCM junctions associated with time division multiplex switching at a tandem exchange. The procedure is first to estimate the original transmission performance and then to compare with this an estimate of the performance to be expected with the proposed arrangement.

Because up to date statistics of telephone plant are very often not available, it is frequently necessary to rely on somewhat historical information; for the present example the original situation will be taken as that existing in about 1955*. Table 8.2, col. 7 offers information on the distribution of subscribers' line plant and Table 8.5, cols 3 and 4 show the converted distributions of SRE and RRE for the type of telephone set in most extensive use at the base date. The distributions of transmission loss of tandem junctions is shown in Table 8.6 for a date of about 1967; at the earlier date the losses might well have been slightly smaller but this will be ignored here. An allowance of 1 dB will be made for transmission losses within the tandem exchange and a mismatch loss of $\frac{3}{4}$ dB will be allocated for impedance mismatches between junction plant of different image impedances. Table 8.7 shows a summary of the mean and standard deviations of the various transmission losses in Figure 7.5b. Each junction will be assumed to generate circuit noise at a level, referred to its input (Points A and E in the figure), of −60 dBmp (1000 pW) and it will be assumed that attenuation/frequency distortion equivalent to an 'impairment' (as defined in Section 4.2.1.2) of 2 dB is attributable to the section A to F of the connection between the local exchanges.

Table 8.7 contains also the corresponding information for the

* More recent statistics do exist to permit estimates based at about 1966 to be made but the earlier date has been chosen for this example because it enables two different types of subscriber's set to be introduced.

Table 8.7. SUMMARY OF MEANS AND STANDARD DEVIATIONS OF TRANSMISSION LOSSES ASSOCIATED WITH EACH ELEMENT OF THE CONNECTIONS

Item	*Original situation*			*Hypothetical future situation*		
	Mean dB	*Standard deviation* dB	$(s.d^2)^2$ dB	*Mean* dB	*Standard deviation* dB	$(s.d.)^2$ dB
x_S (SRE)	+6.6	2.02	4.0804	+6.5	0.61	0.3721
y_S	4.6	1.87	3.4969	3.0	0	0
Switching and mismatches	1.8	0	0	0	0	0
y_R	4.6	1.87	3.4969	3.0	0	0
x_R (RRE)	−1.6	0.93	0.8649	−4.5	0.61	0.3721
Sum	16.0	6.69	11.9391	8.0	1.22	0.7442
Standard deviation of nominal overall reference equivalent			3.46			0.86

hypothetical future situation. The distribution of subscribers' line plant is taken from col. 10 of Table 8.2; the 16-4T-31 (with regulator) type of subscriber's set has been assumed so that the distributions of SRE and RRE are from cols 5 and 6 of Table 8.5. The junction PCM transmission systems are assumed to be lined up to losses of 3 dB each with great precision so that the loss of each remains constant. Switching loss will be negligible and there will be no impedance mismatches attributable to the junctions.

Examination of the probability density histograms from which the above statistics have been drawn will show that these are far from being Gaussian and so they cannot be completely described by the mean and standard deviation; nevertheless, as will be shown now, such skewness and abnormal kurtosis as they possess individually has no appreciable effect upon the distribution of nominal overall reference equivalent which is practically Gaussian with the expected values of mean and standard deviation. Where less than four distributions have to be combined, this result may not apply, in which case proper (for example using the following method) convolution of the probability density functions is necessary.

Table 8.8 shows an approximate, numerical, procedure for combining two probability density histograms; the distributions of SRE and RRE shown in cols 3 and 4 of Table 8.5 are used in this

Table 8.8. AN APPROXIMATE NUMERICAL METHOD FOR COMBINING TWO PROBABILITY DENSITY HISTOGRAMS RELATING TO INDEPENDENT DISTRIBUTIONS

Class Interval dB	SRE f_1	RRE f_2	*Combined class intervals of the two primary distributions*, dB 0–2	1–3	2–4	3–5	4–6	5–7	6–8	7–9	8–10	9–11	10–12	11–13	12–14	13–15	Grand total
(1)	(2)	(3)	(4)	(5)	(6)	(7)	(8)	(9)	(10)	(11)	(12)	(13)	(14)	(15)	(16)	(16)	
−3 to −2		38															
−2 to −1		36															
−1 to 0		19															
0 to +1		6															
+1 to +2		1															
2 to 3																	
3 to 4	7		266	252	133	42	7										
4 to 5	16			608	576	304	96	16									
5 to 6	21				798	756	399	126	21								
6 to 7	18					684	648	342	108	18							
7 to 8	13						494	468	247	78	13						
8 to 9	9							342	324	171	54	9					
9 to 10	8								304	288	152	48	8				
10 to 11	5									190	180	95	30	5			
11 to 12	2										76	72	38	12	2		
12 to 13	1											38	36	19	6	1	
Sum	100	100	266	860	1507	1786	1644	1294	1004	745	475	262	112	36	8	1	10 000

Class Interval	0–1	1–2	2–3	3–4	4–5	5–6	6–7	7–8	8–9	9–10	10–11	11–12	12–13	13–14	14–15
Frequency per 10 000	133	563	1184	1646	1715	1469	1149	874	610	368	187	74	22	4	0

Note: Mean and standard deviations of primary distributions are: SRE; mean +6.6 dB s.d. 2.02 dB. RRE; mean −1.6 dB s.d 0.93.

example with the frequencies rounded to the nearest 1%. In the table, col. 1 contains class intervals of 1 dB extending from the lowest to the highest needed to accommodate the two primary distributions. These class intervals must be uniform and exhaustive. Cols 2 and 3 contain the frequencies within each class interval for each of the two primary distributions. Col. 4 contains the product of the top entry in col. 2 and the top entry in col. 3; col. 5 contains, first, the product of the top entry in col. 2 and the second entry in col. 3 and then the product of the second entry in col. 2 and the first entry in col. 3. The procedure continues assembling in each successive column the products of entries of which the sum of the pair of class intervals is the same. For example, in col. 6 the class intervals concerned are 3 to 4 for f_1 coupled with -1 to 0 for f_2; 4 to 5 and -2 to -1; 5 to 6 and -3 to -2; each pair of class intervals giving the same totals when the respective limits are summed, namely 2 to 4 which range is entered at the head of each column from col. 4 onwards. All the entries in each column from col. 4 onwards are summed respectively and the grand total calculated to serve as an arithmetical check. The probability density histogram representing approximately the distribution of the sums of the variates of the primary distributions is then obtained as follows:

The frequency for the class interval 0 to 1 is given by dividing the sum for the column headed 0 to 2 by two and then dividing by 100 to restore the units to percentages. The frequency for the class interval 1 to 2 is given by taking the arithmetic mean of the sums for the colums headed 0 to 2 and 1 to 3 and then dividing by 100. The results are assembled, together with the probability density histogram percentages for the primary distributions, in Table 8.9. The results have been rounded to the nearest 1%. (Where this does not result in a total of 100% some suitable adjustment should be made.) As abbreviations the symbols S and R are used to represent the two primary distributions and $S{*}R$ denotes the result of performing the convolution procedure.

Table 8.9 also contains the distribution appropriate for each tandem junction (denoted by J_S and J_R) and the result of convolution of this distribution with itself. This is denoted by $J_S^* J_R$. The two binary convolved distributions have then been convolved together to form the quarternary convolved distribution $S^* J_S^* J_R^* R$ which is shown in Table 8.9. It should be noted that the order of performing each binary convolution is not important.

The mean values of the four primary distributions total 14.2 dB and their combined variances amount to 11.94 (*see* Table 8.7)

Table 8.9. PROBABILITY DENSITY HISTOGRAMS FOR SRE, RRE AND JUNCTION TRANSMISSION LOSS AND CONVOLUTIONS OF THESE DISTRIBUTIONS

Class mark dB	*Class Interval* dB	*S*	*R*	*S*R*	J_S J_R	J_S* J_R	$S*J_S*$ J_R*R	*Gaussian approximation*
(1)	(2)	(3)	(4)	(5)	(6)	(7)	(8)	(9)
−2.5	−3 to −2		38					
−1.5	−2 to −1		36					
−0.5	−1 to 0		19					
+0.5	0 to 1		6	1	1			
1.5	1 to 2		1	6	7			
2.5	2 to 3			12	15	1		0.1
3.5	3 to 4	7		16	18	2		0.1
4.5	4 to 5	16		17	16	4	0.2	0.3
5.5	5 to 6	21		15	18	7	0.5	0.5
6.5	6 to 7	18		11	16	10	1.1	1.0
7.5	7 to 8	13		9	6	12	2.1	1.9
8.5	8 to 9	9		6	2	14	3.5	3.0
9.5	9 to 10	8		4	1	14	5.2	4.7
10.5	10 to 11	5		2		12	7.1	6.4
11.5	11 to 12	2		1		10	8.8	8.0
12.5	12 to 13	1				7	10.1	10.2
13.5	13 to 14					4	10.8	11.2
14.5	14 to 15					2	10.7	11.5
15.5	15 to 16					1	9.9	10.7
16.5	16 to 17						8.6	9.3
17.5	17 to 18						6.9	7.4
18.5	18 to 19						5.3	5.3
19.5	19 to 20						3.7	3.6
20.5	20 to 21						2.5	2.2
21.5	21 to 22						1.5	1.2
22.5	22 to 23						0.9	0.8
23.5	23 to 24						0.4	0.3
24.5	24 to 25						0.2	0.2
25.5	25 to 26						0.1	0.1
	Sum	100	100	100	100	100	100.1	100.0
	Mean	6.59	−1.55	5.15	4.58	9.00	14.14	14.2
	s.d.	2.02	0.93	2.25	1.87	2.70	3.59	3.46

corresponding to a standard deviation for the combination of 3.46 dB. Col. 9 of Table 8.9 shows the frequencies within each class interval calculated for a Gaussian distribution having mean 14.2 dB and standard deviation 3.46 dB. It will be seen that these agree within 1 % with those obtained by the convolution procedure.

The distribution produced by the foregoing convolution process (given in col. 8 or the approximation in col. 9) accounts for items x_S, y_S, y_R and x_R listed in Table 8.7; the remaining item (switching and mismatch losses) is being assumed constant and so it can be combined with the others by mere translation along the axis of loss, i.e. by increasing the values of all the class marks by 1.8 dB. The final distribution of nominal ORE therefore has mean 16.0 dB and standard deviation 3.46 dB; assuming this to be Gaussian

Table 8.10. ASSUMED PARTITIONING OF NOMINAL OVERALL REFERENCE EQUIVALENTS (Means and standard deviations of component distributions are given in Table 8.7)

Nominal ORE, dB		Relative frequency	*Assumed partitioning of x*			
Class mark x	*Class Interval* x_i *to* x_{i+1}	$\phi(x)$ %	x_S	x_R	$y_S = y_R$	*Switching and mis-matches*
$\Delta = 2$		$\bar{x} = 16.0$, $\sigma_x = 3.46$	$\Delta = 0.604$	$\Delta = 0.278$	$\Delta = 0.559$	$\Delta = 0$
(1)	(2)	(3)	(4)	(5)	(6)	(7)
4	3 to 5	0.1	+2.9	−3.3	+1.3	+1.8
6	5 to 7	0.4	+3.5	−3.0	+1.8	+1.8
8	7 to 9	1.7	+4.1	−2.7	+2.4	+1.8
10	9 to 11	5.3	+4.7	−2.4	+2.9	+1.8
12	11 to 13	11.7	+5.4	−2.2	+3.5	+1.8
14	13 to 15	19.4	+6.0	−1.9	+4.0	+1.8
16	15 to 17	22.8	+6.6	−1.6	+4.6	+1.8
18	17 to 19	19.4	+7.2	−1.3	+5.2	+1.8
20	19 to 21	11.7	+7.8	−1.0	+5.7	+1.8
22	21 to 23	5.3	+8.5	−0.8	+6.3	+1.8
24	23 to 25	1.7	+9.1	−0.5	+6.8	+1.8
26	25 to 27	0.4	+9.7	−0.2	+7.4	+1.8
28	27 to 29	0.1	+10.3	+0.1	+7.9	+1.8
Sum		100.0	—	—	—	—
Mean			+6.6	−1.6	+4.6	+1.8

yields the relative frequencies shown in Table 8.10, col. 3. In this table class marks of 4, 6, 8, etc. have been used, corresponding to class intervals of 2 dB which are sufficient in this example.

Any particular value of x (nom. ORE) shown in col. 1 of Table 8.10 can, of course be composed of an infinite variety of sets of individual values of x_S, y_S, y_R and x_R and the level of circuit noise reaching the listener from the sources A and E (Figure 7.5b) will depend upon the particular values of y_S, y_R and x_R. To simplify the calculations, a single set of values has been chosen as follows. When x is at its mean value (16.0 dB in this case) all component losses are taken at their respective mean values; for any given departure of x from its mean value, the loss of each component item has been taken to have departed from its mean value by an amount proportional to its standard deviation. In this way, large values of x are made up of correspondingly large contributions by each of its components and correspondingly for small values of x. For example, the intervals between adjacent values of x_S (in col. 4 of Table 8.10) are equal to:

$$\frac{2.02}{6.69}(2 \text{ dB}) = 0.604 \text{ dB}$$

where the denominator is the sum of the standard deviations shown in Table 8.7. The various differences between successive entries are indicated by Δ at the head of each column. The resulting values of x_S, x_R and $y_S = y_R$ are shown in cols 4, 5 and 6 of Table 8.10. Including the switching and mismatch losses (shown in col. 7), it will be noted that the resulting totals correctly add up to each respective value of x.

The chosen sets of values of y_S, y_R and x_R are then used to determine, for each value of x, a representative value of circuit noise level reaching the listener's ear; this calculation is shown in Table 8.11. To avoid having to deal with acoustical quantities, the noise level reaching the listener will be referred to the electrical input to a (hypothetical) receive end having 0 dB receive reference equivalent (RRE); such a point will be referred to by the symbol P. The various values of loss from each noise source to P are shown in cols 2 and 5; the corresponding levels of circuit noise at P attributable to each source are shown in cols 3 and 6; cols 4 and 7 show their conversions to units of power (which are additive). Col. 8 shows the total noise power accumulated at P and this is restored to noise level units in col. 9, this will be denoted by N.

Table 8.11. CALCULATION OF THE LEVEL OF TOTAL NOISE POWER REFERRED TO *P*, THE INPUT TO A RECEIVE END OF 0 dB RRE

Nominal ORE *x* dB	*Loss from A to P* dB	*Noise level at P due to source at A* dBmp	*Noise power at P due to source at A* pWp	*Loss from E to P* dB	*Noise level at P due to source at E* dBmp	*Noise power at P due to source at E* pWp	*Total noise power at P due to both sources* pWp	*Level N, fo total noise power at P due to both sources* dBmp
(1)	(2)	(3)	(4)	(5)	(6)	(7)	(8)	(9)
4	+1.1	−61.1	780	−2.0	−58.0	1580	2360	−56.3
6	+2.4	−62.4	580	−1.2	−58.8	1320	1900	−57.2
8	+3.9	−63.9	410	−0.3	−59.7	1070	1480	−58.3
10	+5.2	−65 2	300	+0.5	−60.5	890	1190	−59.3
12	+6.6	−66.6	220	+1.3	−61.3	740	960	−60.2
14	+7.9	−67.9	160	+2.1	−62.1	620	780	−61.1
16	+9.4	−69.4	120	+3.0	−63.0	500	620	−62.1
18	+10.9	−70.9	80	+3.9	−63.9	410	490	−63.1
20	+12.2	−72.2	60	+4.7	−64.7	340	400	−64.0
22	+13.6	−73.6	40	+5.5	−65.5	280	320	−65.0
24	+14.9	−74.9	30	+6.3	−66.3	230	260	−65.9
26	+16.4	−76.4	20	+7.2	−67.2	190	210	−66.8
28	+17.7	−77.7	20	+8.0	−68.0	160	180	−67.5

Notes:
Hypothetical reference connection: Figure 7.5b
Noise sources
at A 1000 pWp, i.e. −60 dBmp
at E 1000 pWp, i.e. −60 dBmp
Loss from A to P = $y_S+1.8+y_R+x_R$
Loss from E to P = y_R+x_R
The appropriate values of y_S, y_R and x_R are given in Table 8.10
The point P is hypothetical; the level at P is x_R dB lower than that at F in Figure 7.5b

The combinations of nominal ORE and level of total circuit noise power are assembled in cols 1 and 2 of Table 8.12. Col. 3 of this table shows the percentage of telephone users who might be expected to indicate, if interrogated, that they had experienced difficulty in conversing over a connection having the given combination of *x* and N. Strictly, the information entered here ought to include consideration of the distribution of room noise levels to be expected at customers' telephone locations; to avoid this complication, results have been used which are applicable for an average level (50 dBA) and typical frequency spectrum (Hoth) of room noise. Table 4.14 gives such information for values of *x* from 16 dB upwards; use has been made of some provisional estimates for lower values of *x*. Information suitable for the whole range of *x*

Table 8.12. PROBABILITY OF 'DIFFICULT' OPINIONS AS A FUNCTION OF NOMINAL OVERALL REFERENCE EQUIVALENT

Nominal ORE x dB	*Level N of total circuit noise power at P,* dBmp	*Percentage probability of 'difficult'* *Due to loss and noise,* P_{ln} %		
			Due to loss and 2 dB *distortion impairment,* P_{ld} %	*Due to loss alone,* p_l %
(1)	(2)	(3)	(4)	(5)
4	−56.3	4.1	3.9	3.3
6	−57.2	3.4	3.5	2.9
8	−58.3	3.2	3.5	2.8
10	−59.3	3.2	3.8	2.8
12	−60.2	3.5	4.1	3.0
14	−61.1	3.9	4.8	3.2
16	−62.1	4.6	5.7	4.0
18	−63.1	5.8	6.8	5.1
20	−64.0	7.2	8.4	6.2
22	−65.0	9.0	10.4	7.9
24	−65.9	11.5	13.2	10.0
26	−66.8	14.8	16.8	12.8
28	−67.5	18.7	20.9	16.4

Note:
Results in cols 3, 4 and 5 for x = 16 dB and greater are taken from Tables 4.14 and 4.15 which appear in Section 4.2.1.4. For values of x less than 16 dB use has been made of some provisional estimates based on more recent information than that incorporated in those tables.

and applicable for modern types of telephone set is at present lacking.

In certain studies, for example when different types of telephone set are being compared, it is necessary to repeat the estimates of percentage difficult with the assumption of somewhat more severe room noise conditions. Instead of assuming an average level of room noise, a level of perhaps 60 dBA might then be taken in addition; by this means a realistic assessment of the adequacy of the sidetone suppression so far as listening is concerned would be obtained. In the present example, attention is confined to estimates with 50 dBA room noise are sufficient.

In the present example, it is being assumed that the two tandem junctions and the tandem switching centre introduce some attenuation/frequency distortion amounting to the equivalent of an 'impairment' as defined in Section 4.2.1.2 of I = 2 dB. The effect on the percentage 'difficult' opinions of this 'impairment' associated with the overall loss, x, but in the absence of the circuit noise is shown in col. 4 of Table 8.12; these are taken from Table 4.15

with provisional extrapolation below $x = 16$. It is now necessary to estimate the effect of combining loss, noise and attenuation/frequency distortion; for this it is first necessary to have available estimates of the effects of loss alone and these are shown in col. 5. The procedure for estimating the combined effect of all three factors is explained in Section 4.2.4 and Table 8.13 illustrates this for the present example.

Table 8.13. ESTIMATION OF THE PROBABILITIES OF 'DIFFICULT' OPINIONS ATTRIBUTABLE TO THE EFFECTS OF NOISE AND DISTORTION SEPARATELY AND TO THE EFFECTS OF LOSS, NOISE AND DISTORTION COMBINED

Nominal ORE x dB	*Percentage probability of 'difficult'*			
	Due to loss alone, p_l %	*Due to noise alone,* p_n %	*Due to distortion alone,* p_d %	*Due to loss, noise and distortion combined,* P_{lnd} %
(1)	(2)	(3)	(4)	(5)
4	3.3	0.8	0.6	4.7
6	2.9	0.5	0.6	4.0
8	2.8	0.4	0.7	3.9
10	2.8	0.4	1.0	4.2
12	3.0	0.5	1.1	4.6
14	3.2	0.7	1.7	5.5
16	4.0	0.6	1.8	6.3
18	5.1	0.7	1.8	7.5
20	6.2	1.1	2.3	9.4
22	7.9	1.2	2.7	11.5
24	10.0	1.7	3.6	14.7
26	12.8	2.3	4.6	18.7
28	16.4	2.8	5.4	23.1

Note:
Cols 3, 4 and 5 are obtained from the information in Table 8.12 by application of the formulae given in Section 4.2.4 ($p_q = 0$)

The estimates of percentage probability of 'difficult' opinions for the combined effects of loss, noise and attenuation/frequency distortion have been transferred to col. 4 of Table 8.14 which shows the calculation of the total percentage averaged over all values of x according to the relative frequencies with which each interval of x occurs. Two features are worthy of comment: first, the greatest incidence of 'difficulty' is associated with values of x only slightly greater than the mean (16.0 dB) and the extreme values (e.g. the nominal maximum shown in Figure 7.5b as $27\frac{3}{4}$ dB) occur so

Table 8.14. CALCULATION OF THE TOTAL PROPORTION OF 'DIFFICULT' OPINIONS (ORIGINAL SITUATION)

Nominal ORE, dB		Relative frequency $\phi(x)$ %	Probability, $p(x)$, %, that opinion obtained will be 'difficult', given that nom. ORE $= x$; $p(x) = p_{Ind}$ from Table 8.13	$\phi(x).p(x)$ per 10^4
Class mark x	*Class interval* x_i *to* x_{i+1}			
(1)	(2)	(3)	(4)	(5)
4	3 to 5	0.1	4.7	0.47
6	5 to 7	0.4	4.0	1.60
8	7 to 9	1.7	3.9	6.63
10	9 to 11	5.3	4.2	22.26
12	11 to 13	11.7	4.6	53.82
14	13 to 15	19.4	5.5	106.70
16	15 to 17	22.8	6.3	143.64
18	17 to 19	19.4	7.5	145.50
20	19 to 21	11.7	9.4	109.98
22	21 to 23	5.3	11.5	60.95
24	23 to 25	1.7	14.7	24.99
26	25 to 27	0.4	18.7	7.48
28	27 to 29	0.1	23.1	2.31
	Sum	100.0	—	686.33
				6.9%

infrequently that they give rise to negligible volume of 'difficulty'; secondly, the percentage of 6.9% averaged over the whole range of x is very little greater than the value applicable at the mean value of x, namely $x = 16.0$ dB when the percentage = 6.3%. It becomes clear from this exercise that the dispersion of losses has very little influence unless the standard deviations involved are considerably greater than those of the present example.

The task now remains to perform corresponding estimations for the case assumed to apply in the future. Table 8.7 contains the assumed mean and standard deviation values and these could be used, as for Tables 8.10 and 8.14, to construct a distribution of nominal ORE by assuming this to be Gaussian. In this case, however, only two distributions require to be combined (SRE and RRE, the other items having constant values of loss) and it seems prudent to perform the convolution process as shown in Table 8.8; taking the two primary distributions from cols 5 and 6 of Table 8.5 yields the result shown in Table 8.15, cols 1, 2 and 3. The class marks are, however, not convenient integral values and so it is desirable to reclassify the relative frequencies so that the class marks are integers; the result is shown in cols 4, 5 and 6 of Table 8.15. The means and standard deviations are indicated in the table and it is useful to compare them with values to be expected from consideration of the means and standard deviations of the primary distributions shown in Table 8.5; these yield values of 2.0 dB for the mean and 0.86 dB for the standard deviation. The approximations involved in the various numerical processes do not affect the mean value very much but they do slightly increase the standard deviation but the accuracy achieved is quite acceptable for the present purpose.

The calculation procedure then follows the same principles as

Table 8.15. RELATIVE FREQUENCIES FOR THE COMBINED DISTRIBUTION, $S*R$, OF SRE AND RRE FOR 16-4T-31 (WITH REGULATOR) SUBSCRIBERS' SETS

Direct from Table 8.5			*Classes rearranged*		
Class mark dB	*Class Interval* dB	*Relative frequency* %	*Class mark* dB	*Class Interval* dB	*Relative frequency* %
(1)	(2)	(3)	(4)	(5)	(6)
−0.5	−1 to 0	0.7	−1	−1.5 to −0.5	0.3
+0.5	0 to 1	9.9	0	−0.5 to +0.5	5.3
1.5	1 to 2	41.8	1	+0.5 to +1.5	25.8
2.5	2 to 3	37.0	2	1.5 to 2.5	39.4
3.5	3 to 4	6.4	3	2.5 to 3.5	21.7
4.5	4 to 5	2.8	4	3.5 to 4.5	4.6
5.5	5 to 6	1.1	5	4.5 to 5.5	2.0
6.5	6 to 7	0.3	6	5.5 to 6.5	0.7
			7	6.5 to 7.5	0.2
Sum		100.0			100.0
Mean		2.03 dB			2.03 dB
Standard deviation		0.99 dB			1.10 dB

Note: The relative frequencies in col. 6 were obtained by taking mean values of appropriae adjacent pairs of frequencies in col. 3.

was illustrated in Tables 8.11, 8.12 and 8.13 to yield the percentage probabilities, P_{Indq}, that a 'difficult' opinion will be associated with each value of nominal ORE but with the following differences:

1. A level of −67.8 dBmp is assumed for the idle channel noise attributable to the pair of PCM codecs*; this can be considered to arise at point A in Figure 7.5b.
2. Because PCM transmission is assumed in the junctions, the quantising distortion caused by the encoding/decoding process must be considered. In the present case, 7 bit encoding would yield values of Q well in excess of 30 dB* and so the influence of quantising distortion will here be quite negligible.
3. y_S and y_R are constant.

The values of $p(x)$ (= P_{Indq}) have been entered in Table 8.16 together with the relative frequencies in each class interval. The total, weighted, proportion of 'difficult' opinions is then calculated as for the original situation (*see* Table 8.14).

Table 8.16. CALCULATION OF THE TOTAL PROPORTION OF 'DIFFICULT' OPINIONS (HYPOTHETICAL FUTURE SITUATION)

Nominal ORE, dB		*Relative frquency* $\phi(x)$ %	$p(x)$	$\phi(x).p(x)$ *per* 10^4
Class mark x	*Class Interval* x_i *to* x_{i+1}			
(1)	(2)	(3)	(4)	(5)
5	4.5 to 5.5	0.3	3.6	1.08
6	5.5 to 6.5	5.3	3.4	18.02
7	6.5 to 7.5	25.8	3.2	82.56
8	7.5 to 8.5	39.4	3.1	122.14
9	8.5 to 9.5	21.7	3.2	69.44
10	9.5 to 10.5	4.6	3.4	15.64
11	10.5 to 11.5	2.0	3.5	7.00
12	11.5 to 12.5	0.7	3.6	2.52
13	12.5 to 13.5	0.2	3.8	0.76
	Sum	100.0		319.16
				3.2%

* Section 6.2 and Section 4.2.2.1 provide the evidence on which these statements are based.

According to this example, the improvement in performance due to complete replacement of all tandem junctions by PCM transmission systems with integrated switching is shown by the reduction from 6.9% to 3.2% in the proportion of customers likely to experience difficulty in conversation. As can be seen by comparison of Tables 8.14 and 8.16, the improvement given to the relatively infrequent higher loss connections is even greater; considering only the 5% most lossy connections, for the original situation, these are such that about 14% of customers using them experienced difficulty; for the hypothetical future situation, this value is only about $3\frac{1}{2}$%.

Quite clearly, the method illustrated here for estimating the proportion of customers using particular classes of connections who are likely to experience difficulty contains many approximations. The method was evolved in 1951 and has been used in the form described until about 1966; particular applications include those published in References* 4, 5, 6 and 7. With the advent of computers having sufficiently large storage capacity, it has now become possible to employ more elaborate statistical models to represent the various situations to be examined and to make use of a simulation technique whereby the multivariate distributions can be sampled according to any desired random process. Such methods require more extensive fundamental subjective data than are at present available in, for example, Tables 4.14 and 4.15; the need for extension of these data has already been illustrated in Section 4.2.2.1 (where the effects of quantising distortion were examined) and in the present section (where extrapolation below $x = 16$ dB was needed). With the increased computer facilities now available, the opportunity can be taken to extend the scope of such estimation to include consideration of the following which have been treated here rather crudely or not at all:

1. Effects of different values of room noise. As explained in Section 4.1.1, three kinds of factors need consideration (X, Y and Z factors) but room noise (a Y factor) has been treated here as having only a fixed value, namely 50 dBA, Hoth spectrum.
2. Effects of sidetone on talking and listening. These have been treated largely as fixed quantities independent of the subscriber's line; while perhaps acceptable for old type sets (e.g. 13-2P-27), this approximation is no longer accept-

* *See* Section 8.4 of the Bibliography

able for modern, high sensitivity, sets furnished with regulators. Unfortunately the problem of dealing with sidetone is very difficult to treat because sidetone sensitivity depends not only on the subscriber's line but also on the impedance facing the local exchange on its junction side. Furthermore, the effect of sidetone on listening interacts strongly with the level and other characteristics of the room noise.

3. Increased attention will need to be given to the employment in telephone networks of special types of subscriber's station equipment, e.g. loudspeaking telephone sets, PBXS, extension plans; so far the network design for telephony has been based almost entirely on consideration of handset telephones.
4. Telephone networks have, until now, been designed for satisfactory operation for telephony; other services, such as telegraphy and data transmission have had to adapt themselves to this situation. In the future, the specific requirements for these other services may well impose more severe restrictions on degradations than are imposed for telephony, e.g. group delay/frequency distortion and certain forms of nonlinear distortion.

APPENDIX 1

CHARTS FOR DETERMINATION OF MISMATCH AND RETURN LOSSES

Mismatch loss is a measure of the effect on transmitted voltamperes (apparent power) of a mismatch between the impedance of a source and that of a load connected to the source. Mismatch loss is given in decibels by the expression:

$$10 \log_{10} \left| \frac{W_1}{W_2} \right| \text{ dB}$$

where W_1 is the apparent power (complex voltamperes) from the

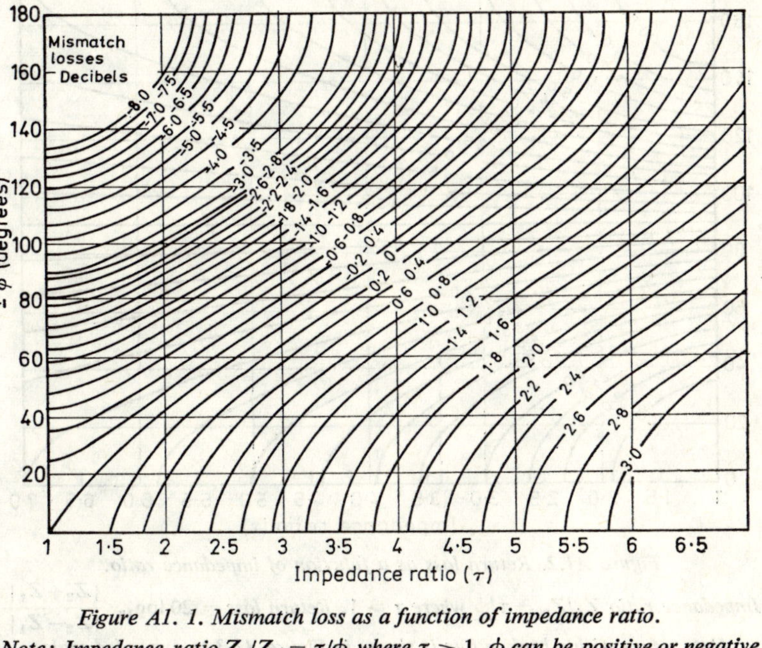

Figure A1. 1. Mismatch loss as a function of impedance ratio.
Note: Impedance ratio $Z_1/Z_2 = \tau/\underline{\phi}$ where $\tau > 1$, ϕ can be positive or negative

source into a load whose impedance is equal in magnitude and angle to that of the source, and W_2 is the apparent power (complex voltamperes) delivered from the source to the load that is mismatched in impedance.

If Z_1 and Z_2 are the impedances respectively of source and mismatched load, then the mismatch loss is numerically equal to:

$$20 \log_{10} \frac{|Z_1+Z_2|}{2(|Z_1Z_2|)^{1/2}} \text{ dB}$$

Where Z_1 and Z_2 are in the form $Z\underline{/\theta}$, the mismatch loss can be expressed as a function of the impedance ratio $\tau\underline{/\phi}$.

$$\tau = \frac{|Z_1|}{|Z_2|} \quad \text{or} \quad \frac{|Z_2|}{|Z_1|} \quad \text{whichever is greater than 1}$$

$$\text{and} \quad \phi = |\theta_1 - \theta_2|$$

$\tau\underline{/\phi}$ can be charted, as shown in Figure A1.1, in the form of contours of constant mismatch loss on the impedance ratio plane. Such a chart is sufficiently accurate for many purposes and can be read to 0.1 dB in the most commonly required area.

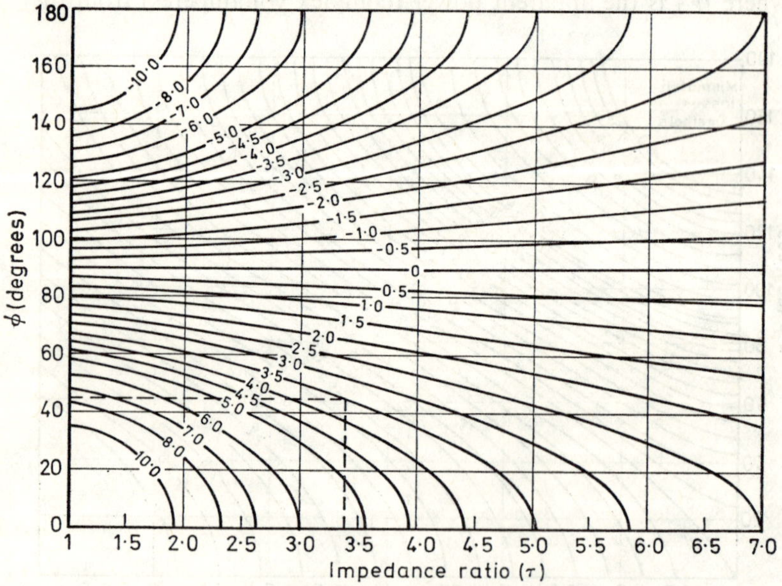

Figure A1.2. Return loss as a function of impedance ratio.

Impedance ratio $Z_1/Z_2 = \tau\underline{/\phi}$ *where* $\tau > 1$. *Return loss* $= 20 \log_{10} \dfrac{|Z_2+Z_1|}{|Z_2-Z_1|}$

Note: Area within broken line is shown in Figure A1.3 to expanded scale

Return loss is another measure of the degree of mismatch in impedance between a source and its load and is given in decibels by the expression:

$$10 \log_{10} \left| \frac{W_1}{W_1 - W_2} \right| \text{ dB}$$

where W_1 and W_2 are defined as above for mismatch loss.

If Z_1 and Z_2 are also defined as above, then the return loss is numerically equal to:

$$20 \log_{10} \frac{|Z_1 + Z_2|}{|Z_1 - Z_2|} \text{ dB}$$

By making use of the impedance ratio $\tau/\underline{\phi}$ defined as above, the numerical values of return loss can also be charted as contours of constant return loss on the impedance ratio plane. Figure A1.2 shows the result for the same range of τ and ϕ as Figure A1.1; Figure A1.3 shows the return loss contours for ranges of τ and ϕ respectively restricted to 1.0–3.4 and 0–45°.

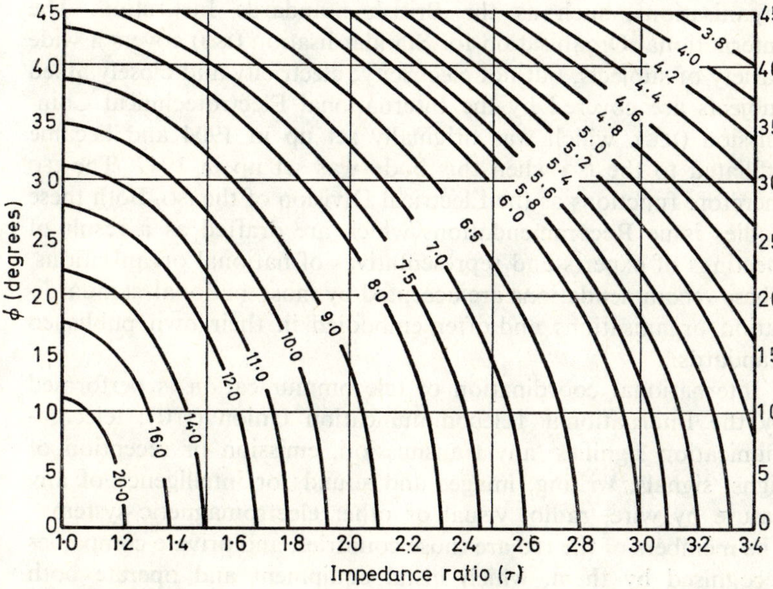

Figure A1.3. Return loss as a function of impedance ratio (narrow range) Impedance ratio $Z_1/Z_2 = \tau/\underline{\phi}$ *where* $\tau > 1$. *Return loss* $= 20 \log_{10} \dfrac{|Z_2+Z_1|}{|Z_2-Z_1|}$

APPENDIX 2
INTERNATIONAL ORGANISATIONS

The work of certain international organisations is of particular importance to the matters treated in this book and so a brief statement of their organisation and functions is given below.

Two international bodies exist for setting up standards to facilitate international exchange of goods and services and to promote cooperation in intellectual, scientific, technological and economic affairs. These bodies mainly function by facilitating coordination of the standardisation work by the various national standardisation organisations, such as the British Standards Institution. The International Organisation for Standardisation (ISO) covers a wide variety of subjects, but not electricity; electricity and closely allied subjects are covered by the International Electrotechnical Commission (IEC), which was originally set up in 1904 and became affiliated to the ISO when this body was set up in 1947. The IEC therefore functions as the Electrical Division of the ISO. Both these bodies issue Recommendations which are drafted as a result of meetings of experts and representatives of national organisations. These recommendations are accepted by most national standardisation organisations and often embodied in their own published standards.

International coordination of telecommunication is performed by the International Telecommunication Union (ITU); telecommunication signifies 'any transmission, emission or reception of signs, signals, writing, images and sounds or intelligence of any nature by wire, radio, visual or other electromagnetic systems'. The members of the ITU are those countries, and private companies recognised by them, which instal equipment and operate both national and international telecommunication services.

The ITU is the oldest of the intergovernmental organisations which, in 1947 (at Atlantic City) were embodied into the structure of the United Nations as specialised agencies. There are now about 12 specialised agencies of the United Nations (some set up since 1947) and these include such organisations as the Food Agriculture Organisation (FAO), United Nations Educational, Scientific and Cultural Organisation (UNESCO), World Health Organisation (WHO), International Civil Aviation Organisation (ICAO), World Meteorological Organisation (WMO), Universal Postal Union (UPU) and International Monetary Fund.

The ITU dates back to 17 May 1865, when the International Telegraph Union was founded in Paris. The Paris Convention was revised in 1868 (Vienna), 1872 (Rome) and 1875 (St Petersburg), the 1868 revision setting up a permanent international bureau, which had its headquarters, until 1948, in Berne. The first provisions for international telephony were drawn up in 1885 (Berlin) and in 1906 (Berlin) the first International Radiotelegraph Convention was signed by twenty-seven maritime states who were interested in ship to shore and ship to ship radio telegraphy. In 1927 (Washington) the first Table of Frequency Allocations was devised. In 1932 (Madrid), the International Telegraph Union and the International Radiotelegraph Union were merged to form the International Telecommunication Union which then published four sets of Regulations: Telegraph, Telephone, Radio and Additional Radio Regulations. In 1948, after the Atlantic City Plenipotentiary Conference at which the present constitution under the United Nations was set up, the headquarters of the ITU was moved from Berne to Geneva where it is still located. The present formal basis of the ITU is the International Telecommunication Convention, signed by plenipotentiaries of 117 countries, to which are annexed four sets of regulations under the same headings as the 1932 Madrid Convention. These regulations (and the convention) are binding on Members and Associate Members of the ITU, the convention ranking as a Treaty between the Governments concerned.

The ITU is now composed of four permanent organs, namely as follows:

The General Secretariat, which is the organ of liaison between telecommunication Administrations throughout the world, is entrusted with all the administrative and financial services of the ITU. It is responsible to the Administrative Council, composed of eighteen members of the ITU elected by the Plenipotentiary Conference which takes place about every six years. The Administrative Council normally meets once a year and functions only when in session.

The International Frequency Registration Board (IFRB), created in 1947, records the frequency assignments made by the various countries and provides Members and Associate Members of the ITU with advice with a view to the operation of as many radio channels as possible in those parts of the spectrum where harmful interference may arise. The work is done by five independent members, one from each Region of the world, who are elected by a World Administrative Radio Conference; while they are working for the IFRB, they act independently of their national Administrations to which they belong and are responsible only to the IFRB.

The International Telegraph and Telephone Consultative Committee (CCITT), which, in 1957, replaced the International Telegraph Consultative Committee (CCIT) and the International Telephone Consultative Committee (CCIF), is called upon to study and issue recommendations on technical, operating and tariff questions connected with telegraphy, facsimile and telephony. The CCIT and CCIF were created, respectively in 1925 and 1924.

The International Radio Consultative Committee (CCIR), created in 1927, is called upon to study and issue recommendations on technical questions connected with radio and on operational matters the solution of which largely depends on considerations bound up with radio technique.

The CCITT and CCIR have, as members, the Administrations of all Members and Associate Members (non-voting members) of the ITU, together with the recognised private operating agencies which, with the assent of the governments recognising them, ask to take part in their work. (On 1 January 1967 there were 129 Members of the ITU, but, at that time there were no Associate Members). The CCI's work largely through their study groups, set up by the plenary assemblies, which normally meet every three years and lay down the program of studies. The study groups undertake the studies, draw conclusions therefore which the following plenary assembly examines, discusses and finally sets forth its conclusions in the form of recommendations. These recommendations are published after each plenary assembly* as a 'Book' in several volumes; the 'colour' of the 'Book' containing the CCITT Recommendations is changed at each plenary assembly, the current version being the White Book, issued after the plenary assembly at Mar del Plata in 1968. The next plenary assembly will be held in 1972. Although these pronouncements (unlike the 'Regulations' of the Plenipotentiary Con-

* References to the Reports of all the Plenary Assemblies are given in Section 1.1 of the Bibliography.

ferences) rank only as 'recommendations', they are formulated with such care and represent unanimous decisions after considerable argument and discussion between the parties concerned that they do in fact form the basis for practically all international telecommunication. Furthermore, although they do not formally apply to national telecommunication, national networks must form parts of international connections and so they are very commonly used also as bases for equipment in national connections. Certain information on national practices is, in any case, assembled as part of the tasks of study groups to assist Members and Associate Members of the ITU, particularly those of the less developed countries.

The work, and the recommendations resulting therefore of certain CCITT Study Groups are of particular relevance to the material included in this book and for the study period 1968/72, the study groups most concerned are as follows:

Study Group XII, which deals with the performance of telephone sets, local lines and junctions and assessment of the overall transmission performance of complete telephone connections; this study group, through its Laboratory Working Party, also controls the work of the CCITT Laboratory which is equipped to conduct (on payment) certain telephonometric measurements for Members and Associate Members.

Study Group XV, which deals with transmission systems, mainly formulation of recommendations applicable to the performances of specific items of equipment.

Study Group XVI, which deals with telephone transmission in a rather general manner, formulating recommendations concerning the desirable performance and characteristics of telephone circuits to satisfy appropriate criteria of overall performance most of which have been recommended as a result of the work proper to Study Group XII.

Study Group Special C, which is a joint CCITT/CCIR study group, was originally set up to unify the clauses in specifications concerned with circuit noise; its work now also includes questions of reliability of service.

Study Group Special D, which deals with pulse code modulation and questions of digital transmission, is charged with the job of ensuring orderly development of a hierarchy of digital transmission systems compatible with the requirements of telephony, telephone switching and signalling (Study Group XI), telegraphy (Study Group VIII, IX and X), facsimile (Study Group XIV), data (Study

Group Special A), television (for which a joint CCIR/CCITT commitee* exists) and broadcast program transmission (CCIR Study Group X).

The current recommendations and the list of questions for study relating to the work of Study Groups XV, XVI, Special C and Special D are published in Volume III of the CCITT White Book and those for Study Group XII are in Volume V.

A number of Autonomous working parties (Groupe Autonome Specialisé, GAS) were set up by the plenary assembly to which they reported directly. One of these (GAS 1) had the task of preparing a handbook on National Automatic Networks, Chapter V of which is concerned with Transmission and contains information relevant to the material of this book. Another (GAS 2) had the corresponding task for Local Networks and Chapter V of that handbook is concerned with Transmission. Both these handbooks† contain information on current practices in the more telephonically developed countries and are intended to assist other less developed countries, particularly in developing their telephone networks and in dealing with the manufacturers of equipment they need. It is also worth noting that the ITU has responsibility for supplying experts, seconded from developed countries, to work in underdeveloped countries to help them build up their telephone networks.

* Joint Commission for Television Transmission (CMTT), administered by CCIR.

† The transmission chapters of both handbooks are now (1972) being revised and will be published separately as a single CCITT Handbook on Transmission.

BIBLIOGRAPHY

1. *GENERAL BACKGROUND*

1.1 *Telephone networks*

(1) Poole, J., *The practical telephone handbook*, Pitman, London, First published 1895, extensively revised in later editions, e.g. Sixth Edition, 1918; Seventh Edition, 1930; Eighth Edition (completely revised) by Knight, N. V. and Prickett, W.), (1944).

(2) Gavey, J., 'The telephone trunk line system in Great Britain', *J.IEE*, **25**, 624–680 (including discussion) (1896).

(3) Gavey, J., 'Inaugural Address', *J.IEE*, **36**, 4–37 (1905).

(4) Hill, J. G., 'Telephone transmission', Instn P. O. Elect. Engrs, Printed Paper No. 3 (1906).

(5) Cohen, B. S. and Shepherd, G. M., 'Telephonic transmission measurements', *J.IEE*, **39**, 503–565 (including discussion) (1907).

(6) Lee, J., 'The Zones and the trunk lines', *P. O. Elect. Engrs, J.*, **1**, 42–45 (1908).

(7) Cohen, B. S., 'Standard cable measurements', *P. O. Elect. Engrs, J.*, **6**, 1–20 (1913).

(8) Slingo, W., 'An address on telephones', *Instn P. O. Elect. Engrs*, Printed Paper No. 64 (1915).

(9) Cohen, B. S. and Hill, J. G., 'Long-distance and cable telephony (Underground and submarine)', *P. O. Elect. Engrs, J.*, **9**, 1–21 and 135–155 (1916).

(10) Cohen, B. S., 'Telephonometry', *Instn P. O. Elect. Engrs*, Printed Paper No. 70 (1916).

(11) Noble, W., 'The long-distance telephone system of the United Kingdom', *J.IEE*, **59**, 389–426 (including discussion) (1921).

(12) Elston, J. S., 'Applied telephone transmission', *Instn P. O. Elect. Engrs*, Printed Paper No. 88 (1922).

(13) Gill, F., Inaugural Address', *J.IEE* **61**, 1–15 (1922/23).

(14) Comité Consultatif International des Communications Téléphoniques a grande distance (CCI), Reports of Plenary Assemblies:

1st Plenary Assembly,	1924, Paris
2nd	1925, Paris
3rd	1926, Paris
4th	1927, Como
5th	1928, Paris
6th	1929, Berlin
7th	1930, Brussels
8th	1931, Paris
9th	1932, Madrid

After 1932, the name of the organisation was changed to 'Comité Consultatif International Téléphonique' which is abbreviated to CCIF: this change was to distinguish the organisation from the Comité Consultatif International Télégraphique (CCIT) which was, at that time, separate.

10th Plenary Assembly,	1934, Budapest
11th	1936, Copenhagen
12th	1938, Cairo
13th	1945, London
14th	1946, Montreux
15th	1949, Paris
16th	1951, Florence
17th	1954, Geneva
18th	1956, Geneva

After the 18th Plenary Assembly of the CCIF, a joint organisation was set up to embrace also the CCIT. The new organisation operates under the International Telecommunication Union (ITU) and is called 'Comité Consultatif International Télégraphique et Téléphonique', CCITT for short. The first Plenary Assembly of the CCITT was purely formal, taking place immediately on conclusion of the 18th Plenary Assembly of the CCIF.

1st Plenary Assembly,	1956, Geneva
2nd	1960, New Delhi
3rd	1964, Geneva
4th	1968, Mar del Plata

(15) BALDWIN, F. G. C., *History of the telephone in the United Kingdom*, Chapman and Hall, London, (a cheaper printing was published in 1938) (1925).

(16) STRATTON, J. and LUXTON, W. G., 'Modern developments in telephone transmission over lines', *Instn P. O. Elect. Engrs*, Printed Paper No. 153 (1933).

(17) BARNES, E. J., WOOD, A. E. and RICHARDS, D. L., 'Standards of transmission and local line limits', *P. O. Elect. Engrs, J.*, **39,** 151–155 and **40,** 8–13 (1947).

(18) FUDGE, G. A. E., 'Introduction and application of transmission performance ratings to subscribers' networks', *Instn P. O. Elect. Engrs*, Printed Paper No. 198 (1949).

(19) SWAFFIELD, J. and RICHARDS, D. L., 'Transmission performance of the telephone network—A survey of some basic problems', *Instn P. O. Elect. Engrs*, (unpublished) (1951).

(20) Various authors: 'Jubilee Number', *P. O. Elect. Engrs, J.*, **49,** 147–274 (see also Jubilee Address, Printed Paper No. 212) (1956).

(21) SWAFFIELD, J. and RICHARDS, D. L., 'Rating of speech links and performance of telephone networks', *Proc. IEE*, **106B,** 65–76 (1959).

(22) RICHARDS, D. L. and SWAFFIELD, J., 'Assessment of speech communication links', *ibid.*, 77–89.

(23) TOBIN, W. J. E. and STRATTON, J., 'A new switching and transmission plan for the inland trunk network', *P. O. Elect. Engrs, J.*, **53,** 75–79 (1960).

(24) CCITT Handbook on National Automatic Networks, ITU, Geneva, (1964.)

(25) WILLIAMS, H., 'Overall survey of transmission performance planning', *Proc. IEE*, **111,** 1–17 (1964).

(26) RICHARDS, D. L., 'Transmission performance assessment for telephone network planning', *ibid.*, 931–940.

(27) HAMSHER, D. H. (Ed.), 'Communication System Engineering Handbook', McGraw-Hill, New York, (Especially: Chapter 3; 'Speech characteristics and acoustic effects' by MEEKER, W. F. and Chapter 6; 'Transmission engineering of switched systems', by JOHNSON, A. C.) (1967).

1.2 *Transmission of Information*

(1) NYQUIST, H., 'Certain factors affecting telegraph speed', *Bell Syst. Tech. J.* **3**, 324–346 (1924).

(2) HARTLEY, R. V. L., 'Transmission of information', *Bell Syst. Tech. J.*, **7**, 535–563 (1928).

(3) GABOR, D., 'Theory of communication', *J. IEE*, **93**, Part III, 429–457 (1946).

(4) SHANNON, C. E., 'A mathematical theory of communication', *Bell Syst. Tech. J.*, **27**, 379–424 and 623–657 (1948).

(5) WIENER, N., *Cybernetics*, John Wiley, (1948).

(6) SHANNON, C. E., and WEAVER, W., *The mathematical theory of communication*, University of Illinois Press, Urbana (1949).

(7) SHANNON, C. E., 'Communication in the presence of noise', *Proc. Inst. Radio Engrs*, **37**, 10–21 (1949).

(8) SHANNON, C. E., 'Prediction and entropy of printed English', *Bell Syst. Tech. J.*, **30**, 50–64 (1951).

(9) GABOR, D., 'A summary of communication theory', *Communication Theory*, Butterworths Scientific Publications, 1–23 (1953).

(10) RICHARDS, D. L. and SWAFFIELD, J. 'Measurement of the communication rate of a speech link—A preliminary appraisal of the problem', *Communication Theory*, Buttterworths Scientific Publications, 450–459 (1953).

(11) QUASTLER, H. (editor) *Information theory in psychology*, The Free Press, Glencoe, Illinois (1955).

(12) QUASTLER, H. 'Studies of human channel capacity', *Information Theory: Third London Symposium (1955)* Editor: Colin Cherry, Butterworths Scientific Publications (1956).

(13) BROADBENT, D. E., *Perception and communication*, Pergamon Press, London (1958).

(14) CHERRY, C., *On human communication*, MIT Press and John Wiley (1959). Also, Science Editions, New York (1961).

(15) SCHWARZ, M., *Information transmission, modulation and noise*, McGraw-Hill (1959).

(16) BELL, D. A., *Information theory and its engineering applications*, Pitman, London (1962).

(17) HARMAN, W. W., *Principles of the statistical theory of communication*, McGraw-Hill (1963).

(18) ROSIE, A. M., *Information and communication theory*, Blackie (1966).

(19) HAMSHER, D. H. (Ed.), *Communication System Engineering Handbook*, McGraw-Hill, New York (1967). Chapter 4: 'Programming, Probability and Information Theory' by NANNI, L. F.

1.3 *Subjective assessment*

(1) MICHELS, W. C. and HELSON, H., 'A reformulation of the Fechner Law in terms of adaptation-level applied to rating-scale data', *Amer. J. Psychol.*, **62,** 355–368 (1949).
(2) NASH, M. C., 'An experimental test of the Michels–Helson theory of judgement', *Amer. J. Psychol.*, **63,** 214–220 (1950).
(3) RICHARDS, D. L., 'Design and analysis of subjective acoustical experiments which involve a quantal response', *Acustica*, **2,** 83–91 (1952).
(4) GUILFORD, J. P., *Psychometric methods*, McGraw-Hill (1954).
(5) HELSON, H., MICHELS, W. C. and STURGEON, A., 'The use of comparative rating scales for the evaluation of psychophysical data', *Amer. J. Psychol.*, **67,** 321–326 (1954).
(6) WOODWORTH, R. S. and SCHLOSBERG, H., *Experimental psychology*, Henry Holt, New York (1955).
(7) MICHELS, W. C. and DOSER, B. T., 'Rating scale method for comparative loudness measurements', *J. Acoust. Soc. Amer.*, **27,** 1173–1180 (1955).
(8) KARLIN, J. E., 'Consideration of the user in telephone research', *Ergonomics*, **1,** 77–83 (1957).
(9) HARRIS, W. F., 'A revised law of comparative judgement', *Psychometrika* **22,** 189–198 (1957).
(10) LADEFOGED, P., *The mechanization of thought processes*, H.M. Stationery Office, London (1959).
(11) STEVENS, S. S. (Editor, Various authors), *Handbook of experimental psychology*, John Wiley, New York (1960).
(12) DODD, S. C. and GERBRICK, T. R., 'Word scales for degrees of opinion', *Lang. & Speech*, **3,** 18–31 (1960).
(13) HELSON, H., *Adaptation level theory*, Harper and Row, New York (1964).
(14) PROSSER, R. D., ALLNATT, J. W. and LEWIS, N. W., 'Quality grading of impaired television pictures' *Proc. IEE*, **111,** 491–502 (1964).
(15) LEWIS, N. W. and ALLNATT, J. W., 'Subjective quality of television pictures with multiple impairments', *Electronics Letters*, **1,** 187–188 (1965).
(16) NUNNALLY, J. C., *Psychometric Theory*, McGraw-Hill, Inc., New York (1967).
(17) GLEISS, N., 'Subjective assessments in telephony', *Tele (*English Edition*)*, **20,** 29–44 (1968).
(18) CORBETT, J. M., 'Effect of observer adaptation on the results of television quality-grading tests', *J. IEE*, **117,** 512–514 (1970).
(19) DUNCANSON, J. P., 'Context effects in categorical judgements', *Fifth International Symposium on Human Factors in Telecommunications*, Post Office Research Department, London, Session 6.4, 1–9 (1970).

1.4 *Mathematics and Statistics*

(1) RENSHAW, A. S., 'The methods of statistics', Instn. P. O. Elect. Engrs, Printed Paper No. 38 (1911).
(2) ADDEY, F., 'Telephone transmission', *P. O. Elect. Engrs, J.*, **5,** 29–40, 220–231 and 337–350 (1912).
(3) FISHER, R. A., 'On the mathematical foundations of theoretical statistics', *Phil. Trans. R. Soc.*, **A222,** 309–368 (1922).

(4) STATTERS, J. E., 'Telephone transmission', *Instn P. O. Elect. Engrs*, Printed Paper No. 101 (1924).

(5) THURSTONE, L. L., *A simplified multiple factor method*, University of Chicago Press (1933).

(6) GADDUM, J. H., *Reports on biological standards. III Methods of Biological assay depending on a quantal response*, Special Report Series, No. 183 Medical Research Council. H.M. Stationery Office, London (1933).

(7) HOLZINGER, K. J., *Student manual of factor analysis*, University of Chicago, Dept. of Education (1937).

(8) THOMSON, G. H., *The factorial analysis of human ability*, University of London Press (1939).

(9) BURT, C., *The factors of the mind*, University of London Press (1940).

(10) HOLZINGER, K. J. and HARMAN, W. W. *Factor analysis*, University of Chicago Press, 1941.

(11) SCHUMACHER, F. X. and CHAPMAN, R. A., *Sampling method in forestry and range management*, Duke University (1942).

(12) FISHER, R. A., *The design of experiments*, Oliver and Boyd, Edinburgh (3rd edition) (1942).

(13) WORTHING, A. G. and GEFFNER, J., *Treatment of experimental data*, John Wiley, New York (1943).

(14) FISHER, R. A., *Statistical methods for research workers*, Oliver and Boyd, Edinburgh (9th Edition) (1944).

(15) RICE, S. O., 'Mathematical analysis of random noise', *Bell Syst. Tech. J.*, **23,** 282–332 (1944) and **24,** 46–156 (1945).

(16) SNEDECOR, G. W., *Statistical methods*, Iowa State College Press (1946).

(17) AITKEN, A. C., *Statistical mathematics*, Oliver and Boyd Edinburgh (5th Edition) (1947).

(18) DAVIES, O. L., *Statistical methods in research and production*, Oliver and Boyd, Edinburgh (1947).

(19) FINNEY, D. J., *Probit analysis*, Cambridge University Press (1947).

(20) HOEL, P. G., *Introduction to mathematical statistics*, John Wiley, New York (1947).

(21) WEATHERBURN, C. E., *A first course in mathematical statistics*, Cambridge University Press (1947).

(22) EISENHART, C., 'The assumptions underlying the analysis of variance', *Biometrics* **3,** 1–21 (1947).

(23) COCHRAN, W. G., 'Some consequences when the assumptions for the analysis variance are not satisfied', *Biometrics*, **3,** 22–38 (1947).

(24) BARTLETT, M. S., 'The use of transformations, *Biometrics*, **3,** 39–52 (1947).

(25) KENDALL, M. G., *The advanced theory of statistics*, (2 vols.), Charles Griffin, London (1948).

(26) EMMENS, C. W., *Principles of biological assay*, Chapman and Hall (1948).

(27) WIENER, N., *The extrapolation, interpolation and smoothing of stationary time series*, Technology Press of M.I.T. and John Wiley (1949).

(28) BROWNLEE, K. A., *Industrial Experimentation*, H.M. Stationery Office (4th Edition) (1949).

(29) YULE, G. U. and KENDALL, M. G., *An introduction to the theory of statistics*, Charles Griffin, London (14th edition) (1950).

(30) GOOD, I. J., *Probability and the weighing of evidence*, Charles Griffin, London (1950).

(31) RICHARDS, D. L., 'An outline of the principles of design and analysis of experiments', *P. O. Elect. Engrs, J.*, **44,** 55–60 (1951).

(32) FINNEY, D. J., *Statistical method in biological assay*, Charles Griffin, London (1952).

(33) FINNEY, D. J., *Experimental design and its statistical basis*, Cambridge University Press (1955).

(34) MILLER, K. S., *Engineering Mathematics*, Constable, London (1956).

(35) RENTON, R. N., *Telecommunications principles*, Isaac Pitman, London (1967).

(36) *The Royal Signals Handbook of Line Communication*, Her Majesty's Stationery Office, London (1962).

(37) FRY, T. C., *Probability and its engineering uses*, van Nostrand, New York, 2nd edition (1965).

2. *MECHANISM OF HUMAN COMMUNICATION*

2.1 *Speech*

2.1.1 *The production mechanism (including artificial mouths)*

(1) PAGET, R., *Human speech*, Harcourt, London, 1930.

(2) INGLIS, A. H., GRAY, C.H. G. and JENKINS, R. T., 'A voice and ear for telephone measurements', *Bell Syst. Tech. J.*, **11,** 293–317 (1932).

(3) DUNN, H. K. and FARNSWORTH, D. W., 'Exploration of the pressure field around the human head during speech', *J. Acoust. Soc. Amer.*, **10,** 184–199 (1939).

(4) STEIN, L., *Speech and voice—their evolution, pathology and therapy*, Methuen and Co., London (1942).

(5) WARD, I. C., *The phonetics of English*, Heffer, Cambridge, England (4th Ed.) (1945).

(6) BODMER, F., *The loom of language*, Allen and Unwin, London (1946).

(7) JONES, D., *An outline of English phonetics*, Heffer, Cambridge, England (1947).

(8) CHAVASSE, P., 'Notes on phonetic statistics of the French language and their application to the study of conversation intelligibility', *Ann. Telecommun.*, **3,** 5–23 (in French) (1948).

(9) STEIN, L., *The infancy of speech and the speech of infancy*, Methuen, London (1949).

(10) MACCARTHY, P. A. D., *English pronunciation*, Heffer, Cambridge (1950).

(11) HAWLEY, M. E. and KETTLER, A. H., 'The apparent source of speech in the mouth', *J. Acoust. Soc. Amer.*, **22,** 365–369 (1950).

(12) LOTZ, J., 'Speech and language', *J. Acoust. Soc. Amer.*, **22,** 712–716 (1950).

(13) MEILLET, P. and COHEN, M. (Eds.), *Les langues du monde*, (CNRS), Paris (1952).

(14) FLETCHER, H., *Speech and hearing in communication*, van Nostrand, New York (1953).

(15) VAN DEN BERG, J. W., 'Transmission of the vocal cavities', *J. Acoust. Soc. Amer.*, **27,** 161–168 (1955).

(16) FLANAGAN, J. L., 'Analog measurements of sound radiation from the mouth', *J. Acoust. Soc. Amer.*, **32,** 1613–1620 (1960).

(17) IBBA, G., 'Test of the characteristics of an artificial mouth by means of audibility and perception', *Alta Frequenza*, **3**, 127–133 (1962).

(18) *CCITT Red Book, 1962, Volume V.* Annexes as follows: 'Artificial mouths used by various telephone administrations and operating organizations' (1962).

Annex 8; France
Annex 9; Italy
Annex 10; Federal German Republic
Annex 11; United Kingdom: the Post Office
Annex 12; Switzerland
Annex 13; Czechoslovakia
Annex 14; Union of Soviet Socialist Republics
Annex 15; Chile
Annex 16; FATME Co. (Italy)

(19) *CCITT Red Book, Volume V bis* (1965).

Annex G; Artificial mouth used by the Swedish Administration

(20) *CCITT Recommendation P.* **51**, 'Artificial voices, artificial mouths, artificial ears', White Book, Volume V. (1968).

2.1.2 *The specification of speech*

(1) DEWEY, G., *Relative frequency of English speech sounds*, Harvard University Press (1923).

(2) POTTER, R. K. and PETERSON, G. E., 'The representation of vowels and their movements', *J. Acoust. Soc. Amer.*, **20,** 528–535 (1948).

(3) NEWMAN, E. P., 'The pattern of vowels and consonants in various languages', *Am. J. Psychol.*, **64,** 369–379 (1951).

(4) PETERSON, G. E. and BARNEY, H. L., 'Control methods used in a study of the vowels', *J. Acoust. Soc. Amer.*, **24,** 175–184 (1952).

(5) PETERSON, G. E., 'Information-bearing elements of speech', *J. Acoust. Soc. Amer.*, **24,** 629–637 (1952).

(6) FLETCHER, H., *Speech and hearing in communication*, van Nostrand, New York (1953).

(7) BERRY, J., 'Some statistical aspects of conversational speech', *Communication Theory*, Butterworths Scientific Publications, 392–401 (1953).

(8) STEVENS, K. N. and HOUSE, A. S., 'Development of a quantitative description of vowel articulation', *J. Acoust. Soc. Amer.*, **27,** 484–493 (1955).

(9) BROADBENT, D. E., LADEFOGED, P. and LAWRENCE, W., 'Vowel sounds and perceptual constancy', *Nature*, **178,** 815–816 (letter) (1956).

(10) LADEFOGED, P. and BROADBENT, D. E., 'Information conveyed by vowels', *J. Acoust. Soc. Amer.*, **29,** 98–104 (1957).

(11) MOLES, A., 'On the "description" of speech and style', *Ann. des Telecommun.*, **12,** 21–32 (1957).

(12) DUNN, H. K. amd BARNEY, H. L., 'Artificial speech in phonetics and communications', *J. Speech Hear. Res.*, **1,** 23–29 (also as Bell Mono. 3020) (1958).

(13) CRABBE, H. J. F., 'Electronics and the phonetician', *Wireless World*, **65,** 289–294 (1959).

(14) FOULKES, J. D., 'Computer identification of vowel types', *J. Acoust. Soc. Amer.*, **33,** 7–11 (1961).

(15) DAVID, E. E., 'Computer-catalyzed speech research', *Fourth Int. Cougress on Acoustics* (Denmark, 1962) also Bell Mono. 4532.

(16) RICHARDSON, E. G. and MEYER, E. (Eds), *Technical aspects of sound, Vol III*, Elsevier Publishing Co., Amsterdam. Chapter 1: FRY, D. B. and DENES, P., 'The role of acoustics in phonetic studies' (1962).

(17) DENES, P. B., 'On the statistics of spoken English', *J. Acoust. Soc. Amer.*, **35,** 892–904 (1963).

(18) GUELKE, R. W. and SMITH, E. D., 'Distribution of information in stop consonants', *Proc. IEE,* **110,** 680–689 (1963).

(19) GUELKE, R. W., 'The transmission of information by means of speech', *Proc. IREE Australia,* **26,** 192–197 (1965).

(20) OLSEN, H. F., BELAR, H. and ROGERS, E. S., 'Speech processing techniques and applications', *IEEE Trans. on Audio and Electroacoustics,* AU-15, 120–126 (1967).

(21) BERGER, K., 'The most common words used in conversation', *J. Communs Disorders,* **1,** 201–214 (1968).

2.1.3 *Physical characteristics important in telephony*

2.1.3.1 *General*

(1) CRANDALL, I. B., 'The sounds of speech', *Bell Syst. Tech. J.*, **4,** 586–626 (1925).

(2) POTTER, R. K., KOPP, G. A. and GREEN, H. C., *Visible speech,* van Nostrand, New York (1947).

(3) PETERSON, G. E. and BARNEY, H. L., 'Control methods used in a study of the vowels', *J. Acoust. Soc. Amer.*, **24,** 175–184 (1952).

(4) FLETCHER, H., *Speech and hearing in communication,* van Nostrand, New York (1953).

(5) RICHARDSON, E. G. (Ed.), *Technical aspects of sound, Vol. I,* Elsevier Publishing Co., Amsterdam, Chapter 10: 'Speech' (1953).

2.1.3.2 *Amplitude structure*

(1) SACIA, C. F. and BECK, C. J., 'The power of fundamental speech sounds', *Bell Syst. Tech. J.*, **5,** 393–403 (1926).

(2) CRANDALL, I. B., 'Dynamic study of the vowel sounds, Part II', *Bell Syst. Tech. J.*, **6,** 100–116 (1927).

(3) SIVIAN, L. J., 'Speech power and its measurement', *Bell Syst. Tech. J.*, **8,** 646–661 (1929).

(4) LOYE, D. P. and MORGAN, K. F., 'Sound picture recording and reproducing characteristics', *J. Soc. Motion Pict. Engrs,* **32,** 631–647 (1939).

(5) DUNN, H. K. and WHITE, S. D., 'Statistical measurements on conversational speech', *J. Acoust. Soc. Amer.*, **11,** 278–288 (1940).

(6) DAVENPORT, W. B., 'An experimental study of speech-wave probability distributions', *J. Acoust. Soc. Amer.*, **24,** 390–399 (1952).

(7) SUBRIZI, V., 'A speech volume survey on telephone message circuits', *Bell. Lab. Record,* **31,** 292–295 (1953).

(8) *American Standards Association, Standard C16.5*–1954, 'American standard practice for volume measurements of electrical speech and program waves'. (1954).
(9) SHEARME, J. N. and RICHARDS, D. L., 'The measurement of speech level', *P. O. Elect. Engrs, J.*, **47,** 159–161 (1954).
(10) YAMOTO, J., 'Statistical characteristics of telephone speech', Reports of the Electrical Communication Laboratory, *Nippon Telegraph and Telephone Public Corporation*, **3,** 15–18 (1955).
(11) BRAUN, K. and SCHOBEL, W., 'The measurement of the statistical distribution of the speech volume in telephone channels' *NTZ* **12,** 291–296 (1959)
(12) HUTTER, J. and ARCHER, E. C., 'The electronic Speech Voltmeter Type 5A', *P. O. Elect. Engrs, J.*, **53,** 92–94 (1960).
(13) *CCITT Red Book, Volume V,* (1962) *Annex 18:* 'SFERT Volume Indicator'
(14) MCADOO, K. L., 'Speech volumes on Bell System massage circuits – 1960 survey', Trans AIEE, Part I – *Commun. Electron.*, **59,** 8–13 (1962) also, with slight changes in text, *Bell. Syst. Tech. J.*, **42,** 1999–2012, (1963) also Bell Mono. 4647.
(15) RICHARDS, D. L., 'Statistical properties of speech signals', *Proc. IEE,* **111,** 941–949 (1964).
(16) BRADY, P. T., 'A statistical basis for objective measurement of speech levels', *Bell Syst. Tech. J.*, **44,** 1453–1486 (1965).
(17) BRICKER, P. D., 'Technique for objective measurement of speech levels', *J. Acoust. Soc. Amer.*, **38,** 361–362 (1965).
(18) FISCHER, G. and RASCH, J., 'Electrical power in telephone and broadcast program transmissions and the influence of preemphasis and compandor', *NTZ-CJ,* **5,** 159–163 (1966).
(19) *CCITT Recommendation P.52:* 'Volume meters', White Book, Volume V (1968).
(20) *Supplement No. 10:* 'ARAEN Volume meter or Speech Voltmeter', White Book, Volume V (1968).
(21) *Supplement No. 11:* 'Volume meter standardised in the United States of America, termed vu-meter', White Book, Volume V (1968).
(22) *Supplement No. 14,* 'Comparison of the readings given on conversational speech by different types of volume meter' (Tests carried out by the British administration), White Book, Volume V (1968).
(23) BRADY, P. T., 'Objective measurement of telephone speech levels and on-off patterns', *Fourth International Symposium on Human Factors in Telephony,* VDE-Verlag GMBH Berlin, 117–126 (1970).
(24) BERRY, R. W., 'Speech-volume measurements on telephone circuits', *Proc. IEE,* **118,** 335–338 (1971).

2.1.3.3 *Time structure*

(1) NORWINE, A. C. and MURPHY, O. J., 'Characteristic time intervals in telephonic conversation', *Bell Syst. Tech. J.*, **17,** 281–291 (1938).
(2) DREW, R. O. and KELLOG, E. W., 'Starting characteristics of speech sounds', *J. Soc. Motion Pict. Engrs,* **34,** 43–58 (1940).
(3) DAVENPORT, W. B., 'An experimental study of speech-wave probability distributions', *J. Acoust. Soc. Amer.*, **24,** 390–399 (1952).
(4) GOLDMAN-EISLER, F., 'The determinants of the rate of speech output and their mutual relations', *J. Psychosom. Res.*, **1,** 137–143 (1956).

(5) Fairbanks, G. and Miron, M. S., 'Effects of vocal effort upon the consonant-vowel ratio within the syllable', *J. Acoust. Soc. Amer.* **29**, 621–626 (1957).
(6) Goldman-Eisler, F., 'Speech production and the predictability of words in context', *Quart. J. Exp. Psychol.* **10**, 96–106 (1958).
(7) Goldman-Eisler, F., 'The predictability of words in context and the length of pauses in speech', *Lang. & Speech*, **1**, 226–231 (1958).
(8) Goldman-Eisler, F., 'Hesitation and information in speeech' *Information Theory*, Butterworths Scientific Publications, 162–174 (1961).
(9) Goldman-Eisler, F., 'Continuity of speech utterance, its determinants and its significance', *Lang. & Speech*, **4**, 220–231 (1961).
(10) Goldman-Eisler, F., 'The distribution of pause durations in speech', *Lang. & Speech*, **4**, 232–237 (1961).
(11) Velichkin, A. I., 'Statistical investigation of the speech process', *Elektrosvyaz'*, **8**, 3–10 (Pergamon translation) (1961).
(12) Boomer, D. S. and Dittman, A. T., 'Hesitation pauses and juncture pauses in speech', *Lang. & Speech*, **5**, 215–220 (1962).
(13) Brady, P. T., 'Technique for investigating on-off patterns of speech', *Bell Syst. Tech. J.* **44**, 1–22 (also as Bell Mono. 4920) (1965).
(14) Henderson, A., Goldman-Eisler, F. and Skarbek, A., 'Sequential temporal patterns in spontaneous speech', *Lang. & Speech* **9**, 207-216 (1966).
(15) Goldman-Eisler, F., 'Sequential temporal patterns and cognitive proesses in speech', *Lang. & Speech*, **10**, 122–132 (1967).
(16) Schwartz, J. and Jaffe, J., 'Markovian prediction of sequential temporal patterns in spontaneous speech', *Lang. & Speech* **11**, 27–30 (1968).

2.1.3.4 *Frequency structure*

(1) Loye, D. P. and Morgan, K. F., 'Sound picture recording and reproducing characteristics', *J. Soc. Motion Pict. Engrs*, **32**, 631–647 (1939).
(2) Dunn, H. K. and White, S. D., 'Statistical measurements on conversational speech', *J. Acoust. Soc. Amer.*, **11**, 278–288 (1940).
(3) Potter, R. K., Kopp, G. A. and Green, H. C., *Visible speech*, van Nostrand, New York (1947).
(4) Stevens, S. S., Egan, J. P. and Miller, G. A., 'Methods of measuring speech spectra', *J. Acoust. Soc. Amer.*, **19**, 771–780 (1947).
(5) Rudmore, H. W. *et al.* 'Voice measurements with an audio spectrometer', *J. Acoust. Soc. Amer.*, **20**, 503–512 (1948).
(6) Fletcher, H., *Speech and hearing in communication*, van Nostrand, New York (1953).
(7) Licklider, J. C. R., Hawley, M. E. and Walkling, R. A. 'Influences of variations of speech intensity and other factors upon the speech spectrum, *J. Acoust. Soc. Amer.*, **27**, 207 (1955).
(8) Fant, G., 'Acoustic analysis and synthesis of speech with applications to Swedish', *Ericsson Tech. No. 1* (1959).
(9) Velichkin, A. I., 'Statistical investigation of the speech process', *Elektrosvyaz'*, **8**, 3–10 (1961) (English translation by Pergamon Press, 843–855)
(10) Leiberman, P. and Michaels, S. B., 'Some aspects of fundamental frequency and envelope amplitude as related to the emotional content of speech', *J. Acoust. Soc. Amer.*, **34**, 922–927 (1962).

(11) VELICHKIN, A. I., 'Correlation function and spectral density of quantized speech', *Soviet Phys. Acoust.*, **9,** 13–18, English translation, 10–14 (1963)

2.1.4 *Acoustical description relevant to recognition and synthesis*

(1) DUDLEY, H., 'Remaking speech', *J. Acoust. Soc. Amer.*, **11,** 169–177 (1939)

(2) DUDLEY, H., 'The carrier nature of speech', *Bell Syst. Tech. J.*, **19,** 495–515 (1940).

(3) POTTER, R. K. and STEINBERG, J. C., 'Toward the specification of speech', *J. Acoust. Soc. Amer.*, **22,** 807–820 (1950).

(4) CHANG, S. H. *et al.*, 'Representations of speech sounds and some of their statistical properties', *Proc. Inst. Radio Engrs*, **39,** 147–153 (1951).

(5) SMITH, C. P., 'The analysis and automatic recognition of speech sounds', *Electron. Engng*, **24,** 368–372 (1952).

(6) KOCK, W. E., 'Problems of selective voice control', *J. Acoust. Soc. Amer.*, **24,** 625–628 (1952).

(7) POLLACK, I., PICKETT, J. M. and SUMBY, W. H., 'On the identification of speakers by voice', *J. Acoust. Soc. Amer.*, **26,** 403–406 (1954).

(8) WEIBEL, E. S., 'Vowel synthesis by means of resonant circuits', *J. Acoust. Soc. Amer.*, **27,** 858–865 (1955).

(9) FLANAGAN, J. L., 'Estimates of the maximum precision necessary in quantising certain "dimensions" of vowel sounds', *J. Acoust. Soc. Amer.*, **29,** 533–534 (1957).

(10) DAVID, E. E., MATTHEWS, M. V. and MACDONALD, H. S., 'Description and results of experiments with speech using digital computer simulation', *Proc. Nat. Electronics Conference*, 766–776 (1958).

(11) FANT, G., 'Modern instruments and methods for acoustic studies of speech', *Acta Polytech. Scand.* (1958).

(12) FANT, G., 'Acoustic analysis and synthesis of speech with applications to Swedish', *Ericsson Tech. No. 1* (1959).

(13) FANT, G., 'The acoustics of speech', *Proceedings of the Third International Congress on Acoustics, (Stuttgart, 1959)* Elsevier Publishing Co., Amsterdam

(14) FOULKES, J. D., 'Computer identification of vowel types', *J. Acoust. Soc. Amer.*, **33,** 7–11 (1961).

(15) SIMMONS, P. L., 'Automation of speech, speech synthesis and synthetic speech. A bibliographical survey from 1950–1960', *IRE Trans. Audio*, **AU--9,** 191–196 (1961).

(16) UHR, L. and VOSSLER, C., 'Recognition of speech by a computer program that was written to simulate a model for human visual pattern recognition', *J. Acoust. Soc. Amer.*, **33,** 1426 (1961).

(17) PETERSON, G. E., 'Automatic speech recognition procedures', *Lang. & Speech*, **4,** 200–219 (1961).

(18) KERSTA, L. G., 'Voiceprint identification', *Nature*, **196,** 1253–1257 (also as Bell Mono. 4485) (1962).

(19) FANT, C. G. M., 'Descriptive analysis of the acoustic aspects of speech', *Logos*, **5,** 3–17 (1962).

(20) FANT, G., FINTOFT, K., LILJENCRANTS, J., LINDBLOM, B. and MARTONY, J., 'Formant amplitude measurements', *J. Acoust. Soc. Amer.*, **35,** 1753–1761 (1963).

(21) FLANAGAN, J. L. *Speech analysis, synthesis and perception*, Springer-Verlag, Berlin and New York (1965).

(22) SCHROEDER, M. R., 'Vocoders: Analysis and synthesis of speech—A review of 30 years of applied speech research', *Proc. IEEE*, **54**, 720–734 (1966)

2.2 *Hearing*

2.2.1 *The reception mechanism (including artificial ears)*
Outer ear

(1) WEST, W., 'Measurements of the acoustical impedances of human ears', *P. O. Elect. Engrs, J.*, **21**, 293–300 (1929).

(2) WEST, W., 'An artificial ear', *P. O. Elect. Engrs, J.*, **22**, 260–263 (1930)

(3) WEBER, H., 'Contribution to the make-up of the orthotelephonic transmission system', *Tech. Mitt. Telegr.-u TelephVerw.*, **24**, 145–149 (1946)

(4) WIENER, F. M., 'On the diffraction of a progressive sound wave by the human head', *J. Acoust. Soc. Amer.*, **19**, 143–146 (1947).

(5) BROGDEN, W. G., and MILLER, S. A., 'Physiological noise generated under earphone cushions', *J. Acoust. Soc. Amer.*, **19**, 620–623 (1947).

(6) *American Standards Association, Standard Z24.9–1949*, 'American standard method for the coupler calibration of earphones'.

(7) WEVER, E. G., *Theory of hearing*, John Wiley, New York (1949).

(8) BERANEK, L. L., *Acoustic measurements*, John Wiley, New York (1949)

(9) BARDUCCI, I., 'Experimental comparison of the average human ear and some artificial ears', *Telecommunicazioni*, **4**, 567–570 (In Italian) (1951).

(10) BARDUCCI, I., 'Contribution to the problem of the artificial ear for use in calibrating telephones', *Ann. des Telecomm.*, **6**, 165–169 (In French) (1951).

(11) ARCHBOLD, R. B., 'An experimental probe microphone for the measurement of sound pressures', *P. O. Elect. Engrs, J.*, **45**, 145–148 (1953).

(12) *British Standard 2042:* 1953, 'An artificial ear for the calibration of earphones of the external type' (1953).

(13) WEVER, E. G., and LAWRENCE, M., *Physiological acoustics*, Princeton University Press, Princeton, New Jersey, (1954).

(14) *CCITT Red Book*, Volume V (1962), Annexes as follows: 'Artificial ears and couplers used by various telephone administrations and operating organizations'.
Annex 8; France
Annex 11; United Kingdom: the Post Office
Annex 12; Switzerland
Annex 14; Union of Soviet Socialist Republics
Annex 17; United States of America

(15) WANSDRONK, C., 'On the mechanism of hearing', *Philips Research Reports Supplement No. 1*, 1–140 (1962).

(16) RICHARDSON, E. G. (Ed.) *Technical aspects of sound, Vol. III*, Elsevier Publishing Co., Amsterdam, Chapter 2: LITTLER, T. S. 'The measurement of hearing loss and the alleviation of deafness' (1962).

(17) ITHELL, A. H., 'The measurement of the acoustical input impedance of human ears', *Acustica*, **13**, 140–145 (1963).

(18) ITHELL, A. H., 'A determination of the acoustical input impedance characteristics of human ears', *Acustica*, **13**, 311–314 (1963).

(19) LITTLER, T. S., *The physics of the ear*, Pergamon Press (1965).

(20) ITHELL, A. H., JOHNSON, E. G. T. and YATES, R. F., 'The acoustical impedance of human ears and a new artificial ear', *Acustica*, **15**, 109–116 (1965).
(21) STEVENS, R. W. B. and BATE, A. E., *Acoustics and vibrational physics*, Edward Arnold, London (1966).
(22) WILLIAMS, F. E. and ITHELL, A. H., 'A new artificial ear', *P. O. Elect. Engrs, J.*, **60**, 47–51 (1967).
(23) *CCITT Recommendation P.51*, 'Artificial voices, artificial mouths, artificial ears', White Book, Volume V (1968).
(24) *International Electrotechnical Commission (IEC) Report Publication 318*, 'An artificial ear of the wide-band type for the calibration of earphones used in audiometry' (1970)

Middle and Inner ear

(25) FURRER, W., 'The ear and hearing—one of the bases of telecommunication technique', *Tech. Mitt. Telegr.-u TelephVerw.*, **22**, 41–57 (1944).
(26) RUEDI, L. and FUARE, W., 'Physics and physiology of acoustic trauma', *J. Acoust. Soc. Amer.*, **18**, 409–412 (1946).
(27) KOBRAK, H. G., 'Observation of cochlear structures during the process of hearing', *J. Acoust. Soc. Amer.*, **19**, 328–331 (1947).
(28) CAUSSE, R., 'The ear as an electro-acoustic organ', *Ann. des Telecommun.*, **2**, 72–76 (1947).
(29) WEVER, E. G., *Theory of hearing*, John Wiley, New York (1949).
(30) BEKESY, G. V., 'The structure of the middle ear and the hearing of one's own voice by bone conduction', *J. Acoust. Soc. Amer.* **21**, 217–232 (1949).
(31) BEKESY, G. V., 'The vibration of the cochlear partition in the anatomical preparations and in models of the inner ear', *ibid.*, 233–245.
(32) BEKESY, G. V., 'On the resonance curve and the decay period at various points on the cochlear partition', *ibid.* 245–254.
(33) PETERSON, L. C. and BOGERT, B. P., 'A dynaminal theory of the cochlea', *J. Acoust, Soc. Amer.* **22**, 369–381 (1950).
(34) HUGGINS, W. H., 'A theory of hearing', *Communication Theory*, 363–380, (Willis Jackson, Ed.) Butterworth Scientific Publications (1953).
(35) RICHARDSON, E. G. (Ed.) *Technical aspects of sound, Vol. I*, Elsevier Publishing Co., Amsterdam, Chapter 11, 'Hearing' (1953).
(36) ONCHI, Y. 'Mechanism of the middle ear', *J. Acoust. Soc. Amer.* **33**, 794–805 (1961).

2.2.2 *Perception*

(1) HIRSH, I. J., *The Measurement of Hearing*, McGraw-Hill, New York (1952).
(2) DAVID, E. E., 'Artificial auditory recognition in telephony', *IBM J. Res. Dev.*, **2**, 294–309 (1958).
(3) BROADBENT, D. E., *Perception and Communication*, Pergamon Press, London (1958).
(4) HIRSH, I. J. 'Bekesy's audiometer', *J. Acoust. Soc. Amer.* **34**, 1333–1336 (1962).

2.2.3 *Characteristics relevant to telephony*

2.2.3.1 *Steady state conditions*

(1) SIVIAN, L. J. and WHITE, S. D., 'Minimum audible sound fields', *J. Acoust. Soc. Amer.*, **4**, 288–321 (1933).

(2) FLETCHER, H. and MUNSON, W. A., 'Loudness, its definition, measurement and calculation', *J. Acoust. Soc. Amer.*, **5,** 82–108 (1933).

(3) FLETCHER, H. and MUNSON, W. A., 'Relation between loudness and making', *J. Acoust. Soc. Amer.*, **9,** 1–10 (1937).

(4) CHURCHER, B. G. and KING, A. J., 'The performance of noise meters in terms of the primary standard', *J. IEE*, **81,** 57–90 (1937).

(5) KNAUSS, H. P., 'An empirical formula for the loudness of a 1000-cycle tone; *J. Acoust. Soc. Amer.*, **9,** 45–46 (1937).

(6) FLETCHER, H., 'Loudness, masking and their relation to the hearing process and the problem of noise measurement', *J. Acoust. Soc. Amer.*, **9,** 275–293 (1938).

(7) INGLISS, A. H., 'Transmission features of the new telephone sets', *Bell Syst. Tech. J.*, **17,** 358–380 (1938).

(8) STEINBERG, J. C. and GARDNER, M. B., 'On the auditory significance of the term hearing loss', *J. Acoust. Soc. Amer.*, **11,** 270–277 (1940).

(9) STEINBERG, J. C., MONTGOMERY, H. C. and GARDNER, M. B., 'Results of the World's Fair hearing tests', *J. Acoust. Soc. Amer.* **12,** 291–301 (1940) also *Bell Syst. Tech. J.*, **19,** 533–563 (1940).

(10) STEVENS, S. S. and VOLKMAN, J., 'The relation of pitch to frequency: a revised scale', *Am. J. Psychol.*, **53,** 329–353 (1940).

(11) BRAUN, K., 'The gain in loudness when listening with both ears and its importance', *Telegr.-Fernspr.-Funk.-u Fernseh-Tech.*, **29,** 343–345 (1940).

(12) STEVENS, S. S., MILLER, J. and TRUSCOTT, I., 'The masking of speech by sine waves, square waves, and regular and modulated pulses', *J. Acoust. Soc. Amer.*, **18,** 418–424 (1946).

(13) WEINER, F. M., 'On the diffraction of a progressive sound wave by the human head', *J. Acoust. Soc. Amer.*, **19,** 143–146 (1947).

(14) HARRIS, J. D., 'Effect of signal/noise ratio on pitch discrimination', *J. Acoust. Soc. Amer.*, **19,** 816–819 (1947).

(15) MILLER, G. A., 'Sensitivity to changes in the intensity of white noise and its relation to masking and loudness', *J. Acoust. Soc. Amer.*, **19,** 609–619 (1947).

(16) MATHER, R. C. and MILLER, R. L., 'Phase effects in monaural perception', *J. Acoust. Soc. Amer.*, **19,** 780–797 (1947).

(17) MOL, H., 'The connection between masking, perception of pitch and loudness', *Het PTT-Bedrijf*, **1,** 24–27 (1947/48).

(18) GOSEWINKEL, M., 'Loudness measurement', *Arch. Tech. Messen*, Issue 155 (1948).

(19) POLLACK, I., 'The atonal interval', *J. Acoust. Soc. Amer.*, **20,** 146–149 (1948).

(20) HAWKINS, J. E. and STEVENS, S. S., 'The masking of pure tones and of speech by white noise', *J. Acoust. Soc. Amer.*, **22,** 6–13 (1950).

(21) MUNSON, W. A. and WEINTER, F. M., 'Sound measurements for psychophysical tests', *J. Acoust. Soc. Amer.*, **22,** 382–386 (1950).

(22) WEBSTER, J. C. *et al.*, 'Individual differences in noise-masked thresholds', *J. Acoust. Soc. Amer.*, **22,** 483–490 (1950).

(23) SCHAFER, T. H. *et al.*, 'The frequency selectivity of the ear as determined by masking experiments', *J. Acoust. Soc Amer.*, **22,** 490–499 (1950).

(24) BEASLEY, W. C. and ROSENWASSER, H., 'Determining factors in composing and analysing speech-hearing tests. Part 1', *Laryngoscope*, **60,** 658–679 (1950).

(25) POLLACK, I. 'Comfortable listening levels for pure tones in quiet and noise', *J. Acoust. Soc. Amer.*, **24,** 158–162 (1952)
(26) DADSON, R. S. and KING, J. H., 'A determination of the normal threshold of hearing and its relationship to the standardisation of audiometers', *J. Lar. Otol.*, **66,** 366–378 (1952).
(27) WHEELER, L. J. and DICKSON, E. D. D., 'The determination of the theshold of hearing', *J. Lar. Otol.*, **66,** 379–395 (1952).
(28) FLETCHER, H., *Speech and hearing in communication*, van Nostrand, New York (1953).
(29) RICHARDSON, E. G. (Ed.), *Technical aspects of sound, Vol. I.* Elsevier Publishing Co., Amsterdam Chapter 11: 'Hearing' (1953).
(30) HIRSH, I. J. and BOWMAN, W. D., 'Masking of speech by bands of noise', *J. Acoust. Soc. Amer.*, **25,** 1175–1180 (1953).
(31) PRITCHETT, J., 'The subjective free-field calibration of an external telephone receiver by the equal-loudness method', *Acustica*, **4,** 544–546 (1954).
(32) *British Standard 2497:* 1954, 'The normal threshold of hearing for pure tones by earphone listening'.
(33) BERANEK, L. L., *Acoustics*, McGraw-Hill, New York (1954).
(34) STEVENS, S. S., 'The measurement of loudness', *J. Acoust. Soc. Amer.*, **27,** 815–829 (1955).
(35) ROBINSON, D. W. and DADSON, R. S., 'A re-determination of the equal-loudness relations for pure tones', *Brit. J. Appl. Phys.*, **7,** 166–181 (1956).
(36) STEVENS, S. S., 'Calculation of the loudness of complex noise', *J. Acoust. Soc. Amer.*, **28,** 807–832 (1956).
(37) STEVENS, S. S., 'Calculating loudness', *Noise Control* 11–22 (1957).
(38) ROBINSON, D. W., 'The subjective loudness scale', *Acustica*, **7,** 217–233 (1957).
(39) ZWICKER, E., FLOTTORP, G. and STEVENS, S., 'Critical bandwidth in loudness summation', *J. Acoust. Soc. Amer.*, **29,** 548–557 (1957).
(40) STEVENS, S. S., 'Concerning the form of the loudness function', *J. Acoust. Soc. Amer.*, **29,** 603–606 (1957).
(41) ROBINSON, D. W. and DADSON, R. S., 'Threshold of hearing and equal-loudness relations for pure tones, and the loudness function', *J. Acoust. Soc. Amer.*, **29,** 1284–88 (1957).
(42) *British Standard 2980:* 1958, 'Pure-tone audiometers'.
(43) *British Standard 3045:* 1958, 'The relation between the sone scale of loudness and the phon scale of loudness level'.
(44) ZWISLOCKI, J., 'Effect of the transmission characteristic of the ear on the threshold of audibility', *J. Acoust. Soc. Amer.* **30,** 430–432 (1958).
(45) SCHARF, B., 'Critical bands and the loudness of complex sounds near threshold', *J. Acoust. Soc. Amer.*, **31,** 365–370 (1959).
(46) SCHARF, B., 'The loudness of complex sounds as a function of the number of components', *J. Acoust. Soc. Amer.*, **31,** 783–785 (1959).
(47) STEVENS, S. S., 'On the validity of the loudness scale', *J. Acoust. Soc. Amer.*, **31,** 995–1003 (1959).
(48) *International Organization for Standardization (ISO) Recommendation R. 131* 'Expression of the physical and subjective magnitudes of sound or noise' (1959).
(49) *International Organization for Standardization (ISO) Recommendation R. 226* 'Normal equal loudness contours for pure tones and normal threshold of hearing under free-field listening conditions' (1961).

(50) ROBINSON, D. W., WHITTLE, L. S. and BOWSHER, J. M., 'The loudness of diffuse sound fields', *Acustica*, **11**, 397–404 (1961).
(51) POLLACK, I., 'Hearing', *Annual Review of Psychology*, **12**, 335–362 (bibliography of 204 items to April 1960) (1961).
(52) GREENWOOD, D. D., 'Auditory masking and the critical band', *J. Acoust. Soc. Amer.*, **33**, 484–502 (1961).
(53) GREENWOOD, D. D., 'Critical bandwidth and the frequency co-ordinates of the basilar membrane', *J. Acoust. Soc. Amer.*, **33**, 1344–1355 (1961).
(54) DE BOER, E., 'Note on the critical bandwidth', *J. Acoust. Soc. Amer.*, **34**, 985–986 (letter) (1962).
(55) BOWSHER, J. M. and ROBINSON, D. W., 'Calculation of Zwicker phons on a digital computer', *Nature*, **200**, 553–555 (1963).
(56) *American Standards Association, Standard S 1.2/1962* 'Physical measurement of sound'; *Standard S 1.4/1961* 'General purpose sound level meters'; *Standard S 3.4* (dated 1963 but not then approved) 'Computation of loudness of noise'.
(57) TILLMAN, T. W., JOHNSON, R. M. and OLSON, W. C., 'Earphone versus sound-field threshold sound-pressure levels for spondee words', *J. Acoust. Soc. Amer.*, **39**, 125–133 (1966).
(58) *International Organization for Standardization (ISO) Recommendation 532* 'Method for calculating loudness level' (1966).
(59) *British Standard 4198: 1967*, 'Method for calculating loudness' (1967).
(60) BURNS, W., 'Noise and Man', John Murray, London (1968).
(61) *British Standard 2497: Part I: 1968*, 'A reference zero for the calibration of pure-tone audiometers' Part 1. Data for earphone coupler combinations maintained at certain standardising laboratories. (Technically identical with ISO Recommendation R. 389–1964 'Standard reference zero for the calibration of pure-tone audiometers'.)
(62) FREIDIN, A. A., 'Role of the critical bands in the processing of information by the human auditory system (Review)', *Sov. Phys., Acoust.*, **14**, 271–283 (1969).
(63) *British Standard 2497: Part 2: 1969*, 'Data for certain earphones used in commercial practice' (1969).

2.2.3.2 *Transient conditions (including speech)*

(1) WEGEL, R. L. and LANE, C. E., 'Auditory masking and the dynamics of the inner ear', *Phys. Rev.*, **23**, 266–285 (1924).
(2) SHOWER, E. G. and BIDDULPH, R., 'Differential pitch sensitivity of the ear', *J. Acoust. Soc. Amer.*, **3**, 275–287 (1931).
(3) BURCK, W., KOTOWSKI, P. and LICHTE, H., 'Development of pitch sensations', *Elektrische Nachrichten Technik*, **12**, 326–333 (1935).
(4) BURCK, W., KOTOWSKI, P. and LICHTE, H., 'Audibility of delays', *ibid.*, 355–362.
(5) HUNT, R. S., 'Damping and selectivity of the inner ear', *J. Acoust. Soc. Amer.*, **14**, 50–57 (1942).
(6) MUNSON, W. A., 'Relation between the theory of hearing and the interpretation of speech sounds', *J. Acoust. Soc. Amer.*, **17**, 103A (1945).
(7) GABOR, D., 'Theory of communication; Part 2. The analysis of hearing', *J. IEE*, **93**, Part III, 442–445 (1946).
(8) HUGHES, J. W., 'The threshold of audition for short periods of stimulation', *Proc. Roy. Soc.*, B, **133**, 486–490 (1946).

(9) Pumphrey, R. J. and Gold, T., 'Transient reception and the degree of resonance in the human ear', Nature, **160**, 124–125 (1947).
(10) Munson, W. A., 'The growth of auditory sensation', *J. Acoust. Soc. Amer.*, **19**, 584–591 (1947).
(11) Garner, W. R., 'Auditory thresholds of short tones as a function of repetition rates', *J. Acoust. Soc. Amer.*, **19**, 600–608 (1947).
(12) Miller, R. L., 'Masking effect of periodically pulsed tones as a function of time and frequency', *J. Acoust. Soc. Amer.*, **19**, 798–807 (1947).
(13) Garner, W. R., 'The effect of frequency spectrum and temporal integration of energy in the ear', *J. Acoust. Soc. Amer.*, **19**, 808–815 (1947).
(14) Rosenblith, W. A. *et al.* 'An auditory after-image', *Science*, **106**, 333–335 (1947).
(15) Miller, G. A., 'The perception of short bursts of noise', *J. Acoust. Soc. Amer.*, **20**, 160–170 (1948).
(16) Miller, G. A. and Taylor, W. G., 'The perception of repeated bursts of noise', *J. Acoust. Soc. Amer.*, **20**, 171–182 (1948).
(17) Garner, W. R., 'The loudness of repeated short tones', *J. Acoust. Soc. Amer.*, **20**, 513–527 (1948).
(18) Miller, G. A. and Garner, W. R., 'Masking of tones by repeated bursts of noise', *J. Acoust. Soc. Amer.*, **20**, 691–696 (1948).
(19) Fletcher, H., 'A method of calculating hearing loss for speech from an audiogram', *J. Acoust. Soc. Amer.*, **22**, 1–5 (also as Bell Mono. 1731) (1950).
(20) Hawkins, J. E. and Stevens, S. S., 'The masking of pure tones and of speech by white noise', *J. Acoust. Soc. Amer.*, **22**, 6–13 (1950).
(21) Haas, H., 'The influence of the single echo on the hearing qualities of speech', *Acustica*, **1**, 49–58 (1951).
(22) Jacobson, H., 'Information and the human ear', *J. Acoust. Soc. Amer.*, **23**, 463–471 (1951).
(23) Dubout, P., 'The perceptability of artifically produced echoes', *Acustica*, **8**, 371–378 (1958).
(24) Plomb, R. and Bowman, M. A., 'Relation between hearing theshold and duration of tone pulses', *J. Acoust. Soc. Amer.*, **31**, 749–758 (1959).
(25) Hirsh, I. J., 'Auditory perception of temporal order', *J. Acoust. Soc. Amer.*, **31**, 759–767 (1959).
(26) Broadbent, D. E. and Ladefoged, P., 'Auditory perception of temporal order', *J. Acoust. Soc. Amer.*, **31**, 1539 (1959).
(27) Lochner, J. P. A. and Burger, J. F., 'Optimum reverberation time for speech rooms based on hearing characteristics', *Acustica*, **10**, 394–399 (1960).

2.2.3.3 *Use of both ears*

(1) Keys, J. W., 'Binaural versus monaural hearing', *J. Acoust. Soc. Amer.*, **29**, 629–631 (1947).
(2) Licklider, J. C. R., 'The influence of interaural phase relations upon the masking of speech by white noise', *J. Acoust. Soc. Amer.*, **20**, 150–159 (1948).
(3) Hirsh, I. J., 'The influence of interaural phase on interaural summation and inhibition', *J. Acoust. Soc. Amer.*, **20**, 536–544 (1948).
(4) Koenig, W., 'Subjective effects in binaural hearing', *J. Acoust. Soc. Amer.* **22**, 61–62 (letter) (1950).
(5) Cherry, E. C. and Taylor, W. K., 'Some further experiments upon the recognition of speech with one and with two ears', *J. Acoust. Soc. Amer.*, **26**, 554–559 (1954).

(6) EGAN, J. P., CARTERETTE, E. C. and THWING, E. J., 'Some factors affecting multi-channel listening', *J. Acoust. Soc. Amer.*, **26,** 774–782 (1954).

(7) CHERRY, E. C. and SAYERS, B. MCA 'Human "Cross-correlator" A technique for measuring certain parameters of speech perception', *J. Acoust. Soc. Amer.*, **28,** 889–895 (1956).

(8) SCHUBERT, E. D., 'Some preliminary experiments on binaural time delay and intelligibility', *J. Acoust. Soc. Amer.*, **28,** 895–901 (1956).

(9) SAYERS, B. MCA. and CHERRY, E. C., 'Mechanism of binaural fusion in the hearing of speech', *J. Acoust. Soc. Amer.*, **29,** 973–987 (1957).

(10) CLARK, H. A. M., DUTTON, G. F. and VANDERLYN, P. B., 'The "Stereophonic" recording and reproducing system', *IRE Trans. Audio,* **AU-5,** 96–111 (1957).

(11) POLLACK, I. and PICKETT, J. M., 'Stereophonic listening and speech intelligibility against voice babble', *J. Acoust. Soc. Amer.*, **30,** 131–133 (1958).

(12) LOCHNER, J. P. A. and BURGER, J. F., 'The binaural summation of speech signals', *Acustica,* **11,** 313–317 (1961).

(13) HARVEY, F. K. and SCHROEDER, M. R., 'Subjective evaluation of factors affecting two-channel stereophony', *J. Audio Engng Soc.*, **9,** 19–28 (also as *Bell Monograph 3869*) (1961).

(14) SCHROEDER, M. R. and LOGAN, B. F., '"Colorless" artificial reverberation', *J. Audio Engng Soc.*, **9,** 192–197 (also as *Bell Monograph 3996*) (1961).

(15) WHITWORTH, R. H. and JEFFRESS, L. A., 'Time vs intensity in the localization of tones', *J. Acoust. Soc. Amer.*, **33,** 925–929 (1961).

(16) SCHROEDER, M. R., 'Improved quasi-stereophony and "colorless" artificial reverberation', *J. Acoust. Soc. Amer.*, **33,** 1061–1064 (Also as Bell Monograph 3996) (1961).

2.3 *Conversation*

(1) NORWINE, A. C. and MURPHY, D., 'Characteristic time intervals in telephonic conversation', *Bell Syst. Tech. J.*, **17,** 281–291 (1938).

(2) RICHARDS, D. L., 'Some aspects of the behaviour of telephone users as affected by the physical properties of the circuit', *Communication Theory,* Butterworths Scientific Publications, London, 442–449 (1953).

(3) RICHARDS, D. L., 'Dynamics of telephone conversation: Study of the alternation of talker and listener roles', *Paper read at 1st Symposium on Human Factors in Telephony, Cambridge, England,* (unpublished) (1961).

(4) RICHARDS, D. L., 'Conversation performance of speech links subject to long propagation times' *International Conference on Satellite Communication (IEE, London)*, 247–251 (1962).

(5) BOERYD, A., 'Some reactions of telephone users during conversation', *Teleteknik, English Edition,* **7,** 97–103 (1963).

(6) BRADY, P. T., 'Technique for investigating on-off patterns of speech', *Bell Syst. Tech. J.*, **44,** 1–22 (also as Bell Mono. 4920) (1965).

(7) BRADY, P. T., 'A statistical analysis of on-off patterns in 16 conversations', *Bell Syst. Tech. J.*, **47,** 73–91 (1968).

(8) RICHARDS, D. L. and HUTTER, J., 'Echo suppressors for telephone connections having long propagation times', *Proc. IEE,* **116,** 955–964 (1969).

(9) BRADY, P. T., 'Objective measurement of telephone speech levels and on-off patterns', *Fourth International Symposium on Human Factors in Telephony,* VDE-Verlag, Berlin, 117–126 (1970).

3. *METHODS OF MEASURING THE COMMUNICATION EFFICIENCY OF A SPEECH LINK*

3.1 *General*

(1) Cohen, B. S., 'Apparatus standards of telephonic transmission and the technique of testing microphones and receivers', *J. IEE* **66,** 165–203 (including discussion) (1928).

(2) Martin, W. H. and Gray, C. H. G., 'Master Reference System for telephone transmission', *Bell Syst. Tech. J.*, **8,** 536–559 (1929).

(3) Pocock, L. C., 'A survey of the telephone transmission-rating problem', *J. IEE* **95,** Part III, 253–270 (1948).

(4) Swaffield, J. and deWardt, R. H., 'A reference telephone system for articulation tests', *P. O. Elect. Engrs, J.*, **43,** 1–7 (1950).

(5) Swaffield, J. and Richards, D. L., 'Transmission performance of the telephone network—A survey of some basic problems', *Paper read to Inst. Post Office Elec. Engrs* (unpublished) (1951).

(6) Richards, D. L., 'Design and analysis of subjective acoustical experiments which involve a quantal response', *Acustica*, **2,** 83–91 (1952).

(7) Richards, D. L. and Swaffield, J., 'Measurement of the communication rate of a speech link—A preliminary appraisal of the problem', *Communication Theory*, Butterworths Scientific Publications, 450–459 (1953).

(8) Pritchett, J., 'The subjective free-field calibration of an external telephone receiver by the equal-loudness method', *Acustica*, **4,** 544–546 (1954).

(9) Slavik, J. B. and Vachek, J., 'A new method for testing the intelligibility of speech', *Slabopr. Obz.*, **19,** 464–466 (1958).

(10) Swaffield, J. and Richards, D. L., 'Rating of speech links and performance of telephone networks', *Proc. IEE*, **106**B, 65–76 (1959).

(11) Richards, D. L. and Swaffield, J., 'Assessment of speech communication links, *Proc. IEE*, **106B,** 77–92 (1959).

(12) Richards, D. L., 'Transmission performance assessment for telephone network planning', *Proc. IEE*, **111,** 931–939 (1964).

(13) Gleiss, N., 'Subjective assessments in telephony', *Tele* (English Edition), **20,** 29–44 (1968).

(14) *CCITT Recommendation P. 41*, 'Description of the ARAEN', White Book, Volume V (1968).

(15) *CCITT Recommendation P. 42*, 'Systems for the determination of reference equivalents', White Book, Volume V (1968).

(16) *CCITT Recommendation P. 74*, 'Methods for subjective determination of transmission quality', White Book, Volume V (1968).

(17) *Supplement No. 8*, 'Methods used by the British telephone administration for rating telephone speech links', White Book, Volume V (1968).

(18) *Supplement No. 9*, 'Absolute calibration of the ARAEN at the CCITT Laboratory', White Book, Volume V (1968).

(19) *Supplement No. 10*, 'ARAEN volume meter or speech voltmeter', White Book, Volume V (1968).

3.2 *Loudness rating*

(1) Braun, K., 'Reference equivalent and its calculation from the frequency response curve of a transmission system', *Telegr.–Fernspr.–Funk.–u Fernseh-Tech.*, **28,** 311–318 (1939).

(2) Braun, K., 'Theoretical and experimental examination of reference

equivalent and sound intensity', *Telegr.–Fernspr.–Funk.–u Fernseh–Tech.*, **29,** 31–37 (1940).

(3) Braun, K., 'A new reference equivalent meter with objective excitation and indicating device', *Telegr.–Fernspr.–Funk.–u Fernseh–Tech.*, **29,** 223–227 (1940).

(4) Munson, W. A., 'Relation between the theory of hearing and the interpretation of speech sounds', *J. Acoust. Soc. Amer.*, **17,** 103A (1945).

(5) Pollack, I., 'The effect of white noise on the loudness of speech of assigned average level', *J. Acoust. Soc. Amer.*, **21,** 225–258 (1949).

(6) Fletcher, H., 'A method of calculating hearing loss for speech from an audiogram', *J. Acoust. Soc. Amer.*, **22,** 1–5 (also *Bell Monograph 1731*) (1950).

(7) Hawkins, J. E. and Stevens, S. S., 'The masking of pure tones and of speech by white noise', *J. Acoust. Soc. Amer.*, **22,** 6–13 (1950).

(8) Fletcher, H. and Galt, R. H., 'The perception of speech and its relation to telephony', *J. Acoust. Soc. Amer.*, **22,** 89–151 (especially Appendix 1) (1950).

(9) Braun, K. and Koshel, H., 'A direct-reading reference equivalent meter and its effect on the improvement of telephone communications' *Fernmeldetech. Z.*, **10,** 447–455 (1952).

(10) Fletcher, H., *Speech and hearing in communication*, von Nostrand, New York (1953).

(11) Blye, P. W., Coolidge, O. H. and Huntley, H. R., 'A revised telephone transmission rating plan', *Bell Syst. Tech. J.*, **34,** 453–472 (also *Bell Monograph 2424*) (1955).

(12) *United Kingdom Administration:* 'The design and analysis of loudness efficacy measurements', CCITT Red Book, Volume V, Annex 7, 232–241 (1962).

(13) *Japanese Telephone Administration*: 'The loudness function and the calculation of relative equivalents for the Japanese language', CCITT Red Book, Volume V, Annex 4 to Question 7/XII, 637–662 (1962).

(14) *CCITT Recommendation P.72:* 'Measurement of reference equivalents and relative equivalents', White Book, Volume V (1968).

(15) *Supplement No. 15:* 'Extract from a study of the differences between results for individual crew members in loudness balance tests', (Contribution by the British administration), White Book, Volume V (1968).

(16) Richards, D. L., 'Loudness ratings of telephone speech paths', *Proc. IEE,* **118,** 423–436 (1971).

(17) Sullivan, J. L., 'A laboratory system for measuring loudness loss of telephone connections', *Bell Syst. Tech. J.*, **50,** 2663–2739 (1971).

3.3 *Articulation*

(1) Campbell, G. A., 'Telephonic intelligibility', *Phil. Mag.*, **19,** 152–159 (1910).

(2) Fletcher, H. and Steinberg, J. C., 'Articulation testing methods', *Bell Syst. Tech. J.*, **8,** 806–854 (1929).

(3) Collard, J., 'A theoretical study of the articulation and intelligibility of a telephone circuit'. *Electl Commun.*, **7,** 168–186 (1929).

(4) Collard, J., 'Calculation of the articulation of a telephone circuit from the circuit constants', *ibid.*, **8,** 141–163 (1930).

(5) Collard, J., 'The accurate measurement of articulation', *P. O. Elect. Engrs'. J.*, **23,** 25–35 (1930).

(6) Pocock, L. C., 'The calculation of articulation for effective rating of telephone circuits', *Electl Commun.*, **18,** 120–132 (1939).

(7) French, N. R. and Steinberg, J. C., 'Factors governing the intelligibility of speech sounds', *J. Acoust. Soc. Amer.*, **19,** 90–119 (1947).

(8) Beranek, L. L., 'The design of speech communication systems', *Proc. Inst. Radio Engrs.*, **35,** 880–890 (1947).

(9) Radley, W. G., 'Speech communication under conditions of deafness and loud noise', *J. IEE,* **95,** Part I, 201–217 (With Appendix by J. O. Ackroyd) (1948).

(10) Egan, J. P., 'Articulation testing methods', *Laryngoscope, St Louis,* **58,** 955–991 (1948).

(11) Fletcher, H. and Galt, R. H., 'The perception of speech and its relation to telephony', *J. Acoust. Soc. Amer.*, **22,** 89–151 (1950).

(12) Swaffield, J. and deWardt, R. H., 'A reference telephone system for articulation tests', *P. O. Elect. Engrs'. J.*, **43,** 1–7 (1950).

(13) deWardt, R. H., 'The conduct of articulation measurements', *P. O. Elect. Engrs'. J.*, **44,** 159–163 (1952).

(14) Markman, F., 'Assessment of transmission properties of telephone instruments based on articulation tests', *Ericsson Tech.*, **9,** 8–50 (1953).

(15) Miller, G. A. and Nicely, P. A., 'An analysis of perceptual confusions among some English consonants', *J. Acoust. Soc.Amer.*, **27,**338–352(1955).

(16) Richards, D. L. and Archbold, R. B., 'A development of the Collard principle of articulation calculation', *Proc. IEE,* **103B,** 679–691 (1956).

(17) Black, J. W., 'Multiple-choice intelligibility tests', *J. Speech Hear. Disorders,* **22,** 213–235 (1957).

(18) Fairbanks, G., 'Test of phonemic differentiation. The rhyme test', *J. Acoust. Soc. Amer.*, **30,** 596–600 (1958).

(19) *American Standards Association, ASA S3.2-1960:* 'American standard method for measurement of monosyllabic word intelligibility',

(20) Black, J. W., 'Speech intelligibility: A summary of recent research' *J. Commun.*, **11,** 87–94 (1961).

(21) Stevens, K. N., 'Simplified nonsense-syllable tests for analytic evaluation of speech transmission systems', *J. Acoust. Soc. Amer.*, **34,** 729 (abstract) (1962).

(22) Kryter, K. D., 'Methods for the calculation and use of the articulation index', *J. Acoust. Soc. Amer.*, **34,** 1689–1697 (1962).

(23) Kryter, K. D., 'Validation of the articulation index', *J. Acoust. Soc. Amer.*, **34,** 1698–1702 (1962).

(24) *CCITT Recommendation P. 45:* 'Measurement of the AEN value of a commercial telephone system (sending and receiving) by comparison with the SRAEN', Red Book, Volume V, 69–114 (This version of the recommendation is in greater detail than that published in 1968) (1962).

(25) Hecker, M. H. L. and Phyte, D. L., 'Diagnostic articulation testing in the design of speech-processing devices', *J. Acoust. Soc. Amer.*, **35,** 1899 (abstract) (1963).

(26) Black, J. W. and Haagen, C. H., 'Multiple-choice intelligibility tests, Forms A and B', *J. Speech Hear. Disorders,* **28,** 77–86 (1963).

(27) House, A. S. *et al.*, 'Articulation-testing methods: Consonantal differentiation with a closed-response set', *J. Acoust. Soc. Amer.*, **37,** 158–166 (1965).

(28) Kryter, K. D. and Whitman, E. C., 'Some comparisons between rhyme

and PB-word intelligibility tests', *J. Acoust. Soc. Amer.*, **37,** 1146 (letter) (1965).

(29) *CCITT Recommendation G. 112 (P.12):* 'Articulation reference equivalent (AEN)', White Book, Volume III (V) 1968).

(30) *CCITT Recommendation P.41:* 'Description of the ARAEN', White Book, Volume V (1968).

(31) *CCITT Recommendation P.44:* 'Description and adjustment of the reference system for the determination of AEN (SRAEN)', White Book, Volume V (1968).

(32) *CCITT Recommendation P.45:* 'Measurement of the AEN value of a commercial telephone system (sending and receiving) by comparison with the SRAEN', White Book, Volume V (An abbreviated version of the 1962 text) (1968).

(33) *Supplement No. 10:* 'ARAEN volume meter or speech voltmeter', White Book, Volume V (1968).

3.4 *Listening methods using sentence material*

(1) FLETCHER, H. and STEINBERG, J. C., 'Articulation testing methods', *Bell Syst. Tech. J.*, **8,** 806–854 (1929).

(2) GRINSTED, W. H., 'The statistical assessment of standards of telephone transmission', *Engineering Supplement to Siemens Magazine*, **No. 140,** 1–10 (1937).

(3) HUGHES, J. R., 'Progress in the study of effective transmission by "Immediate Appreciation" tests and statistical assessment', *ibid.*, **No. 162,** 1–8 (1938).

(4) HUGHES, J. R., 'The comprehensive assessment of telephone communication efficiency', *J. IEE*, **89,** 195–206 (1942).

(5) CODD, W. A., 'Transmission rating of telephone systems', *Trans. Amer. IEE.*, **65,** 694–698 (1946).

(6) GRAY, D. A., 'An immediate appreciation technique for rating the performance of telephone transmission systems', *Telecomm. J. Aust.*, **8,** 352–360 (1952).

(7) RICHARDS, D. L. and BONES, R. A.: 'Measurement of the intelligibility of speech links in terms of a "reference" distortion', *Symposium on Analysis, Recognition and Synthesis of Speech*, Imperial College, London, (Unpublished) (1954).

(8) ROBERTS, W. R., 'A judgemental method for a voice communication system evaluation', *IRE International Convention Record*, **1961**, Part 5, Human Factors in Electronics, 303–308.

(9) *Australian Telephone Administration*, 'Immediate appreciation testing method', CCITT Red Book, Volume V, Annex 32, 525–534 (1962).

(10) MUNSON, W. A. and KARLIN, J. E. 'Isopreference method for evaluating speech transmission circuits', *J. Acoust. Soc. Amer.* **34,** 762–774 (1962).

(11) HECKER, M. H. L. and WILLIAMS, C. E., 'Choice of reference conditions for speech preference tests', *J. Acoust. Soc. Amer.*, **39,** 946–952 (1966).

3.5 *Talking methods*

(1) MCKOWN, F. W. and EMLING, J. W., 'A system of effective transmission data for rating telephone circuits', *Bell Syst. Tech. J.* **12,** 331–346 (1933).

(2) Barnes, E. J., Wood, A. E. and Richards, D. L., Standards of transmission and local line limits', *P. O. Elect. Engrs'. J.*, **39,** 151–155 and **40,** 8–13 (1947).
(3) Richards, D. L. and Buck, G. A., 'Telephone echo tests', *Proc. IEE*, **107B**, 553–556 (1960).
(4) Lane, H. L., Catania, A. C. and Stevens, S. S., 'Voice level: autophonic scale, perceived loudness, and the effects of sidetone', *J. Acoust. Soc. Amer.*, **33,** 160–167 (1961).
(5) Noll, M. A., 'Effects of head and air-leakage sidetone during monaural-telephone speaking', *J. Acoust. Soc. Amer.*, **36,** 598–599 (1964).
(6) Noll, M. A., 'Subjective effects of sidetone during telephone conversation', *AIEEE Trans. Commun. Electron.*, 228–231 (1964).

3.6 *Examination of conversations*

(1) Martin, W. H., 'Rating the transmission performance of telephone circuits', *Bell Syst. Tech. J.*, **10,** 116–131 (1931).
(2) McKown, F. W. and Emling, J. W., 'A system of effective transmission data for rating telephone circuits', *Bell Syst. Tech. J.*, **12,** 331–346 (1933).
(3) Pocock, L. C., 'A survey of the telephone transmission-rating problem', *J. IEE*, **95,** Part III, 253–270 (1948).
(4) Richards, D. L., 'Transmission performance of telephone connexions having long propagation times', *Het PTT–Bedrijf*, **15,** 12–24 (1967).
(5) deJong, C., 'Observations of telephone calls between the Netherlands and the USA', *ibid*, 32–36.

3.7 *Examination of participants during or after conversation*

(1) Irvin, H. D., 'Sibyl: A laboratory for simulation studies of man-machine systems', *IRE Wescon Convention Record*, Pt.4 (1958).
(2) *CCITT Annex 1:* 'Methods used by the British Telephone Administration for rating telephone speech links', Red Book, Volume V, 167–173 (1962).
(3) Riesz, R. R. and Klemmer, E. T., 'Subjective evaluation of delay and echo suppressors in telephone communications', *Bell Syst. Tech. J.* **42,** 2919–2941 (1963).
(4) Richards, D. L., 'Transmission performance assessment for telephone network planning', *Proc. IEE*, **111,** 931–940 (1964).
(5) Helder, G. K., 'Customer evaluation of telephone circuits with delay', *Bell Syst. Tech. J.*, **45,** 1157–1191 (1966).
(6) Karlin, J. E., 'Measuring the acceptability of long delay transmission circuits during the 'Early Bird', transatlantic tests in 1965', *Het PTT-Bedrijf*, **15,** 25–31 (1967).
(7) *Supplement No. 8 (to CCITT Recommendation P.74):* 'Methods used by the British telephone administration for rating telephone speech links', White Book, Volume V (1968).
(8) Duncanson, J. P., 'Measurement of user opinion of telephone transmission quality', *Fourth International Symposium on Human Factors in Telephony*, VDE-Verlag Gmbh, Berlin, 47–62 (1970).
(9) Holloway, C. M., 'Cognitive indices of speech communication efficiency',

Fifth International Symposium on Human Factors in Telecommunications, *P. O. Res. Dept*, London, Session 6.3, 1–7 (1970).

3.8 *Methods based on instrumental measurements*

(1) COLLARD, J., 'Calculation of the articulation of a telephone circuit from the circuit constants', *Electl Commun.*, **8,** 141–163 (1930).
(2) COLLARD, J., A new criterion for circuit performance', *Electl Commun.*, **11,** 226–233 (1933).
(3) FLETCHER, H. and MUNSON, W. A., 'Loudness, its definition, measurement and calculation', *J. Acoust. Soc. Amer.*, **5,** 82–108 (1933).
(4) FLETCHER, H. and MUNSON, W. A., 'Relation between loudness and masking', *J. Acoust. Soc. Amer.*, **9,** 1–10 (1937).
(5) POCOCK, L. C., 'The calculation of articulation for effective rating of telephone circuits', *Electl Commun.*, **18,** 120–132 (1939).
(6) FRENCH, N. R. and STEINBERG, J. C., 'Factors governing the intelligibility of speech sounds', *J. Acoust. Soc. Amer.*, **19,** 90–119 (1947).
(7) BERANEK, L. L., 'The design of speech communication systems', *Proc. Inst. Radio Engrs.*, **35,** 880–890 (1947).
(8) RADLEY, W. G., 'Speech communication under conditions of deafness and loud noise', *J. IEE*, **95,** Part I, 201–216 (With Appendix by J. O. Ackroyd) (1948).
(9) FLETCHER, H. and GALT, R. H., 'The perception of speech and its relation to telephony', *J. Acoust. Soc. Amer.*, **22,** 89–151 (1950).
(10) BRAUN, K. and KOSCHEL, H., 'A direct-reading instrument for measuring reference equivalent, and its influence on the improvement of telephone communication', *Fernmeldetech. Z.* **10,** 447–455 (1952).
(11) BLYE, P. W., COOLIDGE, O. H. and HUNTLEY, H. R., 'A revised telephone transmission rating plan', *Bell Syst. Tech. J.*, **34,** 453–472 (Also as Bell Mono 2424) (1955).
(12) RICHARDS, D. L. and ARCHBOLD, R. B., 'A development of the Collard principle of articulation calculation', *Proc. IEE*, **103B,** 679–691 (1956).
(13) ZAITSEV, T. E., 'The correlation method of determining fidelity and intelligibility of speech transmitted along communication channels', *Elektrosvyaz*, **10,** 38–46 (1958).
(14) LICKLIDER, J. C. R., BISBERG, A. and SCHWARZLANDER, H., 'An electronic device to measure the intelligibility of speech', *Proc. Nat. Electronics Conf.*, **15,** 329–334 (1959).
(15) SCHWARZLANDER, H., 'Intelligibility of voice communications', *Electronics*, **32,** 88–91 (1959).
(16) KRYTER, K. D., 'Methods for the calculation and use of the Articulation Index', *J. Acoust. Soc. Amer.*, **34,** 1689–1697 (1962).
(17) KRYTER, K. D., 'Validation of the Articulation Index', *ibid.*, 1698–1702 (1962).
(18) STUCKEY, C. W. and ROBERTSON, D. W., 'Investigation of an automatic voice intelligibility concept', *IEEE Trans. Audio*, **AU-12,** 3–8 (1964).
(19) RICHARDS, D. L., 'Distortion of speech by quantizing', *Electronics Letters*, **3,** 230–231 (1967).
(20) *CCITT Recommendation P.62:* 'Measurements on subscribers' telephone equipment, White Book, Volume V (1968).

(21) RICHARDS, D. L. and HUTTER, J. H., 'Communication efficiencies of speech links: Possibilities of automatic measurement', *Fourth International Symposium on Human Factors in Telephony*, VDE-Verlag, Berlin, 77–102 (1970).

(22) ERICSSON, E. A., 'Automatic monitoring of subscriber conversations' Fourth International Symposium on Human Factors in Telephony, VDE-Verlag, Berlin, 103–116 (1970).

(23) DIXON, N. R., and TAPPERT, C. C., 'Toward objective phonetic transcription—An on-line interactive technique for machine-processed speech data', *IEEE Trans. Man-machine Systems*, **MMS-11**, 202–210 (1970).

4. *EFFECTS OF SPECIFIC FACTORS*

4.1 *Introduction*

(1) FLETCHER, H., *Speech and hearing in communication*, van Nostrand, New York, (1953).

(2) BERANEK, L. L. *Acoustics*, McGraw-Hill, New York (1954).

(3) HAMSHER, D. H. (Ed.) *Communication system engineering handbook*, McGraw-Hill, New York, (especially Chapter 5; 'Communication system modeling' by FURTH, E. and Chapter 6; 'Transmission engineering of switched systems' by JOHNSON, A. C.) (1967).

(4) *CCITT Recommendation G. 101* 'The transmission plan', White Book, Volume III (1968).

(5) *CCITT Recommendation G. 120 (P. 20)* 'Transmission characteristics of national networks', White Book, Volume III (V) (1968).

(6) *CCITT Recommendation G. 141* 'Transmission losses, relative levels and attenuation distortion', White Book, Volume III (1968).

(7) *CCITT Recommendation G. 142*, 'Transmission characteristics of an international centre (CT)', White Book, Volume III (1968).

(8) *CCITT Recommendation G. 151* 'General characteristics applicable to all modern international circuits and national extension circuits', White Book, Volume III (1968).

4.2 *Listening Factors*

4.2.1 *Loss, noise and attenuation/frequency distortion*

(1) HARBOTTLE, H. R., 'Circuit noise-meter (psophometer) and its applications', *J.IEE*, **83**, 261–274 and (discussion) 275–288 (1938).

(2) SEACORD, D. F., 'Room noise at telephone locations', *Elect. Engng*, **58**, 255–257 (1939).

(3) HOTH, D. F., 'Room noise spectra at subscribers' telephone locations', *J. Acoust. Soc. Amer.*, **12**, 499–504 (1941).

(4) STRECKER, F. and VON SUSANI, C., 'Investigations of the most suitable range of reference equivalent in telephone operating', Elekt. Nachr. Tech., **19**, 241–252 (1942).

(5) *ASA Z24.3–1944* 'Sound level meters for measurement of noise and other sounds', American Standards Association.

(6) CHINN, H. A. and EISENBERG, P., 'Tonal-range and sound-intensity prefer-

ences of broadcast listeners', *Proc. Inst. Radio Engrs*, **33**, 571–580 (1945),

(7) FROST, P. B. and GOULD, E. F. H., 'Practical aspects of telephone interference arising from power systems', *J. IEE*, **93**, 255–274 (1946).

(8) KRYTER, K. D., 'Effects of ear protective devices on the intelligibility of speech in noise', *J. Acoust. Soc. Amer.*, **18**, 413–417 (1946).

(9) PETTERSSON, G. A. and SWEDENBORG, G., 'Interference in communication systems from high-power transmission networks', *Tek. Tidskr.*, **77**, 323–328 (in Swedish) (1947).

(10) MOIR, J., 'Perfect v. pleasing reproduction', *Electron. Engng*, **19**, 23–26 (1947).

(11) ENGEL, H., 'Selector noises in telephone exchangers', *Tech. Mitt. Telegr.-u. TelephVerw.*, **25**, 92–96 (1947).

(12) OLSON, H. F., 'Frequency range preference for speech or music', *J. Acoust. Soc. Amer.* **19**, 549–555 (1947).

(13) ROMANOW, F. F. and HAWLEY, M. S., 'Proposed method of rating microphones and loudspeakers for systems use', *Proc. Inst. Radio Engrs.*, **35**, 953–960 (1947).

(14) RADLEY, W. G., 'Speech communication under conditions of deafness and loud noise', *J. IEE*, **95**, Part 1, 201–217 (1948).

(15) SOMERVILLE, T. and BROWNLESS, S. F., 'Listeners' sound-level preferences', *BBC Quarterly*, **3**, 245–250 (1949).

(16) BONVALLET, G. L., 'Traffic, industrial and residential noise', *J. Acoust. Soc. Amer.*, **24**, 435–439 (1951).

(17) BERANEK, L. L., 'Criteria for office quieting based on questionnaire rating studies', *J. Acoust. Soc. Amer.* **28**, 833–852 (1956).

(18) RICHARDS, D. L. and ARCHBOLD, R. B., 'A development of the Collard principle of articulation calculation', *Proc. IEE*, **103B**, 679–691 (1956).

(19) LICKLIDER, J. C. R. and GUTTMAN, N., 'Masking of speech by line-spectrum interference', *J. Acoust. Soc. Amer.*, **29**, 287–296 (1957).

(20) POLLACK, I. and PICKETT, J. M., 'Masking of speech by noise at high sound levels', *J. Acoust. Soc. Amer.*, **30**, 127–130 (1958).

(21) PICKETT, J. M., 'Limits of direct speech communication in noise', *J. Acoust. Soc. Amer.*, **30**, 278–281 (1958).

(22) KOSHIKAWA, T., 'Scaling of the naturalness for rating the transmission quality', *J. Inst. Elec. Commun. Engrs. Japan*, **41**, 482–488 (1958).

(23) VOLF, V. and KOPIC, A., 'The determination of the effects of intermittent noise on the quality of telephone communications', *Slabopr. Obz.*, **20**, 71–76 (1959).

(24) COOLIDGE, O. H. and REIER, G. C., 'An appraisal of received telephone speech volume', *Bell Syst. Tech. J.*, **38**, 877–897 (also as Bell Monograph 3255) (1959).

(25) RICHARDS, D. L. and SWAFFIELD, J., 'Assessment of speech communication links', *Proc. IEE*, **106B**, 77–92 (1959).

(26) COCHRAN, W. T. and LEWINSKI, D. A., 'A new measuring set for message circuit noise', *Bell Syst. Tech. J.*, **39**, 911–931 (1960).

(27) *IEC Recommendation Publication 123* 'Recommendations for sound level meters' (1961).

(28) Annexes 1 (Japanese Telephone Company), 2 (Federal German Republic Post Office), 3 (Siemens and Halske A.G.) and 4 (Czechoslovak Telephone Administration) to question 3 of CCITT Study Group XII 'Measurement of the disturbing effect of clicks', Red Book, Volume V, 543–571 (1962).

(29) Annex 5 to question 3 of CCITT Study Group XII 'Extracts from the specification for CISPR radio interference measuring apparatus for the frequency range 0.15 to 30 MHz', Red Book, Volume V, 572–576 (1962).

(30) Question 4 of CCITT Study Group XII—'Effect of circuit noise on transmission performance', Comments by SG12 and Annex 1, Red Book, Volume V, 576–594 (1962).

(31) Boeryd, A., 'Some reactions of telephone users during conversation, *Teleteknik* (English Edition), **7,** 97–103 (1963).

(32) Richards, D. L., 'Transmission performance assessment for telephone network planning', *Proc. IEE,* 111, 931–940 (1964).

(33) Webster, J. C., 'Speech communications as limited by ambient noise', *J. Acoust. Soc. Amer.*, **37,** 692–699 (1965).

(34) Annex 4 to question 4 of CCITT Study Group XII 'Contribution by the American Telephone and Telegraph Company' (Information on conversion of readings with various circuit noise meters), Red Book, Volume V bis, 151–152 (1965).

(35) *IEC Recommendation Publication 179* 'Precision sound level meters' (1965).

(36) *CCITT Recommendation G.111 (P.11)* (Reference equivalents in an international connexion', White Book, Volume III (V) (1968).

(37) *CCITT Recommendation G.112 (P.12)* 'Articulation reference equivalent (AEN)', White Book, Volume III (V) (1968).

(38) *CCITT Recommendation G.113 (P.13)*, 'Transmission impairments and Noise', White Book, Volume III (V) (1968).

(39) *CCITT Recommendation G.132*, 'Attenuation distortion', White Book, Volume III (1968).

(40) *CCITT Recommendation P.53*, 'Psophometers (Apparatus for the objective measurement of circuit noise)', White Book, Volume V (1968).

(41) *CCITT Recommendation P.54*, 'Apparatus for the objective measurement of room noise', White Book, Volume V (1968).

(42) Koop, E. J., 'The transmission performance preferences and tolerances of telephone users', *Telecomm. J. Aust.*, **18,** 28–39 (1968).

(43) Gleiss, N., 'The effect of bandwidth restriction on speech transmission quality in telephony', *Fourth International Symposium on Human Factors in Telephony*, VDE-Verlag, Berlin, 1–6 (1970).

4.2.2 *Nonlinear distortion*

(1) von Braunmuhl, H. J. and Weber, W., 'The disturbance capability of non-linear distortion', *Akust. Z.*, **2,** 135 (1937).

(2) Lickider, J. C. R., 'Effects of amplitude distortion on the intelligibility of speech, *J. Acoust. Soc. Amer.*, **18,** 429–434 (1946).

(3) Cunningham, W. J. *et al.*, 'The influence of amplitude limiting and frequency selectivity upon the performance of radio receivers in noise', *Proc. Inst. Radio Engrs*, **35,** 1021–1025 (1947).

(4) Kryter, K. D., Licklider, J. C. R. and Stevens, S. S., 'Premodulation clipping in AM voice communication', *J. Acoust. Soc. Amer.*, **19,** 125–131 (1947).

(5) Bennett, W. R., 'Spectra of quantized signals', *Bell Syst. Tech. J.*, **27,** 446–472 (1948).
(6) Licklider, J. C. R. and Pollack, I., 'Effects of differentiation, integration and infinite peak clipping upon the intelligibility of speech, *J. Acoust. Soc. Amer.*, **20,** 42–51 (1948).
(7) Miller, G. A. and Licklider, J. C. R., 'The intelligibility of interrupted speech', *J. Acoust. Soc. Amer.*, **22,** 167–173 (1950).
(8) Licklider, J. C. R., 'The intelligibility of amplitude-dichotomized, time-quantized speech waves', *J. Acoust. Soc. Amer.*, **22,** 820–823 (1950).
(9) Pollack, I., 'On the effect of frequency and amplitude distortion on the intelligibility of speech in noise', *J. Acoust. Soc. Amer.*, **24,** 538–540 (1952).
(10) van't Groenewout, H. W. F., 'Non-linear distortion in pulse code modulation systems', *Communication Theory*, Butterworths Scientific Publications, 153–160 (1953).
(11) Booton, R. C., Jr., 'The analysis of non-linear control systems with random inputs', *Symposium on Non-linear Circuit Analysis, Polytechnic Institute of Brooklyn,* April 23–24 (1953).
(12) van de Weg, H., 'Quantizing noise of a single integration delta modulation system with an N-digit code', *Philips Res. Rep.*, **8,** 367–385 (1953).
(13) Dukes, J. N. C., 'The effect of severe amplitude limitation on certain types of random signals. A clue to the intelligibility of infinitely clipped speech', *Proc. IEE*, **102C,** 88–95 (1955).
(14) Zetterberg, L. H., 'A comparison between Delta and Pulse Code Modulation', *Ericsson Techn. No. 1,* 95–154 (1955).
(15) Peterson, G. E., Silvertsen, E. and Subrahmanyam, D. L., 'Intelligibility of diphasic speech', *J. Acoust. Soc. Amer.*, **28,** 404–411 (1956).
(16) David, E. E., 'Naturalness and distortion in speech processing devices', *J. Acoust. Soc. Amer.*, **28,** 586–589 (1956).
(17) Smith, B., 'Instantaneous companding of quantized signals', *Bell Syst. Tech. J.*, **36,** 653–709 (1957).
(18) Wathen-Dunn, W. and Lipke, D. W., 'On the power gained by clipped speech', *J. Acoust. Soc. Amer.*, **30,** 36–40 (1958).
(19) Max, J., 'Quantizing for minimum distortion', *IRE Trans. Inf. Theory,* **IT-6,** 7–12 (1960).
(20) Wigan, E. R., 'New distortion criterion'. Electronic Technology (formerly *Wireless Engr.*, now *Indust. Electron.*), **38,** 128–137 and 163–174 (1961).
(21) Purton, R. F., 'A survey of telephone speech-signal statistics and their significance in the choice of a P.C.M. companding law', *Proc. IEE,* **109B,** 60–66 and (discussion) 485–487 (1962).
(22) Law, H. B. and Seymour, R. A., 'A reference distortion system using modulated noise', *Proc. IEE,* **109B,** 484–485 (1962).
(23) Velichkin, A. I., 'Amplitude clipping of speech', *Soviet Phys. Acoust.*, **8,** 130–134 (1962).
(24) Velichkin, A. I., 'Correlation function and spectral density of quantized speech', *Soviet Phys. Acoust.*, **9,** 10–14 (1963).
(25) Velichkin, A. I., 'Optimum characteristics of quantizers', *Radio Engng. Electron. Phys.*, **18,** 1–7 (1963).
(26) Richards, D. L., 'Transmission performance assessment for telephone network planning', *Proc. IEE,* **111,** 931–940 (1964).
(27) Keneko, H., 'PCM distortion due to encoding without holding the sampled value', *Electron. Commun. Japan*, **47,** 38 (1964).

(28) Richards, D. L., 'Speech transmission performance of P.C.M. systems', *Electron. Lett.*, **1,** 40–41 (1965).

(29) Panter, P. F., *Modulation, Noise and Spectral Analysis*, McGraw-Hill, New York (1965).

(30) Velichkin, A. I., 'Some aspects of decoding of messages in communication systems with pulse-code modulation', *Rádióteknika*, **20,** 127–129 (English translation) (1965).

(31) O'Neal, J. B., 'Delta modulation quantizing noise analytical and computer simulation results for Gaussian and television signals', *Bell Syst. Tech. J.*, **45,** 117–142 (also with title: 'Quantizing noise in delta modulation systems for Gaussian and TV signals', as Bell Mono. 5145) (1966).

(32) McDonald, R. A., 'Signal-to-noise performance and idle channel performance of differential pulse code modulation systems, with particular applications to voice signals', *Bell Syst. Tech. J.*, **45,** 1123–1151 (1966).

(33) O'Neal, J. B., 'A bound on signal-to-quantizing noise ratios for digital encoding systems', *Proc. IEEE,* **55,** 287–292 (1967).

(34) Williams, G., 'Quantizing for minimum error with particular reference to speech', *Electron. Lett.*, **3,** 134–135 (1967).

(35) Protonotarios, E. N., 'Slope overload noise in differential pulse code modulation', *Bell Syst. Tech. J.*, **46,** 2119–2162 (1967).

(36) Richards, D. L., 'Distortion of speech by quantizing', *Electron. Lett.*, **3,** 230–231 (1967).

(37) Decina, M. and Giacomelli, L., 'Signal to noise ratio dependence on quantization characteristics in PCM telephone systems', *Note, Recensioni e. Notizie*, **17,** 321–367 (in Italian) (1968).

(38) Richards, D. L., 'Transmission performance of telephone networks containing p.c.m. links', *Proc. IEE,* **115,** 1245–1258 (1968).

(39) Richards, D. L. and Berry, R. W., 'Impairment of telephone speech by quantizing' *4th International Symposium on Human Factors in Telephony*, 29–46 (1968).

(40) *CCITT Recommendation G.711*, 'Characteristics of pcm systems', White Book, Volume III (1968).

(41) Johnson, F. B., 'Calculating delta modulation performance', *IEEE Trans. Audio*, **AU-16,** 121–129 (1968).

(42) Wang, P. P., 'Idle channel noise of delta modulation, *IEEE Trans. Commun. Technol.*, **COM-16,** 737–742 (1968).

(43) Aaron, M. R., Fleishman, R. A. and Protonotarios, E. N., 'Response of delta modulation to gaussian signals', *Bell Syst. Tech. J.*, **48,** 1167–1195 (1969).

(44) Goodman, D. J., 'Delta modulation granular quantizing noise', *Bell Syst. Tech. J.*, **48,** 1197–1218 (1969).

(45) Iwersen, J. E., 'Calculated quantization noise of single integration delta-modulation coders', *Bell Syst. Tech. J.*, **48,** 2359–2389 (1969).

(46) Cattermole, K. W., *Principles of pulse code modulation*, Iliffe Books, London (1969).

(47) Laane, R. R., 'Measured quantizing noise spectrum for single integration delta modulation coders', *Bell Syst. Tech. J.*, **49,** 191–195 (1970).

(48) Cartmale, A. A. and Steele, R., 'Calculating the performance of syllabically companded delta-sigma modulators', *Proc. IEE*, **117,** 1915–1921 (1970).

(49) Hooijkamp, C., 'Comparison of the transmission quality of speech of a number of transmission systems employing PCM and Delta Modulation by means of a listening opinion test', *Fifth International Symposium on Human Factors in Telecommunications*, Post Office Research Department, London, Section 2.1, 1–8 (1970).

4.2.3 *Group delay/frequency distortion*

(1) Steinberg, J. C., 'Effect of phase distortion on telephone quality', *Bell Syst. Tech. J.*, **9**, 550–566 (1930).
(2) Flanagan, J. L., 'Effects of delay distortion upon intelligibility and quality of speech', *J. Acoust. Soc. Amer.*, **23**, 303–307 (1951).
(3) Lochner, J. P. A., and Burger, J. F., 'The subjective masking of short-time-delayed echoes by their primary sound and their contribution to the intelligibility of speech', *Acustica*, **8**, 1–10 (1958).
(4) Lochner, J. P. A., and Burger, J. F., 'The intelligibility of speech under reverberant conditions', *Acustica*, **11**, 195–200 (1961).
(5) Gatfield, A. G., 'Delay distortion in telephone lines', *IRE International Convention Record, Part 8, Communications Systems — Techniques*, 109–118 (1961).
(6) *CCITT Recommendation G.133 (P.15)* 'Group-delay distortion' White Book, Volume III (V) (1968).
(7) *CCITT Recommendation G.232* '12-channel terminal equipments' White Book, Volume III (1968).

4.2.4 *Combination of the effects of different listening factors*

(1) Richards, D. L., 'Transmission performance assessment for telephone network planning', *Proc. IEE*, **111**, 931–940 (1964).
(2) Gleiss, N., 'Subjective measurement in telephony', *Tele* (English Edition), **20**, 29–44 (1967).
(3) Allnatt, J. W., 'Subjective studies concerning the Viewphone Project', *4th International Symposium on Human Factors in Telephony*, VDE-Verlag, Berlin 259–268 (1968).

4.3 *Talking factors*

(1) Black, J. W., 'Some effects of auditory stimuli upon voice', *Aviat. Med.*, 251–255 and 277 (1950).
(2) Black, J. W., 'The effects of room characteristics upon vocal rate and intensity', *J. Acoust. Soc. Amer.*, **22**, 174–176 (1950).
(3) Lee, B. S., 'Some effects of sidetone delay', *J. Acoust. Soc. Amer.*, **22**, 639 (letter) (1950).
(4) Black, J. W., 'The effect of delayed sidetone upon vocal rate and intensity', *J. Speech Hear. Disorders*, **16**, 56–60 (1951).
(5) Richards, D. L., 'Some aspects of the behaviour of telephone users as affected by the physical properties of the circuit', *Communication Theory*, Butterworths Scientific Publications, 442–449 (1953).

(6) PHILLIPS, G. M., 'Echo and its effects on the telephone user', *Bell Lab. Record*, **32**, 281–284 (1954).

(7) PICKETT, J. M., 'Effects of vocal force on the intelligibility of speech sounds', *J. Acoust. Soc. Amer.*, **28**, 902–905 (1956).

(8) DREHER, J. J. and O'NEILL, J. J., 'Effects of ambient noise on speaker intelligibility for words and phrases', *J. Acoust. Soc. Amer.*, **29**, 1320–1323 (1957) (also *Laryngoscope*, **68**, St Louis, 539–548 (1958).

(9) RICHARDS, D. L., and BUCK, G. A., 'Telephone echo tests', *Proc. IEE*, **107B**, 553–556 (1960).

(10) LANE, H. L., CATANIA, A. C., and STEVENS, S. S., 'Voice level: Autophonic scale, perceived loudness and effects of sidetone', *J. Acoust. Soc. Amer.*, **33**, 160–167 (1961).

(11) RICHARDS, D. L., 'Theoretical study of the functioning of echo suppressors', *Teletechnik* (English Edition), **7**, 71–74 (1963).

(12) PFLEIDERER, F., 'Results of investigations for determining the most desirable sidetone (return loss) for subscriber sets in connection with different circuit attenuations', *Teleteknik* (English Edition) **7**, 95–96 (1963).

(13) NOLL, A. M., 'Effects of head and air-leakage sidetone during monaural-telephone speaking', *J. Acoust. Soc. Amer.*, **36**, 598–599 (1964).

(14) NOLL, A. M., 'Subjective effects of sidetone during telephone conversation', *IEEE Trans. Commun. Electron.*, **83**, 228-231 (1964).

(15) ALLANSON, J. J., *et al.*, 'Delayed auditory feedback. An assessment of speech disturbance as a function of the filtration and distortion of the fed-back signal', *J. Sound Vib.*, **2**, 409–422 (1965).

(16) SINGH, S., and BLACK, J. W., 'A study of nonsense syllables spoken by two language groups in varying conditions of sidetone and reading rate', *Lang. & Speech*, **8**, 208–213 (1965).

(17) *CCITT Recommendation G.122* 'Influence of national networks on stability and echo in international connections', White Book, Volume III (1968).

(18) *CCITT Recommendation G.131* 'Stability and echo', White Book, Volume III (1968).

4.4 *Conversational factors*

4.4.1 *Asymmetry*

(1) RICHARDS, D. L., 'Some aspects of the behaviour of telephone users as affected by the physical properties of the circuit', *Communication Theory*, Butterworths Scientific Publications, 442–449 (1953).

(2) BOERYD, A., 'Subscriber reaction due to unbalanced transmission levels', *Het PTT-Bedrijf*, **15**, 39–43 (1967).

(3) GLEISS, N., 'Subjective measurements in telephony', *Tele* (English Edition) **20**, 29–44 (1968).

4.4.2 *Propagation time*

(1) BLACKWELL, O. B., 'The time factor in telephone transmission', *Bell Syst. Tech. J.*, **11**, 53–66 (1932).

(2) RICHARDS, D. L., 'Conversation performance of speech links subject to

long propagation times', *International Conference on Satellite Communication* (IEE, London) 247–251 (1962).

(3) RIESZ, R. R., and KLEMMER, E. T., 'Subjective evaluation of delay and echo-suppressors', *Bell Syst. Tech. J.*, **42,** 2919–2941 (1963).

(4) ANDREWS, F. T., 'The effect of long delays and echo-suppressors on telephone communications', *Teleteknik* (English Edition), **7,** 75–86 (1963).

(5) HALSTROM, H. L., 'Can long propagation delays be rendered less noticeable by acoustical masking?', *Teleteknik* (English Edition), **7,** 87–94 (1963).

(6) GOULD, R. C., 'Echo-suppression and time delay, A Subjective evaluation', *IEEE Trans. Commun. Technol.*, **COM-12,** No. 3, 74–82 (1964).

(7) HELDER, G. K., 'Customer evaluation of telephone circuits with delay', *Bell Syst. Tech. J.*, **45,** 1157–1191 (1966).

(8) VARTABEDIAN, A. G., 'The effects of transmission delay in four-wire teleconferencing', *Bell Syst. Tech. J.*, **45,** 1673–1688 (1966).

(9) RICHARDS, D. L., 'Transmission performance of telephone connexions having long propagation times', *Het PTT-Bedrijf* (English Edition), **15,** 12–14 (1967).

(10) KARLIN, J. E., 'Measuring the acceptability of long-delay transmission circuits used during the 'Early Bird' transatlantic tests in 1965', *Het PTT-Bedrijf* (English Edition), **15,** 25–31 (1967).

(11) KLEMMER, E. T., 'Subjective evaluation of transmission delay in telephone conversations', *Bell Syst. Tech. J.*, **46,** 1141–1147 (1967).

(12) KRAUSS, R. M. and BRICKER, P. D., 'Effects of transmission and access delay on the efficiency of verbal communication' *J. Acoust. Soc. Amer.*, **41,** 286–292 (1967).

(13) HUTTER, J., 'Customer response to telephone circuits routed via a synchronous-orbit satellite', *P. O. Elect. Engrs'. J.*, **60,** 181–186 (1967).

(14) *CCITT Recommendation G. 114 (P. 14)* 'Mean one-way propagation time', White Book, Volume III (V) (1968).

(15) *Supplement No. 1* 'Subscriber tolerance of lengthened propagation time, echo and echo suppressors', (Contribution by the Telephone Association of Canada), White Book, Volume V (1968).

(16) *Supplement No. 2* 'Subjective evaluation of transmission performance on telephone connexions with long propagation time', (Contribution by K. D. D.–Japan), White Book, Volume V (1968).

(17) *Supplement No. 3* 'Subscribers' reaction to "Early Bird" circuits, (Contribution by the administrations of Denmark, Norway and Sweden), White Book, Volume V (1968).

(18) *Supplement No. 4* 'Telephone tests on satellite HS303', (Contribution by Telespazio, Italy), White Book, Volume V (1968).

(19) *Supplement No. 5* 'Results of measurements and traffic observations of satellite circuits', (Contribution by the Federal Republic of Germany), White Book, Volume V (1968).

(20) *Supplement No. 6* 'Correlation of the telephone transmission impairment due to long propagation time and that due to noise', (Contribution by the Telephone Association of Canada), White Book, Volume V (1968).

(21) HOHNE, H. D., 'Influence of long transmission delays and reverberation on the telephone conversations of testpersons', *Fourth International Symposium on Human Factors in Telephony*, VDE-Verlag, Berlin, 7–28 (1970).

(22) WILLIAMS, G., 'Subjective evaluation of unsuppressed echo in simulated

long delay telephone communications', *Fifth International Symposium on Human Factors in Telecommunications*, Post Office Research Department, London, Session 2.2, 1–8 (1970).

(23) HOHNE, H. D., 'Final results of tests investigating the influence of transmission delay', *ibid.* Session 2.2, 1–10.

4.4.3 *Voice-operated devices*

(1) CALDWELL, S. E., *Switching theory and logical design*, John Wiley and Sons, Inc., New York (1958).

(2) BUSALA, A., 'Fundamental considerations in the design of a voice-switched Speakerphone', *Bell Syst. Tech. J.*, **39,** 265–294 (also as Bell Mono. 3512) (1960).

(3) MIEDMA, H. and SCHACHTMAN, M. G., 'TASI quality—Effect of speech detectors and interpolation', *Bell Syst. Tech. J.*, **41,** 1455–1473 (1962).

(4) BRADY, P. T. and HELDER, G. K., 'Echo-suppressor design in telephone communications', *Bell Syst. Tech. J.*, **42,** 2893–2918 (also as Bell Monograph 4675) (1963).

(5) GARDNER, M. B. and NELSON, J. R., 'Combating echo in speech circuits with long delay', *J. Acoust. Soc. Amer.*, **35,** 1762–1772 (1963).

(6) RICHARDS, D. L., 'Theoretical study of the functioning of echo-suppressors', *Teleteknik* (English Edition), **7,** 71–74 (1963).

(7) SHANKS, P. H., 'A new echo-suppressor for long-distance communications', *P. O. Elect. Engrs'*, *J.*, **60,** 288–292 (1968).

(8) *CCITT Recommendation G. 161* 'Definitions relating to echo suppressors and characteristics of a far-end operated, differential, half-echo suppressor' White Book, Volume III (1968).

(9) RICHARDS, D. L. and HUTTER, J., 'Echo suppressors for telephone connections having long propagation times', *Proc. IEE*, **116,** 955–964 (1969).

4.4.4 *Handsfree telephony*

(1) EMLING, J. W., 'General aspects of hands-free telephony', *AIEE Trans.* Part 1, **76,** 201–205 (1957).

(2) GARDNER, M. B., 'A study of talking distance and related parameters in hands-free telephony', *Bell Syst. Tech. J.*, **39,** 1529–1551 (1960).

(3) GARDNER, M. B., 'Effect of noise, system gain and assigned task on talking levels in loudspeaking communication', *J. Acoust. Soc. Amer.*, **40,** 955–965 (1966).

(4) HEBERLE, W., 'Loudspeaking telephone versus standard handset', *Fourth International Symposium on Human Factors in Telephony*, VDE-Verlag, Berlin, 63–76 (1970).

(5) FLOHRER, W., 'Human aspects of loudspeaking telephones', *Fifth International Symposium on Human Factors in Telecommunications*, Post Office Research Department, London, Session 2.4, 1–7 (1970).

4.5 *Crosstalk*

(1) HAWKINS, J. E. and STEVENS, S. S., 'The masking of pure tones and of speech by white noise', *J. Acoust. Soc. Amer.*, **22,** 6–13 (1950).

(2) *CCITT Recommendation G. 134* 'Linear crosstalk', White Book, Volume III (1968).

(3) SEN, T. K., 'Masking of crosstalk by speech and noise', *Bell Syst. Tech. J.*, **49,** 561–584 (1970).

4.6 *Miscellaneous*

Training

(1) BLACK, J. W. and MASON, H. W., 'Training for voice communication', *J. Acoust. Soc. Amer.*, **18,** 441–445 (1946).

(2) MOSER, H. M. and DREHER, J. J., 'The effects of training on listeners in intelligibility studies', *J. Acoust. Soc. Amer.*, **27,** 1213–1219 (1955).

Context and language

(3) MILLER, G. A., HEISE, G. A. and LICHTEN, W., 'The intelligibility of speech as a function of the context of the test materials', *J. Exp. Psychol.* **41,** 329–335 (1951).

(4) HIRSH, I. J., REYNOLDS, E. G. and JOSEPH, M., 'Intelligibility of different speech materials', *J. Acoust. Soc. Amer.*, **26,** 530–538 (1954).

(5) HOWES, D., 'On the relation between the intelligibility and frequency of occurrence of English words', *J. Acoust. Soc. Amer.*, **29,** 296–305 (1957).

(6) POLLACK, I., 'Message procedures for unfavourable communication conditions', *J. Acoust. Soc., Amer.*, **30,** 196–201 (1958).

(7) CARTERETTE, E. C., 'Message repetition and receiver confirmation of messages in noise', *J. Acoust. Soc. Amer.*, **30,** 846–855 (1958).

(8) LANE, H., 'Foreign accent and speech distortion', *J. Acoust. Soc. Amer.*, **35,** 451–453 (1963).

(9) STOWE, A. N., HARRIS, W. P. and HAMPTON, D. B., 'Signal and context components of word-recognition behaviour', *J. Acoust. Soc. Amer.*, **35,** 639–644 (1963).

(10) TRAUL, G. N. and BLACK, J. W., 'The effect of context on aural perception of words', *J. Speech Hear. Res.*, **8,** 363–369 (1965).

Voice Recognition

(11) DAVID, E. E., 'Artificial auditory recognition in telephony', *IBM J. Res. Dev.* **2,** 294–309 (1958).

Vision

(12) SUMBY, W. H. and POLLACK, I., 'Visual contribution to speech intelligibility in noise', *J. Acoust. Soc. Amer.*, **26,** 212–215 (1954).

(13) HALL, A. D., 'Developing Picturephone service', *Bell Teleph. Mag.*, **43,** 14–21 (1964).

(14) POLLACK, I., 'Interaction between auditory and visual information sources in word identification', *Lang.* and *Speech*, **7,** 79–83 (1964).

(15) REID, A. A. L., 'Comparisons between telephone and face-to-face conversation', *Fifth International Symposium on Human Factors in Telecommunications*, Post Office Research Department, London, Session 6.6, 1–9 (1970).

Teleconferencing

(16) GARDNER, M. B., 'Effect of noise on listening levels in conference telephony', *J. Acoust. Soc. Amer.*, **36,** 2354–2362 (1964).

(17) VARTABEDIAN, A. G., 'The effects of transmission delay in four-wire teleconferencing', *Bell Syst. Tech. J.*, **45,** 1673–1688 (1966).

5. *TELEPHONE SETS*

5.1 *Handset telephones*

(1) CAMPBELL, G. A. and FOSTER, R. M., 'Maximum output networks for telephone substation and repeater circuits' *Trans. AIEE*, **39,** 231–280 (1920).

(2) ALDRIDGE, A. J., BARNES, E. J. and FOULGER, E., 'A new CB microtelephone', *P. O. Elect. Engrs'. J.*, **22,** 185–193 (1929).

(3) JOHNSON, K. S., *Transmission circuits for telephonic communication*, van Nostrand, New York (1929).

(4) *CCIF Red Book*, 'Special precautions to be taken when measuring the reference equivalents of hand-microtelephones', 53–60 (1931).

(5) JONES, W. C. and INGLIS, A. H., 'Development of a handset for telephone stations', *Bell Syst. Tech. J.*, **11,** 245–263 (1932).

(6) WIGAN, E. R., 'The effect of sidetone on the efficiency of telephone systems and the principles governing sidetone control', *P. O. Elect. Engrs'. J.*, **25,** 197–204 (1932) (See also Letter, *ibid.*, **27,** 69 (1934).

(7) *CCIF White Book*, Volume IV, 183–190, 'Measurements by voice and ear. Precautions to be taken during telephonometric measurements. Position of the lips with respect to the microphone' (1934).

(8) MCMILLAN, D., 'Some performance characteristics of the subscriber's telephone transmitter', *P. O. Elect. Engrs'. J.*, **28,** 167–178 and 313–317 (1935).

(9) INGLIS, A. H., 'Transmission features of the new telephone sets', *Bell Syst. Tech. J.*, **17,** 358–381 (1938).

(10) HOTH, D. F., 'Room noise spectra at subscribers' telephone locations', *J. Acoust. Soc. Amer.*, **12,** 499–504 (1941).

(11) ACKROYD, J. O., 'Theoretical consideration of the optimum frequency response characteristic of a hearing aid', Appendix to RADLEY, W. G., 'Speech communication under conditions of deafness or loud noise', *J. IEE*, **95,** 201–217 (1948).

(12) *CCIF Yellow Book*, Volume IV, 68, 'Method proposed for the determination of AEN in Europe: program of the 8th Series of Experiments of the CCIF Laboratory. Volume of speech sounds and speaking distance to be used' (1949).

(13) INGLIS, A. H. and TUFFNELL, W. L., 'An improved telephone set', *Bell Syst. Tech. J.*, **30,** 239–270 (1951).

(14) BENNETT, A. F., 'An improved circuit for the telephone set', *Bell Syst. Tech. J.*, **32,** 611–626 (1953).

(15) ZWISLOCKI, J., 'Factors determining the sound attenuation produced by earphone sockets', *J. Acoust. Soc. Amer.*, **27,** 146–154 (1955).

(16) ZWISLOCKI, J., 'Development of a semiplastic earphone socket', *J. Acoust. Soc. Amer.*, **27,** 155–161 (1955).

(17) Zwislocki, J., 'Design and testing of earmuffs', *J. Acoust. Soc. Amer.*, **27,** 1154–1163 (1955).

(18) Spencer, H. J. C., 'Some principles of antisidetone telephone circuits', *P. O. Elect. Engrs', J.*, **48,** 208–211 (1956).

(19) Roberton, J. S. P., 'The rocking armature receiver', *P. O. Elect. Engrs'. J.*, **49,** 40–46 (1956).

(20) Spencer, H. J. C. and Wilson, F. A., 'The new 700-type telephone' *P. O. Elect. Engrs. J.*, **49,** 69–80 (1956).

(21) Williams, F. E. and Wilson, F. A., 'Design of an automatic sensitivity control for a new subscriber's telephone set', *Proc. IEE*, **106B,** 361–371 (1959).

(22) Spencer, H. J. C. and Wilson, F. A., 'The new 700-type table telephone —Telephone No. 706', *P. O. Elect. Engrs'. J.*, **52,** 1–12 (1959).

(23) *CCITT Recommendation P.45*, 'Measurement of the AEN value of a commercial telephone system (sending and receiving) by comparison with the SRAEN', Section B. 'Conditions of measurement', Red Book, Volume V 70–77 (1962).

(24) *United Kingdom telephone administration*, 'Artificial mouth and ear used by the United Kingdom telephone administration', CCITT Red Book, Volume V, Annex 11, 302–326 (1962).

(25) *National Automatic Networks*, Handbook published by CCITT, Chapter V, Transmission (1964).

(26) Troke, F. E. I., 'Field trial of the Trimphone', *P. O. Elect. Engrs'. J.*, **58,** 8–10 (1965).

(27) Wilson, F. A., 'Speech-level control in telephone instruments', *Proc. IEE*, **114,** 907–915 (1967).

(28) Seeman, E., 'New telephone set', *PTT Tech. Mitt.*, No. 5, 248–254 (1967).

(29) *Local Telephone Networks*, Handbook published by CCITT, Chapter V, Transmission and signalling aspects of local networks.

(30) *CCITT Recommendation P.31*, 'Conditions which should be satisfied by subscribers' stations used with international circuits rented temporarily for private purposes', White Book, Volume V (1968).

(31) *CCITT Recommendation P.72*, 'Measurement of reference equivalents and relative equivalents', Section C. 'Precautions to be taken during telephonometric measurements', White Book, Volume V (1968).

(32) *Supplement No. 8 (to CCITT Recommendations P.20 and P.74)* 'Methods used by the British telephone administration for rating telephone speech links', White Book, Volume V (1968).

5.2 *Loudspeaking telephones*

(1) Lochner, J. P. A. and Burger, J. F., 'Optimum reverberation time for speech rooms based on hearing characteristics', *Acustica*, **10,** 394–399 (1960).

(2) Larson, R. J. and Adducci, A. J., 'Transient distortion in loudspeakers', *IRE International Convention Record, Part 2, Electroacoustics* (Session 12), 23–31 (1961).

(3) Griffiths, J. D., 'Apparent bass and non-linear distortion', *ibid.* 32–39.

(4) Clemency, W. F. and Goodale, W. D., 'Functional design of a voice-switched Speakerphone', *Bell Syst. Tech. J.*, **40,** 649–668 (1961).

(5) Copping, B. *et al.* 'Designing a voice-switched loudspeaking telephone

—Loudspeaking Telephone No. 4', *P. O. Elect. Engrs. J.*, **60,** 65–71 (1967).

(6) *CCITT Recommendation P.33* 'Subscriber telephone sets containing either loudspeaking receivers or microphones associated with amplifiers', White Book, Volume V (1968).

(7) *CCITT Study Group XII, Question 17/XII,* 'Loudspeaker telephones', White Book, Volume V and particularly the following Annexes to this Question (1968).

Annex 2, 'A method for measuring the sensitivity of a loudspeaking telephone set' (Contribution of the United Kingdom Administration).

Annex 3, 'Effect of room noise and reverberation time in the room in which the loudspeaking telephone is used', (Contributions by the United Kingdom, Swedish Administration and American Telephone and Telegraph Co.).

Annex 4, 'Contribution by the Administration of the Federal Republic of Germany' (Printed in Red Book, Volume V, 705–706 (1962).

Annex 5, 'Contribution by the Swedish Administration' (Printed in Red Book, Volume V, 706–708) (1962).

Annex 6, 'Contribution by the Polish Administration' (Printed in Red Book, Volume Vbis, 284–286) (1965).

5.3 *Headset telephones*

(1) SPENCER, H. J. C. and Roberton, J. S. P., 'A lightweight headset for telephone operators', *P. O. Elect. Engrs'. J.*, **53,** 177–180 (1960).

(2) *CCITT Recommendation P.22* 'Manual trunk exchanges', White Book, Volume V (1968).

6. *LINE TRANSMISSION SYSTEMS*

6.1 *Subscribers' lines and local telephone circuits*

(1) HAYES, H. C. S., SEYMOUR, R. A. and BRAY, P. R., 'Crosstalk', Inst. Post Office Elec. Engrs, Printed Paper No. 169 (1938).

(2) MEEK, L., 'Subscribers' cable distribution—Recent developments', Inst. Post Office Elec. Engrs., Printed Paper No. 170 (1939).

(3) SUMMERS, F., 'Subscribers' cable distribution—Some further considerations', Inst. Post Office Elec. Engrs., Printed Paper No. 180 (1944).

(4) CALAME, L. E., 'Cable distribution by means of large outdoor terminal pillars', *Telecomm. J. Aust.*, **4,** 328–333 and **5,** 31–39 (1944).

(5) TAYLOR, R. E., 'Unit type cable for subscribers' distribution schemes', *P. O. Elect. Engrs'., J.* **39,** 81–84 (1946).

(6) EDWARDS, J. J. and HARDING, J. P., 'Flexibility units for local line networks–Part 1.—Pillars, cabinets and combined units', *P. O. Elect. Engrs'. J.*, **39,** 100–104 (1946).

(7) HARDING, J. P. and HUMPHRIES, A. J., 'Flexibility units for local line networks. Part 2.—Terminal strip assemblies', *P. O. Elect. Engrs. J.* **39,** 159–163 (1947).

(8) BARNES, E. J., WOOD, A. E. and RICHARDS, D. L., 'Standards of transmis-

sion and local line limits. Part 1.—Factors affecting the transmission of speech over a telephone circuit', *P. O. Elect. Engrs.' J.*, **39,** 151–155, 'Part 2. The effective transmission technique', **40,** 8–13 (1947).

(9) SIDENMARK, N., 'Distribution system in local telephone networks with through connected basic load', *Ericsson Rev.*, No. 2, 26–35 (1947).

(10) DE LUCA, L., 'Optimum conductor diameter for urban telephone cables', *Poste e Telecomm.*, **No. 3,** 97–106 (1948).

(11) FUDGE, G. A. E., 'Introduction and application of transmission performance ratings to subscribers' networks', Inst. Post Office Elec. Engrs. Printed Paper No. 198 (1949).

(12) BOGAN, L. B. and YOUNG, K. D., 'Simplified transmission engineering in exchange cable plant design', *Commun. Electron.*, **No. 15,** 498–502 (1954).

(13) MACWHIRTER, R. and MASON, H. J. S., 'A review of basic factors in local line planning', Inst. Post Office Elec. Engrs. Printed Paper **No.** 21 (1956).

(14) INOSE H., YASUDA, Y., KAWAI, Y. and TAKAGI, M., 'Subscriber line circuits for an experimental time division, multiplex exchange system, featuring Delta Modulation techniques', *J. Elect. Commun. Engrs Japan*, **44,** 1322–1328 (1961).

(15) ANDERSON, E. W., 'Local networks', *Proc. IEE*, **111,** 715–726 (1964).

(16) JONES, R. A., 'A prototype conference repeater using a new voice-switching technique', *P. O. Elect. Engrs*'. *J.*, **59,** 246–249 (1967).

(17) MITCHELL, D., 'General transmission considerations in telephone conference systems', *IEEE Trans. Commun. Technol.*, **COM-16,** 163–167 (1968).

(18) KUEBLER, W. P., 'A multiport telephone conference bridge', *IEEE Trans. Commin. Technol.*, **COM-16,** 168–172 (1968).

(19) *Supplement No. 8 (to CCITT Recommendations P.20 and P.74)* 'Methods used by the British telephone administration for rating telephone speech links', White Book, Volume V (1968).

(20) *Local Telephone Networks*, Handbook published by CCITT, Chapter V, Transmission and Signalling Aspects of Local Networks.

(21) GRESH, P. A., 'Physical and transmission characteristics of customer loop plant', *Bell Syst. Tech. J.*, **48,** 3337–3385 (1969).

(22) SPENCER, H. J., 'A new method of local transmission planning', *P. O. Elect. Engrs*'. *J.*, **63,** 84–85 (1970).

6.2 *Junctions*

(1) GOODALL, W. M., 'Telephony by pulse code modulation', *Bell Syst. Tech. J.*, **26,** 395–409 (1947).

(2) LIBOIS, L. J., 'A new process of code modulation "Delta Modulation"', *L'Echo des Recherches*, No. 3, 8–15 (1951) (also appears in *Onde Elect.*, **32,** 26–31 (in French) (1952)).

(3) SCHOUTEN, J. F., DE JAGER, F., and GREEFKES, J. A., 'Delta Modulation, a new modulating system for telecommunication', *Philips Tech. Rev.*, **13,** 237–245, (Articles in Dutch have appeared earlier by the same authors) (1952).

(4) DE JAGER, F., 'Delta modulation—A method of P. C. M. transmission using the 1-unit code', *Philips Res. Rep.*, **7,** 442–466 (1952) (also in *Communication Theory*, Butterworths Scientific Publications, 119–137 (1953)).

(5) WINDELER, A. S., 'Polyethylene-insulated telephone cable', *Trans. AIEE, Part 1 — Commun. Electron.*, **73,** 106–111 (1954).

(6) BOWERS, F. K., 'What use is Delta Modulation to the transmission engineer?', *Trans. AIEE, Part 1 — Commun. Electron.*, **76,** 142–147 (1957).

(7) TURNER, D. and NEILL, T. B. M., 'Principles of negative impedance 2-wire repeater', *P. O. Elect. Engrs'. J.*, **51,** 206–211 (1958).

(8) EAGER, G. S. *et al.* 'Transmission properties of polyethylene insulated telephone cables at voice and carrier frequencies', *Trans. AIEE, Part 1 — Commun. Electron.*, **78,** 618–640 (1959).

(9) WHYTE, J. S., JARVIS, J. R., and HUTT, B. J., 'Pulse code modulation', *P. O. Elect. Engrs'. J.*, **54,** 86–91 and 189–195 (1961).

(10) HENNEBERGER, T. C., and FAGEN, M. D., 'Comparative transmission characteristics of polyethylene insulated and paper insulated communication cables', *Trans. AIEE, Part 1 — Commun. Electron.*, **81,** 27–33 (1962).

(11) DAVIS, C. G., 'An experimental pulse code modulation system for short-haul trunks', *Bell Syst. Tech. J.*, **41,** 1–24 (1962).

(12) SHENNUM, R. H., and GRAY, J. R., 'Performance limitations of a practical PCM terminal', *Bell Syst. Tech. J.* **41,** 143–171 (1962).

(13) MANN, H. STRAUBE, H. M., and VILLARS, C. P., 'A companded coder for an experimental PCM terminal', *Bell Syst. Tech. J.*, **41,** 173–226 (1962).

Note The above three articles, with others relating to PCM, appear in Bell Mono. 4085.

(14) GRAVIS, H. and CRATER, T. V., 'Engineering of T1 Carrier System repeatered lines', *Bell Syst. Tech. J.*, **42,** 431–486 (1963).

(15) INOSE, H. L., and YASUDA, Y., 'A unity bit coding method by negative feedback', *Proc. IEEE,* **51,** 1524–1535 (1963).

(16) WINKLER, M. R., 'High information Delta Modulation', *IEEE Int. Conv. Rec.*, **11,** Part 8 260–265 (1963).

(17) FRANKLIN, R. H., 'Transmission systems for trunk and junction circuits', *Proc. IEE,* **111,** 700–712 (1964).

(18) BARTLETT, C. J. C., and GRESZCZUK, J., 'Companding in a PCM system', *Conference on Transmission Aspects of Communication Networks, IEE,* London, 183–186 (1964).

(19) CARTER, R. O., 'Application of pulse-code modulation to speech transmission in civil networks', *Electron. Power*, 2–5 (1964).

(20) BOYD, R. C., *et al.* 'N2 Carrier System', *Bell Syst. Tech. J.*, **44,** 731–822 (1965).

(21) FULTZ, K. E., and PENICK, D. B., 'The T1 Carrier System', *Bell Syst. Tech. J.*, **44,** 1405–1451 (1965) (also Bell Mono. 5055).

(22) SHTEIN, V. M., 'Some problems in PCM system design', *Telecomms Radio Engng*, 14–20 (Mar. 1966).

(23) EINARSSON, G., 'An experimental 24-channel system with pulse code modulation', *Ericsson Rev.*, **43,** 42–48 (1966).

(24) BOLTON, L. J., and BENNETT, G. H., 'Pulse code modulation', Inst. P. O. Elec. Engrs. Paper read in 1966–67.

(25) CATTERMOLE, K. W., 'PCM junction telephone transmission', *Communs*, Nov, 22–26 and Dec. 22–25 (1966).

(26) MAEDA, K., 'A 24-channel PCM system for short trunks — (Part 1) — The PCM-24 System', *Japan Telecomms Rev.*, **8,** 3–13 (1966).

(27) O'NEAL, J. B., 'Predictive quantizing systems (differential pulse code modulation) for the transmission of television signals', *Bell Syst. Tech. J.*,

45, 689–721 (1966).

(28) Jousset, A., and Lachaise, J., 'An experimental PCM junction: Bonne-Nouvelle to Chaville', *L'Echo des Recherches,* **No. 49,** 29–33 (1967).

(29) Chandler, T. W., Cleobury, D. J., and Payne, A. J., 'Pulse Code Modulation', A collection of six articles reprinted from *GEC Telecommuns,* **No. 35,** 2–29 (1967).

(30) Abate, J. E., 'Linear and adaptive delta modulation', *Proc. IEE,* **55,** 298–308 (1967).

(31) Bello, P. A., Lincoln, R. N., and Gish, H., 'Statistical Delta Modulation', *Proc. IEEE,* **55,** 308–318 (1967).

(32) Greefkes, J. A., and de Jager, F., 'Continuous delta modulation', *Philips Res. Rep.,* **23,** 233–246 (1968).

(33) Tomozawa, A., and Kaneko, H., 'Companded delta modulation for telephone transmission', *IEEE, Trans. Commun. Technol.,* **COM-16,** 149–157 (1968).

(34) Brolin, S. J., and Brown, J. M., 'Companded delta modulation for telephony', *IEEE Trans. Commun. Technol.,* **COM-16,** 157–162 (1968).

(35) Nasell, I., Ellison, Jr., C. R., and Holmstrom, R., 'The transmission performance of Bell System intertoll trunks', *Bell Syst. Tech. J.,* **47,** 1561–1613 (1968).

(36) *CCITT Recommendation G.125* 'Characteristics of national circuits on carrier systems over very short distances', White Book, Volume III (1968).

(37) *CCITT Recommendation G.711* 'Characteristics of P.C.M. Systems', White Book, Volume III (1968).

(38) Bolton, L. J., and Bennett, G. K., 'Design features and application of the British Post Office 24-channel pulsecode-modulation system', *P. O. Elect. Engrs. J.,* **61,** 95–101 (1968).

(39) Richards, D. L., 'Transmission performance of telephone networks containing p.c.m. links', *Proc. IEE,* **115,** 1245–1258 (1968).

(40) Richards, D. L., and Berry, R. W., 'Impairment of telephone speech by quantizing' *4th International Symposium on Human Factors in Telephony,* 29–46 (1968).

(41) Wing, P. A., 'Wisdom — Self synchronizing delta modulation orthogonal channel modulator', *Electron. Lett.,* **4,** 211–212 (1968).

(42) Cattermole, K. W., *Principles of pulse code modulation,* Iliffe Books, London (1969).

(43) Goodman, D. J., 'The application of delta modulation to analog-to-PCM encoding', *Bell Syst. Tech. J.,* **48,** 321–343 (1969).

(44) Daguet, J., *et al.* (60-channel telephony multiplex using Delta-modulation with controlled slope', *Cables et Transmission,* No. **2,** 178–197 (1969).

(45) Michel, A. N., 'Necessary conditions for a given mode of oscillations in delta modulated control systems', *IEEE Trans. Aerospace Syst.* **AES-5,** 351–353 (1969).

(46) Wing, P. A., 'Adaptive delta modulation', *Electron. Lett.,* **5,** 191–192 (1969).

(47) Donaldson, R. W., and Douville, R. J., 'The performance capabilities of PCM, DPCM, ΔM and PM voice communication systems', *IEEE Trans. Commun. Technol.,* **COM-17,** 421–431 (1969).

(48) McKibbin, J., and Stralding, J. B. M., 'The use of delta modulation in pulse transmission systems', *Electrical Engineering Transactions,* The Institution of Engineers, Australia, 35–39 (1970).

(49) JAYANT, N. S., 'Adaptive delta modulation with a one-bit memory', *Bell Syst. Tech. J.*, **49,** 321–342 (1970).

(50) NEWTON, C. M. B., 'Delta modulation with slope-overload prediction', *Electron. Lett.*, **6,** 272–274 (1970).

(51) CLARKE, C. M., and PETFORD, B., 'A companded delta-sigma speech digitiser', *IEE Conf. Publ.* **64,** 59–63 (1970).

(52) GREEFKES, J. A., 'A digitally companded delta modem for speech transmission', *IEEE Internat. Conf. Comm. Rec.*, 7–33 - 7–48 (1970).

(53) SCHINDLER, H. R., 'Better communications are possible when the signals produced by the human voice are changed into a series of binary digits. One conversion method offers special advantages', *IEEE Spectrum*, **7,** 69–78 (1970).

6.3 *Trunk circuits*

(1) HOLBROOK, B. D., and DIXON, J. T., 'Load rating theory for multichannel amplifiers', *Bell Syst. Tech. J.*, **18,** 624–644 (1939).

(2) JACOBSEN, B. B., 'Effect of non-linear distortion in multichannel amplifiers', *Elect. Commun.*, **20,** 29–34 (1940).

(3) HALSEY, R. J., 'Modern submarine cable telephony and the use of submerged repeaters', *J. IEE*, **91,** Part III, 218–236 (1944).

(4) BROCKBANK, R. A., and WASS, C. A. A., 'Non-linear distortion in transmission systems', *J. IEE*, **92,** Part III, 45–56 (1945).

(5) STANESBY, H., 'A review of line and radio-relay communication systems', *Proc. IEE*, **103B,** 11–17 (1956).

(6) PRINGLE, J. B., and GANDET, G., 'A statistical approach to telephone transmission maintenance', *Trans. AIEE, Part 1 — Commun. Electron.*, **76,** 663–669 (1957).

(7) FRANKLIN, R. H., and WILLIAMS, M. B., 'Recent developments in line transmission systems', *Inst. Post Office Elec. Engrs'. Printed Paper No. 213* (1958).

(8) JACOBSEN, B. B., 'Thermal noise in multi-section radio links', *Proc. IEE*, **105C,** 139–150 (1958).

(9) RICE, S. O., 'Distribution of the duration of fades in radio transmission — Gaussian noise model', *Bell Syst. Tech. J.*, **37,** 581–635 (1958).

(10) PARRY, C. A., 'CCITT Recommendations for multichannel radio relays and white noise', *Trans. AIEE, Part 1 — Commun. Electron.*, **78,** 107–117 (1959).

(11) POLLACK, L., 'Radio communication using earth-satellite repeaters', *Elect. Commun.*, **36,** 180–188 (1960).

(12) KITCHENN, R. C., 'Stability and echo in the trunk network', *Telecomm. J. Aust.*, **13,** 49–61 (1961).

(13) EMLING, J. W., and MITCHELL, D., 'Effects of time delay and echoes on telephone conversations', *Bell Syst. Tech. J.*, **42,** 2869–2892 (1963).

(14) BRADY, P. T., and HELDER, G. K., 'Echo suppressor design in telephone communications', *Bell Syst. Tech. J.*, **42,** 2893–2918 (1963).

Note The above two articles appear as Bell Mono. 4675.

(15) LEWINSKI, D. A., 'A new objective for message circuit noise', *Bell Syst. Tech. J.*, **43,** 719–740 (1964).

(16) FRANKLIN, R. H., 'Transmission systems for trunk and junction circuits', *Proc. IEE*, **111,** 700–712 (1964).

38

(17) WILLIAMS, H., 'Overall survey of transmission performance planning', *Proc. IEE*, **111**, 727–743 (1964).

(18) RICHARDS, D. L., 'Statistical properties of speech signals', *Proc. IEE*, **111**, 941–949 (1964).

(19) PEARSON, K. W., 'Method for the prediction of the fading performance of a multi-section microwave link', *Proc. IEE*, **112**, 1291–1300 (1965).

(20) MAYO, J. S., 'Experimental 224 Mb/s PCM terminals', *Bell Syst. Tech. J.*, **44**, 1813–1842 (1965).

(21) BLEISCH, G. W., 'The N3 Carrier System; transmission aspects and objectives', *IEEE Trans. Commun. Technol.* **COM-13**, 355–360 (1965).

(22) OLIVER, W., 'White noise loading of multi-channel communication systems', *Electron. Engng*, **37**, 714–717 (1965).

(23) CYR, M. J., and THURSWALDNER, A. 'Multichannel load calculation using the Monte Carlo method', *IEEE Trans. Commun. Technol.* **COM-14**, 177–181 (1966).

(24) HOLMAN, E. W., and SUHOCKI, V. P., 'A new echo suppressor', *Bell Lab. Rec.*, **44**, 139–142 (1966).

(25) BECKER, F. K., and RUDIN, H. R., 'Application of automatic transversal filters to the problem of echo suppression', *Bell Syst. Tech. J.*, **45**, 1847–1850 (1966).

(26) SONDHI, M. M., and PRESTI, A. J., 'A self-adaptive echo canceller', *Bell Syst. Tech. J.*, **45**, 1851–1854 (1966).

(27) MOFFATT, A. M., 'Stability and echo in telephone networks', *N. Z. Engng*, **15**, 12–17 (1967).

(28) MUNDAY, S., 'New international transmission and switching plan recommended by the CCITT for public telephony', *Proc. IEE*, **114**, 619–627 (1967).

(29) SONDHI, M. M., 'An adaptive echo-canceller', *Bell Syst. Tech. J.*, **46**, 497–511 (1967).

(30) FONTANELLAZ, G., PFYFFER, H. K., and EMMENEGGER, H., 'Load measurements on carrier frequency telephone systems', *PTT Tech. Mitt.*, No. **5**, 237–241 (1967).

(31) BRUGIA, O., GIACOMELLI, L., and ROVERI, A., 'The method of automatic echo neutralization', *Note, Recensioni e Notizie*, **16**, 577–598 (1967).

(32) THIES, A. W., and ZMOOD, R. B., 'New ways of echo suppression' *Aust. Telecomm. Res.*, **1**, 14–19 (1967).

(33) SHANKS, P., 'A new echo-suppressor for long-distance communications', *P. O. Elect. Engrs. J.*, **60**, 288–292 (1968).

(34) *CCITT Recommendation G.121* (*P.21*) 'Reference equivalents of national systems', White Book, Volume III (V) (1968).

(35) *CCITT Recommendation G.123* 'Circuit noise in national networks', White Book, Volume III (1968).

(36) *CCITT Recommendation G.124* 'Characteristics of long-distance loaded-cable circuits liable to carry international calls', White Book, Volume III (1968).

(37) *CCITT Recommendation G.143* 'Circuit noise and the use of compandors', White Book, Volume III (1968).

(38) *CCITT Recommendation G.161* 'Echo suppressors suitable for circuits having either short or long propagation times', White Book, Volume III (1968).

(39) *CCITT Recommendation G.162* 'Characteristics of compandors for telephony', White Book, Volume III (1968).

(40) *CCITT Recommendations G.211* 'Make-up of a carrier link', White Book, Volume III (1968).

(41) *CCITT Recommendation G.212* 'Hypothetical reference circuits', White Book, Volume III (1968).

(42) *CCITT Recommendation G.222* 'Noise objectives for design of carrier-transmission systems of 2500 km, White Book, Volume III (1968).

(43) *CCITT Recommendation G.223* 'Assumptions for the calculation of noise on hypothetical reference circuits for telephony', White Book, Volume III (1968).

(44) *CCITT Recommendation G.226* 'Noise on a real link', White Book, Volume III (1968).

(45) *CCITT Recommendation G.232* '12-channel terminal equipments', White Book, Volume III (1968).

(46) *CCITT Recommendation G.233* 'Recommendations concerning translating equipments', White Book, Volume III (1968).

(47) *CCITT Recommendations in Section 3* 'Individual characteristics of international carrier telephone systems on metallic lines', White Book, Volume III (1968).

(48) *CCITT Recommendations in Section 4* 'General characteristics of international carrier telephone systems on radio-relay or satellite links and interconnexion with metallic links', White Book, Volume III (1968).

(49) *CCITT Recommendations in Section 5* 'Audio-frequency circuits', White Book, Volume III (1968).

(50) *CCITT Recommendations in Section 6* 'Co-ordination of radio-telephony and line telephony', White Book, Volume III (1968).

(51) *CCITT Recommendations in Section 7* 'Digital transmission systems', White Book, Volume III (1968).

(52) MIURA, A., *et al.* 'Blockless echo suppressor', *IEEE Trans. Commun. Technol.* **COM-17,** 489–495 (1969).

6.4 *Overall transmission loss of a connexion*

(1) *CCITT Recommendation G.111* (*P.11*) 'Reference equivalents in an international connexion', White Book, Volume III (V) (1968).

(2) *CCITT Recommendation G.112* (*P.12*) 'Articulation reference equivalent (AEN)', White Book, Volume III (V) (1968).

6.5 *Economy systems*

Amplitude compression or limiting

(1) DE JAGER, F., and GREEFKES, J. A., "FRENA", a system of speech transmission at high noise levels', *Philips Tech. Rev.*, **19,** 73–83 (1957–58).

(2) RIZZONI, E. M., 'Compandor loading and noise improvement in frequency division multiplex radio-relay systems', *Proc. IRE*, **48,** 208–220 (1960).

(3) Daguet, J. and Gilabert, K., 'Constant-level speech in single-sideband transmitters', *Onde Elect.*, **16,** 498–509 (in French) (1961).

(4) Rupp, H., 'Improvement of signal-to-noise ratio in speech transmission with pre-emphasis and using compandors', *NTZ-CJ*, 178–182 (1963).

(5) Thomson, D., 'A compandor using junction transistors', *Proc. IEE*, **106B,** Suppl. No. 16, 619–624 (1959).

(6) Carter, R. O., 'Theory of syllabic compandors', *Proc. IEE,* **111,** 503–513 (1964).

(7) Richards, D. L., 'Transmission performance assessment for telephone network planning', *Proc. IEE,* **111,** 931–940 (1964).

(8) Watt-Carter, D. E., and Wheeler, L. K., 'The Lincompex System for the protection of HF radio-telephone circuits', *P. O. Elect. Engrs. J.*, **59,** 163–167 (1966).

(9) *CCITT Recommendation G.162* 'Characteristics of compandors for telephony', White Book, Volume III (1968).

Time compression

(10) Gabor, D., 'New possibilities in speech transmission', *Proc. IEE*, **94,** Part III, 369–390 (1947).

(11) Bullington, K., and Fraser, J. M., 'Engineering aspects of TASI', *Bell Syst. Tech. J.*, **38,** 353–364 (1959).

(12) Clinch, C. E. E., 'Time assignment speech interpolation (TASI)', *P. O. Elect. Engrs'. J.*, **53,** 197–200 (1960).

(13) Fraser, J. M., Bullock, D. B., and Long, N. G., 'Over-all characteristics of a TASI system', *Bell Syst. Tech. J.*, **41,** 1439–1454 (1962).

(14) Miedema, H., and Schachtman, M. G., 'TASI quality–Effect of speech detectors and interpolation', *Bell Syst. Tech. J.*, **41,** 1455–1473 (1962).

(15) *CCITT Recommendation G.163* 'Call concentrating systems', White Book, Volume III (1968).

Frequency compression

(16) Bogert, B. P., 'The Vobanc — A two-to-one speech bandwidth reduction system', *J. Acoust. Soc. Amer.*, **28,** 399–404 (1956).

(17) Marcou, P. and Daguet, J., 'New methods of speech transmission', *Third London Symposium on Information Theory, London* (1956). Butterworth's Scientific Publications, 231–244 (also, in French, *Ann. Telecommun.*, **11,** 118–126 (1956)).

(18) Kryter, K. D., 'Speech bandwidth compression through spectrum selection', *J. Acoust. Soc. Amer.*, **32,** 547–556 (1960).

(19) Cherry, E. C., and Phillips, V. J., 'Some possible uses of single sideband signals in formant tracking systems', *J. Acoust. Soc. Amer.*, **33,** 1067–1077 (1961).

(20) Kryter, K. D., 'Bandwidth compression by spectrum sampling' *J., Acoust. Soc. Amer.*, **33,** 1663 (1961).

(21) Schroeder, M. R., Flanagan, J. L., and Lundry, E. A., 'Bandwidth compression of speech by analytic-signal rooting', *Proc. IEEE*, **55,** 396–401 (1967).

Analysis/synthesis

(22) SWAFFIELD, J., 'The potentialities of the vocoder for telephony over very long distances', *P. O. Elect. Engrs'., J.*, **41,** 22–28 (1948).
(23) HALSEY, R. J. and SWAFFIELD, J., 'Analysis-synthesis telephony with special reference to the vocoder', *Proc. IEE*, **95,** Part III, 391–411 (1948).
(24) CAMPANELLA, S. J., 'A survey of speech bandwidth compression techniques', *IRE Trans. Audio*, **5,** 104–116 (1958).
(25) FLANAGAN, J. L., 'Resonance vocoder and baseband complement: Hybrid speech transmission', *Wescon Convention Record,* (also as Bell Monograph 3407) (1959).
(26) POLYKOVSKII, A. M., 'A new method of synthesizing the speech spectrum', *Telecomms (U.S.S.R.)* No. 12, 1403–1410 (1960).
(27) SCHROEDER, M. R., and DAVID, E. E., 'A vocoder for transmitting 10 kc/s speech over a 3.5 kc/s channel', *Acustica*, **10,** 35–43 (1960).
(28) STEAD, L. G., 'Narrow-bandwidth speech communication', *J. IEE*, **8,** 454–457 (1962).
(29) FLANAGAN, J. L., *Speech analysis, synthesis and perception*, Springer-Verlag, Berlin (1965).
(30) OLSON, H. F., BELAR, H., and ROGERS, E. S., 'Research towards a high efficiency voice communication system', *J. Audio Engng Soc.*, **14,** 233–239 (1966).
(31) SCHROEDER, M. R., 'Vocoders: Analysis and synthesis of speech — A review of 30 years of applied speech research', *Proc. IEEE*, **54,** 720–734 (1966).

7 PLANNING OF TELEPHONE NETWORKS

7.1 *A look backwards*

(1) GAVEY, J., 'The telephone trunk line system in Great Britain', *J. IEE*, **25,** 624–680 (including discussion) (1896).
(2) GAVEY, J., 'Inaugural Address', *J. IEE*, **36,** 4–37 (1905).
(3) HILL, J. G., 'Telephone transmission', *Inst. Post Office Elect. Engrs, Printed Paper No.* 2 (1906).
(4) PLUMMER, T., 'Improvements in telephonic transmission over underground circuits by the insertion of inductance coils', *Inst. Post Office Elect. Engrs, Printed Paper No.* 4 (1906).
(5) SLINGO, W., 'An address on telephones', *Inst. Post Office Elect. Engrs, Printed Paper No.* 64 (1915).
(6) COHEN, B. S., 'Telephonometry', *Inst. Post Office Elect. Engrs, Printed Paper No.* 70 (1916).
(7) NOBLE, W., 'The long-distance telephone system of the United Kingdom', *J. IEE*, **59,** 389–426 (including discussion) (1921).
(8) ELSTON, J. S., 'Applied telephone transmission', *Inst. Post Office Elect. Engrs, Printed Paper No.* 88 (1922).
(9) MARTIN, W. H., 'The transmission unit and telephone transmission reference systems', *Bell Syst. Tech. J.*, **3,** 400–408 (1924).
(10) SMITH, C. W., 'Practical application of the recently adopted transmission unit', *Bell Syst. Tech. J.*, **3,** 409–413 (1924).
(11) COHEN, B. S., ALDRIDGE, A. J. and WEST, W., 'The frequency characteristics of telephone systems and audio-frequency apparatus', *J. IEE*, **64,** 1023–1064 (1926).

(12) COHEN, B. S., 'Apparatus standards of telephonic transmission and the technique of testing microphones and receivers', *J. IEE*, **66**, 165–203 (1928).

(13) STRATTON, J. and LUXTON, W. G., 'Modern developments in telephone transmission over lines', *Inst. Post Office Elect. Engrs, Printed Paper No. 153* (1933).

(14) RYALL, L. E., 'A few recent developments in telephone transmission apparatus', *Inst. Post Offiice Elect. Engrs, Printed Paper No. 155* (1934).

(15) BARNES, E. J., WOOD, A. E. and RICHARDS, D. L., 'Standards of transmission and local line limits', *P. O., Elect. Engrs, J.*, **39**, 151–155 and **40**, 8–13 (1947).

(16) FUDGE, G. A. E., 'Introduction and application of transmission performance ratings to subscribers' networks', *Inst. Post Office Elect. Engrs, Printed Paper No. 198* (1949).

(17) *CCITT Red Book, Volume V, Annex 1*, 167–173 'Methods used by the British telephone administration for rating telephone speech links' (1962).

(18) *CCITT Red Book, Volume V bis, Annex 4*, 85–118 'Methods applied by various administrations in inland local and trunk networks, with a view to providing satisfactory performance for national calls' (1965).

7.2 *The network planning process*

(1) SWAFFIELD, J. and RICHARDS, D. L., 'Transmission performance of the telephone network—A survey of some basic problems', *Paper read to Institution of Post Office Electrical Engineers*, (unpublished) (1951).

(2) HUNTLEY, H. R., 'Where we are and where we are going in telephone transmission', *Commun. Electron.*, No. 29, 54–63 (1957).

(3) FLOWERS, T. H., 'Transmission network planning with electronic exchanges', *Proc. IEE*, **107B**, Supp. 20, 364–370 (1960).

(4) TOBIN, W. J. E. and STRATTON, J., 'A new switching and transmission plan for the inland trunk network', *P. O. Elect. Engrs, J.*, **53**, 75–79 (1960).

(5) *CCITT Handbook on National Automatic Networks*, Chapter V, Transmission, International Telecommunication Union, Geneva (1964).

(6) WILLIAMS, H., 'Overall survey of transmission performance planning', *Proc. IEE*, **111**, 727–743 (1964).

(7) DUERDOTH, W. T., 'The possibility of an integrated P.C.M. switching and transmission network', *Paper read to Inst. Post Office Elect. Engrs* (1966).

(8) MUNDAY, S., 'New international transmission and switching plan recommended by the CCITT for public telephony', *Proc. IEE*, **114**, 619–627 (1967).

(9) KITCHENN, R. G., 'Telephone transmission objectives', *Telecomm. J. Aust.*, **18**, 15–27 (1968).

(10) *CCITT Recommendation G. 101*, 'The transmission plan', White Book, Volume III (1968).

(11) *CCITT Recommendation G. 103*, 'Hypothetical reference connections' White Book, Volume III (1968).

7.3 *Notes on specific aspects of the network planning process*

(1) POCOCK, L. C., 'A survey of the telephone transmission-rating problem', *J. IEE*, **95**, Part III, 253–270 (1948).

(2) SWAFFIELD, J. and RICHARDS, D. L., 'Rating of speech links and performance of telephone networks', *Proc. IEE*, **106B**, 65–76 (1959).

(3) RICHARDS, D. L. and SWAFFIELD, J., 'Assessment of speech communication links', *Proc. IEE*, **106B**, 77–92 (1959).

(4) RICHARDS, D. L., 'Transmission performance assessment for telephone network planning', *Proc. IEE*, **111**, 931–940 (1964).

(5) HUTTER, J., 'Customer response to telephone circuits routed via a synchronous-orbit satellite', *P.O. Elect. Engrs, J.*, **60**, 181–186 (1967).

(6) DAVIS, H. T., MOOT, G. and WALKER, J. M., 'Engineering aspects of service standards in the Australian telephone network', *Telecomm. J. Aust.*, **17**, 181–194 (1967).

(7) RICHARDS, D. L., 'Transmission performance of telephone networks containing p.c.m. links', *Proc. IEE*, **115**, 1245–1258 (1968).

(8) *CCITT Recommendation G. 103*, 'Hypothetical reference connections', White Book, Volume III (1968).

(9) *CCITT Recommendation G. 120 (P. 20)*, 'Transmission characteristics of national networks', White Book, Volume III (V) (1968).

(10) *Supplement No.* **7**, 'Methods applied by various administrations in inland and trunk networks, with a view to providing satisfactory performance for national calls', CCITT White Book, Volume V (1968).

(11) *Supplement No. 8*, 'Methods used by the British telephone administration for rating telephone speech links', CCITT White Book, Volume V.

(12) RICHARDS, D. L. and MUNDAY, S., 'The planning process and assembling fundamental information' and 'A survey of the transmission plans of the United Kingdom Post Office', *International Seminar on National Telephone Transmission Planning, Melbourne, Australia* (1970).

(13) SPENCER, H. J., 'A new method of local transmission planning', *P. O. Elect. Engrs, J.*, **63**, 84–85 (1970).

8. *TRANSMISSION PERFORMANCE OF TELEPHONE NETWORKS*

8.1 *Some estimates of the telephonic performance of the telephone network of the United Kingdom since 1905*

(1) GAVEY, J., 'Inaugural Address', *J. IEE*, **36**, 4–37 (1905).

(2) SLINGO, W., 'An address on telephones', *Inst. P. O. Elect. Engrs*, Printed Paper No. 64 (1915).

(3) ELSTON, J. S., 'Applied telephone transmission', *Inst. P. O. Elect. Engrs* Printed Paper No. 88 (1922).

(4) STRATTON, J. and LUXTON, W. G., 'Modern developments in telephone transmission over lines', *Inst. P. O. Elect. Engrs*, Printed Paper No. 153 (1933).

(5) BARNES, E. J., WOOD, A. E. and RICHARDS, D. L., 'Standards of transmission and local line limits; *P. O. Elect. Engrs, J.*, **39**, 151–155 and **40**, 8–13 (1947).

(6) FUDGE, G. A. E., 'Introduction and application of transmission performance ratings to subscribers' networks', *Inst. P. O. Elect. Engrs*, Printed Paper No. 198 (1949).

(7) RICHARDS, D. L. and SWAFFIELD, J., 'Assessment of speech communication links', *Proc. IEE*, **106B**, 77–92 (1959).

(8) SPENCER, H. J. C. and WILSON, F. A., 'The new 700-type table telephone—Telephone No. 706', *P. O. Elect. Engrs, J.*, **52,** 1–12 (1959).
(9) TOBIN, W. J. E. and STRATTON, J., 'A new switching and transmission plan for the inland trunk network', *P. O. Elect. Engrs, J.*, **53,** 75–79 (1960).
(10) RICHARDS, D. L., 'Transmission performance assessment for telephone network planning', *Proc. IEE*, **111,** 931–940 (1964).
(11) RICHARDS, D. L., 'Transmission performance of telephone networks containing p.c.m. links', *Proc. IEE*, **115,** 1245–1258 (1968).

8.2 *Distribution of transmission losses*

(1) MARKMAN, F., 'Determination of cumulative distribution of reference equivalents and transmission performance ratings', *Ericsson Tech.*, **No. 2,** 192–252 (1954).
(2) KITCHENN, R. J., 'A transmission survey of subscribers' telephone services', *J. Inst. Engrs, Aust.*, 195–209 (1968).

8.3 *Frequencies of calls over various classes of connection*

(1) *Annex A*, 'Statistical distribution of reference equivalents of international calls complying with the old CCITT transmission plan and the new one', CCITT Red Book, Volume V bis, 41–45 (1965).
(2) MUNDAY, S., 'A survey of the transmission plans of the United Kingdom Post Office', *International Seminar on National Teephone Transmssion Planning*, Melbourne, Australia, 1–24 – 1–77 (1970).
(3) 'The World's Telephones', *American Telephone and Telegraph Company, New York*, published yearly (1970).
(4) 'Post Office Report and Accounts, October 1969 to March 1970', *Her Majesty's Stationary Office, London* (1970).

8.4 *Estimation of future network transmission performance*

(1) SWAFFIELD, J. and RICHARDS, D. L., 'Transmission performance of the telephone network—A survey of some basic problem', *Paper read to Inst. P. O. Elect. Engrs* (unpublished) (1951).
(2) SWAFFIELD, J. and RICHARDS, D. L., 'Rating of speech links and performance of telephone networks', *Proc. IEE*, **106B,** 65–76 (1959).
(3) MARKMAN, F., 'Transmission performance in presence of circuit noise', *Ericsson Tech.*, No. 2, 166–206 (1960).
(4) WILLIAMS, H., 'Overall survey of transmission performance planning' *Proc. IEE*, **111,** 727–743 (1964).
(5) RICHARDS, D. L., 'Transmission performance assessment for telephone network planning', *Proc. IEE*, **111,** 931–940 (1964).
(6) *Annex B*, 'Effects of certain levels of circuit noise on percentage of unsatisfactory calls', CCITT Red Book, Volume V bis, 45–49 (1965).
(7) *Annex D*, 'Determination of a limit for circuit noise on an international circuit used for public telephony and not fitted with a compandor', CCITT Red Book, Volume V bis, 50–56 (1965).

(8) RICHARDS, D. L., 'Transmission performance of telephone networks containing p.c.m. links', *Proc. IEE*, **115,** 1245–1258 (1968).

(9) MUNDAY, S., 'A survey of the transmission plans of the United Kingdom Post Office', *International Seminar on National Telephone Transmission Planning Melbourne, Australia*, 1–24 – 1–77 (1970).

INDEX